W9-AQD-466

An Introduction to Transportation Engineering

Robert T. Aangeenbrug

Board of Advisors, Engineering

A. H-S. Ang
University of Illinois

Civil Engineering–Systems and Probability

Donald S. Berry
Northwestern University

Transportation Engineering

James Gere
Stanford University

Civil Engineering and Applied Mechanics

J. Stuart Hunter
Princeton University

Engineering Statistics

T. William Lambe
R. V. Whitman
Massachusetts Institute
of Technology

Civil Engineering–Soil Mechanics

Perry L. McCarty
Stanford University

Environmental Engineering

Don T. Phillips
Texas A & M University

Industrial Engineering

Dale Rudd
University of Wisconsin

Chemical Engineering

Robert F. Steidel, Jr.
University of California–
Berkeley

Mechanical Engineering

R. N. White
Cornell University

Civil Engineering–Structures

R. T. Aangeenbrug

An Introduction to Transportation Engineering

Robert T. Aangeenbrug

Second Edition

William W. Hay

**Professor of Railway Civil Engineering, Emeritus
University of Illinois, Urbana Campus**

JOHN WILEY & SONS
New York   Santa Barbara
London   Sydney   Toronto

Copyright © 1961 and 1977, by John Wiley & Sons, Inc.

All rights reserved. Published simultaneously in Canada.

No part of this book may be reproduced by any means, nor transmitted, nor translated into a machine language without the written permission of the publisher.

Library of Congress Cataloging in Publication Data:

Hay, William Walter, 1908–
  An introduction to transportation engineering.

  Includes bibliographical references and index.
  1. Transportation. I. Title.
TA1145.H35   1977   380.5   77-9293
ISBN 0-471-36433-9

Printed in the United States of America

10 9 8 7 6 5 4 3 2 1

To Mary, Bill, and Mary Elisabeth, who have
contributed more than they realize

Preface to the First Edition

I have long been aware of a gap in transportation textbook material regarding those factors and principles that have to do with the technological utility of the various modes of transport in moving persons and goods. Many excellent books have been written on the structural design and formation of plant and equipment. Possibly even more have been written on the economics and regulatory aspects of the industry. The intermediate area of technological use and utility, by which the two are united, has been largely ignored except for portions of an occasional textbook within the area of one mode—such as Wellington's classic *Economic Theory of the Location of Railroads.*

This book has been written, at an introductory level, to fill that gap and to bridge from structural design to economic functioning. It is related to the structural area (including the design of prime movers and equipment) by the conditions of roadway and motive power that must bear and pull the loads imposed by a given traffic pattern and system of operation. It merges in the other direction into the economic and regulatory aspects through the effects of technological characteristics on cost.

The subject matter of this intermediate area is, therefore, the effects of technological factors on movement and the principles involved. Appropriate topics for such a study include the propulsive resistance encountered by all modes of transport and the propulsive force that must be exerted to overcome it. Operating characteristics and criteria determine the suitability of a particular mode of transport for a given situation and traffic. Obviously both route and traffic capacity are of prime importance in determining transport utility. The frequently overlooked factors of terminals, coordination, and operational control must also be considered.

vii

The effects of the foregoing characteristics on costs follow in logical sequence.

All of these find their ultimate significance in planning, an element of fundamental importance in developing economical and useful transportation. Planning should, even if it has not always in the past, consider suitability and utility as principal factors in making decisions as to the types of transport to be utilized and developed. Consideration of the nation's transportation resources is another factor frequently overlooked in planning.

Such matters are worthy of the attention of everyone. Not only the engineer but every citizen needs to have some understanding of the significance of these factors to the transportation system, which is the lifeblood of his nation's economy and for which, in one way or another, he must pay.

The student and transportation engineer must be acquainted with these principles as the basis upon which their specialization in one particular mode or area—railroads, waterways, urban transit, etc.—is founded.

An additional value in this presentation is available to the student in engineering. A study of the technological characteristics of transport systems is, in effect, a review of many of the major elements in the engineering sciences and the way these are combined to achieve a desired purpose, a form of systems analysis.

I have attempted to develop a unified approach to these problems, cutting across the conventional boundaries that have separated railways, highways, airways, etc., into unrelated compartments. I have tried to show a common core of problems and principles present in all transport systems.

I am humbly aware of my shortcomings in dealing adequately with any of the foregoing approaches to transportation. It is my hope, however, that this initial attempt will serve to inspire and guide others to a fuller development of the possibilities within this approach. If it does no more than this, the effort will have been worth while.

In the pages that follow, I have endeavored to give full credit to the findings and work of others where these have been knowingly utilized. However, in the extensive reading, study, and contacts that preceded this effort, many ideas and expressions were undoubtedly acquired that I unwittingly now assume as my own. For any such lapses, I hope to receive understanding pardon.

Much of the chronological data on railroad development was obtained from *A Chronology of American Railroads*, compiled and published by the Association of American Railroads. I gratefully acknowledge permission to use these data.

Numerous friends, associates, and students of mine have contributed encouragement and ideas for this work. Space prohibits listing them by name but I am extremely grateful for their valued assistance. I do wish especially to acknowledge the inspiring help and suggestions received from Professor A. S. Lang of the Massachusetts Institute of Technology. Professor Lang conducted a thorough and searching review of the text as it was prepared and made many helpful recommendations for improving the content as well as uncovering an occasional "goof."

Many other mistakes were eliminated through the reading of manuscript and proof by my wife. I am especially grateful for her loyal encouragement and inspiration.

Any errors that remain in the text are of my own doing, and for these I assume full responsibility.

William W. Hay

Urbana, Illinois
January, 1961

Preface

Transportation has experienced many significant developments in the 15 years since the first edition was published. The Federal System of Interstate Highways is nearing completion. Research and development have taken innovative concepts from the Sunday supplements and put them onto drawing boards and test tracks. Great new systems of rapid transit are recently in operation or under construction in our largest cities.

But not all aspects of transportation are favorable. Our once magnificent railroad plant, equipment, and service have deteriorated to an alarming degree. The ground time for air travel often equals or exceeds the in-flight time. Human beings have rocketed to the moon but still fight bumper-to-bumper congestion going to and from their daily work. The Federal Interstate's promise of mobility has been dimmed by the shortage of petroleum energy and the outcry against pollution. Pipelines and super-tankers, in the attempt to make more energy available, have been tagged as environmental hazards. Innovative systems, too, have lost their first luster as money and effort have been directed toward renovating and improving existing systems and technologies to meet the immediate needs of the present.

Just trying to delineate these needs warrants the effort to update the materials contained in the first edition. There is still a need to examine transportation systems and conduct transportation planning in the light of modal utility; this will achieve a balanced, economical, and productive transportation system for the nation and its individual regions and communities. Students must be given the whole, not just a part of the picture.

The reasons for our present predicament do not lie with transportation technology but with the use we have made of that technology. The primary concepts of mobile utility must govern the planning process. For that reason I have continued to present a unified approach instead of one based on modes; my approach stresses factors and principles that are common and significant to all systems of transport. Classroom presentation may be slightly more difficult, but that, in my opinion, is a small price for guiding students to an overall view. There have and will be plenty of books on the techniques and problems of individual carrier modes.

As with the first edition, a great debt is owed to other people. I note especially the courtesy extended in permitting me to incorporate techniques and concepts of others that contribute needed depth and breadth to my presentation. I sincerely hope no one has failed to receive full credit. I extend thanks to the people who provided illustrative materials. I also thank Betty Wise for her typewritten interpretation of my original copy. Most of all I thank my devoted wife, Mary, for her critical reading of the manuscript and her understanding and encouragement. With so many diverse features to encompass there undoubtedly remain errors and omissions. For these I fully acknowledge my personal responsibility.

William W. Hay

Champaign-Urbana, Illinois, 1977

Contents

Part 1
The Transportation System

Chapter 1

Transportation Function and Development

Transportation is an essential part of today's life. It has been an essential part of most societies of the past. One can hardly conceive of a future society in which it would not continue to be essential. The adequacy of its transportation system is a fair index of a country's economic development.

FUNCTION

Transportation is the movement of people and goods and the facilities used for that purpose. The movement of people assumes first importance in the minds of many, especially in urban areas, but the movement of goods, that is, freight transportation, is probably of greater significance to the functioning of our economic society. Both must be considered essential.

Significance Transportation exhibits characteristics and attributes that determine its specific functions and significance. A prime function is *relating population to land use*. As an *integrating and coordinating factor* in our highly complex and industrialized society, transportation is involved heavily in the movement of goods. Goods have little value unless given *utility*, that is, the capacity for being useful and satisfying wants. Transportation contributes two kinds of utility: *place utility* and *time utility*, economic terms that simply mean having goods where they are wanted

3

FIGURE 1.1 PRODUCT REQUIREMENTS FOR ONE INDUSTRY. DEPENDENCE OF AN EAST SAINT LOUIS INDUSTRY UPON DISTANT AREAS. (COURTESY OF RICHARD L. DAY, *FREIGHT TRAFFIC PATTERNS OF EAST ST. LOUIS,* PH.D. THESIS, 1959, DEPARTMENT OF GEOGRAPHY, UNIVERSITY OF ILLINOIS, URBANA.)

4

when they are wanted, essential functions that can also be applied to the movement of people. The dependence of a particular industry for materials used in its manufacturing processes, that is, the giving of place utility to those materials, is illustrated in Figure 1.1.

In the urban areas especially, transportation provides the connecting link between dwelling units and work opportunities. As much as 50 percent or more of all urban travel is likely to be work trips. Shopping, entertainment, travel to and from school and a variety of other reasons for making trips depend on transportation to make corresponding land uses accessible to urban dwellers. The movement of people represents a vital transportation service that calls for the use of streets and highways, buses, car pools, and other forms of transit in the most efficient manner possible. Intercity travel for business and recreation takes place between all parts of the nation and of the world.

Transportation as a cost factor and as a system will be defined later in the chapter.

FRAMES OF REFERENCE

Systems Characteristics A *system* can be defined as a group or assemblage of parts or elements used for a common purpose so interrelated that a change in one component has an effect or feedback upon the other components. The popular concept of individual modes or operating companies as systems is seen when one speaks of the Burlington Northern System or of the United Airlines system. The national transportation system refers to the entirety of the nation's transportation lines and facilities. One may also refer to regional, state, or urban transport systems.

Transportation systems, however the term is applied, are also *subsystems* in the socioeconomic system. Transportation, for example, is only one of several subsystems, albeit an important one, along with sewage, lighting water supply and other components of the socioeconomic land use organization known as the urban system or community. The advent of the electronic computer and a variety of system analysis techniques are assisting in obtaining solutions for problems that involve the interrelation between transportation, population, and land use. More will be said on this later.

Technological Systems Transportation as a *technological system* is a major frame of reference for this book. A transport system is composed of five major components: vehicle, motive power, roadway, terminals, and control systems. The way in which these components combine to give service and utility is indicated in Figure 1.2. Hauling capacity depends in part on the size and speed of the vehicle. Size and capacity of the vehicle determine needed motive power (and vice versa). Vehicle size and load are also related to the bearing capacity and stability of the roadway, and the number of vehicles per hour (route capacity) is a function of the number of paths (lanes, tracks, and channels) in the roadway and the system of operational control (rules, signals, switching arrangements, and communications). Route or roadway capacity is of little value if the terminals are not designed for prompt handling of traffic through the terminal or to and from the line. Labor, supervision, and management must be present to shape the technological components into a viable performing transportation service. These, in turn, are dependent upon the availability of working, replacement, and expansion capital to finance the operation. Figure 1.3 shows the technological system as part of the national socioeconomic

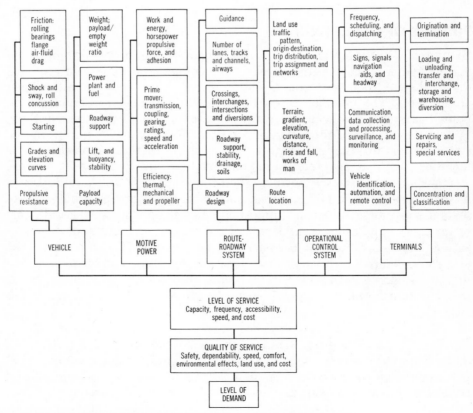

FIGURE 1.2 TRANSPORT TECHNOLOGY-DEMAND SYSTEM.

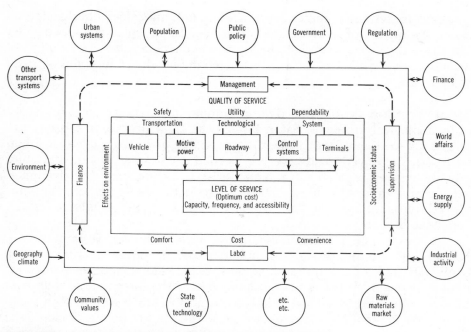

FIGURE 1.3 TRANSPORTATION SYSTEM ENVIRONMENT.

system, subject to impacts and feedbacks to and from a multiplicity of external forces and factors.

Technoeconomic Characteristics   An additional frame of reference is a group of technological characteristics that have a pronounced effect upon costs. The several *technoeconomic characteristics* combine to give a particular transport system or mode its degree of usefulness or utility for a particular type of traffic or service. These characteristics include propulsive resistance and propulsive force in their various forms, payload to empty weight ratios, thermal efficiency and fuel consumption, power loading ratios (horsepower per ton), route capacity, stability, speed, safety, dependability, flexibility, productivity, effects on the environment, and others. The significance of these characteristics should be understood and the effects on utility, land use, and cost included in the planning process.

Environmental Factors   Transportation has a demonstrated major impact upon the environment. The automobile is known as a principal contributor to air pollution, but all modes of transport contribute to air, water, visual, noise, and land pollution in varying degrees. The advent of transport routes and facilities can consume land, divide neighborhoods, and lessen the quality of living. Land values can also be enhanced and a fuller life made available. Foresight in the selection of a mode and design of location and facilities can add to or reduce the incidence of pollution in all of its forms.

Transportation Problems   Transportation engineering has its full share of technological problems. Safer, faster vehicles with greater capacity are sought for road haul, and smaller, more flexible units are sought for urban transport. Improved roadways are needed to bear imposed loads and provide greater capacities. Problems exist in vehicle-roadway dynamics, vehicle suspensions, roadway stability, reduction of pollution, and improved terminal systems—all within the basic requirements of safety, dependability, and economy.

Problems in policy and implementation have technological implications. Should reliance be placed on public carriage, private carriage, or a combination? Should the system be designed for individual transport or for mass movement? What is the optimum mix of the foregoing as well as of the several modes to achieve a balanced transportation system with maximum utility? What are the roles and responsibilities of public agencies and private industry in conducting transportation and in carrying on research and development? What kinds and degrees of public regulation are needed to assure adequate nondiscriminatory service and safety? Who will own and operate the various units of the system and, of fundamental importance, how will the nation's transport needs be financed and by whom?

Financing bears directly on the answers to all the foregoing questions and problems. The mix of transport modes may depend on the amount of financing received by each rather than on a strict analysis of technoeconomic utility. Private funding of for-hire transport has not been entirely adequate. The need for public support has been expressed and sometimes it is given. In the public realm, however, transportation must compete for funds with other public activities: education, health, welfare, national defense, and other needs. Priorities are often decided on a political basis between these alternatives and also between conflicting demands within the transportation area. As such, these decisions are subject to all the vagaries of any political decision. Various interests press their demands for public funds—for more aerial navigation devices, for expansion of the interstate highway system, for urban mass transit, and for aid in meeting more exacting standards of track maintenance and passenger service. Special interest groups and lobbies weigh the impact of each public decision according to their own situations while environmentalists eye the effects upon the national or regional ecology.

Transportation Planning   The planning process is itself a problem area. The engineer can be usefully involved at all stages of transport development—in planning, design, construction, and operation. So complex are the planning problems it is no longer desirable for the engineer to work alone. He or she serves, rather, as one of a team, along with urban planners, sociologists, ecologists, architects, legal and financial experts, and with engineers from associated disciplines.

But final decisions are seldom made either by the engineer or by the planning group. The planners, instead, must develop the possible alternatives with the consequences of each—costs, advantages, and disadvantages —from which the real decision makers—boards of directors, financial institutions, town councils, state legislatures, or the Congress—make their selection or reject all.

Devising a solution to meet an identified demand involves selecting a *level and quality of service* that will satisfy the demand within the limits of feasible funding. The level of service relates to the capacity, frequency, and accessibility of the service. Quality of services involves, for example, safety, dependability, comfort, privacy, speed, and effects on the community and the environment. A cost is attached to each.

Specific goals for suitable transport should include providing sufficient capacity, ready accessibility, a minimum door-to-door travel time, all-weather safety and dependability, reasonable comfort, convenience, and minimal adverse effects on neighborhoods and the environment—all at a reasonable cost. Such goals are not always achieved or even possible, but that is no reason for failing to make an effort toward their attainment.

FACTORS IN TRANSPORTATION DEVELOPMENT

Transportation develops because of several and frequently overlapping factors. The National Road in the United States was built not only to overcome a mountain barrier but also to aid in economic development, political unity, and national defense. For purposes of study only, the several factors are considered under individual headings of economic, geographic, political, military, technological, competitive, and urban, but any of these or all of these may work together.

Economic Factors   Almost all transport development is economic in origin. The chief preoccupation of the first human beings was the procurement of food, shelter, and sometimes clothing. As they became more highly developed their wants increased, often beyond what their local economy could supply. Means of transporting goods from distant places had to be

TABLE 1.1. IMPROVED TRANSPORT TECHNIQUES VERSUS COST AND OUTPUT[a]

Type of Carrier	Output per Man-Ton-Miles per Day	Value of Vehicular Equipment (in dollars)	Accessories Required	Costs per Day[b] (a) Accessory (b) Operating (c) Interest (d) Wages (in dollars)		Total Cost per Day (in dollars)	Costs per Ton-Mile (in dollars)
Man's back (100 lb carried 20 miles)	1	0	Trail and pack rack	(a) (b) (c) (d)	0.01 — — 0.20	0.21	0.21
Pack horse (200 lb carried 40 miles)	4	80	Trail and pack saddle	(a) (b) (c) (d)	0.02 0.20 0.01 0.40	0.63	0.16
Wheelbarrow (400 lb moved 20 miles)	4	10	Path	(a) (b) (c) (d)	0.04 0.02 0.01 0.30	0.37	0.093
Cart, best conditions (1000 lb moved 20 miles)	10	10	Pavement	(a) (b) (c) (d)	0.08 0.02 0.01 0.30	0.41	0.041
Team and wagon (3 net tons moved 40 miles)	120	600	Good road	(a) (b) (c) (d)	0.53 0.36 0.12 3.60	4.61	0.038
Motor truck (20 tons moved 200 miles)	4000	36,000	Pavement and structures	(a) (b) (c) (d)	7.89 40.80 3.38 40.00	92.07	0.021
Railroad train (2000 net tons moved 50 miles)	100,000	800,000	Tracks and structures	(a) (b) (c) (d)	288.00 500.00 147.00 100.00	1035.00	0.0103

[a] Based on a suggestion in unpublished papers of the late Dr. E. G. Young, former professor, Railway Mechanical Engineering, Univeristy of Illinois.

[b] Costs are those of time and locality where type of carrier was most prevalent. (a) Includes all costs of maintiaining and operating facilities, capital costs not included. (b) Fuel (or feed), oil, water, maintenance, etc., except labor. (c) Includes interest on vehicle only, plus simple yearly amortization charge. (d) Direct labor cost of operating vehicle only. (e) Mileage moved is statistical average for all railroads in the United States; obviously a railroad train could run 320 miles ± in 8 hr.

devised, adding to the costs of the goods thereby secured. The need for transporting individuals over a wider area also arose.

Today, as much as 10 to 15 percent of the costs of any product are transport and distribution costs. Transportation costs in terms of automobile ownership and commutation can represent as much as 10 to 20 percent of the budget in a modern household.

Increasing transportation productivity and lower unit costs have occurred over the years as the systems of transportation become more highly developed and complex. Table 1.1 shows in a generalized form the increases in transport productivity and lowered unit costs as the technology is improved. Costs must be considered at all stages of development.

Geographic Factors   Geography is closely related to economics. The geographical location of natural resources determines the transport routes that give access to those resources and create economic utility, that is, time and place utility, by taking them from a location where they have little value to processing and consuming areas where their value is vastly increased.

The small, waterbound area of the British Isles forced those islanders to look to other lands for food and raw materials and for markets in which to sell their industrial production. It made Britain a seafaring nation with a citizenry skilled in seamanship, and focused world ship building on the banks of the River Clyde. (Japan has gained seafaring and shipbuilding prominence for similar reasons.) Britain established sea routes to her colonial possessions, built ports and fueling stations, and developed bases for a mighty navy to protect those routes and facilities.

The Americas were discovered in an attempt to find a new all-water route for trade with the Indies as an improvement upon overland caravan routes. In the United States, a major objective of early highway, canal, and railway construction was the exploitation and economic development of new territory, especially after the Louisiana Purchase. The ore fields in Michigan, Wisconsin, and Minnesota brought forth one of the most efficiently coordinated transportation systems in the world, the rail-water-rail movement down the Great Lakes via specially designed railroad cars, ore boats, and ore-handling facilities.

The United States economy suffered extensive repercussions when petroleum tankers ceased to ply between the Eastern Seaboard and the Middle East during the winter of 1973–1974. Individual citizens felt the impact in waiting lines at the gasoline pumps.

Great Lakes shipping is a phenomenon of geography in the contiguous placement of natural resources and natural waterways. The Erie Canal and its successor, the New York State Barge Canal, pass through the only natural break in the Appalachian Mountains. The Rocky Mountains yielded only a few passes through which rail lines and highways might be built and led to the development of powerful locomotives and of tunnels, snowsheds, snowplows, and other devices to overcome the mountain defenses. River valleys offered natural waterways and provided easy construction for railroads and highways. Trade winds determined the routes of sailing vessels, and today the winds of the stratosphere affect airline routes and schedules.

The shores of rivers, lakes, ocean harbors, and the crossing points of land routes were the natural choices for centers of population and the problems of urban transport that those centers engendered. In the United States, the Eastern Seaboard and the navigable portions of rivers attracted early settlement. Population followed the westward development of roads, waterways, and railroads.

Political Policies   Political policies frequently play a deciding role in transport development. Note, for example, Russia's constant search for a warm water port. The Japanese desire for economic self-sufficiency, The Greater East Asia Co-Prosperity Policy, would have meant little without the Korean and South Manchurian railways.

The size of the United States has prompted action by the federal government, such as construction of the National Road and the Land Grant Act for railroad construction, to bind its isolated areas with lines of communication. The intimate relation between political policies and the development of the United States highway system, the automotive industry, and the relation of all of these to urbanization and suburbanization is one of the more interesting and sometimes frustrating phenomena of the century. Construction of the Federal System of Interstate (and Defense) Highways illustrates a combination of need, pressures from special interests, and the effectiveness of defense as a motivating factor for Congressional action. The use of Interstate funding to extend the system's routes into and through cities has provided expressways between suburbs and central areas, thereby expanding the urban sprawl into the suburbs and beyond.

Military   The military might of a nation is primarily intended to support its political policies and to provide for national defense. Concurrently, military strategy and tactics often have a direct influence on transport development. The Roman roads were built as routes of conquest for Roman armies. The Civil War provided convincing proof of the vital part transport plays in military operations and highlighted the need for a standard gauge of railroad track. Behind the fighting zones, men and materials of war must be trained, produced, and transported. The historian, Alfred Thayer Mahan, interpreted history largely as a continuing struggle for control of the seas.[1] Sir Halford Mackinder, on the other hand, saw land transportation as enabling land power to outflank sea power.[2] Airplanes and rockets nullified the protection afforded by oceans and forced the United States from a policy of isolation.

A wartime speed-up in research and development enabled both highway and air transport to emerge full-fledged from World War I and to grow into principal modes of transport before World War II. World War II established air transport on a global basis. Rocketry and jet propulsion, radar, and the electronic computer are typical war-inspired developments that have contributed to improved transportation technology.

Technological Factors   Progress in direct and supporting technologies has played an obvious role in transportation.

Early mariners were forced to remain close to shore where the land contours served as their guide. Invention of the astrolabe permitted the mariner to hold to a true course when beyond the sight of land. Application of the compass principal to navigation about 1400 A.D. simplified the problem of keeping on course. The sundial, astrolabe, and compass enabled the navigator to determine his latitude. The problem of determining longitude was not satisfactorily solved until about 1669 when the Italian astronamer, Cacini, suggested that the movements of the satellites of the planet Jupiter be used as an independent clock. More accurate maps and charts were now possible, but errors still existed because of the crude approximations of the earth's circumference. An accurate chronometer was achieved in 1750 by the English watchmaker, John Harrison, in the form of a pendulum clock. Today, a ship or survey party can obtain correct time anywhere on the globe by radio. A system of standard clocks is essential in train operation.

Early in the 1800s, George Stephenson adapted Watt's principles of the steam engine to a steam locomotive. His *Rocket*, Figure 1.4, that incorporated correct design features, accounted in no small part for the rapid growth of rail transportation immediately thereafter.

[1] Alfred Thayer Mahan, *The Influence of Sea Power upon History, 1660-1783*, Little, Brown, Boston, 1890, 1911.

[2] Sir Halford J. Mackinder, *Democratic Ideals and Democracy*, Henry Holt, New York, 1919, 1942.

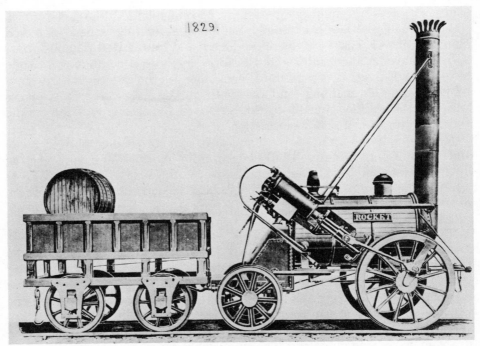

FIGURE 1.4 STEPHENSON'S *ROCKET*. (COURTESY OF THE ASSOCIATION OF AMERICAN RAILROADS, WASHINGTON, D.C.)

Even a successful locomotive was not enough. The economies of long, heavy trains moving at high speeds could not have been realized without the development of Bessemer and open hearth steel for safe, dependable rails; without automatic couplers and draft gears to hold trains together and absorb impacts; and without the automatic air brake to control the enormous momentum and kinetic energy stored in a moving train.

Today's system of highway transport exists because of the invention of powerful, dependable, lightweight gasoline engines, the development of pneumatic tires, and the use of concrete and bituminous materials for highway surfaces. Consider where highway transport would be today without a dependable storage battery, or the Edison electric light, or without Kettering's self-starter. Automobiles might still be a "rich man's toy" had not Henry Ford designed an inexpensive car and the assembly-line process for its low-cost, mass production. The future of automotive transport rests in part upon the ability to design safe and ecologically acceptable vehicles.

The discovery of oil created the technological need for pipelines. The drawing of seamless steel tubing and electric welding of joints made possible modern pipelines capable of withstanding high pumping pressures and tight enough to hold highly viscous refined products.

Airplane development has been largely a story of obtaining a powerful, light weight, and dependable engine. The expansion of air transport is further indebted to lightweight metals, jet engines, and the various electronic navigation and communication aids, plus a knowledge of soils engineering and pavement design. These last permit airport runways capable of withstanding the wheel loads, jet exhausts, vibration, and impacts of big airliners.

The list of technological innovations that have increased the role of transport in our daily life is endless and cumulative.

Competition   The competitive urge in Western capitalistic society has given a powerful impetus to transport development. Railroads compete with railroads and also with trucks, barges, pipelines, and airlines. Freight services compete with express and express with parcel post. Airlines have counted heavily on speed but have also been forced to greater safety and dependability to meet ground transport competition.

No less real is the competition between products and industries tributary to transport. Bituminous materials compete with concrete as a road surface. Diesel locomotives have won over steam but may face competition from electric locomotives in the next decade, especially as petroleum energy resources decline.

Communities and geographical areas are also in competition. Competition between East and West, North and South has brought about traffic patterns that demand and get new transport routes or improvements of the old.

Urbanization   The rapid growth of urban areas by an even more rapidly expanding population is a phenomenon that cannot be overlooked among transport development factors. Accessibility to land and the intensity of land use are closely related to transport availability. Experts will argue as to whether urbanization has created the problems and demands on transportation or vice versa; perhaps it is both. The urban sprawl that accommodates so much of the urban population could not have occurred without the automobile to give personalized mobility to that population and access to land in almost any location.

Population growth and movement from rural to urban sites calls increasingly for improvement of intracity transport. A concurrent "flight to the suburbs" has been abetted (if not initiated) by expressway construction and has led to the building of high-speed automated suburban rail systems such as the Lindenwold line out of Philadelphia and the BART system in the San Francisco Bay area. The expansion of urban areas into metro complexes and then into urban corridors as on the Eastern Seaboard (coupled with concurrent air corridor congestion) has brought efforts at developing high-speed ground transportation (HSGT) systems to permit traveling these corridors at speeds of 100 to 300 mph or more. The detailed configuration of future solutions is not entirely clear, but the problems are already at hand for solution.

Other Factors   The foregoing are not the only factors that determine the extent and type of transport development. Availability of or lack of adequate financing has meant life for some routes and systems, failure for others. Railroads were forced to inferior initial construction standards because shortages of capital precluded more expensive methods of construction and locations. A similar lack of funding is causing problems in the quality of plant and service offered to the public today. Labor has played its part in advancing standards of safety, determining levels of service through the medium of work rules, and aided or hindered modal development through the productivity that comes from its effort. The many facets of man's complex nature and society have all played some part, large or small, in transportation development.

OTHER ASPECTS OF TRANSPORTATION

Sociological   This has been called an "age on wheels." Transportation has changed the customs and patterns of American life. Trailer courts, camper trucks, hitchhikers, motels, and the omnipresent service station are aspects of a migrant population that have accompanied highway transport development and have brought new and difficult problems of health, sanitation, changes in moral standards, law enforcement, and property values. Aerial hijacking is a crime unknown a decade ago.

Urban patterns are changing. Shoestring communities border and blight the edges of highways. Shopping centers appear in outlying locations and drive-in services proliferate. The thrust of expressways into the inner city permits the more affluent to maintain business and employment in the core area while living amid the amenities of the suburbs. The less affluent remain in the decaying central city. Work opportunities for them can only be realized if public transportation is available to job sites. The tragic rioting of 1968 in the Watts area has been attributed, in part, to a lack of transportation between that area and job opportunity sections of Los Angeles. The problem worsens as commercial and industrial enterprises seek less expensive land, labor, and tax costs in the suburbs, thereby creating a need for reverse commuting.

Cultural Transportation's contribution to cultural patterns include a decrease in provincialism. Worldwide differences have diminished through contact by travel and dissemination of printed matter, pictures, and products of industry. Only political restraints deter the process. In the United States, cultural differences and distinctions are disappearing. No part of the country need persist in isolated ignorance.

Not all effects are advantageous. With the disappearance of regional differences, much of the local color and character also disappear, leaving instead a sameness or a drab imitation of genuine local characteristics. The easing of tensions from overcrowding in one region too often means land congestion and the creation of new tensions where migration has occurred. National competitive animosities become international.

The Transportation Industry Private carriage provides transport for the individual and small business. Large industries own and operate private transportation systems as exemplified by United States Steel's railroads and Great Lakes Fleet. Beyond these private carriers is a vast industry of for-hire and common-carrier operations—railroads, trucklines, airlines, and pipeline companies providing service to all or a part of the general public.

These agencies help to support an array of supply and equipment manufacturers—steel mills producing rails, pipe, highway pavement reinforcing bars, and bridge steel; rubber plants making tires for trucks, autos, and airplanes; and copper manufacturers producing wire for signals, motors, and generators, and communications for all carriers. Other industries produce automobiles, barges, locomotives, trucks, ships, and airplanes. All of these must be fueled and lubricated by the products of petroleum production and manufacture. For better or for worse, much of the nation's economy is presently dependent upon the automotive industry. Chapter 3 describes in more detail the present United States system of inland transport and the transportation industry.

Transportation and the Individual The reader might well be amazed at the close interrelationship between transportation and his or her personal necessities, comforts, and luxuries; goods and services available only because of the transport industries that serve his community. He or she and several million more people, may be directly employed by transportation agencies; other millions are indirectly employed in the industries that supply transport equipment.

The cost of almost everything he or she buys is determined in part by the cost and availability of transportation. A large percentage of Federal, state, and local taxes, which supply funds for roads, schools, fire and police protection, and countless other services, is provided by the transportation industry. As an engineer, the reader may design routes or equipment for transport agencies or plan the use of transport in carrying out some other project. As a manufacturer, he will be required to select the types of transport that will give the most efficient and economical access to raw materials and parts and to markets for his finished products. As a financier, he will find transportation a ready field for investment but one beset with many problems. As a lawmaker, he must consider the needs and problems of the country in relation to transportation and the formulation of regulatory policies and measures for transportation services. As a military leader, he must be aware of the essential role of transportation in military logistics and national defense. As a city planner, he must know that a city without adequate transportation will wither and decay or strangle in its own traffic congestion. As sociologist, he must consider the effect upon patterns of life and culture that the freedom to move readily from one locality to another has engendered. As a citizen, he is concerned with all of these manifold functions and relations that are so vital to the individual's and the country's well-being. Finally, as a student, he has the obligation to understand the fundamental principles and relations that govern this essential element in society. Imperfect understanding of these principles has led at times to misuse of transport potential and to economic and social loss. Effects are cumulative; the more highly complex and industrialized a society, the more it is dependent upon transportation.

Engineering students can find in the study of transportation an introduction to the realm of engineering and applications of almost all of the engineering sciences and many of the natural sciences. The design of stable subgrades for railroads, highways, and airports draws upon soils engineering and the mechanics of elastic deformation. Roadbed drainage applies the principles of hydrology, hydraulics, and fluid mechanics. Running surfaces involve the student in the field of engineering materials—steel, concrete, bitumens, and timber—and their behavior under load and varying conditions of temperature, moisture, and support. A study of the roadway and roadway structure presents problems of bridge, tunnel, and general-structure design. Motive power applies the principles of thermodynamics, electricity, chemistry, and the specialized fields of fuels and lubricants. Aerodynamics and fluid mechanics govern much of the equipment design and construction of ships and aircraft. The applications of electronics and radar to operational control, signaling, and communications are numerous.

Summary The transportation system may be considered as a coordinating and integrating agency. It relates population to land use and environment and integrates the United States and, indeed, the entire world into one vast industrially productive unit. It unites the nation and makes it essentially one people in its economy and culture. It could so unite the world, except for political and social barriers.

QUESTIONS FOR STUDY

1. Explain how you, as an individual, and your local community are dependent upon transportation.
2. What transport facilities do you require and which could you do without? Answer this question first with regard to your immediate pattern of living and then in relation to the overall pattern of your society.
3. Discuss the dependence of the United States on world trade and world trade routes.
4. How might the coordinating and integrating functions of transportation be shown graphically?
5. Develop the cost-per-ton-mile relation, similar to that of Table 1.1, for one specific type of transport, that is, trucks, ships, airplanes, pipelines, conveyors, or cableways.
6. Cite examples of regional specialization and show how transportation helps to make that specialization possible.
7. Prepare an outline or diagram to show the scientific principles that are applied in (a) railroads, (b) highway transport, (c) airlines, (d) waterways, (e) pipelines.
8. Describe the role of transportation in the production and distribution of (a) flour, (b) gasoline, (c) fresh fruits and vegetables, (d) an automobile, (e) other specific commodities.
9. (a) Why did the Allied bombers work so intensively upon the destruction of transport facilities during World War II?
   (b) Why did the United States bombing have so little effect in halting the flow of enemy men and matériel during the Vietnam war?
10. What alternatives are there to the services and functions performed by transportation in moving people and goods (a) within an urban area, (b) between urban areas or between urban and rural areas?
11. (a) If you were deprived of your driver's license today, what problems would you experience?
    (b) What groups of people in your community face similar problems?
12. What part has transportation played in (a) growth of the suburbs and (b) inner city problems?

13. What might be the United States' foreign policy today if the airplane and rocketry had not been invented?
14. Using a flow diagram or outline presentation, indicate the various stages at which transportation costs are added to the production of foods stuffs and to the manufacture and sale of automobiles.
15. Explain the extent to which the economy of the United States is related (*a*) to the highway transportation industry and (*b*) to all transportation.

SUGGESTED READINGS

1. Kent T. Healy, *The Economics of Transportation in the United States*, Ronald Press, New York, 1940, Chapters 2 and 3, pp. 41–46.
2. D. Phillip Locklin, *Economics of Transportation*, 6th edition, Richard D. Irwin, Homewood, Illinois, 1966.
3. *Economic Characteristics of the Urban Transportation Industry*, Institute for Defense Analysis for the U. S. Department of Transportation, Washington, D. C., February 1972.
4. A. M. Vorhees, "The Changing Role of Transportation in Urban Development," *Traffic Quarterly*, November 1969, pp. 527–536.
5. *Transportation: Principles and Perspectives*, edited by Stanley J. Hille and Richard F. Poist, Interstate Printers & Publishers, Danville, Illinois 1974.

Chapter 2

Historical Development

PRETWENTIETH CENTURY

Before 1800   Prior to the nineteenth century, human beings depended on unprocessed nature for their means of locomotion. Wind, currents, gravity, animal, and manpower were the means of propulsion. No one really knows when the wheel was invented, but four-wheeled wagons were used by the Mesopotanians as early as 3000 B.C. In the period from 360 B.C. to 360 A.D. the Romans developed the technique of massive construction for roads. Layers of heavy stone were placed in several tiers and topped with large flat stones, bonded together with a lime mortar. Footways and chariotways were separated by low curbing. Attention to drainage and the stone arch bridge were further Roman contributions.

Tresaguet, Telford, and MacAdam, in the late 1700s to early 1800s, developed light road building processes that emphasized drainage, crushed stone subgrades, and covers of finely crushed stone bound with water or oil, giving the familiar macadam type of design.

1800–1900   The most notable early roadbuilding effort in the United States was the National Road extending from Cumberland, Maryland to Vandalia, Illinois (now the route of U. S. 40). Its construction (1806–1830) featured a 20-ft (6.09-m) roadway on a 66-ft (20.1-m) right of way. The road had a 12-in. (30.5-cm) base covered with 6 in. (15.2 cm) of broken stone well-compacted, stone arch bridges, and toll houses every 15 miles

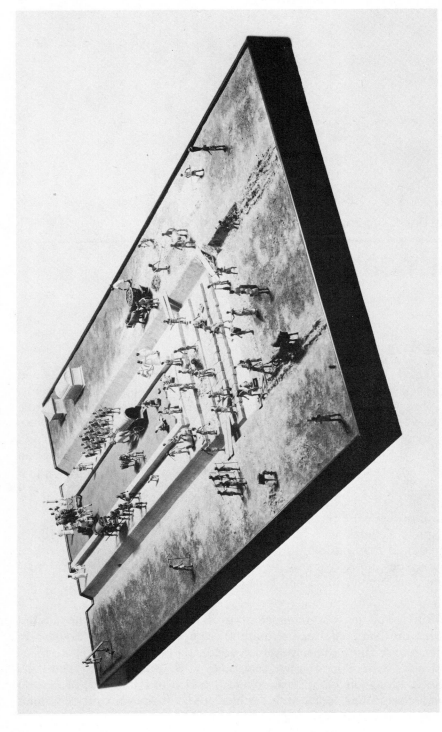

FIGURE 2.1 MODEL CROSS SECTION OF THE APPIAN WAY. MODEL IS PERMANENTLY LOCATED IN THE NATIONAL MUSEUM IN WASHINGTON, D.C. (PREPARED BY THE BUREAU OF PUBLIC ROADS, DEPARTMENT OF COMMERCE.)

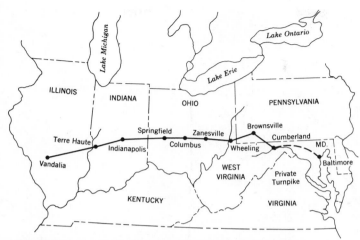

FIGURE 2.2 ROUTE OF THE NATIONAL ROAD, 1830

(24.1 km) with heavy iron gates. A strict interpretation of the Constitution (road building was not mentioned specifically) by the Andrew Jackson administration halted further federal road building enterprise until 1916.

The solution to the problem of establishing latitude and longitude for navigation in open water has already been described. Thus was advanced the many possibilities of navigation—first by current and oar power, later by sail, and, in the early 1800s, by steam.

Aerial transport barely got underway prior to 1900 with balloon ascensions and with dirigibles—envelopes containing several balloonlike bags of a buoyant gas (hot air or hydrogen) and propelled by a light weight motor.

The principles of locks and canals were known and used at an early date, the eighth century in China and the seventeenth century in Europe. In the United States, the Erie Canal, completed in 1825, connected New York via the Hudson River to Troy with the Great Lakes at Buffalo thereby opening a low cost system of freight movement to the newly developing West.

The year 1800 is an approximate date for the coming of steam propulsion with the invention of a steam locomotive by Richard Trevethick and its further development from 1814 to 1830 by George Stephenson, Timothy Hackworth, and others. The first United States commercial railroad, the Baltimore and Ohio, was opened in 1830. By 1860, a network of railroads had spread over the eastern United States. A series of railway reconnaisance surveys, the Land Grant acts, and destruction of the Southern economy by the Civil War marked the decline of river steamboating and permitted the railroads to assume a major role in the westward expansion

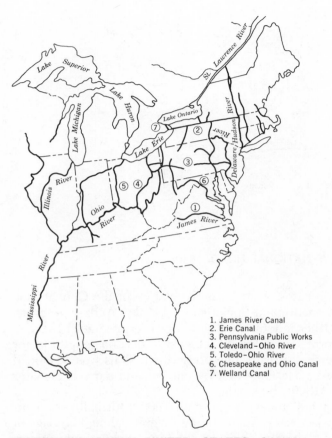

FIGURE 2.3 EARLY UNITED STATES CANAL
ROUTES, 1850

that followed the end of hostilities. The nation's first "transcontinental"
rail line, the Union Pacific-Central Pacific, was completed in 1869 between
Council Bluffs, Iowa and San Francisco Bay. By 1900 there were five
major rail lines to the West Coast. In 1887 the Interstate Commerce
Commission was established; it is a Federal agency intended to regulate
rates and services of the railraods and supplementing agencies. The need
for such regulation arose from the unfair monopolistic attitudes, rates, and
services extended to the shippers and to cutthroat competition among the
rail carriers that too often led to bankruptcy of individual companies and
to poor service to the public.

FIGURE 2.4 RAIL CONSTRUCTION FOR WESTWARD EXPANSION. A CONSTRUCTION SCENE DURING THE EARLY DAYS OF RAILROADING. (COURTESY OF PARAMOUNT PICTURES.)

Between 1860 and 1897 a practical design of the gasoline motor-driven highway vehicle was developed. This radical innovation in highway transport technology evolved slowly and gave little early indication of the important role it would play in the twentieth century.

Petroleum, destined to play a major role in highway transportation, was brought to the surface from the first oil well in 1859. The first petroleum pipeline went on-line in 1865.

AFTER 1900

Twentieth Century Rail Transport   For the railroads, the twentieth century has been a time of combination, consolidation and, more recently, of retrenchment. During the first two decades large systems such as the Pennsylvania, Southern, and Union Pacific, grew by consolidating smaller companies. Some mergers ran afoul of the antitrust laws and were disbanded by court orders. During the 1960s and 1970s, a more liberal view permitted merger of the Burlington, Northern Pacific, and Great Northern into the successful Burlington Northern, the Seaboard Coast Line from the Atlantic Coast Line and the Seaboard Air Line, and the less successful Penn Central combination (from the Pennsylvania, New York Central, and the New Haven), now reorganized with other lines into Conrail.

Some of the mileage constructed prior to 1930 has become superfluous because of mergers, competition from other modes, or the shifting and dwindling of traffic sources. A slow but steady reduction in such mileage is underway, thereby relieving the railroads of operating unprofitable branch lines and permitting a concentrated effort in upgrading high density lines.

Highway trailers hauled on railroad freight cars in the early 1930s had only limited use until the introduction of a two-trailer flat car and corresponding rates in the 1950s. Trailer-on-Flatcar (TOFC) or "flatback" combines the railroads' long haul efficiency with the speed and flexibility of highway carriers in making origin-destination pickups and deliveries and in performing feeder service.

A new type of motive power, the diesel-electric locomotive, was placed in yard switching service on the Central of New Jersey, 20 October 1925. With the exception of a relatively few miles of electrified line, most railroads today are 100 percent dieselized.

Immediately before and after World War II numerous railroads attained speeds of 90 to 100 mph (145 to 161 kph) with passenger trains using steam, diesel, and electric locomotives. In the 1960s trains have attained speeds of over 120 mph (193 kph) (the Japanese Railways's *Takaido Express*) and as high as 186+ mph (299 kph) in the United States. The electrified Metroliners between New York and Washington are targeted for eventual average speeds of 125 mph (201 kph) with spot speeds as much as 150 mph (241 kph). The world's speed record is still held by the French SNCFs *Mistral*, which attained a speed of more than 200 mph (322 kph) on the Paris-Lyon-Marseilles line.[1]

A system of centralized traffic control (CTC) was placed in operation on the New York Central on 25 July 1923 between Toledo and Berwick, Ohio. Its application during World War II gave a much-needed increase in traffic capacity to many single-track lines, which otherwise would have been required, in a time of steel shortage, to build additional track miles.

Other major technological innovations of the century have included air conditioning, mechanization of maintenance work, stabilization of roadbeds through soils engineering, and the use of continuous welded rails (CWR) in track. High capacity cars and locomotives, improved brake

[1]The rail-mounted Linear-Induction Motor Vehicle (LIMV) recently obtained a speed of 255 mph (410.3 kph) at the Department of Transportation's Pueblo Test Track.

systems and draft gears, and the use of remotely controlled slave or drone locomotive units spaced throughout the consist permit trains of 100 to 200 cars. Unit trains hauling train loads of a single commodity in shuttle service exploit the high capacity–low cost capability of rail service but have also introduced a need for extended research in track/train dynamics that is now in progress under AAR, FRA, and railway suppliers auspices.

Radio has improved communication and the safety and efficiency of railroad and train operations. Microwave is coming into use for dispatching, especially in areas where line wires are subject to wind and ice storm damage. Electronic computers are essential for payroll and car accounting, for automated classification yards, for automatic car identification and for providing real time and other data to assist in supervision and managerial decision making.

Following an all-time high during the automobile fuel rationing of World War II, rail passenger traffic dwindled to a money-losing low in 1970. At that point, 30 October 1970, Congress enacted legislation creating a National Railroad Passenger Corporation, popularly known as AMTRAK, to assume ownership of intercity rail passenger equipment and to contract with the railroads for operating AMTRAK trains. Rail service for commuters remained with the railroads and has maintained a moderate level of patronage.

Financial difficulties of railroads in the Northeastern United States, especially those of the bankrupt Penn Central, led to the Rail Reorganization Act of 1973 that became law in January 1974. The Act provides for various loans and grants to maintain temporary operation of existing services and sets up the United States Rail Corporation to function as a planning and finance agency to establish a system of viable rail lines in the Northeastern United States. It has created the Consolidated Rail Corporation (Conrail) to operate the system restructured from viable and necessary portions of existing rail lines in the Northeast as a private for-profit rail system.

The Automotive Age   The Federal Road Act of 1916 provided Federal funds for up to 50 percent of the cost of improving state roads. Initiative was left with the states to determine the roads to be built and to carry on actual construction and maintenance.

The motor truck gave a convincing demonstration of its usefulness in World War I. To preserve order, competition, and adequate service both for the public and the rapidly expanding trucking industry, the Motor Carrier Act of 1935 (later—in 1940—Part II of the Interstate Commerce Act) was enacted by the Congress. Lack of funds and the heavy demands placed on highways by World War II traffic added to the congestion, rapid deterioration, and obsolescence of all roads.

An intensified effort was therefore made to build highways of modern design with four- and six-lane roadways, divided as to direction, with limited access and with no cross traffic or intersections at grade. Financing in many instances depended on tolls as security for revenue bonds. The Federal Highway Act, passed June 29, 1956, and supplemented by an amendment of 16 April 1958, authorized construction with Federal funds based on these high standards of more than 41,000 miles (65,969 km) of interstate highways to connect principal centers of population. Funding was provided by Federal taxes on automotive fuel and other supplies and was segregated in a trust fund for highway construction purposes only.

The heavy trucking in use after World War I made earlier construction inadequate and led to the use of concrete as a paving material. There are now more than 85,000 miles (136,765 km) of concrete-paved highways among the more than 3 million miles (4.8 million km) of United States roads, much of it in urban areas. The superhighways of today call forth the best in cement and asphaltic-concrete, heavy-duty pavement construction with a full application of the principles of soils engineering and drainage. See Figure 2.5.

An ample supply of fuel seemed assured with the bringing in of the fabulous Spindletop oil well near Beaumont, Texas on 10 January 1901. This supply coupled with the lightness, relative cleanliness, simplicity, and ruggedness of the internal combustion engine made gasoline the favored means of automobile propulsion. In recent days, a possible shortage of such fuel, and ensuing higher prices, has aroused concern and created a growing demand for smaller cars with greater fuel economy.

By 1904 there were 55,000 cars on the road, and the automotive industry was tending to center around Detroit. Important manfacturing techniques led to the large-scale production of a car at a price the general public could afford. In 1908 Henry Leland introduced Eli Whitney's interchangeable parts in his Cadillacs, and Henry Ford designed a simple standardized car produced on a moving assembly line. The invention of the self-starter in 1910 by Charles F. Kettering helped to popularize motor cars. Standardization of parts was introduced in 1910 by the Society of Automotive Engineers (SAE), a factor that contributed economy of manfacture and interchangeability of parts.[2]

By the end of World War I in 1918 there were 5.5 million passenger cars registered. Registration has since risen to more than 93 million passenger

[2]*The Automobile Story*, General Motors Corporation, Detroit, Michigan, 1955, p. 3.

FIGURE 2.5 MODERN SUPERHIGHWAY AND INTERCHANGE—EAST SHORE EXIT FROM SAN FRANCISCO'S OAKLAND BAY BRIDGE. (COURTESY OF *HIGHWAY MAGAZINE*, ARMCO DRAINAGE AND METAL PRODUCTS, INC., MIDDLETOWN, OHIO.)

vehicles and 18 million trucks. The latter have increased in capacity from the small 5- to 10-ton (4.5- to 9.1-tonne) truck of the 1920s to tractor-trailer combinations hauling 20 to 40 tons (18.1 to 36.2 tonnes) on the highways and to specialized vehicles of 80 to 150-tons (72.6- to 136.1-tonnes) capacity used in mine hauls and earth moving.

Flight   Count Ferdinand von Zeppelin's first rigid dirigible, a cigar-shaped craft, made its first successful flight in 1900. Buoyancy was obtained from 16 hydrogen-gas-filled bags contained in aluminum compartments within the 416 ft (126.8-m) frame. A speed of 20 mph (32.2 kph) was attained from the thrust of two 16-hp Daimler engines. The total weight was 9 tons. During World War 1 the zeppelins engaged in aerial reconnaissance and bombardment of England. Following the war, Germany built a few zeppelin-type dirigibles for the United States as a part of war reparations. The United States Navy, in cooperation with the Goodyear Tire and Rubber Company, also built rigid and semirigid dirigibles. Others were constructed by the British. Although numerous intercontinental flights were made, including one over the North Pole and another around the world, the lighter-than-air craft proved expensive to build and operate and difficult to control in rough weather. After a series of spectacular disasters had wrecked all but one of these airships, efforts to develop the design were abandoned.

The nonrigid types of dirigible, developed by A.S. Dumont during the first years of the century, have had a continuing minor use in naval operations and advertising.

Successful flight in a heavier-than-air craft was first attained December 17, 1903 at Kitty Hawk, North Carolina, when the Wright brothers, Orville and Wilbur, launched and flew a heavier-than-air craft for three wobbly minutes. It was powered by a 16-hp gasoline engine designed by the Wrights and weighed only 7 lb (3.18 kg) per horsepower. The pilot lay stretched out full length on the lower ring. See Figure 2.6.

Airplanes were used for observation and reconnaissance, bombing attack, and pursuit in World War 1. The box-kite design took on a fuselage and aerodynamic design was improved, but probably the most important development was the designing of dependable, lightweight motors.

At the close of the War, coast-to-coast flights in the United States, intercontinental, and around-the-world flights became realities. The most notable of these was Charles Lindbergh's solo flight across the Atlantic in his single-engined Ryan monoplane, "The Spirit of St. Louis" on May 20–21 1927.

FIGURE 2.6 THE WRIGHT BROTHERS AIRPLANE. THE START OF IT ALL! THE FIRST FLIGHT TOOK OFF FROM THE SANDS OF KITTY HAWK, NORTH CAROLINA, ON DECEMBER 17, 1903. THE FLIGHT COVERED 120 FT (36.6 M) IN 12 SECONDS. (COURTESY OF UNITED AIR LINES.)

Airmail service was established in 1916 by the Post Office Department. Scheduled transcontinental airmail between New York and San Francisco became operative in 1924. The Kelly Mail Act of 1925 authorized mail-carrying contracts between the Post Office Department and commercial airline opertors, which led to the establishment in 1926 of 14 domestic airmail lines, forerunners of today's commercial carriers of passenger, mail, express, and freight.

The Air Commerce Act of 1926 provided, among other things, for safety, licensing, and air traffic control. The Civil Aeronautics Act of 1938 combined and placed under the Civil Aeronautics Authority (CAA) all of the functions of aid and regulation of the aircraft and air-transportation industries. Establishment of airways and air-traffic control were among the important aspects of CAA supervision. Under the revisions of the Federal Aviation Act of 1958 operational control of United States navigable air space was conducted by a Federal Aviation Agency that, in 1968 became a part of the Department of Transportation as the Federal Aviation Administration. Economic regulation has continued since 1928 under the Civil Aeronautics Board (CAB) with authority over rates, mergers, and other economic matters.

By 1940 improved fuels and engines, radio communications and electronic navigational aids, lightweight metals, and other developments made possible a dependable and comfortable airplane for commercial service, best exemplified by the still-active DC-3. The DC-3 with two engines, had a capacity of 21 passengers and 3 crew members, and a cruising speed of 180 mph (290 kph).

Planes that were designed to carry World War II heavy bomb loads for long distances held equal opportunity as cargo carriers. Troop carriers presaged the high-load capacities of the DC-6, DC-9, and Constellation classes. High-octane fuels, superchargers, and pressurized cabins made flight in the stratosphere a reality. Radar added safety and comfort. Improved knowledge of soils and paving materials made possible the runways for these craft.

The War also introduced jet propulsion that is now adapted to commercial use in such modern airlines as the 300± passenger Boeing 747 and Douglas DC-10. The German V-1 and V-2 rockets of World War II were forerunners of the modern rockets with which man traveled to the moon and now probes outer space.

Speeds exceeding that of sound have been achieved by military and experimental airplanes, but the United States' effort to build a supersonic commercial transport (SST), successfully achieved by the French-British *Concorde* and the Russian *Tupolo*, have run afoul of environmental impacts and high design, construction, and operating costs.

A more successful innovation is the helicoptor, which has performed successfully in commercial transport service, as an adjunct to construction in difficult places, and for troop carrying and other military purposes in the war in Southeast Asia.

Waterways   An International Waterways Convention, held in Cleveland in 1900, and the first Rivers and Harbors gathering, held that same year in Baltimore, created a new interest in inland-water transportation. Passage by the Congress of the Rivers and Harbor Act in 1902 established a Board of Army Engineers to receive all recommendations for inland-waterways projects.

By 1903 New York State had voted $101 million for the relocation and improvement of the Erie Canal, renaming it the New York State Barge Canal. October 23, 1929 marked the official completion of a 9-ft (2.74-m) channel in the Ohio River (begun in 1907) from Pittsburgh, Pennsylvania, to Cairo, Illinois and the Mississippi River. In later years there was increased slack-water development of the upper Mississippi and its tributaries, notably the Missouri and Illinois rivers.

In the late 1930s diesel-powered, propellor-driven towboats or rams began to replace the steam-driven stern-wheelers. The propellers are mounted in grooves or tunnels in the ship's hull to permit operation in shallow water. Diesal craft are of 1000- to 5000-hp, 117 to 202 ft (35.7 to 61.6m) long, 30 to 55 ft (9.1 to 16.8m) in beam, and drawing 7.6 to 8.6 ft (2.32 to 2.62m) of water. Craft of 10,000 hp are under construction. See Figure 4.2. Modern vessels are equipped with radio telephone and with radar that permits safe operation when visibility is impaired by rain or fog. Mechanization of loading and unloading at terminals and wharves added speed and economy. Completion of the Calumet Sag project in Chicago integrates the inland-river system with the Great Lakes-St. Lawrence Seaway System.

The relatively minor passenger traffic on the Great Lakes has virtually disappeared, but carriers of ore, grain, and coal have become larger and numerous, especially during the world wars when military needs required the utmost in productive capacity. Today's steel freighters, 700 ft (213.4m) long, carry 15,000 to 50,000 tons of cargo. High level gravity ore docks and conveyor systems for loading and bridge cranes and Hulett unloaders, familiar at southern lake ports, permit vessels to be loaded and unloaded and turned around in a few hours instead of several days.

Pipelines   The introduction of welded joints to replace screw joints permitted the development of high pressures and long-distance pumping. The first oxyacetylene welded line for gas flow was 1000 ft (304.8m) long, in Philadelphia in 1911, but the first successful high-pressure, electrically welded line was not completed until 1920. In 1928 electric welding and the manufacture of seamless steel tubing reached a state of technical excellence that permitted a combination of the two in high-pressure, long-distance lines. The tight joints and pipe necessary to retain high-viscosity liquids and gases became a reality. There has been a rapid expansion of high-pressure, long-distance pipelines not only for the transportation of petroleum but for the flow of gasoline, kerosene, and natural gas.

Pumping facilities have reflected improvements in line and related technologies. High-pressure turbine and centrifugal pumps and electric and diesel-fueled prime movers have replaced cumbersome steam equipment in most stations, along with centralized and computerized control of pumping stations and flow. Various forms of radio and microwave transmission facilitate communication between stations, a vital part of pipeline operation.

The "solids" lines have had some success. A "solids" pipeline, built in 1957, carries Gilsonite in suspension 72 miles (115.8km) from Bonanza, Utah to Gilsonite, Colorado.[3]

A 108-mile (174-km) pipeline, placed in operation in April 1958, to carry powdered coal in suspension from Georgetown, Ohio to Cleveland, Ohio proved the technological feasibility of the "solids" pipeline, but experienced an economic setback with the advent of the unit-train type of rail haul. See Chapter 5 for a more recent installation.

Innovative Systems   The third quarter of the century has been marked by considerable interest, research and development, in new and exotic transport systems. Vehicles supported on cushions of air, guided by center or side rails, or by directed jets of air and propelled by linear-induction motors have been proposed for very high-speed (over 300mph) (483 kph) intercity and regional passenger travel. Slower-speed modifications of these and other technologies are being proposed as "people movers" within the

[3]A.S. Lang, "The Use of Pipe Lines for the Long Distance Transportation of Solids," *Proceedings of the A.R.E.A.*, Vol. 60, American Railway Engineering Association, Chicago, 1959, pp. 240-244.

crowded, congested urban areas. Several experimental projects have been constructed but only small intra-airport systems and the air-jet supported Hovercraft have had extended commercial service, the latter primarily across the English Channel.

Summary   The twentieth century has seen the rise of rail transportation and its loss of significance, perhaps temporarily only, as the automobile technology and widespread, public-supported highway construction and the publicly assisted air transportation technology offered competitive alternatives, especially in the movement of passengers. The railroads' unit train operation is in turn offering stiff competition to water transport in the movement of bulk freight. Pipeline technology provides another alternative, for solids as well as for liquids. An outstanding feature of later years is the problem of urban transportation. Exotic transport systems are undergoing research and development to satisfy both urban and interurban needs and to provide freight as well as passenger movement.

QUESTIONS FOR STUDY

1. What were the contributions of the Romans to transportation engineering?
2. Develop significant illustrations, other than those used in the text, for each of the several factors affecting transport development.
3. Contrast the contributions of Tresaguet, Telford, and MacAdam. What engineering principles did their work exemplify?
4. Why does the beginning of the nineteenth century stand as an approximate milestone and dividing line in the development of transportation?
5. Early United States railroad development was said to be "pioneering" in character. Explain the technological meaning and significance of the statement and the factors that permitted and/or required that type of development.
6. Using a modern transport unit (automobile, airplane, ship, etc.), outline the technological developments not connected with transportation that make that unit possible.
7. Why did the end of the Civil War mark the decline of steamboating and the rise of western railroads?
8. What technological developments make possible a 150-car freight train running at 50 mph?
9. Account for the rapid growth of gasoline, kerosene, and benzine pipelines after the late 1920s.
10. On what technological basis can one account for the rapid popularization and use by the general public of private automobiles?

11. Prepare a graph showing the change in average and maximum speeds from the year 5000 B.C. to the present.
12. Relate the rate and extent of urban and suburban growth to the development of transport technology.

SUGGESTED READINGS

1. Caroline MacGill and a staff of collaborators, *History of Transportation in the United States before 1860*, Carnegie Institution of Washington, Washington, D.C., 1917.
2. Nicholas Woods, *Treatise on Railroads*, Longman, Orme, Brown, Green, and Longman's, London, 1838.
3. Charles Francis Adams, Jr., *Railroads: Their Origin and Problems*, G. P. Putnam's Sons, New York, 1878.
4. Michel Chevalier, *Histoire et description des voies de communication aux Etats Unis et des travaux d'art qui en dependent*, Librairie de Charles Gosselin, Paris, 1840.
5. Harlan Hatcher, *Lake Erie*, and others in his Lakes and Rivers series, Bobbs-Merrill, New York, 1945.
6. Kent T. Healy, *The Economics of Transportation in America*, Ronald Press, New York, 1940, Chapters 1 through 9.
7. Franz von Gerstner, *Report on the Interior Communication of the United States*, Vienna, 1843.
8. Harlan Hatcher, *A Century of Iron & Men*, Bobbs-Merrill, New York, 1950.
9. Charles Francis Adams, Jr., *A Chapter of Erie*, Fields, Osgood and Company, Boston, 1869.
10. Mark Twain (Samuel L. Clemans), *Life on the Mississippi*, Harper & Brothers, New York.
11. Charles Edgar Ames, *Pioneering the Union Pacific*, Appleton-Century-Crofts, New York, 1969.
12. Christy Borth, *Mankind on the Move*, Automotive Safety Foundation, Washington, D.C., 1969.
13. Frank J. Taylor, *High Horizons*, McGraw-Hill, New York, 1959.

Chapter 3

The Transportation System

There is no United States transportation system in the sense of a formally organized entity. The term must refer to a loosely fashioned group of individual systems and facilities, some publicly owned, others privately owned or privately owned but with common carrier status and subject to varying degrees of public regulation.[1] This system has developed with a minimum of planning. Hence there is a lack of uniform integration and coordination and a considerable disproportion in the quality and extent of services offered and with wide variations in the degree of private and public funding.

Only two modes, rail and pipeline, combine ownership of the five system elements (see Chapter 1) in private operating companies. With the others —highway, waterway, airway—the roadway-guidance and terminal elements of the systems are provided by public agencies. There are 58 Class I railroad companies, 12 major trunk and 14 local service airlines, 71 large bus lines, 1355 major motor freight lines, 56 large carriers by water, and numerous urban transit companies. The largest system of passenger carriage resides in the individual private ownership of automobiles driven upon publicly owned and maintained streets and highways. There are over 101 million of these.

[1]This is in contrast to an arrangement in many foreign countries where, as in South Africa for example, the ownership and operation of rail, air, highway, port, and coastal shipping facilities are vested in a ministry of transport.

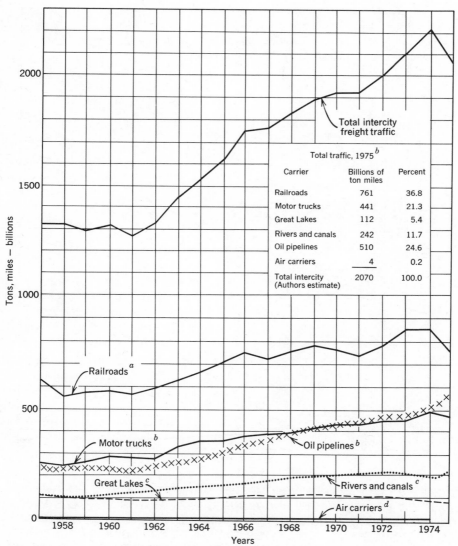

Total intercity freight traffic

Total traffic, 1975[b]

Carrier	Billions of ton miles	Percent
Railroads	761	36.8
Motor trucks	441	21.3
Great Lakes	112	5.4
Rivers and canals	242	11.7
Oil pipelines	510	24.6
Air carriers	4	0.2
Total intercity (Authors estimate)	2070	100.0

Railroads[a]

Motor trucks[b]

Oil pipelines[b]

Great Lakes[c]

Rivers and canals[c]

Air carriers[d]

Tons, miles — billions

Years

FIGURE 3.1 COMPARATIVE VOLUMES OF INTERCITY FREIGHT TRAFFIC. (*a*) *TRANSPORT STATISTICS IN THE UNITED STATES* FOR THE YEAR ENDING 31 DECEMBER 1974, INTERSTATE COMMERCE COMMISSION. (*b*) *MOTOR VEHICLE FACTS AND FIGURES 1976*, MOTOR VEHICLES MANUFACTURERS ASSOCIATION, DETROIT, MICHIGAN, 1976, P. 56. (*c*) *INLAND WATERBORNE COMMERCE STATISTICS*, AMERICAN WATERWAYS OPERATORS, INC., 1971, P. 4. (*d*) *AIR TRANSPORT 1976*, AIR TRANSPORT ASSOCIATION OF AMERICA, 1976, P. 26.

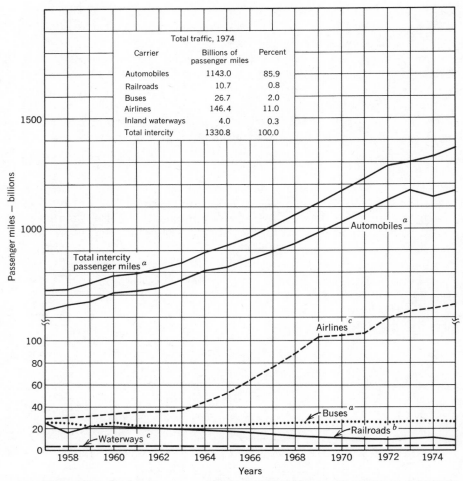

FIGURE 3.2 COMPARATIVE VOLUMES OF INTERCITY PASSENGER TRAFFIC. (*a*) *MOTOR VEHICLE FACTS AND FIGURES 1976*, MOTOR VEHICLES MANU-FACTURERS ASSOCIATION, DETROIT, MICHIGAN, P. 54. (*b*) *TRANSPORT STATISTICS IN THE UNITED STATES* FOR THE YEAR ENDING 31 DECEMBER 1974, INTERSTATE COMMERCE COMMISSION, WASHINGTON, D.C. (*c*) *AIR TRANSPORT 1976*, AIR TRANSPORT ASSOCIATION OF AMERICA, WASHING-TON, D.C., 1976, P. 26.

Each category of society has its own characteristic transport requirements. A family unit will utilize one or more automobiles for shopping, pleasure, work trips, and personal business. These are serviced and maintained at the family's expense. A large corporation may require and own or contract for fleets of automobiles and airplanes for its personnel. It may also finance, operate, and maintain its own private railroad or steamship line and have its own fleet of trucks on the publicly owned highways. Each unit of society—family, farm, or business—will also make some use of for-hire or public transportation, whether by special contract or by utilizing common carrier services offered to all at established and government-regulated rates.

The extent to which use is made of major units in the "system" is indicated in Figures 3.1 and 3.2. The detailed classifications that follow aid in understanding their extent, variety and complexity. These classifications are presented without inquiry into the reasons for the existence or design and operating practices of an individual carrier. An understanding of the technological and economic considerations that are responsible, in part, for the system in its present form is a purpose of this book and will be developed in later chapters.

CARRIER CLASSIFICATIONS

Components of the United System   Major transportation modes operating in the United States can be classified in many ways. One obvious classification is by traffic types, primarily *freight and passenger*. Passenger systems can be subdivided into the intercity and urban. Freight systems can be classified by the commodities handled or as originators, terminators, or transferers of freight. Another possible grouping is the *for-hire* and the *private* transport systems. The for-hire group are further classified as *common carriers*, *contract carriers*, and *exempt* (from regulation) *carriers*. Personal or individual carriage contrasts with mass or quantity movement. Carriers are technologically grouped as to mode—railways, highways, waterways, conveyors, pipelines, and numerous others. Further classification can be by motive power—internal combustion engine, jet propulsion, electric drive, or steampower, etc. Classification also can be by vehicle motive power and carrying arrangement—single unit (carrying space and motive power in the same vehicle), multiple or assembled units (vehicles assembled into trains or tows), and continuous flow systems with the roadway usually serving as cargo container as with pipelines or conveyors.

Note that none of these classifications are absolute or exclusive. One grouping merges into another making precise classification difficult.

Modal Systems The use and distribution of transport by mode are continuing sources of controversy. Six major categories and several fringe and developing groups can be identified.

Railways. This mode utilizes the flanged-wheel-on-rail technology. The rail may be either conventionally rigid or flexible as are aerial tramway cables. The familiar intercity railroads, rapid transit, street cars (light rail), some designs of monorail, aerial tramways, automobile carriers, inclined planes and tracked air cushion vehicles (TACV) are of this mode.

Highways. A rubber-tired wheel on a smooth, firm roadway features the technology used by automobiles, trucks, buses, tractor-trailer combinations, bicycles, motorbikes, some monorails, taxis, dial-a-bus and minibus systems and some portions of so-called personalized rapid transit (PRT) people carriers for urban traffic.

Waterways. Natural or artificial channels and bodies of water serve as roadways. Ships of many designs and purposes, barges, tows, car floats, pleasure craft, hydroplanes, submarines, and towbags float in or on the water.

Airways. The use of air space at a more than nominal height above the ground is basic to this technology. Commercial jet and propellor craft are the usual examples, but the group also includes balloons, rigid and semirigid dirigibles, helicopters, vertical-takeoff-and-landing (VTOL) and steep-take-off-and-landing (STOL) craft, small personalized private planes, rockets, and spacecraft.

Pipelines. Cargo, usually in liquid form, is pumped through long tubes that provide containment as well as roadway. The pipes, normally but not necessarily, located below the ground surface, provide transportation for water, sewage, petroleum and its products, gas, steam, heat, and other liquids and gases. "Solids" pipelines and pneumatic tube systems are in this category.

Conveyors. The belt conveyor on which cargo is carried and propelled by motorized rollers is the usual form of this technology and is customarily associated with the movement of granular bulks. There are other conveyor types but only the belt conveyor has been used for long distance haul. Applications of the belt conveyor principle include moving stairs and sidewalks and the Carveyor and similar types of people movers for urban areas.

Multimodal systems. These combine two or more modes to provide utility and service. Highway trailers and containers on railroad flatcars (TOFC and COFC) combine the long distance speed and economy of rail haul with the flexibility of highway movement within terminals and urban areas. Auto-train systems have recently permitted travelers to take their cars with them on the same train. Buses have been designed to run both on rail and highway.

Exotic systems.   These are mostly design concepts with only a limited amount of pilot model testing. Here included are the tracked air cushion vehicle (TACV), the air cushioned vehicle (ACV), the linear-induction motor vehicle (LIMV), very high-speed tube systems such as the Foa and Edwards, and various designs of urban people movers—the Starrcar, Teletran, and Transit Expressway. None of these except the Transit Expressway have yet been accepted for commercial use in scheduled operation through all kinds of weather, day in, day out.

Quasi-transport.   Transportation substitutes or alternatives are found in this group. The telephone, television, and their variations reduce the need for people travel. Facsimile transmission by wire and radio of letters, pictures, and diagrams reduces the need for mail service. Electrical energy produced by thermal plants at the mine mouth eliminates the need to transport coal long distances through the use of power transmission lines. Mine mouth ore beneficiation and the concentration of frozen juices near the fruit orchards reduce the bulk of raw materials being moved.

Unit of Carriage   Transport systems may be distinguished by the prevailing unit of traffic carried.

People movers.   The *individual* ownership and use of a private means of transport permits a degree of flexibility, privacy, comfort, and freedom that has universal attraction to all peoples and has led to the proliferation of bicycles, motorbikes, and, most obviously, the automobile. The effort to provide mass movement by an individual carrier, the automobile, has led to much of the urban transportation problem with its high costs, congestion, and adverse environmental impact. Minibuses, taxis, and limousines combine the characteristics of individual and mass carriers.

Mass transportation is geared to the economical movement of large numbers of people in a minimum number of vehicle units. Mass transit may involve the use of 25 to 55 passenger buses, 6- to 10-car rapid transit trains, with capacities of 60 to 200+ passengers (including standees) per car, and commuter railroad trains of 6 to 10 cars with seating capacities of 80 to 165 persons. Modern aircraft carry 200 to 400 persons at a time.

Movers of Goods.   Individual or *small lot freight services* are exemplified by the pickup truck serving a single individual, farm, or firm. Less-carload and less-truckload freight give an individualized service. Express, mail, and parcel post services are also more or less individualized.

Mass, bulk, or quantity freight movement is conducted by pipelines, bulk cargo carriers, and barges, and by unit trains of 80 to 150 cars transporting such bulk commodities as coal, ore, grains, and petroleum. Containerization is an effort to obtain the economies of bulk handling for small and individual shipments.

Private vs. For-Hire Systems Increasingly large volumes of traffic are moving in privately owned vehicles. The major mover of people is the privately owned automobile. Privately owned airplanes also carry some intercity traffic. Private freight carriage ranges from the small farm truck or delivery vehicle to large fleets of trucks serving major segments of industry. Private freight carriers, like the automobile exempt from all but vehicle licensing and safety regulations, transport the owner's goods or goods for which the truck owner is lessee or bailee. They are not for hire. The subterfuge of a bill of sale for a bill of lading has sometimes been used to create the appearance of private ownership and thereby escape regulation. The steel and petroleum industries operate private fleets of bulk cargo carriers and barges. They also own and operate private railroads.

For-Hire Systems For-hire systems make the hauling of traffic for revenue purposes their sole activity and are subject to varying degrees of economic and administrative regulation as well as being subject to safety requirements. For-hire carriers in 1972 accounted for 76.3 percent of the ton miles of motor freight on main rural roads; private carriers accounted for 23.7 percent.[2]

Common Carriers These carriers offer their services to any of the public wishing such use. They have the status of public utilities, that is, they are closely endowed with the public interest and are subject to the maximum amount of public regulation, especially of rates, by Federal and state agencies. The conventional railroads, airlines, truck lines, bus companies, and urban transit companies fall into this category. Common carriers may also engage in contract carriage. Common carriers must have a certificate of public convenience and necessity in order to operate over specified routes and cannot terminate service without regulatory permission. They must give adequate service without discrimination as to persons,

[2]*Motor Truck Facts*, 1975 edition, Automobile Manufacturers' Association, Detroit, Michigan, p. 34.

places, or things. Economic regulation extends not only to the setting of maximum and minimum rates but also to the setting of specific rates.

Contract Carriers   These companies do not offer open services to the general public but only to that segment with which they engage under individual contract for transportation. Contract carriers are mostly in the motor freight field. Among these are many small owner-operator one-rig independents. They are subject to modified regulation by the ICC (and the state authority) including control of maximum and minimum rates.

Exempt Carriers   These would fall within the for-hire categories were they not specifically exempted by statute from the economic regulatory features of the Interstate Commerce Act. The exemption is an effort to aid the industry being served. Carriers of farm products, newsprint, fish, and other commodities are exempted.

Ownership of System Components   Railroads, pipelines, and conveyor systems are usually completely owned—right of way, roadway, vehicles, motive power, and terminals—by the operating carrier. The financing of both capital and operating expenses are the carriers' responsibility.* Their properties are taxed by federal, state, and local governments. An exception to this is Amtrak, a quasi-public corporation sponsored with government funds that contracts with rail carriers to move its passenger trains.

Highway carriers own their own trailers, tractors, and freight terminal facilities but utilize the publicly owned and financed highways. They contribute to the costs of the roadways through taxes on fuel, oil (and on tires and accessories for trucks).

Airlines own their own aircraft and maintenance facilities, but the airports are almost universally publicly owned, usually by cities, counties, or the Federal government. The airlines pay landing fees (average in 1973 of $0.95 per 1000 lb) and space rentals for hangars, ticket sales, office space, passenger waiting rooms, and loading ramps. The airways are established and marked with navigational aids by the Federal Aviation Administration. Airways operations are controlled by the FAA's Air Traffic Control Stations. Airlines pay a seat tax that is passed on to the patrons.

The waterways are maintained by the Corps of Engineers, which also build and operate dams and locks to provide canalized and slack-water

*It should be noted that in recent years a considerable sum of money through outright grants and guaranteed loans has been made to some railroads through the Federal Government.

navigation channels. Operators own their own towboats and barges but pay no fees or taxes of any kind for using the waterways and locks. All efforts to impose some kind of a tax have met with strong Congressional opposition. Wharf and terminal facilities may be owned by the ship or barge line or by local municipalities.

PUBLIC POLICY AND REGULATION

U. S. Transportation Policy   The Transportation Act of 1940 amending the Interstate Commerce Act declares it to be the national transportation policy of the Congress:

> To provide fair and impartial regulation of all modes of transportation so administered as to preserve the inherent advantages of each; to promote safe, adequate, economical and efficient service and foster sound economic conditions in transportation and among the several carriers; and to encourage the establishment of reasonable charges without unjust discrimination or unfair or destructive competitive practices.

Opinion varies as to whether these goals, worthy as they are, are sufficient, let alone achieved. Public policy is better indicated by the content of Congressional legislation and by the attitudes and actions of the public itself.

Regulation has been, and is, a part of that policy. The origins of United States regulation go back to the days when railroads, enjoying a transportation monopoly, engaged in a variety of abuses. Among these abuses were the charging of exorbitant freight rates, particularly to the farmers, discrimination, especially through the granting of rebates to a preferred group of large oil shippers, and cutthroat competition among themselves. This last, in which goods were sometimes hauled at rates less than the costs of the service, drove certain roads into bankruptcy or left them in such a weakened financial condition that they were unable to compete or to render adequate service to the public.

Regulation   Regulation of transport companies, exemplified to the fullest extent by the Interstate Commerce Act, was originally intended to assure to the public adequate, dependable service at reasonable rates without discrimination as to persons, places, or things. Specific abuses were also proscribed. The Transportation Act of 1920 advanced the concept that consideration also be given to the welfare of the industry in order for it to have the financial strength to provide service for the publice. Regulation was not unwelcome to the railroads as it offered a means of maintaining order and stability within their industry. Whether public interest or carrier welfare has been the prime concern of regulation as practiced has been a matter of warm debate.

In order to carry out the foregoing purposes, regulatory bodies usually possess most of the following powers and authority.

Rates   Control is exercised over the raising and lowering of rates. Rates, in general, must be fully compensatory and cannot be below the costs of service. The ICC, for example, can establish not only maximum and minimum rates but also specific rates. Division of joint rates between carriers, control of rates for long and short hauls, and the forbidding of rebating in its several forms are among the many powers over rates given to regulatory bodies.

Services   To assure adequate service, authority is given to require carriers to possess a sufficient amount of motive power and equipment, to maintain an adequate number of schedules and a reasonable adherence to them, and to make a fair distribution of equipment among the carriers' patrons. The rendering of service without discrimination, either for or against, is subject to regulatory inquiry and action.

Entrance/Exit   A certificate of public convenience and necessity is required of a carrier before it can offer public service. The purpose is to permit inquiry into the need for the service, the effect upon carriers already giving service, and the physical and financial capability of the applicant to render the service proposed.

Once a service is established, it cannot be discontinued without regulatory permission. Permission is based on (1) the lack of need for the service, (2) the availability of alternative services, and (3) the financial impact of rendering the service upon the applicant.

Records and Accounts   In order to have sufficient and comparable data on which to render decisions in carrying out the foregoing duties, regulatory bodies prescribe the accounting and recordkeeping procedures for carriers within their jurisdiction. Financial and statistical reports of various kinds are also required.

Other Powers   Regulatory powers may also extend over mergers and consolidations, the issuance of securities, safety, and numerous other aspects of transportation.

PUBLIC AGENCIES

A variety of agencies has been established to conduct and administer economic and other regulation of transport companies. Most of these are located in the Departments of Commerce and of Transportation.

Department of Commerce   Four agencies closely associated with transportation are in the Department of Commerce.

The Interstate Commerce Commission   The oldest of the regulatory agencies was created in 1887 through passage of the Act to Regulate Interstate Commerce (the Interstate Commerce Act). The Commission consists of 12 members (increased from nine members) each of whom serves seven years. The Commission was charged with the regulation of rates, services, and other aspects of interstate railroads including routes and operating rights. They prescribe uniform systems of accounts and records, approve (or disapprove) mergers and consolidations, entry and abandonment, and authorize the issuance of securities. They also hold rate hearings. Subsequently, in 1906, express and sleeping car companies, electric railways, and common carrier petroleum pipelines were brought under ICC jurisdiction. Motor carriers by highway were brought under the ICC by the Motor Carriers Act of 1935 (which in 1940 became Part II of the Interstate Commerce Act). Part III of the Act, giving jurisdiction over carriers operating in coastwise, intercoastal, and inland waters, came with the Transportation Act of 1940. Private carriers and carriers of liquid bulks or of three or less dry bulk commodities in a single vessel or tow are exempted, thereby omitting a major part of the water transportation industry. Part IV of the Act gives the Commission jurisdiction over freight forwarders through the Transportation Act of 1942. Revisions of the original Act strengthened the power and authority of the ICC. Its findings may not be appealed on questions of fact, only of law and due process.

Civil Aeronautics Board   The CAB, which grew out of the Civil Aeronautics Act of 1938 and Presidential Reorganization Plans of 1940 and the Federal Aviation Act of 1958, exercises over the airline industry essentially the same regulatory powers as does the ICC over ground transportation. In addition it determines grants and subsidies, assists in developing international air transport, and grants, subject to Presidential approval, foreign certification to United States carriers and United States operating certificates to foreign carriers. The CAB has five members serving six-year terms.

Federal Power Commission   The six-member FPC administers the regulation of interstate gas pipelines and electric power transmission lines.

Federal Maritime Commission   The present FMC, established by the Presidential Reorganization Plan of 1961, has regulatory powers similar to those granted by the Shipping Act (and subsequent amendments) of 1916. These powers include the regulation of services, rates, and practices and of agreements in international trade for vessels carrying the American flag in foreign commerce and for common carriers by water that operate in domestic trade beyond the continental United States. There are five members appointed for four-year terms by the President with Senate approval.

Department of Transportation   The need to coordinate and centralize federal jurisdiction over transportation policies and processes led, through the Department of Transportation Act of 1966, to the creation at cabinet level of a Department of Transportation, which began operation on 1 April 1967. The enabling act calls upon the Department of Transportation (DOT) to coordinate the government's transportation programs and provide leadership in transportation matters, to develop national transportation policies and programs, and to develop compliance with safety laws that pertain to all modes of transport. See Figure 3.3.

There are seven major operating administrations:

The *Federal Aviation Administration* has taken over the duties of the Federal Aviation Agency and develops and operates the airways including Air Traffic Control Centers; administers the Federal Airport Development Aid Program; promotes and guides the long range growth of civil aviation generally including research and development, and enforces safety regulations.

The *United States Coast Guard* provides police duty along the United States coastlines; enforces maritime safety including the management of lighthouses and navigation aids for water and for transoceanic air traffic; approves plans for vessel construction and repair; administers the Great Lakes Pilotage Act of 1960; and has taken over from the Corps of Engineers responsibility for the location and clearance of bridges over navigable waters, vessel anchorages, and drawbridge operation.

The *Federal Highway Administration* assumed the activities of the Bureau of Public Roads including the administration of Federal highway construction, research planning, safety programs, and implementation of the Federal-Aid Highway Program; administers the National Traffic and Motor Vehicle Safety and the Highway Safety Acts of 1966; has assumed for the Corps of Engineers responsibility for the reasonableness of tolls on bridges over navigable waters; coordinating with the Transportation Research Board conducts and administers research in all areas.[3]

The *National Highway and Traffic Safety Administration* develops and promulgates programs for use by the several states for driver performance; develops uniform standards for accident records and accident investigations

[3]The Transportation Research Board prior to 1974 was known as the Highway Research Board.

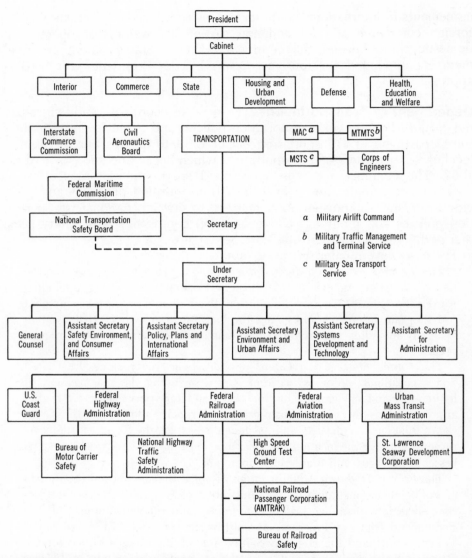

FIGURE 3.3 DEPARTMENT OF TRANSPORTATION ORGANIZATION CHART. [A]MILITARY AIRLIFT COMMAND. [B]MILITARY TRAFFIC MANAGEMENT AND TERMINAL SERVICE. [C]MILITARY SEA TRANSPORT SERVICE.

and for vehicle registration and inspection; it engages in the planning, development, and enforcement of Federal motor vehicle standards in the manufacture of automotive vehicles.

The *Bureau of Motor Carrier Safety*, a division of the FHA, has taken over from the ICC the administration and enforcement of Federal motor carrier safety regulations and those regulations governing the transportation of hazardous materials by highway.

Federal Railroad Administration (FRA) advises the Secretary on matters pertaining to national railroad policy developments; has responsibility through its Bureau of Railroad Safety for the implementation of railroad and pipeline safety laws and regulations, including the recently enacted track inspection legislation; is responsible for operation of the Alaska Railroad; and administers the High Speed Ground Transportation Program.

Within the FRA, the *Office of High Speed Ground Transportation* administers the HSGT research program, which includes high-speed rail research on the Boston–Washington Northeast Corridor, development of auto-on-train experiments, tracked-air-cushion vehicles, and other high-speed surface innovations, and operates the Pueblo Test Tracks at the Research Center near Pueblo, Colorado.

St. Lawrence Seaway Development Corporation administers the rates and tolls, operation, and maintenance of the United States portion of the St. Lawrence Seaway.

Urban Mass Transportation Administration (UMTA) develops comprehensive coordinated mass transit systems for urban and metropolitan areas; provides for research and development including demonstration projects, aid for technical studies, planning and engineering, and designing; administers financial grants to public bodies for modernization, equipment, and training of personnel. Cooperation and consultation are required between the Department of Housing and Urban Development and UMTA to assure compatability between urban plans and urban transportation.

In addition, there is a *National Transportation Safety Board*, reporting directly in part to the Secretary, that investigates and reports on causes and circumstances of transportation accidents; exercises all functions relating to aircraft accident investigations; and reviews on appeal the revocation, suspension, or denial of certificates or licenses issued by the Department.

U. S. Railway Association   The Regional Rail Reorganization Act of 1973 created the USRA as a quasi-public body to plan and finance the reorganization of bankrupt railroads in the Northeast (and their Midwest extensions) into a viable rail company (or companies), the new company to be known as the Consolidated Rail Corporation or Conrail. A part of the proposed reorganization would be a rationalization of the rail network through the abandonment of unprofitable branch and duplicating mileage (except where states or other local interests would wish to retain and subsidize losses sustained through marginal operations). Conrail came into legal existence on 10 November, and began operations on 27 February 1976.

Maritime Administration   This is a branch of the Commerce Department that runs the U. S. Merchant Marine and determines ship requirements, ocean services, routes and lines required for development, and maintenance of the foreign commerce of the United States; it maintains the National Defense Reserve Fleet; grants ship mortgage insurance; develops ship design and other technological concepts. Subsidies are awarded and the degree of service and specific routes of subsidized operators are a function of this Administration's Maritime Subsidy Board.

National Transportation Reports   A series of reports, covering all modes of transportation, to provide information and analyses upon which to base recommendations required by the Congress has been initiated by the Department of Transportation. These reports fulfill the requirements of Sections 2 and 4 of the Department of Transportation Act. The 1972 and 1974 National Transportation Reports present "...(The) state of the nation's transportation system and planning alternatives of Federal, State, and local governments for improving the system over the longer range future."[4] These reports contain (a) a status report of the nation's transportation system performance, operation, finance, and related problems, (b) estimates of future demand for transportation services, (c) estimates of investment needs and priorities for all sectors, (d) studies of system alternatives in urban and intercity transportation, and (e) guidelines for future action by Federal, state, and local governments and by the private sector.

Two additional series supplement the NTR reports. The National Highway Needs Report (NHNR) was issued in 1968, 1970, and 1972. The National Airport System Plan (NASP) came out in June 1973 to fulfill a provision of Section 12(a) of the Airport and Airway Development Act of 1970. In 1974 all three reports were combined into one.

These reports provide a wealth of information and statistical data on the status, development, and future transportation needs of the United States.

[4]*1972 National Transportation Report*, Office of the Assistant Secretary for Policy and International Affairs, U. S. Department of Transportation, Washington, D. C., July 1972, p. 4.

State and Local Agencies The Federal regulatory agencies have jurisdiction over all interstate commerce and transport. These powers are so broadly interpreted by the courts as to touch on almost all aspects of transport. Purely intrastate transport, however, is subject to control by state agencies (e.g., public utility commissions and commerce commissions) that have generally the same powers within their states as the Federal agencies have nationally.

Many states have established departments of transportation to evaluate state transportation needs, plan for and foster future development, and encourage coordination of transportation at the state, Federal, local, and private levels.

Regional authorities have been established within a state or across state lines to perform similar functions and to own, operate, and sometimes impose taxes for transportation facilities. The Regional Transportation Authority of the six-county Chicago Metropolitan Area encompases suburban bus lines, and the Chicago Transit Authority's rapid transit and bus operations; it purchases service agreements with five intercity railroads to provide commuter service. Its charter calls for it,

> ...to provide and facilitate public transportation which is attractive and economical to users, comprehensive, coordinated among its various elements, safe, efficient, and coordinated with area and state plans.

The Authority may plan and construct new facilities, has taxing powers, and makes grants to transportation agencies for operating expenses and other purposes.

In the New York area, the Port of New York and New Jersey Authority owns and operates airports, wharves, motor freight stations, buildings, tunnels, bridges, and a commuter railroad. The list of regional authorities could be expanded.

At a more local level, departments of public works, county and city planning commissions, traffic engineers, and local town councils plan for and/or operate streets, expressways, bus and rapid transit for small communities, cities, and metro areas. These bodies, each with jealously guarded perogatives, are sometimes more of a hindrance than a help in transportation planning.

FREIGHT CARRIERS—PLANT AND TRAFFIC

Railroads Railroads may be grouped initially into *private* and *common carriers*. The private group is relatively small and unimportant. Comprising principally lumbering, mining, and industrial plant facilities, often these

FIGURE 3.4 RAILROAD NETWORK OF THE UNITED STATES, 1975. (COURTESY OF THE ASSOCIATION OF AMERICAN RAILROADS, WASHINGTON, D.C.)

roads establish a common carrier status through which they make their services available to the public.

Railroads are further designated by area of use as *line haul, switching or belt line*, and *terminal*. The first group, handling intercity traffic, is exemplified by the Santa Fe, Burlington Northern, and Seaboard Coast Line. Switching or belt-line carriers offer sidetrack services at wharf and waterfront locations, industries, and warehouses, for example, to line-haul railroads, for a switching charge that is included by the line-haul carrier in its total charges. Examples are the New Orleans Public Belt Railroad owned by the City of New Orleans and the Belt Railway of Chicago owned jointly by 12 or more user railraods.

Terminal railroads supply terminal services and interchange to line-haul carriers and are usually owned by one or more of the railroads served. Several railroads can thereby gain access to city and terminal areas without the expense of providing duplicate terminal facilities. The Kansas City Terminal Company and the Terminal Railraod Association in St. Louis are examples of this kind of rail carrier.

The Interstate Commerce Commission classifies railroads according to gross operating revenues. Class I carriers are those with gross annual operating revenues of $10 million or more. There were 58 Class I line-haul railroads in 1976. These railroads in 1974 hauled more than 99 percent of the United States rail freight traffic (856 billion ton miles), employed 93 percent of all railroad workers (544,400 workers earning 5893 million dollars), and operated 96 percent of total line-haul mileage (195,000 route miles (317,755 km)). There are approximately 431 Class II carriers with revenues less than ten million dollars and some 225 small terminal, switching, and industrial carriers.

The extent and completeness of the railroad network in the United States are illustrated in Figure 3.4. Total mileage in that network and a summary of the equipment owned are given in Tables 3.1 and 3.2. With the minor

TABLE 3.1 RAILROAD MILEAGE IN THE UNITED STATES, 1971[a]

Tracks	Class I	All Others	Total
Road (first main track)	194,772	9,924	204,696
All other main tracks	18,900	100	19,000
Yard track and sidings	97,004	13,168	110,172
Total all tracks	310,676	23,192	333,868
	(1 mile = 1.609 kilometers)		

[a]Transport Statistics in the United States, Year Ended December 31, 1971, Bureau of Accounts, Interstate Commerce Commission, Washington, D. C., Tables 1, 2, and 4, pp. 1, 2, 4.

TABLE 3.2 RAILROAD EQUIPMENT, ALL UNITED STATES RAILROADS, 1975[a]

Kind	All Class I Railroads and Others	Class I Railroads	Other Railroads	Car Companies and Shippers	Amtrak and Auto-Train
Locomotives:					
Diesel-electric	28,289				
Electric	224				
Steam	11				
Total	28,524				
Freight cars:	1,723,605	1,359,459	29,407	334,739	
Passenger cars	6,534	4,651			1,883

[a]*Yearbook of Railroad Facts*, 1976 edition, Economics and Finance Department, Association of American Railroads, Washington, D. C., pp. 50, 51, 54, 62.

TABLE 3.3 REVENUE TONS OF FREIGHT ORIGINATED BY CLASS I RAILROADS[a] 1972

Commodity	Revenue Tons Originated (000s)	Percent of Total
Farm products	130,269	8.82
Metallic ores	110,034	7.45
Coal	374,970	25.45
Nonmetallic minerals except fuel	170,413	11.52
Food and kindred products	106,066	7.47
Lumber and wood products	109,986	7.65
Pulp, paper, and allied products	44,298	3.06
Chemicals and allied products	91,846	6.42
Petroleum and coal products	47,951	3.26
Stone, clay, and glass products	73,487	5.56
Primary metal products	60,901	4.48
Transportation equipment	31,410	2.06
Waste and scrap materials	39,741	2.70
Miscellaneous shipments	55,828	4.05
Subtotal carload traffic	1,447,200	99.95
Small-packaged freight shipments	651	0.05
Total tons originated	1,447,851	100.00

[a]*Based on Statistics of Railroads of Class I in the United States*, Years 1962 to 1972, Economics and Finance Department, Association of American Railroads, Washington, D. C., November 1973.

exception of trackage and joint operation agreed upon between individual roads, only a railroad's own trains operate on these tracks. There is no intermingling of private and public, commercial, and pleasure vehicles. Even when cars or trains (such as Amtrak passenger trains, circus trains, or privately owned refrigerator, tank, or chemical cars) are hauled, these are moved by the railroad's own locomotives, staffed by its crews, and are under full railroad control and direction.[5]

Class I railroads in 1975 represented a net investment of over $29.5 billion. Operating revenues were $16,423 billion; operating expenses $13,123 billion, and taxes paid $1,596 billion. The rate of return for all Class I railroads was 1.20 percent in 1975.[6] Revenue sources and quantities are given in Table 3.3.

Highway Freight Carriers   The variety of trucking operations makes it difficult to select any one or two as typical. As with railroads, a distinction first can be made between private and for-hire operators.

Private Carriers   Private carriers, hauling the transporter's own goods or goods for which the truck owner is lessee or bailee, can assure flexibility of service, prompt delivery of goods without delays enroute or in terminals, and without the excessive crating or packaging required by common carriers.

In 1972 there were six private fleets containing more than 10,000 trucks each of various types and sizes. Over 65,000 fleets contained 10 or more truck vehicles, a total of over seven million vehicles.[7]

Common Carriers   These for-hire agencies serve the public by hauling traffic as offered over regular routes or irregular routes. They are subject to close control by the ICC and must have a certificate of public convenience and necessity from that body in order to engage in interstate business. There were 1398 common carriers of freight reporting to the ICC in 1972.[8] For purely intrastate operations, the state regulatory agencies exercise the same supervision.

Contract Carriers   Certain for-hire agencies engage under individual contracts for motor freight transportation. Contract carriers are subject to

[5]Amtrak is in the process of acquiring, operating, and maintaining its own fleet of locomotives. See Table 3.2.

[6]"Yearbook of Railroad Facts," Economics and Finance Department, Association of American Railroads, Washington, D. C., 1976, p. 9.

[7]*Motor Truck Facts*, 1974 edition, Motor Vehicle Manufacturers Association, p. 40.

[8]*Transportation Statistics in the United States, Part 7, Motor Carriers*, ICC period ending 31 December 1972, p. iv.

modified regulation by the ICC (and/or state authority). Permits to operate were held by 127 reporting contract carriers in 1972.[9] Contract carriers may contract for freight distribution from warehouse or rail or water terminals (pickup and delivery) or may furnish drivers and vehicles for the exclusive use of designated shippers to operate under the shippers' jurisdiction. Others may serve within a specialized commodity area (e.g., haulers of petroleum). By hauling truckload traffic directly between shipper and consignee, few if any terminal facilities need be furnished by the carrier.

Exempt Carriers   Carriers exempt from economic and service regulation are operators of for-hire vehicles that transport agricultural products (not including their manufactured products), farm supplies, livestock, and fish; that distribute newspapers; that are operated by farmers cooperative marketing organizations; that are used in transport incidental to airline operations (freight and express service); that are used in occasional or casual services by persons not regularly engaged in the transportation business. Certain carriers operate over a fixed route, but others move from area to area picking up a load wherever available for any destination suitable to the operator. Runs of 100 to 2000 miles (161 to 3218 km) are not uncommon. About 250,000 vehicles are engaged in exempt service, 50 percent more than are engaged as common and contract carriers.

Revenue Classification   In the ICCs revenue classification for common and contract carriers, Class I highway carriers are those with annual gross operating revenues of $1 million or more over a three-year period. Class II carriers have operating revenues of less than one million but more than $300,000. In 1972 there were 1663 Class I carriers.[10] These moved over 651 million tons (590.5 million tonnes) of intercity highway traffic.

Other Classifications   The ICC further classifies highway carriers as *local* or *intercity*. Local carriers are those receiving 50 percent or more of revenue from local service. Intercity carriers are also identified as common carriers of general freight, common carriers of commodities other than general freight, and as contract carriers.

An earlier, and useful, classification was by routes.

(A)   Regular route, scheduled service (point to point)

(B)   Regular route, nonscheduled service (point to point)

---

[9]Ibid.
[10]Ibid.

(C) Irregular route, radial service (point to point over radial lines from a fixed or base point)

(D) Irregular route, nonradial service (no established pattern)

(E) Local cartage service

A further ICC classification is by type of commodity.

1. Carriers of general freight
2. Carriers of household goods
3. Carriers of heavy machinery
4. Carriers of liquid petroleum products
5. Carriers of refrigerated liquid products
6. Carriers of refrigerated solid products
7. Carriers engaged in dump trucking
8. Carriers of agricultural products
9. Carriers of motor vehicles
10. Carriers engaged in armored truck service
11. Carriers of building materials
12. Carriers of films and associated commodities
13. Carriers of fruit products
14. Carriers of mine ore (excluding coal)
15. Carriers engaged in retail store delivery
16. Carriers of explosives or dangerous articles
17. Carriers of commodities not otherwise designated

By combining these classifications, any carrier could be specifically designated. A contract carrier of building materials over an irregular route, nonradial, would be designated as a Contract Carrier, Class D-11.

Note that a carrier certificated for a certain commodity is not in violation of the "discrimination against things" rule, but must, if a common carrier, accept all traffic offered within the certificated commodity category. Tables 3.4 and 3.5 show the traffic carried by highway and the vehicles used in that carriage.

Highways The highway system of the United States is even more elaborate and ubiquitous than is the rail network. In 1974 the highway was comprised of the mileages given in Table 3.6. These figures include more

TABLE 3.4 INTERCITY MOTOR FREIGHT TONNAGE 1970. CLASSES I AND II. MOTOR FREIGHT CARRIERS OF PROPERTY[a]

Commodity	Tonnage (millions)	Percent
General freight	225	37.7
Household goods	7	1.2
Heavy machinery	6	1.0
Liquid petroleum	154	25.8
Refrigerated liquids and solids	12	2.1
Agricultural commodities	8	1.3
Motor vehicles	16	2.7
Building materials	26	4.4
All other intercity classes	142	23.8
Total all freight	596	100.0
(1 ton = 0.907 tonnes)		

[a]*1971 Motor Truck Facts*, Automobile Manufacturers Association, Detroit, Michigan, 1971, p. 47.

TABLE 3.5 VEHICLES IN HIGHWAY TRANSPORTATION, 1974[a]

Vehicle Type	Private Vehicles	Public Vehicles	Total Vehicles
Automobile registration (all types except military)		104,901,066	
Motor buses			
Commercial and Federal	90,072	2,200	92,272
School and other	128,617	225,669	354,286
Total buses	218,689	227,869	446,558
Freight vehicles			
Trucks	23,471,468	1,118,710	24,590,178
Full trailer	301,033		301,033
Semitrailer	2,189,816		2,189,816
Total trailers	2,490,849		2,490,849
Truck tractors	1,064,600		1,064,600

[a]*Motor Vehicle Facts*, Motor Vehicle Manufacturers Association, Detroit, Michigan, 1976, pp. 29–33.

TABLE 3.6 UNITED STATES HIGHWAY MILEAGE, DECEMBER 31, 1971[a](IN THOUSANDS OF MILES; 1 MILE = 1.609 KILOMETERS)

System	Nonsur-faced Mileage	Surfaced Mileage Low Grade	Medium Grade	High Grade	Total
Rural mileage					
Under state control					
State primary system	3	9	114	282	408
State secondary system	3	8	67	44	122
County roads under state	16	47	60	30	153
State parks, forests, etc.	13	9	2	5	29
Total state control	35	73	243	361	712
Under local control					
County roads	455	805	331	135	1726
Town and township roads	108	283	82	25	498
Other local roads	14	13	4	1	32
Total local control	577	1101	417	161	2256
Under Federal control					
National parks, forests	138	46	7	6	197
Total rural mileage	750	1219	668	528	3165
Municipal mileage					
Under state control					
Extension of state primary system	—	—	6	55	61
Extension of state secondary system	—	—	8	8	16
Total state control	—	—	14	63	77
Under local control					
City streets	26	68	226	196	516
Total municipal mileage	26	68	240	259	593
Total rural and municipal mileage in the United States	776	1287	908	787	3758

[a]*Highway Statistics, Table M-2*, Federal Highway Administration, U. S. Department of Transportation, Washington, D. C., 1971, p. 188

than 35,460 miles (57,055 kilometers) of completed superhighway and 3036 miles (4885 km) under construction.

All primary rural routes are built and maintained by the states. An approximate 156,000 miles (251,004 kilometers) are designated as national highways because the Federal Government contributes to the cost of their construction and maintenance. A total of 7060 million dollars was contributed by the Federal Government in the fiscal years 1917–1955 prior to the Interstate System. National highways bear the Federal shield marker and route number. East-west highways are given even numbers, From U.S. 2 between Boston and Seattle on the north to U.S. 90 between Jacksonville, Florida and El Paso, Texas on the south. North-south highways receive odd numbers beginning with U.S. 1 on the eastern seaboard between Maine and Key West, Florida. Basic numbers are one or two digits (except U.S. 101 on the West Coast). Three-digit numbers are for main highway branches. For example, U.S. 224 is a branch of U.S. 24.

Interstate Highway routes are similarly designated. For example, a primary north-south highway in the Mississippi Valley region is I-57. Highways coming off I-57 providing a regional connection will be designated I-157, I-257, etc., regardless of which direction they run. Thus I-157 can run east-west and I-257 might run north-south but both are connected to I-57.

The Federal Highway Act of 1956 and the supplementary act of 1958 introduced a new dimension to highway planning. Recognizing the inadequacy of the highway system for present as well as future traffic, the Congress of the United States appropriated more than $40 billion to be used principally for the construction in the next 16 to 20 years of more than 42,500 miles of a National System of Interstate and Defense Highways. Of this mileage, approximately 70 percent are new route locations connecting some 90 percent of all cities with populations of 50,000 or more, including 42 state capitals. See Figure 3.5. Construction is to superhighway standards, with easy curves designed for speeds of 70 mph or more, usually with separate lanes for opposing directions with a wide median strip between, and full control of access. The Federal Government pays 90 percent of the costs (collected from gasoline, tire, and other accessories taxes) and the states, 10 percent. The monies thus collected are held in a trust fund for the continuing expansion and improvement of the system. (But efforts continue in the Congress to permit the allocation of a portion of the trust fund to nonhighway transportation development. In late 1973 states were given permission to use a portion of their share of trust fund money for the improvement of public transit.) Initiative for instituting a section of the system rests with the states and, as with other national fuel-tax highways, is carried out by forces under the supervision of

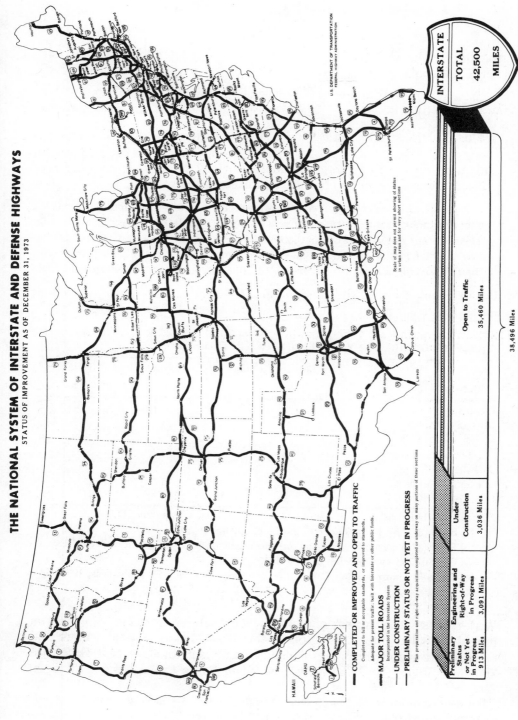

FIGURE 3.5 THE NATIONAL SYSTEM OF INTERSTATE AND DEFENSE HIGHWAYS. (COURTESY OF THE FEDERAL HIGHWAY ADMINISTRATION.)

the states. The Federal Highway Administration establishes general standards and policies. This system carries approximately 20 percent of the nation's highway traffic. A different basis of classification is given in Chapter 17.

Water Carriers   Domestic carriers by water may be divided into three groups: the Great Lakes carriers, those plying coastal waters and canals, and those on inland rivers and canals. Each of these categories includes private, contract, common carrier, and exempt shipping.

In 1971, 177 inland and coastal waterways carriers reported to the ICC, Class A being those with revenues exceeding $500,000, Class B exceeding $100,000 but not exceeding $500,000, and Class C with revenues of $100,000 or less. Eight Class A and seven Class B carriers on the Great Lakes reported to the ICC; so did 18 maritime carriers.[11] Other categories are shown in Table 3.7.

Because only a few of the common and contract carriers are in bulk cargo trade, the major part of waterway traffic is exempt from regulation.

TABLE 3.7 CARRIERS BY WATER IN THE UNITED STATES

Waterway	Common and Contract Carriers	Exempt Carriers	Private Carriers
Great Lakes[a]	11	87	36
Mississippi River System and Gulf Intercoastal system[b]	31	663	238
Atlantic, Gulf, and Pacific coasts[c]	67	503	143
Total	109	1,253	417

[a,b,c]Transportation Series, U.S. Army Corps of Engineers, Washington, D.C.
  (a) Series 3. Transportation Lines on the Great Lakes, 1975, pp. 25–31.
  (b) Series 4. Transportation Lines on the Mississippi River System and Gulf Intercoastal Waterway, 1975, pp. 265–305.
  (c) Series 5. Transportation Lines on the Atlantic, Gulf, and Pacific Coasts, 1975, pp. 161–191.

Waterways   The system of domestic or inland waterways also may be roughly divided into three groups: the Great Lakes, coastal waters and canals, and inland rivers and canals. Components of the Great Lakes are shown in Figure 3.6. The New York State Barge Canal has a 14-ft open

[11]*Transport Statistics of the United States, Part 5, Carriers by Water*, Interstate Commerce Commission, Washington, D.C.; year ending 31 December 1971, p. 1.

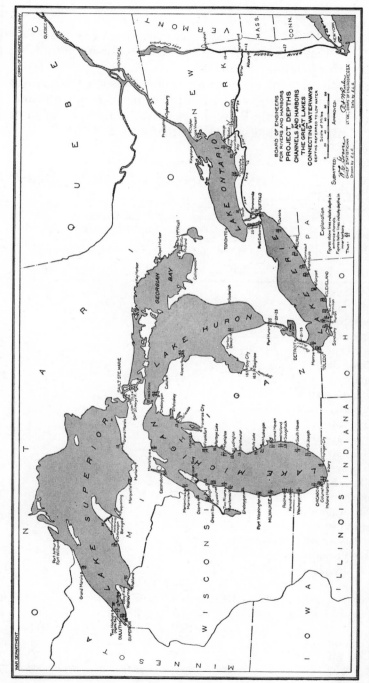

FIGURE 3.6 GREAT LAKES SYSTEM. (COURTESY OF THE U.S. ARMY CORPS OF ENGINEERS.)

canal depth and a 12-ft lock sill depth, and the Great Lakes-St. Lawrence Seaway channel depths have been increased to 27 ft, permitting most ocean shipping to reach inland ports.

Intercoastal shipping in deep water follows a somewhat longer route through the Panama Canal. Part of the trade moves in deep-water vessels with characteristics similar to those in ocean trade. These companies report to the U.S. Maritime Commission rather than to the ICC.

TABLE 3.8 NAVIGABLE LENGTHS AND DEPTHS OF UNITED STATES INLAND WATERWAY ROUTES[a]

Group	Length in Miles of Waterways (1 mile = 1.609 kilometers)					
	Under 6 ft	6 to 9 Ft	9 to 12 Ft	12 to 14 Ft	and Over	Total
Atlantic coast waterways (exclusive of Atlantic intracoastal waterway from Norfolk, Virginia, to Key West, Florida.) but including New York State Barge Canal System	1426	1241	584	918	1581	5768
Atlantic intracoastal waterway from Norfolk, Virginia, to Key West, Florida	—	65	65	1104	—	1,234
Gulf coast waterways (exclusive of Gulf intracoastal waterway from St. Marks River, Florida, to the Mexican border)	2055	647	1133	79	378	4292
Gulf intracoastal waterway from St. Marks River, Florida, to the Mexican border (including the Plaquemine—Morgan City Alternate Route)	—	—	—	1173	—	1,173
Mississippi River system	3020	969	4957	740	268	8,954
Pacific coast waterways	730	498	237	26	2084	3578
Great Lakes	45	89	—	8	348	490
All other waterways (Exclusive of Alaska)	76	7	—	1	7	91
Total of all systems	6352	3516	6976	4033	4666	25,543

[a]*Waterways of the United States*, map and statistical insert, for "Big Load Afloat" The American Waterways Operators, Inc., Washington, D.C., 1973.

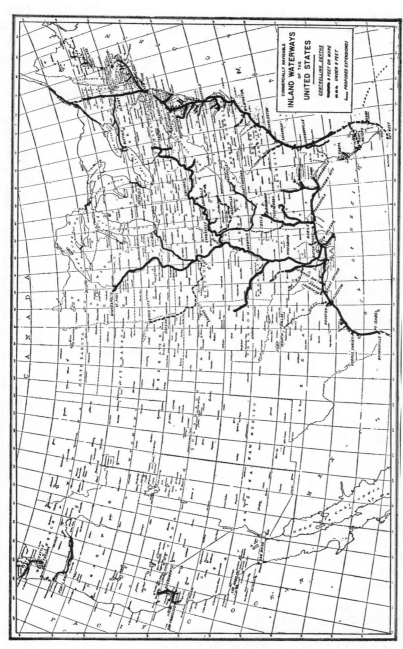

FIGURE 3.7 INLAND AND COASTAL WATERWAYS OF THE UNITED STATES, 1975. (COURTESY OF THE AMERICAN WATERWAYS OPERATORS, INC., WASHINGTON, D.C.)

Canals and rivers form routes of limited and sometimes circuitous extent as seen in Table 3.8 and Figure 3.7. Many parts of the country have no access to navigable waters; few reaches of water have a naturally sufficient channel depth of 6 to 14 ft. Dredging or dam construction is required to permit navigation.

Operators on rivers, lakes, and canals are spared the cost and responsibility of channel and lock construction and maintenance. They make no user payment, equivalent, for example, to the highway fuel tax or the airlines' seat tax. All types of traffic—pleasure and commercial, private, contract, and common carrier—share the canals, rivers, lakes, and coastal waters. An exception to the free use of waterways is the toll charged on the St. Lawrence Seaway, the proper rate for which is a continuing subject of

TABLES 3.9 NUMBER OF VESSELS ON INLAND WATERWAYS OF UNITED STATES FOR TRANSPORTATION OF FREIGHT AS OF JANUARY 1, 1971[b]

Types of Vessels	Mississippi River System and the Gulf Intercoastal Waterway	Atlantic, Gulf, and Pacific Coasts	Great Lakes	Total
Self-propelled				
Towboats and tugs				
Number of vessels	2,344	1,748	138	4,230[a]
Horsepower	2,305,305	1,541,640	108,056	3,995,001
Nonself-propelled				
Dry cargo barges and scows				
Number of vessels	13,310	3,000	121	16,439
Cargo capacity (net tons)	14,863,812	3,296,521	111,681	18,272,014
Tank barges				
Number of vessels	2,581	581	23	3,815
Cargo capacity (net tons)	4,753,480	1,521,222	55,596	6,330,298
Total Nonself-Propelled				
Number of vessels	15,899	3,581	144	19,624
Cargo capacity (net tons)	19,617,292	4,817,743	167,277	24,602,312

[a]Includes only those vessels engaged in performing transportation services.
[b]*Inland Waterborne Commerce Statistics 1971*, The American Waterways Operators, Inc., 1280 Conneticut Avenue, Washington, D.C. 20036, p.2.

TABLE 3.10 COMMODITIES CARRIED BY UNITED STATES INLAND WATERWAYS, 1971 (EXCLUSIVE OF THE GREAT LAKES)[a]

Commodity	Net Tons (Millions)	Percent
Grain and grain products	19.08	3.40
Soy beans	8.62	1.54
Marine shells (unmanufactured)	18.31	3.26
Bituminous coal and lignite	109.53	19.51
Crude petroleum	57.76	10.30
Sand, gravel, and crushed rock	75.45	13.46
Sulphur, dry and liquid	5.07	0.90
Nonmetallic minerals (except fuels)	5.31	0.94
Rafted logs	16.30	2.90
Caustic soda	3.04	0.54
Basic chemicals and products	13.19	2.35
Fertilizers and fertilizer materials	6.06	1.08
Gasoline	44.20	7.88
Jet fuel	6.71	1.19
Other petroleum and coal products	100.17	17.87
Iron and steel products and scrap	7.00	1.25
All other commodities	65.21	11.63
Total—all commodities	560.47	100.00

[a]*1971 Waterborne Commerce Statistics*, The American Waterways Operators, Inc., October 1972, p. 5.

debate. Waterway equipment and revenue tons of freight are shown in Tables 3.9 and 3.10.

Air Cargo Freight, express, and domestic intercity first class mail move by air (over 6 billion ton miles in 1973), but the principal function of airlines is carrying passengers. The reader is referred to the section on passengers carriage for additional comment on air transport, pp. 74–79. Three lines carry only freight traffic.

Pipelines Nearly all pipelines for petroleum products have common carrier status under the law. As a fact of operating practice, most pipelines are gigantic plant facilities, owned by the petroleum industry, which transport the owner's oil from field to refinery or from refinery to market. There were 99 pipeline companies reporting to the ICC in 1971 (including one coal-slurry line). These lines transported in trunk-line movement 4781 million barrels (bbl) of crude petroleum and 3,017 million bbl of refined products. They produced 1,439,195 million barrel-miles of crude-oil transportation and 1095 million barrel-miles of refined-products transportation.

Gas pipelines have common carrier status in some localities, but others operate as private facilities. The status of gas lines is not so clearly defined as that of petroleum lines.

The network of petroleum product pipelines in the United States is shown in Figure 3.8. Petroleum pipelines may be divided into three categories: gathering lines, which carry crude oil from the fields to the initial pumping station, crude trunk lines for pumping crude petroleum, and product trunk lines for handling gasoline, kerosene, and distillate fuel oils. The mileage and traffic of the 99 lines reporting to the ICC are given in Table 3-11.

The pipeline routes generally lead from the oil fields to refining centers or to river or gulf ports for water transport to the eastern seaboard or abroad. Much petroleum also arrives from abroad, especially from Venezuela and the Middle East, via oceangoing tankers, then moves to refineries via pipeline or barge. The Alaska Pipeline extends from Prudhoe Bay on Alaska's north coast to Valdez on the southern shore.

TABLE 3.11 MILEAGE AND TRAFFIC OF UNITED STATES PETROLEUM AND PETROLEUM PRODUCTS PIPELINES, 1971[a] (1 MILE = 1.609 KILOMETERS)

Pipeline mileage	
Trunk lines:	
Crude petroleum	60,946 miles
Refined products	61,525 miles
Total trunk lines	122,471 miles
Gathering lines:	45,759 miles
Total petroleum pipeline mileage	168,230 miles
Pipeline traffic	
Total barrels having trunk line movement:	
Crude petroleum	4,781,043,520
Refined products	3,016,574,466
Total	7,797,617,986
Barrel miles having trunk line movement:	
Crude petroleum	1,439,195,444
Refined products	1,045,399,200
Total	2,484,594,644

[a]*Transport Statistics in the United States, Part 6, Oil Pipelines,* I.C.C., Washington, D. C., for the year ended December 31, 1971, pp. 3 and 4.

Natural-gas lines totaled more than 400,000 miles (643,600 km) in 1957. Principal lines (Figure 3.9) run to population and industrial centers, where the gas constitutes about 24 percent of the nation's fuel supply. An increasingly large amount of gas is used in the production of petrochemicals.

Conveyors   Conveyors have generally been limited to use as plant facilities in factories and mines or as carriers of aggregates for distances up to eight miles in connection with large construction projects. Considerable interest was shown during the 1950s in adapting them to haul coal and ore between the Great Lakes and the Ohio River, indicating a possible future development. Moving stairways and moving sidewalks are applications of the conveyor technology.

Cableways   Aerial tramways, cableways, or ropeways offer a specialized type of transportation. They are used primarily for moving products of mines relatively short distances over terrain that is too rough for the economical construction and operation of other types of transport. They are also used as a plant facility in operating gravel pits and quarries and in moving aggregates from nearby sources of supply to large construction projects.

Cableways have been used extensively in the rugged terrain of the Alpine countries for hauling freight and passengers. It was there that this type of transport originated and has reached its fullest development.

There are only a few major cableway installations in the United States; each of the two largest is 13 miles (20.9 km) long. There is a 20-mile cableway in the Andes, one 54 miles long in Colombia, another 8 miles

TABLE 3.12 PERCENT INTERCITY TON MILES CARRIED IN 1965[a]

Mode	Length of Haul	
	Under 400 Miles	Over 400 Miles
Rail	9.52	27.75
Truck	19.31	2.23
Water	11.47	13.58
Pipeline	10.63	4.45
Air	0.01	0.07

[a]*1972 National Transportation Report*, Office of the Assistant Secretary for Policy and International Development, U.S. Department of Transportation, Washington, D.C., July 1972, Chart III-ll, p. 41.

(12.8 km) long in the Phillippines for hauling gold ore, and still another that moves asbestos a distance of 12 miles (19.3 km) in Swaziland to a railhead of the South African Railways.

PASSENGER CARRIERS—PLANT AND TRAFFIC

Highway Passenger Carriers   In the United States the principal carrier of passengers is the privately owned and operated automobile. Almost every family has a car. In 1972, 79.5 percent of all families owned one or more cars, 30.2 percent owned more than one car, and 20.5 percent had no car at all.[12]

This is not the total use of cars by individuals. Car rental on a drive-it-yourself basis, expecially at airports and other terminals, is steadily increasing. Car rental by commercial and industrial concerns and other agencies for the business use of their employees is commonplace, abetted by the extensive availability of expense accounts. Rental for general private use is a small but growing trend. Intercity travel by automobile rose from 350 billion passenger[13] miles in 1947 to 675 in 1958, 800 in 1965, and 1026 in 1970.

Buses   Over 611,255 buses were registered, largely on a for-hire basis in the United States in 1972.[14] Some large companies, government agencies, institutions, and traveling groups such as dance bands and professional ball teams have their own private buses. More often contracts are made with a bus company (which may also be providing a common carrier operation) for the required service. Intercity bus lines carried over 164 million passengers in 1972 for a total of over 13 billion passenger miles.[15] Sightseeing buses account for a considerable number of them. Approximately 19 million schoolchildren were transported over 2.4 billion vehicle miles in 316,421 registered vehicles in 1972.[16]

The ICC has divided intercity bus companies into two classes, Class I carriers with gross operating revenues of $1 million or more and Class II carriers with revenues of less than $1 million.

[12]*Automobile Facts and Figures*, Motor Vehicle Manufacturers Association, Detroit, Michigan, 1973/1974, p. 30.

[13]*1972 National Transportation Report*, U.S. Department of Transportation, Washington, D.C., July 1972, p. 39, Chart III-6.

[14]Ibid., p. 16.

[15]*Transportation Statistics of the United States, Part 7 Motor Carriers*, Interstate Commerce Commission, for the period ending 31 December 1972, p. 116, Table 27.

[16]*1974 Motor Truck Facts*, Motor Vehicle Manufacturers Association, Detroit, Michigan, pp. 21 and 34.

TABLE 3.13 URBAN REVENUE PASSENGERS BY MODE[a]

Mode	Riders (Millions)	Revenue (Millions)	Percent Change (1960–1970)
Taxicabs	2378	2221	+30
Bus	4058	1194	−24
Trolley coach	128	30	−24
Rail rapid transit	1746	415	−6
Rail commuter	247	206	No change

[a]*1972 National Transportation Report*, U.S. Department of Transportation, Washington, D.C., July 1972, Chart IV-22, p. 47.

Urban Carriers   The automobile is the principal urban passenger carrier. A large portion of highway and urban planning is intended to facilitate movement by automobile. In urban areas the automobile moves annually an estimated billion or more people. The automobile mode, including taxicabs, offers a personalized service, that is, vehicle capacities of 1 to 6 persons not operated on a schedule. Taxicabs are almost universally privately owned with a few large fleets, such as Yellow Cab, and many smaller companies. All others—bus, trolley coach, light rail, rail rapid transit, and commuter railroad—are mass transport. Table 3.13 shows the use made of each in urban areas.

Bus Systems   Bus lines have been the only viable alternative, following the general demise of street railways, for cities with populations from 5000 to 500,000. Both level and quality of service are variable. Bus lines, privately owned, have found costs increasing faster than revenues. Automobile competition has been keen. Since World War II the number of privately owned systems has decreased. From 1959 to 1970, 235 companies have gone bankrupt. Of these, 89 received public ownership. The remaining 146 went out of service.[17]

Urban bus systems may be divided into those operating only within city limits and those operating both within city limits and to the surrounding suburbs. A single fare predominates, but variations by zones may also be in effect. In those cities having commuter railroad or rapid transit, buses may be used to perform a feeder service by delivering from local neighborhoods to transfer points on the rail lines.

[17]*1972 National Transportation Report*, U.S. Department of Transportation, Washington, D.C., July 1972, p. 22.

Rapid Transit   Rail rapid transit is also a mass carrier. Its distinguishing features include an exclusive right of way (on, above, or below ground), frequent service, and numerous stops in an area within 6 to 8 miles (9.7 to 12.9 km) of the central business district (CBD). Speeds are in the 30 to 45 mph (48.3 to 72.4 km) range. Traffic exerts peak demands from 7 to 9 a.m. and from 4 to 6 p.m. ±. Single fares prevail, sometimes coupled with zonal increments. All North American systems are publicly owned. Most of those require public aid to make up the difference between operating costs and fare box revenues. See Table 3.14.

TABLE 3.14 RAPID TRANSIT IN NORTH AMERICA[a]

System	Designation	Route Miles	Riders (Millions)	Stations	Cars	Operating Ratio[b]
		(1 mile = 1.609 kilometers)				
San Francisco[c]	BART	75	45.4	34	225	n.a.
Boston	MBTA	77.3	26.0	n.a.	353	1.32
Chicago[d]	RTA	205	94.2	142	1100	1.11
Cleveland	CTS	19	12.7	17	n.a.	1.34
Lindenwold	PATCO	14.5	7.8	12	75	1.04
Mexico City	Metro	26	72.8	n.a.	537	n.a.
Montreal	Metro	21.0	20.8	n.a.	333	0.84
Newark		4.2	0.7	n.a.	30	0.92
New York	NYCTA	237.2	1300.0	n.a.	6800	1.14
New York	PATH	14.2	7.3	n.a.	299	1.57
Philadelphia	PTC	82.7	66.2	n.a.	1039	1.02
Shaker Heights	SHTS	13.2	0.7	n.a.	55	0.98
Toronto	TTC	23	170.0	45	410	0.76
Atlanta[e]	MARTA	61	—	34	200	—
Baltimore[e]	BMTA	71	—	n.a.	300	—
Washington[f]	WMATA	99.8	7.0 (est.)	86	300	

[a] "1973 Transit: City-by-City," *Modern Railroads*, May 1973. Insert.
[b] "*1972 National Transportation Report*," Office of Assistant Secretary for Policy and International Affairs, Washington, D.C., July 1972, p. 54, Table III.32.
[c] W.D. Middleton, "Trouble-Plagued BART...," *Railway Age*, 12 April 1976, pp. 24–27.
[d] "CTA, statistically", *CTA Quarterly*, Chicago Transit Authority, Summer 1975, p. 24.
[e] Under construction.
[f] 4.6 miles in operation 3 April 1976.

Light Rail Light rail, a modern term for street railways, is characterized by lower speeds and capacities than rapid transit and often by the use of city streets as rights of way. The modern light rail concept includes quiet, rapid, lightweight equipment operating in single or in two- to four-car trains, with a restricted right of way (off-street or in a dedicated portion of the street), on light rails, with a capability of moving mass volumes of riders at lower costs than conventional rapid transit. Once considered obsolete, light rail systems have experienced an awakened public interest and are still in operation in San Francisco, Boston, Toronto, Philadelphia, and Pittsburgh. These, too, impose single fares or use zonal increments and depend on public funding of operating deficits.

Commuter Railroads Commuter trains operate over the tracks and rights of way of intercity line-haul railroads, serving a series of communities spaced along the routes. Fares are based on the length of ride with lowered rates for regular (and peak period) riders. Car capacities and train speeds are usually higher than for rapid transit. Train movements are less frequent, especially during off-peak hours. This operation generally loses money and railroads view it largely as a nuisance.

The service is not a part of Amtrak. In the Chicago and Philadelphia areas the public has, through regional organizations, contracted with the railroads to maintain service, sometimes aided by funds from the Department of Transportation for the purchase of new equipment. Where such assistance is not available, losses suffered must be recouped from freight revenues. Commuter rail service is provided in five metro areas—New York, Philadelphia, Chicago, Pittsburgh, and San Francisco.

Intercity Passenger Railroads Rail passenger service, once the epitome of luxury and excellence, experienced a sharp decline in ridership and service quality following World War II. Air and highway competition (mostly highway) drained traffic from the rails, and railroads made a strong effort to divest themselves of this service because of the high costs associated with it.

The National Railroad Passenger Corporation, usually referred to as Amtrak, was created by an act of Congress signed into law on 30 October 1970 in an effort to maintain a basic network of intercity rail passenger service while relieving the railroads of the burden of passenger train costs. The corporation is not a government agency but is subject to the District of Colombia Business Corporation Act. It receives funds, nevertheless, from Congressional appropriation. Twenty railroads, paying into Amtrak almost $200 million over a three-year period, were allowed thereby to

TABLE 3.15 RAIL PASSENGER TRAVEL[a]

Carrier	Passengers (Millions)		Passenger Miles (Millions)	
	1974	1975	1974	1975
Amtrak	18.3	16.8	4259	3572
Auto-Train	0.32	0.31	273.5	267.5
Commuter rail	214	210	4534	4519

[a]*Yearbook of Railroad Facts*, Association of American Railroads, 1975 edition, pp. 31, 32, 62 and 1976 edition, pp. 30, 32, 62.

discontinue their own passenger service. Fourteen hold contracts to run Amtrak trains over their tracks; five railroads elected to continuing operating their own passenger trains otside of Amtrak. Amtrak has established and "operates" a network of long and medium-distance passenger trains. Commuter service remains under railroad ownership.[18] Amtrak owns its own cars and locomotives and employs its own train crews. The future of Amtrak depends upon its acceptance by the traveling public and on the willingness of the Congress to provide funding during the period of development. A continuing shortage of petroleum fuels is turning more attention and passenger patronage to the railroads as efficient users of energy.

Air Transport   Air transport is generally divided into air carrier, general aviation, and military. Another grouping is civil, military, and general. The private or general category includes pleasure craft and those used by private industry and commerce for transporting equipment, supplies, and personnel. Several companies own fleets of planes for company use and may rent planes for executive use; contract flying by companies solely engaged in this enterprise is a developing alternative to private ownership for industry. A special group of for-hire planes engages in crop dusting, aerial reconnaissance and photography, skywriting, fire fighting, timber inspection, and other nontransport activities.

Airline Classifications   For transportation services, the Civil Aeronautics Board, which has the same relation to air transport as the ICC has to other types of transportation, has divided commercial airlines into seven general categories:

[18]The high-speed Washington-Boston Corridor service has also become Amtrak's responsibility. It is scheduled also to acquire ownership of the tracks.

1. **Domestic trunk lines**   are certified carriers holding permanent operating rights within the continental United States; these 11 major transcontinental carriers enplaned over 144 million passengers in 1973, produced 71.2 percent of all revenue passenger miles, or 115.352 billion, and 49.1 percent of cargo traffic ton miles, or 2962 billion.

2. **Local-service**   carriers operate on routes of light traffic density on short flights between small traffic centers and provide feeder service for trunk lines. Eight domestic local-service carriers in 1973 flew over 32 million passengers 9830 billion revenue passenger miles (15,917 billion km) and produced 114 million revenue cargo ton miles (183 million km) in serving more than 600 cities.

3. **International and territorial**   airlines, 10 in all, are United States flag carriers between the United States and foreign countries other than Canada. Several are also engaged in domestic trunk and local service. In 1973 they flew 35,640 billion revenue miles (57,345 billion km) with 18.9 million passengers and produced 1590 billion cargo ton miles of freight.

4. **Territorial lines**   are in two groups. Two intra-Hawaiian airlines, operating solely within the State of Hawaii produced in 1973 610 million revenue passenger miles with 4.4 million passengers and carried 6.4 million cargo miles of freight. The four intra-Alaska lines, operating within the States of Alaska, enplaned 933,000 passengers in 1973 for 498 million revenue passenger miles and produced 28.2 million ton miles of freight.

5. **All-cargo lines**   are certified to perform freight, mail, and express service between points in the United States. In 1973 three companies flew 496 million ton miles of cargo, express, and priority and nonpriority United States mail. All-cargo international carriers flew 838 revenue cargo ton miles.

6. **Helicopter**   airlines carry mail between the airports and central post offices in New York, Chicago, and San Francisco–Oakland. Helicopters are also used to carry freight and passengers between major airports. In 1973 they flew 613,000 passengers 10.9 million revenue passenger miles (17.6 million km) and produced 14,000 ton miles of cargo.[19]

7. **Supplemental**   air carriers hold temporary certificates of public convenience and necessity and are authorized to perform passenger and cargo charter services supplementing scheduled services of certificated carriers. They may also perform temporary limited scheduled service.[20]

[19]These, and foregoing data, are from *Air Transport Facts and Figures*, 1974, published by the Air Transport Association of America, Washington, D.C., 1974, pp. 20–23, 35.
[20]F.A.A. *Statistical Handbook of Aviation*, U.S. Federal Aviation, Administration, Department of Transportation, Washington, D. C., 1972 edition, p. 277.

Airports   Airports provide the interface between air and ground transport as well as fulfilling the usual terminal functions. Airports are classified by the Federal Aviation Administration as (1) local interest airports, (2) national system of airports, and (3) military airports. The Airport and Airway Development Act of 1970 directs the Secretary of Transportation to prepare "...a National Airport Systems Plan (NASP) for the development of public airports in the United States." Its principal purpose is to provide for expansion and improvement of the nation's airport and airway system to meet actual and anticipated growth in civil aviation.

Airport needs are developed under that Act in terms of Federal Aviation Administration standards for runway length, pavement thickness, gradient, for example, as related to aircraft size and type and to the level of service standards regarding departure delays—of not more than four minutes because of traffic congestion, for example.

The National Airport System Plan Designation arises from a combination of public service (annual number of passengers enplaned) and aeronautical operational density (annual aircraft operation) and furthermore as to primary, secondary, and feeder systems. See Chapter 10 and Table 10.2 for an elaboration of airport classification and function and design features.

Airports generally are developed with public funds—national, state, county, municipal, or a combination. The Federal Airport Act of 1946 gave impetus to airport construction by making available $520 million in matching funds to state and local governments for airport planning and construction.

Airports are only occasionally privately owned and then usually by an industry or flight school. Military airports are under the jurisdiction of the armed forces, especially the U.S. Air Force. Those comprising the National System of Airports may be owned and operated by a municipality (O'Hare Airport is a part of Chicago's Department of Public Works), by a regional authority such as the Port of New York and New Jersey Authority that owns and operates Kennedy, La Guardia, Newark, and Teeterboro airports in the New York area, by a county (Allegheny County Airport at Pittsburgh), or by the Federal Government (Dulles and National airports at Washington, D.C.)

There is always a need for ground transportation to airports, usually provided by private car, taxi, limousine, or bus. Because of air and ground space requirements, airports must be many miles from the inner city. Direct expressways, express bus service, rapid transit, and even suburban railroads are entering the picture to provide rapid access to the central city.

Airways Airway routes are closely related to the ground on which both visual and radio guides are established in designated patterns. A system of Federal Airways was established originally by the Civil Aeronautics Authority of the United States Department of Commerce. These airways are policed, controlled, and maintained by the Federal Aviation Administration. Low- or medium-frequency, radio-range-controlled, color-designated airways extended over 67,738 miles in 1956 including Alaska and Hawaii and consisted of 10-mile-wide air lanes between principals terminals. Only 94 miles of low-frequency routes remained in 1971.[21] Flight separation was provided laterally, horizontally, and vertically [in 1000 ft (304.8 m) layers] by rules and by the supervision of traffic-control centers located in or near principal cities.

Replacing the phased out colored airways are 289,740 nautical miles (536,599 km) of very high frequency (VHF) controlled routes protected by VOR/VORTAC guidance systems. VOR (VHF omnidirectional range) uses electronic en route facilities that provide directional guidance to or from a ground station. VORTAC is a combined VOR and VHF tactical air navigation system that enables a plane to determine its position by reference to only one facility. The system includes 945 miles (1521 km) of low frequency control, 142,093 miles (228,628 km) of direct and 33,274 miles (53,538 km) of alternate direct low altitude [700 up to 18,000 ft (213.4 to 5486.4 m)] and 114,373 miles (184,026 m) of high altitude [18,000 to 60,000 ft (5486.4 to 18,288 m)] or jet routes.[22] There is, in addition, considerable cross-country flying by private operators. Routes of the Federal Airways are shown in Figure 3.10.

Air space is governed by Air Positive Control (APC). Jet routes are under the FAA's Instrument Flight Rules (IFR). Control of flight is exercised through the FAA's 27 air traffic control centers, 346 airport traffic control towers, and nine international flight service stations.[23]

In 1971 there were 133,870 active certified civil aircraft, 2642 of which were owned by certificated airlines and commercial operators.[24] The number of aircraft by number and type of engines is given in Table 3.16. Revenues and expenses of air transport carriers are given in Tables 3.17 and 3.18. United States scheduled airlines employed 311,499 persons with

---

[21]These are located in Alaska but are being phased out. Air routes designated by color are still used in Canada.
[22]ibid., p. 6.
[23]ibid., p. 8.
[24]ibid., p. 101.

TABLE 3.16 ACTIVE AIRCRAFT CERTIFICATED IN THE UNITED STATES IN 1972[a]

Type of Use	Type of Airplane				Total
	Piston	Turbine	Rotorcraft	Other	
Domestic airlines	45	2,215	7	—	2,267
General aviation	156,500	4,500	3,400	2,600	167,000
Totals	156,545	6,715	3,407	2,600	169,267
Other Aircraft[b]					
Gliders	2,516				
Ballons	80				
Blimps	4				
Total aircraft	169,267				

[a]*Air Transport 1976*, Air Transport Association of America, Washington, D.C., 1976, p. 30.
[b]Estimated.

TABLE 3.17 TRAFFIC AND SERVICE: UNITED STATES SCHEDULED AIRLINES[a]

Traffic		Revenue
Revenue passengers enplaned	202,208,000	
Revenue passenger miles	161,957,307,000	10,275,689,000
Charter service		420,763,000
Cargo traffic	Ton miles	
Freight	4,736,729	1,038,510,000
Express	100,497,000	36,175,000
Priority U.S. mail	602,709,000	190,430,000
Nonpriority U.S. mail	595,265,000	103,237,000
Total cargo ton miles	6,035,300,000	
Public service revenue		68,930,000
Other[b]		285,037,000
Total revenue		$12,418,771,000

[a]*Air Transport Facts and Figures*, Air Transport Association of America, Washington, D.C., 1974, pp. 20, 25.
[b]This figure includes excess baggage, foreign mail, incidental revenues, and other transportation.

TABLE 3.18 OPERATING EXPENSES OF UNITED STATES SCHEDULED AIRLINES[a]

Item		Expense	Percent
Flying operations		$3,389,637,000	28.6
Maintenance		1,745,702,000	14.8
General services and administration			
Passenger service	$1,269,402,000		10.8
Aircraft and traffic			
servicing	2,335,695,000		19.7
Promotion and sales	1,424,741,000		12.0
Administration	605,143,000		5.1
Total		5,634,981,000	
Depreciation and amortization		1,064,440,000	9.0
Total operating expenses		$11,834,760,000	100.00

[a]*Air Transport Facts and Figures*, Air Transport Association of America, Washington, D. C., 1974, p. 25.

TABLE 3.19 APPROXIMATE ROUTE MILEAGE OF THE UNITED STATES TRANSPORTATION SYSTEM

Carrier	Route Mileage	Kilometers
Railroads (miles of main line)	204,696	329,356
Highways (paved roads)	2,982,000	4,798,038
Inland waterways (authorized		
navigable lengths)	25,543	41,099
Pipelines (crude oil and petroleum		
products)	168,230[a]	270,682
Natural gas transmission lines	247,000	397,423
Airways (VOR/VORTAC control)	278,822	448,625
Total transport route miles	3,906,291	6,285,223

[a]There are additionally 599,000 miles of distribution lines and 72,000 miles of field and gathering lines.

a payroll of $4,640,370. The average domestic passenger trip in 1973 was 797 miles.

TRANSPORT CORRIDORS

Mileage Summary In summary, Table 3.19 shows the approximate total mileage of the five major elements in the United States transportation system.

Lines of Flow   Certain rather well-defined transportation corridors, usually served by several carrier types, have developed in this country. Such development has arisen from the geographical pathways and barriers formed by lakes, rivers, valleys, mountain ranges, and mountain passes and from the sites of natural resources and patterns of economic activity and population density. Although modern engineering can overcome many natural barriers, population and industrial uses have been established along these corridors to establish their patterns. A full dozen transportation corridors covering most of the continental United States can be identified.

1. The **Atlantic Seaboard Corridor** between the Atlantic Ocean and the Appalachian Ranges extends from Maine to Florida with movement via coastal shipping, rail, air, and highway.

2. The **North and Middle Atlantic-Gulf Coast Corridor** between New York, Philadelphia, Washington and Gulf Coast points, served by rail, air, highway, and pipeline.

3. **Trunk Line Territory** between Boston, New York, Philadelphia, and Washington in the East and Detroit, Chicago, and St. Louis in the Midwest, served by rail, air, highway, pipelines, the Great Lakes, and the New York State Barge Canal.

4. **Pocohontas Routes,** principally coal haulage by rail from West Virginia, Kentucky, and Virginia to tidewater at Norfolk and Baltimore and to the Great Lakes at Toledo, Huron, and Sandusky.

5. **Mississippi Valley Corridor** served by water, air, rail, and highway between the Gulf Coast at Mobile and New Orleans and Chicago and the Twin Cities.

6. **Eastern Gulf Corridor** extends eastward from the Mississippi River as far north as Memphis to eastern and Florida points via water, rail, air, highway, and pipeline.

7. **Pacific Northwest Corridor** originates in the Chicago, Twin Cities, Duluth-Superior region and extends westward via air, rail, highway to the Seattle area through the High Plains.

8. **Central or Overland Corridor** by highway, rail, and air from Chicago and Mississippi-Missouri river crossings and gateways (St. Louis, Kansas City, Peoria, and Omaha) to Los Angeles and San Francisco via Denver, Cheyenne, Salt Lake City and Ogden.

9. **Southwestern Corridor** originating at St. Louis, Memphis, and Kansas City and extending rail, highway, pipeline, and air routes southwestward through Arkansas, Oklahoma, and Texas to the Pacific Coast.

10. The **Gulf-Pacific Coast or Southern Transcontinental Corridor**   extends from Mobile, Birmingham, New Orleans, Baton Rouge, and Shreveport in the Gulf Coast region westward to Texas and West Coast points using rail, air, water, and pipeline.

11. Pacific Coast Corridor   extends north and south from San Diego and Los Angeles through the San Francisco Bay area to Seattle, Vancouver, and northwestward by deep water and air to Alaska. Rail and highway serve the United States coastal portion.

12. Panama Canal routes   connect the Pacific with Gulf and eastern seaboard cities via deep-water shipping.

Gateways   Corridor routes frequently are concentrated through certain limiting zones. These may be geographical but also because of rates and tariffs whereby certain connections are quoted via a few key points. Peoria, St. Louis, Omaha, and the Twin Cities have traditionally served as gateways, and traffic has tended to flow and routes to be concentrated in these areas accordingly.

Terrain and topography are other important limiting factors. A river valley provides easy grades and locations for railways and highways as well as for water-borne commerce. The industry and population that then congregate along the river intensify the traffic needs and production of that area. The Great Lakes and their adjacent coastal plain similarly provide a natural location for shipping, railways, highways, pipelines, and a route for airways to follow. Mountain passes have long been a limiting factor in route location. The Delaware Water Gap, Cumberland Gap, Mohawk Valley, Donner Pass, and others are occupied by one or more railroads and highways.

Urban Corridors   Population and urban growth have led in some areas to expanding metropolitan centers advancing toward each other to form an urban continuum hundreds of miles in length. Problems arise in the movement of people and goods through these urbanized corridors where there is demand for rapid and extensive transportation service but where land is densely populated and expensive. The Northeast Corridor along the Atlantic Seaboard between Washington and Boston, a similar if less densely populated corridor between San Diego and Sacramento on the West Coast, and the several corridors that emanate from Chicago to Milwaukee, St. Louis, Indianapolis–Cincinnati, and to South Bend to form a Midwest Network are typical of this development. See Figure 3.11.

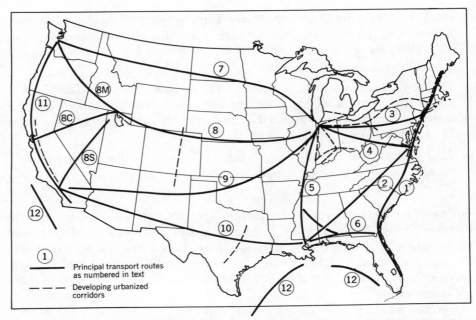

FIGURE 3.11 TRANSPORTATION CORRIDORS IN THE UNITED STATES (AND DEVELOPING URBANIZED CORRIDORS).

QUESTIONS FOR STUDY

1. Present in graphic form the various components of the United States transportation system.
2. Distinguish between the services performed by (*a*) a belt-line or switching railroad, (*b*) a terminal railroad, (*c*) a line-haul railroad.
3. Give a complete ICC designation for (*a*) a Class II contract carrier serving an irregular route, nonscheduled service, and carrying petroleum products, (*b*) a Class I common carrier hauling motor vehicles over a regular route with scheduled point-to-point service.
4. Explain the advantages and disadvantages of private systems of transport for (*a*) private individuals, (*b*) small companies, and (*c*) large companies and corporations.
5. What are the distinguishing features of (*a*) personalized transport, (*b*) mass transport, (*c*) rail rapid transit, and (*d*) commuter railroads?
6. What are the airport classifications for the National Airport Plan and what is the basis for those classifications?
7. What problems are likely to arise in operation, maintenance, and financing when roadway and vehicle are owned by separate entities? What competitive problems are engendered?

8. What factors give rise to transportation corridors (*a*) in general, (*b*) in the case of a selected individual corridor?
9. Described the formation and transportation requirements of an urbanized corridor.
10. What implications for the future does the drive-it-yourself system of automobile and truck rental offer?
11. What kinds of data might be found in the Department of Transportation's *National Transportation Reports*?
12. What are the usual duties and powers of the Federal regulatory agencies? Of state agencies? Of regional, county, and municipal agencies?
13. Which, if any, of the various modes of transportation have promise as growth industries during the next decade? Why?

SUGGESTED READINGS

1. Wilbur G. Hudson, *Conveyors and Related Equipment*, 3rd edition, Wiley, New York, 1954.
2. Harold M. Mayer, *The Port of Chicago and the St. Lawrence Seaway*, University of Chicago Press, Chicago, 1957.
3. *Transportation on the Great Lakes*, Transportation Series No. 1, 1937 revision, Corps of Engineers, U. S. Army, U. S. Government Printing Office.
4. Royale James Heyl, "The Southern Routeway of the United States—A Study of Interregional Surface Freight Movements," doctoral thesis, University of Illinois—Urbana Campus, 1962.
5. "1972 and 1974 National Transportation Reports: Present Status—Future Alternatives," U.S. Department of Transportation, Office of the Secretary for Policy and International Affairs, July 1972 and 1974, Washington, D.C.
6. *Big Load Afloat*, a publication of the American Waterways Operators, Washington, D.C., 1973.
7. *Transportation, The Nation's Life Line*, George M. Harmon, Editor, Industrial College of the Armed Forces, Washington, D. C., 1968.
8. *Competition Between Rail and Truck in Freight Transportation*, Charles River Associates, Cambridge, Massachusetts, December 1969, Contract No. DOTOS-A9-060, Office of Assistant Secretary for Policy and International Affairs.
9. *Inalnd Waterways Transportation*, C. W. Howe et al., Resources for the Future, Washington, D. C., 1969.
10. *National Transportation Policy* (The Doyle Report), Preliminary draft of a report prepared for the Committee on Interstate and Foreign Commerce, U.S. Senate by the Special Study Group on Transportation Policies in the United States, 87th Congress, 1st Session, January 3, 1961.
11. *Preliminary System Plan for Restructuring Railraods in the Northeast and Midwest Region* pursuant to the Regional Rail Reorganizational Act of 1973, U.S. Railway Association, February 26, 1975.

12. *Evaluation of the U.S. Railway Association Preliminary System Plan*, Interstate Commerce Commision, report of the Rail Services Planning Office to the U.S. Railway Association, April 28,1975.
13. Harold L. Gauthier, *Geography of Transportation*, Prentice-Hall, Englewood Cliffs, New Jersey, 1973.
14. *Defining Transportation Requirements*, papers and discussion of the 1968 Transportation Engineering Conference of the American Society of Mechanical Engineers, October 1968, New York Academy of Science, New York.
15. *Bus Use of Highways—State of the Art*, National Cooperative Highway Research Program No. 143, Highway Research Board—National Research Council et al., Washington, D.C., 1973.
16. *Bus Use of Highways—Planning and Design Guidelines*, National Cooperative Highway Research Program No. 155, Transportation Research Board—National Research Council, Washington, D.C., 1975.
17. Donald R. Whitnah, *Super Skyways: Federal Control of Aviation*, Iowa State University Press, Ames, Iowa, 1966.
18. Dudley E. Pegrum, *Transportation Economics and Public Policy*, Richard D. Irwin, Homewood, Illinois, 1968.
19. Noël Mostert, *Supership*, Knopf, New York, 1974.
20. G.M. Smerk, *Urban Transportation, The Federal Role*, Indiana University Press, Bloomington, Indiana, 1965.
21. United States Railway Association *Preliminary System Plan*, Vol. I and II, Washington, D. C., February 1975.
22. Evaluation of the *U. S. Railway Association Preliminary System Plan*, Interstate Commerce Commission, Office of Rail Services Planning Office, 28 April 1975, Washington, D. C.
23. *Final System Plan, Vol. I and II*, U. S. Railway Association, Washington, D. C. 26 July 1975.

Part 2
Transport Technology

Chapter 4

Technological Characteristics

The technological components of a transport system—vehicle, motive power, roadway, terminals, and operational control—shown in Figure 1.2 combine to provide transport capability and utility, and are present in any transport system regardless of modal configuration. In this chapter the vehicle and its characteristics central to transport operation will be considered first, although the separation of vehicle from motive power and roadway cannot always be clearly defined.

MODAL CHARACTERISTICS

Inherent Advantage Each carrier type has closely interrelated technological and economic characteristics that bring both advantages and disadvantages to its operation. The combination gives each mode a particular field of usefulness. Outside that field, its usefulness may be marginal. The marginal carrier may remain in a particular kind of service because of the overall demand for transportation (as in time of war, in a rapidly expanding economy, through user preference, or when no other transport is available) or because of some form of government subsidy or regulatory restriction on competitors.

An efficient domestic or international system of transportation permits and encourages the full utilization of inherent advantages. This is the stated policy of the United States government in the 1940 amendment to the Interstate Commerce Act, which declares it to be the national transportation policy of the Congress (among other things):

To provide fair and impartial regulation of all modes of transportation so administered as to preserve the inherent advantages of each.

Often the advantages of each modal type are best realized in combination with those of other types. The full advantages of waterborne bulk-cargo carriers can be realized, for example, only in combination with a coordinated system of land transportation. Problems of adequate coordination and integration among transport agencies and facilities are considered later. This chapter, and those that follow, presents technological characteristics common to all or nearly all types of transport. These characteristics include units of transport, degrees of freedom, systems of guidance, propulsive resistance and propulsive force, fuel consumption and thermal efficiency, payload to empty weight ratios, suspension, stability and buoyancy, capacity, speed, effects on environment, guidability and maneuverability, loading and unloading capabilities (passenger traffic is normally self-loading and self-transferring), and the effects of curvature, grades, and elevation. Certain technological characteristics are reflected in carrier capacity and in the operational and economic aspects of transport, especially costs. All are combined and interrelated in the selection of mode and in the planning and location of transport systems. Figure 1.2 shows technoeconomic relationships.

UNIT-OF-TRANSPORT CLASSIFICATION

In Chapter 3, transportation agencies were conventionally classified as railroads, highways, airways, waterways, and pipelines and also according to the classification systems of the regulatory bodies. A different classification, indicative of the transportation potential of the carrier, is the unit of carriage. In Chapter 12 a relation between the following groupings and the costs of each type is noted.

Single Units   This type of transport combines motive power (propulsion) and cargo or passenger space in one unit. The single-unit group includes automobiles, trucks, buses, aircraft, ships, rail cars (both dual and mono-rail), street cars, and interurban cars. Most of the exotic transport concepts are of the single-unit type.

Assembled or Multiple Units   The assembled or multiple-unit system of transport normally has a separate motive-power unit that propels, tows, or pushes one or more cargo-carrying units. One railroad locomotive hauls a few or many cars, as the volume of traffic requires. A tow may consist of 1 to 20 or more barges that are picked up and set out by the towboat (usually a pusher or a ram type) at various points along the route. Rapid transit trains composed of conventional multiple-unit (MU) cars or of other types of cars in commuting, rapid-transit, or high-speed corridor service combine elements of the single- and multiple-unit systems. Often with multiple-unit operation, one car is motored and pulls an unmotored trailer, but car pairs may be combined into trains of as many units as traffic requires. The Northeast Corridor Metroliners and some other systems have all cars motored. A highway tractor-trailer combination also has elements of each group. Normally there is only one trailer, but it is separate from the motive-power unit (the tractor) and can be set out and picked up at will. Thus one tractor can keep several trailers in short-haul operation. The operation of so-called double bottoms, that is, two trailers hauled by one tractor, has become commonplace on interstate road systems.

Continuous-flow or Stationary-Propulsion System   A third category includes those systems in which the transport "roadway" channels and guides the traffic, and the propulsive force is supplied by one or several intermediate and stationary power sources. In pipeline operation the cargo moves, but the pumping units and the pipelines remain fixed. There is no "vehicle" as such. In conveyors and moving sidewalks the belt moves, but again the prime mover is stationary. The propulsive unit for an aerial tramway is fixed. Freight usually moved by a continuous-flow system requires no packaging or stowing (as presently operated) but may require positioning (the cargo must be centered on the conveyor belt) or processing (coal must be pulverized and placed in suspension for pipeline transport). The loading and unloading of the system are usually done without manual labor, that is, by means of a valve or a tripper and chute.

From the foregoing it is evident that the single-unit carriers have characteristics especially adaptable to moving individuals or small groups of persons and small quantities of packaged goods, that multiple-unit systems can move people in mass quantities and either packaged or bulk products, and that continuous-flow systems are especially suited to move raw products in bulk quantities but with limited adaptation to moving people.

GUIDANCE AND MANEUVERABILITY

Degrees of Freedom   Operational flexibility and safety are related to the degrees of motion and maneuverability a vehicle has.

One degree.   The vehicle is limited to forward and backward movements by means of a track or other structurally limited guidance.

Two degrees. The vehicle can also move laterally right or left in one plane, exemplified by autos, buses, and ships.

Three degrees. This category can also move up and down. Airplanes, helicopters, dirigibles, and submarines can be so classified.

Closed-System Guidance Pipelines, conveyors, and moving sidewalks offer a confining roadway that prevents departure from the intended route or interference with other traffic. The adherence to route is independent of weather, and complete guidance is inherent in the system but with minimum flexibility.

Lateral Wheel-Rail Guidance A recent innovation is an I-beam positioned in the roadway between the tractive wheels (usually rubber-tired) against which suspended guide wheels bear laterally. See Figure 4.1.

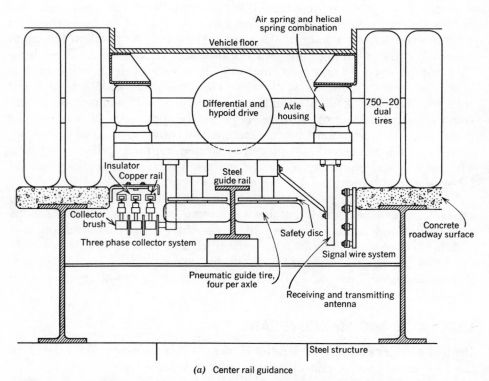

(a) Center rail guidance

FIGURE 4.1 LATERAL WHEEL GUIDANCE SYSTEMS. (*a*) CENTER RAIL GUIDANCE: TRANSIT EXPRESSWAY (COURTESY OF WESTINGHOUSE ELECTRIC CORPORATION).

Guidance is absolute as long as the I-beam-wheel combination remain inteact. A variation has the lateral wheels bear against side boards or guides outside the traction wheels. Montreal and Paris subways use this system, combined with conventional flanged wheels to give guidance through the switches at diverging routes and crossovers.

With center rail guidance, the entire rail or guide beam must be moved, a process much slower than the three seconds or less required to move a railroad switch. One system uses a transfer table that moves not only the vehicle but an entire section of roadway laterally from one path to another, a time-consuming operation. A different system uses arms projecting from the vehicle that ride in grooves of guide channels placed parallel to the roadway. The guide arms follow the channel path and impart guidance to the vehicle. The presence of such a guide channel on one side or the other determines which guide arm, right or left, will receive guidance and impart direction at a diversion point to the vehicle.

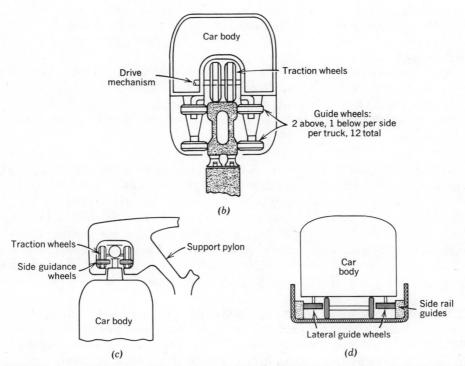

FIGURE 4.1 (CONTINUED) (*b*) CENTER RAIL GUIDANCE: ALWEG SUPPORTED SYSTEM. (*c*) SIDE GUIDANCE—SAFEGE-SUSPENDED SYSTEM. (*d*) SIDE GUIDANCE—RUBBER-TIRED.

Rail-Flange Guidance   Railroad trains and rail cars of all types are guided by flanged wheels bearing against a steel rail. The flange may project little more than an inch (2.54 cm) beyond the wheel tread but, combined with vehicle weight and a smooth track, gives complete safety in guidance with no dependence on human effort. The vehicle cannot be blown or drift off course or become lost regardless of weather. Inclined planes, skips, elevators, monorail systems, and some types of cableways use similar guidance, usually with a double flange on the guide wheels. Rail-flange guidance definitely lacks flexibility. The vehicle can go only where a track is laid in advance. Switches are necessary to divert trains from one track to another. (See the section on geometrics.) However, the ubiquity of tracks and turnouts makes rail flexibility relatively good.

Electronic Guidance   The "automatic pilot" of an aircraft can be "locked on" to the guidance beam of a directional radio transmitter. A ship can have its course set by a similar automatic device that steers the ship on its predetermined way. On an experimental basis, highway vehicles have had the steering mechanism inductively coupled to a cable buried in the center of a pavement lane, thereby offering the possibility of an automated expressway.

The guidance of aircraft, rockets, and other missile types by radio remote control is a familiar event today and offers the possibility of extending that type of control to all other carriers—to ships, trucks, and automobiles. Small bursts fired from jet nozzles give directional control to spaceships.

Driver-Pilot Guidance   The pilot or driver must steer the vehicle or craft continuously to keep it on route or course and to avoid contact with other vehicles. Safety of movement depends largely on the driver–pilot's skill and attention to his task. The system offers maximum flexibility, but the hazards are obvious. In passing or overtaking, the operators of both vehicles must be fully alert to prevent conflicts. The human element bears primary responsibility for the vehicle's safety. The section on geometrics describes the role of lateral friction in guiding highway vehicles on curves.

Confining Roadway-Waterway Guidance   Highways, airport runways, and confining waterway channels indicate the route to follow. There is little danger of losing the way unless heavy storms or fog reduce visibility to zero or the way becomes obliterated by drifting snow or sand.

In a river tow, barges and towboat are lashed together in a single more or less rigid unit (Figure 4.2). Towboats, often referred to as "pushers" or "rams" actually push the barges instead of towing. The towboat's rudders and manipulation of its propellors steer the entire combination. An attempt to pull a string of barges around sharp bends would end in disaster, possibly with the barges piled on the river bank. True towing, with the barges strung out behind the towboat, is usually attempted only in open water or in sea-level intracoastal waterways where currents are less dangerous.

Open Waterways Lakes, oceans, and wide rivers offer no guidance to vessels. Constant pilotage must be observed to keep vessels on course and to avoid contact with obstacles and other vessels. Flexibility of direction is unlimited, but adequate harbors must be available for vessels to reach shore in safety.

Guidance of vessels during storms, when changing course, or when in harbor and in other restricted areas is provided by a helmsman in the wheelhouse. In very small vessels and in older sailing ships the rudder may be controlled directly by hand or by ropes or cables running from the wheel to the rudder post. In most ships of commercial size a small wheel is

FIGURE 4.2 A RIVER TOW. (TOWBOAT *HUMPHREY* WITH 11-BARGE COAL TOW; 1600-HP 132-FT TOWBOAT BUILT BY DRAVO CORPORATION, PITTSBURGH, PENNSYLVANIA, FOR CONSOLIDATED COAL COMPANY.)

used to control a powered device that actually moves the rudder. When on course in calm water, the responsibility for guidance can be delegated to an automatic pilot, which guides the ship automatically, once a course is set, by control of a gyroscopic compass. Large ships have sufficient flexibility of guidance for open water. In harbors and narrow channels their extreme bulk and mass make guidance difficult, especially when the unfavorable tides and currents found in many harbors are encountered. Tugs are usually necessary to effect a safe berthing or putting out. In storms or foggy weather radar plus traffic regulations aid the helmsman to direct the ship.

Airways   Airways offer the same problems as open waterways with the additional requirement of guidance in three dimensions. (This requirement is also present for submarines in open waterways.) Aircraft are maneuvered or guided by movable surfaces—the ailerons, rudder, and elevators. Retractable flaps are also used to increase wing area at the slower speeds of takeoff and landing. Airplanes can be flown and guided, like ships, with an automatic pilot, but human guidance is usually required in takeoff and landing and under adverse weather conditions. An aid in guiding the plane is the radio beam (in various types) that establishes both a sonic and visual (on the instrument panel) indication of proximity to the established route.

Aircraft offer considerable flexibility in direction (although restrained by air traffic regulations), but smooth landing fields are mandatory. Helicopters overcome this difficulty to a considerable degree, as do experimental models of more conventional types of planes, which rise or land vertically, leveling off in full flight.

SUPPORT—BUOYANCY AND STABILITY

Stable support is an obvious requirement for vehicle operation. Land-based vehicles receive solid support from highways, runways, and tracks. These, along with stability requirements for vehicles, are considered in Chapter 6. Water-supported ships and barges are supported through buoyancy. Balloons and dirigibles receive buoyant support, but heavier-than-air craft are dependent on air pressure differentials. Stability is essential to riding comfort and safety for all vehicles and craft. It is a factor in designing the suspension systems of land vehicles—wheels, axles, and truck assemblies—and in designing and maintaining the smoothness of roadway and track and the superelevation for highway and railway curvature.

Ship's Buoyancy   Archimedes' principle states that a body immersed in a liquid is buoyed up by the weight of the volume of liquid displaced. Fresh water weighs 62.4 lb (28.3 kg) per ft^3, salt water 64.0 lb (29.1 kg) per

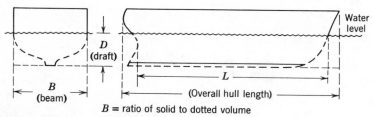

B = ratio of solid to dotted volume

FIGURE 4.3 BLOCK RATIO.

ft³. The weight of a ship in tons is equal to the weight of water its submerged portion displaces. The displacement space equivalent of one ton of salt water is $2240 \div 64 = 35$ ft³ (0.99 m³); of fresh water, $2240 \div 62.4$ $= 35.9$ ft³ (1.02 m³). Both measurements are based on long tons. A rectangular block D ft deep having a wetted length L, breadth B, and draft or submerged depth H would, in salt water, displace $L \times B \times H \div 35$ long tons; in fresh water the displacement becomes $L \times B \times H \div 35.7$ tons.

Ship hulls, however, are not rectangular. They are more or less stream-lined for easier movement against the resistance of water and waves. The ratio between the actual underwater volume and the corresponding LBH block is known as the block coefficient, C_B. For a yacht, C_B may be as low as 0.50; for a river barge C_B ranges from 0.88 to 0.95 and for a towboat or ram from 0.58 to 0.68.[1] The C_B for Great Lakes bulk carriers averages 0.87 to 0.88. It varies with the draft (and therefore with the weight of ballast, fuel, and cargo at any given time) and must be obtained from the designer's table or chart for the draft at any particular loading. The actual displacement (weight in long tons) of a vessel for salt water is $(LBH \times C_B)$ $\div 35$. See Figure 4.3.

The safe depth of submersion is, among other factors, a function of the load, length of vessel, draft, size of wave to be encountered, and depth of channel to be coursed. As allowed by law, it is indicated by the load line or Plimsoll line, so-named in Britain after its advocate of 1890, Samuel Plimsoll. A typical load line takes the form shown in Figure 4.4 and is located exactly amidship. The horizontal arms indicate the point to which a vessel may be legally and safely submerged, depending on its route and season. The Plimsoll mark on United States vessels is inspected regularly by the Coast Guard.

[1]According to a letter received from Braxton B. Carr, President of the American Waterways Association, the block coefficient is not in general use in the design of inland waterways equipment but the relations hold nonetheless. Note also that C_B is usually designated β in European practice.

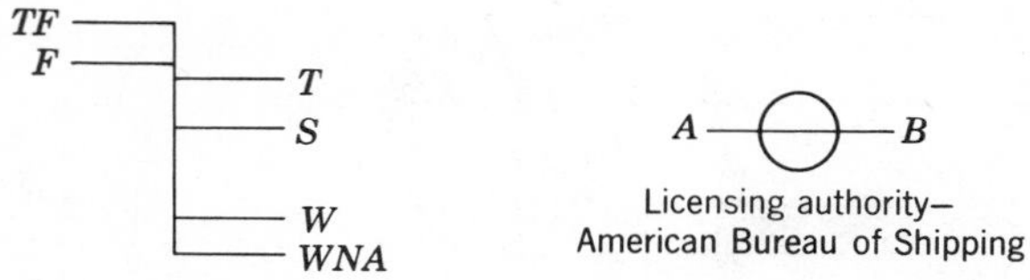

TF — tropical fresh water
F — Fresh water
T — Tropical water
S — Summer
W — Winter
WNA — Winter, North Atlantic

FIGURE 4.4 LOAD LINES.

Draft is the submerged depth of a vessel or barge. The required depth for operation must include additional clearance to prevent contact with the channel bottom. A ship in motion experiences a so-called squat by which the stern is pulled downward by the propellor. Allowance must also be made for wave height, at least 4 ft (1.22 m) for soft bottom channels and harbors, 6 ft (1.83 m) for rock bottoms. Great Lakes bulk cargo carriers normally draw or require 23 ft (7.01 m) of channel depth when loaded and 20 ft (6.1 m) empty. Most ocean shipping requires at least 26 ft (7.93 m), 36 to 40 ft (10.97 to 12.19 m) for the larger ships, and 60 to 90 ft (18.29 to 27.43 m) for the largest intercontinental tankers carrying up to 3,000,000 bbl of oil. River craft require depths of 6 to 12 ft (1.83 to 3.66 m). Freeboard is the height of hull above the water line.

Because the total actual weight of a vessel depends on the weight of what is in it, there are three weights that determine displacement.

Light. Vessel without crew and supplies, cargo, fuel, or passengers.

Loaded. Weight when loaded to maximum draft, to the deep load line.

Actual. Weight of the vessel and tonnage at any time during the voyage, varying with fuel, cargo, and passengers aboard.

Ships are registered by tonnage, either *net* or *gross*.

Vessels have displacement curves showing the displacement tonnage (hence C_B) at any draft. The displacement of a warship is always given as actual weight displacement tons.

Four other weights are here noted as used in maritime circles.

1. Deadweight, the maximum weight of fuel, passengers, and cargo that a vessel can carry when loaded to the deep load line. It is the difference between displacement tonnage light and loaded and is the basis on which payment is made for vessels in charter. The term deadweight for barges (and land transport vehicles) means exactly the opposite, that is, it is the weight of the barge (or vehicle) when empty.

2. Vessel gross weight, expressed in units of space, is based on a seldom-used old assumption that 100 ft³ (2.8 m³) of space accommodates one ton of cargo. It is equal to the closed-in capacity in cubic feet divided by 100 (or the closed-in space in cubic meters divided by 2.83).

3. Net weight is the actual gross tonnage minus the space devoted to the operation of the vessel, that is, with space occupied by machinery and fuel, crew accommodations, stores, and galley deducted. Usually about 30 percent of the closed-in space is so deducted.

4. Cargo tonnage may be expressed either in actual tons or measure (100 ft³) tons. The long or 2240-lb (1017-kg) ton is used in most ocean shipping and increasingly on the Great Lakes, the short or 2000-lb (908-kg) ton in the United States, and the metric or 2204.6-lb ton in France. For bulky lightweight cargos, a measurement ton of 40 ft³ (1.12 m³) is used.

Aircraft Support Only balloons and dirigibles have buoyancy in the same sense as a ship. Archimedes' principle and Boyle's law both find application here. Nevertheless, heavier-than-air craft (omitting rockets) are sustained in flight by pressure from the medium in which they operate, that is, air. Bernoulli's law states that the pressure of any fluid stream is least when velocity is greatest, greatest when velocity is least. If this law is applied to an airfoil (wing) as shown in Figure 4.5, the air stream, having a greater distance to flow across the upper surface, flows faster, exerting less pressure, than it does on the under surface. The results are a difference in pressure beneath the upper and undersides of the airfoil and an upward lift or sustaining force.

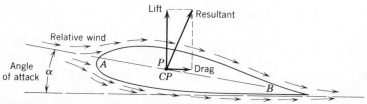

FIGURE 4.5 FORCES ACTING ON AN AIRFOIL.

Empirical expressions for forces acting on a plane's wing, based on Bernoulli's theorem, are

$$L = C_L(\rho/2)Sv^2 = \begin{array}{l}\text{lift or supporting force in pounds} \\ \text{perpendicular to the selective} \\ \text{wind}\end{array}$$

$$D = C_D(\rho/2)Sv^2 = \begin{array}{l}\text{drag or resistive force in pounds} \\ \text{perpendicular to relative wind}\end{array}$$

$$R = C_R(\rho/2)Sv^2 = \begin{array}{l}\text{resultant of the previous forces in} \\ \text{pounds}\end{array}$$

In these equations C_L = lift coefficient = the ratio between the lift of the airfoil and impact pressure of air on a flat plate of the same area. It varies from 0.0 to 3.0 (with flaps), as the angle of attack increases (it is three times greater at 6 degrees than at 2). Figure 4.6 shows C_L, C_D, and other characteristics varying with the angle of attack for an airfoil in frequent use (the N.A.C.A. with an infinite aspect ratio). ρ = air density, which varies with the altitude and temperature but which is usually taken as 0.002378 slugs per cubic foot at sea level; v = relative velocity in feet per second (equal to 1.47 V, where V is the velocity in miles per hour); S = wing area in square feet, including flaps for landing and take-off. The force supporting a plane can be seen in the following problem example.

Problem Example

A light transport plane on the ground is subject to an air pressure of 14.7 psi on both sides of the wings. When in motion at take-off, the air pressure above the wing is 14.523 psi. There is thus an upward difference in pressure of 0.177 psi or 25.5 psf of wing area. With an effective wing area of 987 ft², the total upward pressure on the underside of the wing becomes $987 \times 25.5 = 25{,}169$ lb (11,427 kg).

The resultant force R produces a moment about the leading edge of the wing. The supporting pressures under an airfoil may be considered concentrated at one point known as the center of pressure or center of lift. It is located by the percentage of chord length where it is applied behind the leading edge, where CP = center of pressure = AP/AB. If the moment formed by the lift or resultant tends to raise the leading edge (thereby increasing the angle of attack), there is a tendency toward stalling. If the moment tends to lower the leading edge, decreasing the angle of attack, there is a diving tendency. Values for the lift coefficient C_L and also the drag coefficient C_D, for various angles of attack, are shown in Figure 4.6 for the NACA airfoil. Further explanation of these terms and factors is given later in this chapter.

Robert T. Aangeenbrug

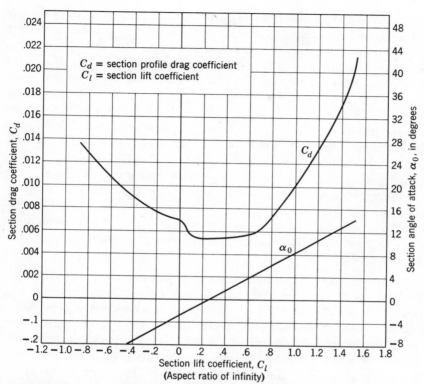

FIGURE 4.6 CHARACTERISTICS OF NACA 63_1-412 AIRFOIL. (NATIONAL ADVISORY COMMITTEE FOR AERONAUTICS, REPORT NO. 824, P. 165, BY IRA H. ABBOTT, ALBERT E. VON DOENHOFF, AND LOUIS S. STIVERS, JR., LANGLEY MEMORIAL AERONAUTICAL LABORATORY, LANGLEY FIELD, VIRGINIA, U.S. GOVERNMENT PRINTING OFFICE, WASHINGTON, D.C.)

The decreased pressure on the upper surface of the wing is based on a smooth flow of air over the surface, which occurs when the angle of attack α is small. Air will continue to flow smoothly as α is increased, until the "burble" angle is reached where the air flow breaks away from the upper surface near the trailing edge and burbles and eddies. Since the support at the trailing edge is decreased, the center of pressure shifts forward and tends to depress the trailing edge still further with an inclination toward stalling. The burble or stalling angle varies with the shape of the airfoil (and with the speed) but has a value of 16 to 20 degrees for most airfoils. Slots in the leading edge of the wing reduce the stalling angle and the chance of falling in a spiral.

Lift increases as the square of the speed, but the critical point occurs at the slow speeds of takeoff and landing. The CAB sets the minimum stalling speed in "approach" for various classes of airplanes. For larger craft, the long runways of modern airports are required. Increased lift at slower speeds is obtained by the use of retractable flaps that increase the wing area at takeoff and landing but that can be retracted to decrease resistance or drag when the plane is in full flight. Jets and jet-assisted propeller planes develop more initial speed and can take off in shorter distances. When a plane lands, the propeller pitch may be reversed to act as a brake (thereby shortening runway requirements once the plane is on the ground).

The lifting power of a plane varies with wing and power loading. Wing loading is the gross weight of the airplane in pounds divided by its total wing area in square feet. Power loading is the gross weight divided by total horsepower. The product of wing and power loading gives a commonly used index for plane classification and runway design. Airplanes must have motion to sustain flight.

For supersonic flight some modification of the foregoing is required. Sound travels at a speed of 760 mph (1222.8 kph) at sea level and a temperature of 59° F (15° C) but loses speed as altitude increases and temperature decreases. At subsonic speeds sound and pressure waves in advance of the aircraft permit airflow particles to adjust to the craft in a more or less streamlined pattern. At transonic speeds where air flow over a part of the plane is around the speed of sound and supersonic speeds (all airflow superflow) the air molecules are not "arranged" in advance but are thrust violently to form a bowlike shock wave just ahead of the craft no thicker perhaps than ten thousandths of an inch. Subsonic flow and turbulence occur behind the shock wave. There are energy losses in heat and an added wave drag factor of 5 to 10 percent that requires additional engine horsepower. The ensuing loss of lift, turbulence, and buffeting create problems in control, stalling, and possible structural damage.

An ideal airfoil design would eliminate the transonic range by keeping the airflow subsonic until Mach 1 is reached. A Mach number is the ratio of true air speed to the speed of sound. Subsonic speeds are those with Mach numbers below 0.75, transonic 0.75 to 1.2 and supersonic speeds have Mach numbers of 1.2 to 5.0+. Such an airfoil would be thin yet have sufficient thickness for structural strength and for carrying devices to increase wing area for safe takeoff and landing speeds. Wing taper, aspect ratio, and sweepback of the wings are significant design elements. Sweptback wings increase the critical Mach number, M_c (where Mach 1 is reached at some point on the wing). The drag coefficient is reduced as the sweep angle is increased. Sweep introduces, however, the problems of lateral control, tip stalling, a forward shift of the center of pressure, and need for higher landing speeds. Solutions include a compromise sweep angle and the use of movable wings for which the angle of sweep can be increased when the plane is in flight.

The shock wave creates a sonic boom that may have adverse environmental effects. It can cause structural damage to ground installations and frighten people and animals. Large quantities of fuel are consumed by the requirements of high horsepower. For these reasons supersonic flight has been limited largely to military and experimental aircraft. Much opposition has arisen to the development of commercial supersonic transports (SSTs).

At this writing, the Congress has discontinued funds for research and development of an American version designed to carry up to 298 passengers at a cruising speed of 1800 mph (2896 kph). The Russian-built Tupolev TU-144 is designed for 130 passengers traveling at a speed of 1550 mph (2494 kph) while the Anglo-French Concorde has a similar seating capability at 1336 mph (2150 kph). As such aircraft become commercially feasible, they are likely to operate under rather severe restrictions as to routes and speeds over land areas.

Fluid Support Systems   The dynamic interaction between roadway and vehicle requires a smooth road or track to keep shock, sway, and concussion to a minimum and for comfort and safety to passengers and goods. Because dynamic effects are intensified by speed, the criteria for roadway smoothness become more exacting as speed increases. Construction and maintenance to meet these criteria become correspondingly costly. Problems with suspension and contact dynamics and wear phenomena of wheels, whether steel or rubber-tired, arise at high speeds and may prove to be limiting factors. There is evidence that the practical limit with present technology for roadway contact systems is in the range of 200 to 250 mph, (322 to 402 kph).

One alternative utilizes a fluid support system that depends upon maintaining a pocket or cushion of air under the vehicle surface or a set of lifting pads. The pocket of air is sealed within the pad by a fluid or mechanical means. A fixed horizontal plane is required as a roadway but, where there is no contact between that plane (roadway) and the vehicle, a lesser degree of precision may be permissible in the regularity of its surface. That degree of precision depends on the clearance between vehicle and roadway arising from three different variations of the fluid support system.

The close clearance design is exemplified by a "shoe" surrounding and sliding on a rail. The separating and supporting medium may be the conventional petroleum lubricant or a very thin layer of air, about 0.001 in. (0.00254 cm), forced through openings in the shoe under high pressure. See Figure 4.7.

Medium pressure systems designated as ground effects machines (GEM) and air cushion vehicles (ACV) maintain a clearance varying from about 0.01 to 0.05 in. (0.0254 to 0.127 cm) with the French designed Aerotrain to 3 to 4 ft (0.91 to 1.22 m) with the Hovercraft. Heavy fabric "skirts" attached to the bottom of the vehicle or to the support pads form the air pocket seal. Pressures are moderate, about 25 gm/cm^2 for the Aerotrain.

The sliding shoe requires low volumes of air under high pressures, the ground effects system utilizes large volumes at low pressures. Future GEM systems may include wheels for switching, station stops, and other slow speed movements. Hovercrafts have had considerable commercial use in Britain including a route crossing the English Channel. The Aerotrain concept, originated in France, is undergoing research and development by the U. S. Department of Transportation. Vehicular design and motive power features of the system will be reviewed in later chapters.

A third type of aerodynamic support is that of low pressure and high clearance exemplified by rotor jet flaps and wing-in-ground effect concepts that have had very little application to practical vehicles. These devices are essentially ground related but require minimum roadway precision. Actually the application has been largely confined to watercraft designed to skim or plane over rather than through the water surface, a type of hydroplane or hydrofoil.

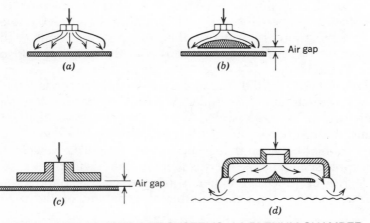

FIGURE 4.7 FLUID SUPPORT SYSTEMS. (*a*) PLENUM CHAMBER. (*b*) PERIPHERAL JET. (*c*) AIR BEARING. (*d*) ANNULAR JET— WATER CRAFT.

An entirely different system is the maglev (magnitude levitation vehicle) in which the vehicle is lifted above a metallic roadway by magnetic "repulsion." An alternative procedure has the vehicle suspended above the roadway by magnetic attraction applied from above. Linear induction motors are proposed for propulsion with both systems. Center rail guidance and relatively close clearance requiring precision in roadway design and maintenance characterize these systems.

Ship Stability   Stability is the ability of a ship to remain in an upright position about its axis or to return to that position if moved from it by an outside force, such as a wave. When a ship is in trim, the deck is level and the center of gravity G and center of buoyancy B are on the same vertical axial line. See Figure 4.8. If disturbed from that position, a stable ship tends to return to its original position of equilibrium. Consider the transverse section of Figure 4.8. The vessel is shown in a heeled position with

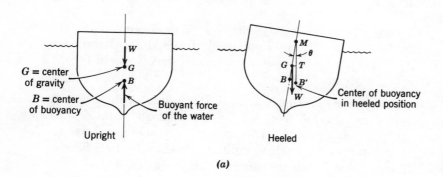

(a)

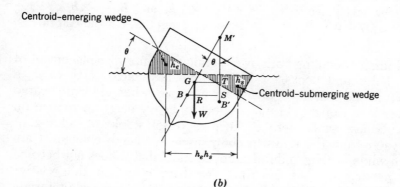

(b)

FIGURE 4.8 SHIP'S STABILITY: (a) ANGLES OF HEEL < 15 DEGREES; (b) ANGLES OF HEEL > 15 DEGREES.

the center of buoyancy shifted to B' and acting vertically upward to intersect the extension of BG at M. A righting couple is produced to overcome the heeling couple whereby

$$W \times GT = W \times GM \sin\theta$$

where W = the weight of the vessel is displacement tons and $GT = GM \sin\theta$ is the righting lever in feet. The point M is the transverse metacenter and for small angles of heel (up to 15 degrees) is approximately fixed in position. GM is the metacentric height. The height of the metacenter above the center of buoyancy is $BM = I/V$, where V is the volume of displacement in cubic feet and I is the moment of inertia of the water plane about the axis of rotation.

The position of centers of gravity and buoyancy can be calculated by knowing the position and weight of the elements that comprise the ship. These will vary with the distribution and weight of cargo, fuel, and ballast. Typical metacentric heights vary from 1 to 2 ft (0.34 to 0.68 m) for tugs and liners to 2 ft (0.68 m) for cargo vessels and up to 5 ft (1.7 m) for battleships.

For heeling angles greater than 15 degrees, M does not remain fixed but varies its position to M'. There is no simple formula for determining the value of BM', but a method frequently followed makes use of the Atwood formula developed about 1795.

V is the total immersed volume, and v is the volume of the emerging or submerging wedge in cubic feet; $h_e h_s$ is the horizontal distance in feet between the centroids g_e and g_s of the emerged and submerged wedges. A moment of $W \times GT$ ft-tons tends to right the vessel. $GT = BS - BR = BS - BG \sin\theta$. The horizontal shift in the center of buoyancy to $B' = BS$. The increase in the submerged wedge volume is opposed by the buoyancy acting through B'; by taking moments, $BS \times V = v \times h_e h_s$ and $BS = v \times h_e h_s / V$; therefore, $W \times GT = W(v \times h_e h_s / V - BG \sin\theta)$. From this, GT and GM' are computed.

Unstable loading, including both overloading and improper placement of cargo in the holds and decks, or ballast in ballast tanks, may cause a dangerous shift in the center of gravity; for example, when unloading cargo from the lower levels of a hold while allowing heavy cargo to remain on upper decks. Ballast tanks help to keep a vessel stable by adding or discharging water from the tanks to compensate for cargo changes. Care must be taken while loading and unloading a vessel that it does not become unstable or get too far out of trim. The trim refers to the relative displacement fore and aft of the longitudinal axis. The cargo is uniformly distributed or removed as the loading or unloading proceeds. A loading plan must be carefully prepared in advance by one skilled in the art. Those items to be off-loaded at the first port of call must be readily accessible. On the other hand, the center of gravity must be kept as low as possible by placing the heavier cargo items near the bottom of the hold, regardless of order of unloading. With bulk cargo ships, a low center of gravity is not so desirable as it contributes to instability by causing the ship to roll too rapidly.

The longitudinal metacenter is not so critical in most ships as is the transverse. Its determination and solution are, however, similar to that of the transverse.

Aircraft Stability   The forces that sustain an airplane in air are roughly analogous to those that affect the stability of a ship. Stability in an aircraft is the ability to fly in a straight line and to remain in or return to the same attitude with respect to the relative wind. The distribution of individual pressure and vacuum forces acting on an airfoil varies with the angle of attack. See Figure 4.5. The center of lift or pressure in an aircraft, where these forces are assumed to be concentrated, corresponds to the center of buoyancy in a ship. As the angle of attack increases, the center of pressure moves forward; as it decreases, the center of pressure moves backward. Consequently, the location of the resultant forces also varies with the angle of attack. Lift is always perpendicular to the direction of the relative wind.

An airplane possesses stability only when the forces acting on it are in equilibrium, that is, when the sum of the vertical forces, the horizontal forces, and the moments about any point are zero; when $\Sigma V = 0$, $\Sigma H = 0$, and $\Sigma M = 0$. The active forces in level flight are the propeller thrust (forward), the drag (backward), the lift (upward), and the tail load, which may act either upward or downward.

An airplane must have stability about three axes of movement.

Longitudinal stability involves the pitching of a plane about the lateral axis and the relations between the center of pressure and the center of gravity. There is no change in the axial position of the center of gravity (where the mass of the plane is considered as being centered), but the position of the center of lift varies with the speed and angle of attack.

As the center of pressure moves forward, the resultant force on the wings exerts a turning effect about the tail as a pivot. See Figure 4.9. The leading wing edge tends to rise, thereby increasing the angle of attack still more and increasing a tendency to stall. Airplane stability is dependent upon the location of the center of gravity with relation to the center of lift. The center of gravity must set in front of the center of pressure or lift for the corresponding flight speed to provide a restoring moment to offset the

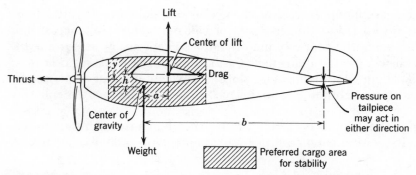

FIGURE 4.9 STABILITY OF A PLANE IN HORIZONTAL FLIGHT.

lift at the center of pressure and its tendency to rotate about the tail. Horizontal tail surfaces or stabilizers aid in reducing pitch and in keeping the center of pressure back of the center of gravity.

A summation of moments counterclockwise about the center of gravity (see Figure 4.9) would be

$$\text{thrust} \times (y) - \text{drag} \times (h) + \text{lift} \times (a) \pm \text{tail load} \times (b) = 0$$

For a high-wing monoplane, the forces normally acting on the tail will be positive (downward) because of downwash from the wing. If the angle of attack is lessened by a gust of wind when the airplane is flying in a stable position, the lift will be decreased and the center of lift moved backward (reducing the angle of attack still more). However, the downward pressure on the elevators will be increased to form a moment that will tend to rotate the plane to stability. For a low-wing plane, the normally acting forces will be upward (negative). The forces acting normally on the tail may also be reinforced upward or downward by raising or lowering the elevators at the trailing edge of the tail.

In designing an airplane for cargo service, the relation between positions of the center of lift and the center of gravity must be carefully considered. A minimum shifting of the center of lift is desirable. Every unit of weight acts with a moment arm, depending on its position, to oppose or reinforce the moment of lift. Weight in the tail has the greatest adverse affect because of the long moment arms involved. The area adjacent to the center of pressure requires minimum weight-distribution planning because of the shorter moment arms. The problems of correct weight distribution in loading and unloading a cargo plane are similar to those for a ship.

With changes in *lateral stability*, the center of gravity does not remain in the same vertical plane. Unstable lateral motions include rolling and side slipping. Yaw and roll are mutually causal, and side slipping produces both. Yaw should turn the plane in the direction in which it is slipping. The fixed vertical fin to which the rudder is attached in the tail assembly or empennage aids in reducing lateral instability. Owing to lateral symmetry along the longitudinal axis, lateral balance is obtained by a symmetrical distribution of weight on each side of a vertical plane through the longitudinal axis. This factor must be kept in mind in loading cargo planes. Setting wings at a dihedral angle also increases lateral stability.

EFFECTS OF RESISTANCE ON TRANSPORT

The initial study will be given to inherent modal characteristics that have a prime influence on operating costs. The amount of resistance to forward movement within any one mode significantly influences the operating costs of that mode.

Propulsive Resistance   Without friction or gravitation, a vehicle and its load, once set in motion, would continue in that state, according to Newton's laws, until acted upon by an outside force. Actually, there are a great many resistive forces acting to retard the movement of a train, truck, ship, or aircraft, caused by the friction of moving parts, by air, wind, or wave pressures, and by turbulence, shock, sway, and concussion. Gravity, or the effects of overcoming elevation, is also a form of resistance. Because additional tractive force is required to overcome these resistive forces, as the speed increases, acceleration may also be considered a kind of resistance. The total of such forces varies with the weight of vehicle and cargo and with vehicle speed. In one way or another, the total propulsive force of the motive power unit is expended in overcoming resistance.

There is much similarity in the cause and effect of resistive forces, whatever the name and units applied. In railroading the term is "tractive" or "train" resistance and is expressed as pounds of resistance, in total, per car, or per ton. Unit resistance, pounds per ton, decreases as the weight of the vehicle increases. With highway vehicles, the opposing force is sometimes referred to as "road" or "tire-rim" resistance. In water transport skin friction and residual resistance are also expressed in pounds, as is drag or the summation of resistive forces acting on an airplane in flight. Pipeline operation encounters flow resistance, usually expressed in pounds per square inch or in feet of head—but the total resistance is expressed in pounds. The frictional and grade resistance for conveyors can be similarly expressed. All resistances can also be expressed in terms of resistive or elevation head and also in terms of the resistance of equivalent distance of level tangent movement.

Vehicles with aerodynamic support such as ground effects and air cushion vehicles do not experience frictional roadway contact but do encounter normal air resistance.

A definite series of relations exists between transportation capacities, propulsive or motive power, and the resistance to movement of the several forms of transport. In all cases the principal effects of resistance are to reduce the speed or the amount of tractive or propulsive force available for moving the vehicle and the pay or revenue load. The effort expended in overcoming resistive forces, which act upon the dead weight of power unit and cargo container, is wasted from a transportation standpoint. Nevertheless, it is a loss inherent in each system, although not to the same extent. Measures of transport efficiency are the pounds of resistance per gross ton and per revenue or net ton and the ratio between the two. A second criterion is directly related, that is, horsepower per gross and per net or revenue ton. These are discussed later.

Tractive and Road Resistance   A vehicle moving upon tangent track or roadway, in still air, and at constant speed encounters resistance that must be overcome by the tractive effort of the locomotive, truck, or automobile engine. These resistances include (*a*) rolling friction between wheel and rail or tire and pavement, which is probably fixed in quantity for a given type of surface but may vary with the condition of the surface, especially highway surfaces; (*b*) axle-bearing friction that varies with vehicle weight and load and with the type of bearing; (*c*) losses that vary with the speed, principally flange friction in the case of railroads but also sway, buff, and concussion, including the effects of bumps on rough roads; and (*d*) air resistance that varies directly with the cross-sectional area, with the length and shape (streamlining) of the vehicle, and with the square of the speed.

The Davis formula[2] for train resistance is generally used by railroad and rapid transit engineers. It combines the foregoing terms into a single empirical formula, whereby for unit resistance

$$R_t = (1.3 + 29/w + bV + CAV^2/wn) \text{ lb/ton}$$

[2]W. J. Davis, Tractive Resistance of Electric Locomotive and Cars, *General Electric Review*, Vol. 29, October 1926, pp. 685–708.

where R_t = train or tractive resistance in pounds for car or locomotive on tangent level track in still air

w = weight in tons per axle of car or locomotive

n = number of axles

b = coefficient of flange friction, swaying, and concussion (0.045 for freight cars and motor cars in trains, 0.03 for locomotives and passenger cars, and 0.09 for single-rail cars)

C = drag coefficient of air; 0.0025 for locomotives (0.0017 for streamlined locomotives) and single- or head-end-rail cars; 0.0005 for freight cars, and 0.00034 for trailing passenger cars including rapid transit

A = cross-sectional area of locomotives and cars (usually 105 to 120 ft^2 for locomotives, 85 to 90 ft^2 for freight cars, 110–120 ft^2 for multiple-unit and passenger cars, and 70 to 110 ft^2 for single- or head-end-rail cars)

V = speed in miles per hour

Unit resistance is multiplied by the weight of the car on rails in tons to obtain resistance of an individual car or by the number of tons in a train to compute the total resistance of the train.

The resistance thus computed may be multiplied by a factor k to recognize the improved rollability of modern track and equipment. Thus k has a value of 0.85 for equipment built since 1950, 1.00 for flat cars carrying highway trailers (TOFC), or containers on flat cars (COFC), 1.33 for loaded auto carriers (bi- and tri-level racks) and 1.90 for empty rack cars.

Committee 16 of the American Railway Engineering Association recommends a modified form of the Davis equation for use with modern specialized equipment.[3]

$$R = 0.6 + \frac{20}{w} + 0.01V + \frac{KV^2}{wn}$$

where

K = 0.07 for conventional equipment

= 0.16 for trailers on flatcars

= 0.0935 for containers on flat cars

For speeds greater than 50 mph (80.5 kph) the experimental values prepared from field tests of Tuthill, which extended the Schmidt tests on

[3]"Manual for Railway Engineering (Fixed Properties)," American Railway Engineering Association, Chicago, Illinois, 1970 revision, p. 16-2-2.

which the Davis equation is based into higher-speed ranges may be used.[4] See Figure 4.10. The foregoing k factors may also be applied to the Schmidt-Tuthill values. Studies with greater precision take into account track flexibility, wind resistance, streamlining, car length, losses due to car axle generators, and condition of the rails. Inertia and cooling and draining of bearing lubricants increase starting resistance to values of 15 to 50 lb/ton depending on temperature, bearing type, and length of time that the car has been standing.

Monorail, rack, and cable systems are subject to similar types of tractive resistance. In the case of rack railways, the friction between gear and rack has to be added. This, however, would be small compared to the total grade resistance of a slope steep enough to require the use of a rack system. Monorail equipment encounters different resistance values, depending on the design of truck and method of suspension. Tests and published data for this type of equipment are lacking. Recognition may be given the resistive losses for a monorail with the Davis formula, using Davis's values for the single-rail car where steel wheels are used and tire-on-pavement rolling resistance values where rubber tires and concrete rails are used.

Propulsive resistance for highway vehicles is directly analogous to tractive resistance for trains. The propulsive resistance (tire-rim or road resistance) is again taken at constant speed on tangent roadway in still air. The variety of factors that affect resistance includes the friction of main and shaft bearings, gears and differential, design and state of tire tread, air pressure in tires, wheel bearings, condition of roadway surface, impact and concussion, air resistance, and gross weight of vehicle.

1. The friction losses in main and shaft bearings and in transmission and differential gears are usually accounted for in the mechanical efficiency of the engine as a part of a 10 to 15 percent internal-loss factor.
2. The condition of the roadway reflects its material, degree of crowning, and state of maintenance. A constant unit resistance in pounds per ton is often combined with a surface condition factor to give a total value for rolling resistance. Based on an average speed of 40 mph (65 kph), concrete in good condition is usually assigned a resistive effect of 20 lb/ton (9.1 kg/ton), 30 lb/ton (13.6 kg/ton), when in fair condition,

---

[4]John K. Tuthill, High Speed Freight Train Resistance, *University of Illinois Engineering Experiment Station Bulletin 376*, Urbana, 1948, and Edward C. Schmidt, Freight Train Resistance, *University of Illinois Engineering Department Station Bulletin 43*, Urbana, 1934.

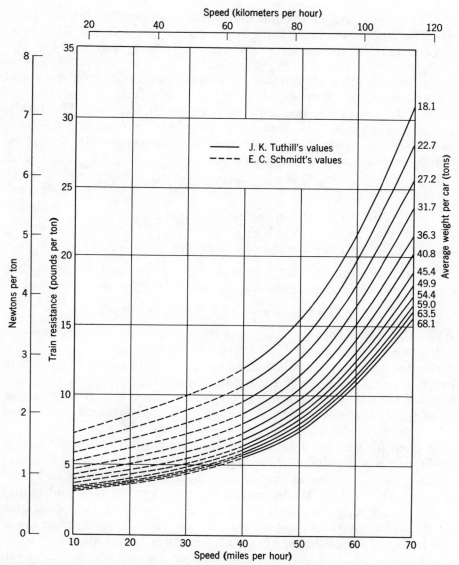

FIGURE 4.10 SCHMIDT-TUTHILL TRAIN RESISTANCE CURVES. (COURTESY OF UNIVERSITY OF *ILLINOIS ENGINEERING EXPERIMENT STATION BULLETIN* 376, 1948, P. 29.)

and 40 lb/ton (18.2 kg/ton), if the paving is in poor condition. Corresponding values for macadam pavements are 30, 45, and 70 lb/ton, respectively, and for earth roads, 70, 90, and 150 lb/ton.

These simplifications do not take into account variations with speed and gross vehicle weight (corresponding to car weight for railroads). Using a series of total road resistance tests made for the Bureau of Public Roads, Starr determined rolling resistance by deducting air resistance from the total resistance for speeds of 4 to 40 mph.[5] To these results, Mr. Starr fitted a straight-line equation having the form[6]

$$R_r = 17.9 + (1.39V - 10.2)/W_v$$

where

R_r = unit rolling resistance in pounds per ton
V = speed in miles per hour
W_v = gross vehicle weight in tons

Total unit tire-rim or road resistance can be obtained by adding the conventional value for air resistance, $R_a = CAV^2/W_v$, whence

$$R = 17.9 + (1.39V - 10.2 + 0.0024AV^2)/W_v$$

Here 0.0024 represents the drag coefficient (sometimes taken simply as 0.002) and A is the cross-sectional area of the vehicle.

Since the tests were conducted on good concrete roads, the effect of other types and qualities of road surfaces can be applied to the value computed for rolling resistance (air resistance excluded) by modifying that part of the total equation by the ratio of good concrete road to the road surface under consideration.

The need for fuel economy directs attention toward reducing air resistance through streamlining. Benefits derived are evaluated in the drag coefficient term of the air resistance portion of the Davis and Starr equations. The goal in streamlining is to attain laminar flow and reduce turbulence and eddy currents that are caused by elements protruding from the vehicle, by blunt head and tail ends, by wheels and under carriage, and space between vehicle units. The rather extensive organization and equations of the Totten revision for the Davis equation air resistance term can

[5]C. C. Saal, Hill Climbing Ability of Motor Trucks, *Public Roads*, Vol. 23, No. 23, U. S. Bureau of Public Roads, May 1942, pp. 33–54.

[6]Millard O. Starr, A Comparative Analysis of Resistance to Motion in Commercial Transportation, unpublished Master of Science thesis, Department of Mechanical Engineering, University of Illinois, Urbana, 1945.

be found in reference 3 at the end of this chapter or on pages 77–78 of Hay's *Railroad Engineering*.[7]

Air resistance, being greater at high speeds, is important to passenger autos, but trucks and buses running 50 to 70 mph will encounter air resistance of marked quantity. Savings in air resistance and fuel can be achieved by rounding the corners of trailers, placing deflectors or cowling between tractor cab and trailer, and by rounding corners of cab and trailer. A reduction in the aerodynamic drag coefficient of 10 to 20 percent may be possible with deflectors that can be attached to the front end of tractors to deflect air upward in a laminar similar flow similar to that achieved with the sloping front ends of early diesel passenger locomotives. Side or quartering winds, it should be noted, can produce high resistance, but quantitative data on these effects are scant.

In both rail and highway resistive expressions one finds an equation of the form $A + (B/W) + CV + DV^2$ that contains a fixed factor for rolling resistance, one that varies with the weight, another varying with the speed, and a fourth term that varies as the square of the speed. See Figures 4.11a and 4.11b for resistance values of typical rail and highway vehicles.

Resistance for monorails depends on whether steel wheels on steel rails are used or rubber tires on a pavement material pad. Average coefficients of friction (adhesion) of 0.20 and 0.80 apply respectively. Thus average resistance for the steel wheel is about 4 lb (1.8 kg) per ton and for the rubber tire 16 lb (7.3 kg) per ton. Taking speed and other practical effects into account, these values will approach the 6 lb (2.72 kg) and 30 lb (13.6 kg) per ton that are often taken as the average rail and highway resistance, respectively.

Ships   Many factors enter into ship's resistance—design and condition of the hull, width and depth of channel, height of waves, and disposition of cargo and ballast. The following greatly simplified discussion presents some of the more important relations.

The most important resistances to be overcome by a ship's propulsive units are (a) skin friction, that is, friction of quiet water on the wetted surface of the hull, (b) streamline resistance, (c) eddy-current resistance, and (d) wave resistance. These resistances are taken at constant speed in still air. In addition, there is air resistance of $R_a = CAV^2$ on that part of the ship that is out of the water. The slow speed and streamlining of most ships keep this from being excessive.

[7]W. W. Hay, *Railroad Engineering, Vol. I*, Wiley, New York, 1953, pp. 77–78.

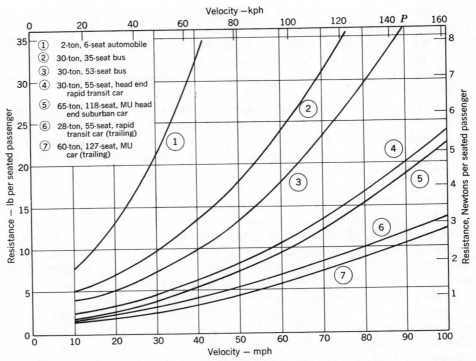

FIGURE 4.11 RESISTANCE PER PASSENGER FOR HIGHWAY AND RAIL TRANSIT VEHICLES.

Skin friction, which constitutes 50 to 85 percent of the total, depends upon the speed and the design of the hull. Durand's commonly used empirical formula for skin resistance is $R_s = fA_w V^{1.83}$, where R_s is the total skin friction in pounds and f is the friction factor varying from 0.01 for hull lengths of 20 ft (6.1 m) to 0.0085 for 600 ft (182.9 m) (longer hulls produce better streamline effects) and increasing 25 to 50 percent for dirty hull bottoms and up to 100 percent for fouled and barnacled hulls. V is the speed in knots (1 knot = 1 nautical mile or 6080 ft per hr). A_w is the wetted surface of the hull in square feet and is equal very approximately to $15.6\sqrt{D_s \times L}$, where D_s is the displacement in tons and L is the length in feet. The remaining resistances (known appropriately as residual resistance), include streamline resistance caused by eddy-current effects, but principally by wave action, and are included in the Taylor empirical equation[8]

$$R_r = 12.5 \times C_B \times D_s \times (V/\sqrt{L})^4$$

where

R_r = resistance in pounds
C_B = block ratio
D_s = displacement in tons
V = speed in knots
L = wetted length of hull in feet

The expression $(V/\sqrt{L})^4$ is the ratio of speed to the square root of the length and relates the effect of stern and bow waves on one another.

The total resistance is $R = R_s + R_r + R_a$. R_a is the air resistance of the portion of the ship out of water, A = cross-sectional area, C = drag coefficient = 0.002, and V = speed in knots. Unit resistance would be R/D_s, in which D_s = gross loaded displacement. Giving each term its detailed components, unit resistance is

$$R = \left(f 15.6\sqrt{D_s L} \; V^{1.83} + CAV^2 \right)/D_s + 12.5C_B V^4/L^2$$

In a narrow canal or channel water is crowded against the confining walls or slopes, increasing the resistance so that $R_c = 8.5R/(2 + a/15.6\sqrt{D_s \times L})$, where R_c = resistance in a confining channel, R = resistance in a large body of water, a = cross-sectional area of the channel in square feet, and D_s and L have the same meaning as before.

Articulated barges help reduce resistance when assembled into tows. Only the ends of leading and trailing barges in the tow have rake (sloping stern and bow). The intermediate barges are square-ended and are lashed tightly together into an almost rigid unit, thereby eliminating the turbulence that occurs when each barge in the tow has rake. Uniform draft and channel depths also help keep resistance to a minimum.

Aircraft The subject of airplane resistance is rather complicated. The following discussion is greatly simplified for inclusion in the scope of this book. It illustrates the principles and demonstrates the similarity of airplane resistance to other types of propulsive resistance.

Airplane resistance, more commonly called "drag," is the resistance of air to the forward motion of the plane. Rolling and bearing friction

[8]D. W. Taylor, *The Speed and Power of Ships*, Wiley, New York, 1910.

resistances are absent. Nevertheless, weight, as represented by the required wing area for a given weight of plane, enters the evaluation. From fluid mechanics, the resistance of a fluid to the passage of a body is $D = C_D(\rho/2)Sv^2$, where C_D is a coefficient depending on the shape of the body with skin friction and eddy components, ρ is the fluid (air) density in slugs per cubic foot at the altitude of flight, S is the wing area in square feet, and v is the velocity in feet per second. This expression is similar to that for lift except for the coefficient; the two expressions for drag and lift are sine and cosine components, respectively, of the total force acting on the wing or airfoil.

Drag is composed of two elements: (a) parasitic drag caused by the frontal pressure and side friction on the parts of the plane and (b) induced drag inherent in the production of lift, due principally to the downdraft vortexes at the ends of the wings. Resistance = drag = $D = (C_{Dp} + C_{Di})(\rho/2)Sv^2$, where C_{Dp} and C_{Di} are parasitic and induced-drag coefficients, respectively. The values of these coefficients vary with the type of airfoil and are usually determined by wind-tunnel tests. These values are presented graphically in Figure 4.6. Values can also be computed.

C_{Dp} is determined as the sum of the component drags on the several parts of the plane structure. The drag-coefficient values in Figure 4.6 are for the wing alone. Parasitic drag also arises from the tail assembly, the fuselage, engine nacelles, and other elements offering frontal areas. These component drags are expressed as the combined equivalent drag of a flat plate perpendicular to the relative wind and having a coefficient of unity. Thus the total C_{Dp} equals the C_{Dp} for the wings (in Figure 4.6 or equivalent) plus an assumed C_{Dp} for the tail area and other elements. The tail area is often taken as 30 percent of the wing area with a C_{Dp} of 00 .01; a value of 0.10 is used for the remaining elements.

The ratio of wing span to average wing chord or width is the aspect ratio $AR = b/c = b^2/S$, where b is the span, c is the chord, and S is again the area. Values of AR may be as low as 2 to 3 for missiles and supersonic craft, 6 to 8 for small planes, and 8 to 15 for long-range commercial transports. Data in Figure 4.6, following the current practice, are based on airfoils of infinite length. Obviously such an airfoil has no induced drag but only the parasitic drag. With airfoils of finite length, air moves under the wing tips to the area of lower pressure above the tip, thereby introducing downwash velocities along the wing span that reduce the lift and the angle of attack. To maintain the same lifting force, the angle of attack of a finite airfoil must thereby be increased by a small induced angle, α_1 creating in turn an additional or induced drag. The induced or additional angle of attack for the finite length of airfoil can be determined, from principles of fluid flow, as $\alpha_1 = C_L/\pi AR$ and C_{Di} as $C_{L\alpha_1}$, whereby $C_{Di} = C_L^2/\pi AR$.

It should be noted that for level flight there is only one speed for each angle of attack, hence only one lift and one C_L for that speed and angle.

Induced drag is dependent only on wing span. An increase in wing span decreases the induced drag. A longer wing creates a greater bending moment. Wing strength, and usually weight, must be increased. Proper design calls for an economic balance between wing span and induced drag.

Drag increases as the square of the speed but decreases with altitude because of lower atmospheric densities at high altitudes. A decrease in wing span decreases parasitic drag but also reduces lift. The designer must determine whether he requires speed or load-carrying capacity for a given thrust and horsepower.

In summary, the gross load that an airplane can support in level flight in still air at constant speed at a given altitude depends on speed, a factor of power plant, on air density, a factor of altitude and temperature, and on drag, a factor of wing span and angle of attack. Drag divided by gross weight gives unit resistance.

Problem Example

Given a light transport airplane with two 48-in. diameter, 1600-hp engines (1300 hp at 8000 ft), fuselage diameter of approximately 8 ft, wing area of 980 ft^2 (N.A.C.A. 63_1–412 section, aspect ratio of infinity), and lift of 25,000 lb. Determine (a) the adequacy of the wing area for landing and takeoff speeds of 80 mph and (b) the drag when flying at an elevation of 8000 ft at 180 mph, air density = 0.0019 slugs.

1. Lift = L = gross weight = 25,000 lb = $C_L(\rho/2)v^2S$. Assume maximum lift; C_L (from Figure 4.6) = 1.56.

$$\rho = 0.0024 \text{ slugs/ft}^3 \text{ at sea level}$$

$$v = 80 \times 5280/3600 = 118 \text{ ft/sec}$$

$$S = \text{required wing area in square feet}$$

$$25,000 = 1.56 \times 0.0024 \times 118 \times 118 \times S/2$$

$S = 959$ ft^2; this is less than 980 ft^2; therefore, the given wing area is adequate.

2. Drag = $C_D\rho v^2S/2$ where $C_D = C_{Dp} + C_{Di}$. At an altitude of 8000 ft and speed of 180 mph,

$$C_L = \text{weight}/\rho v^2S/2 = 25,000/0.0019 \times 264 \times 264 \times 980/2$$

$$C_L = 0.385$$

From Figure 4.6, D_{Dp} for the airfoil = 0.0055. Assume $C_{Dp} = 0.01$ for tail drag and 0.10 for frontal areas; also assume the tail area as approximately 30 percent of wing area and aspect ratio $(AR) = 8$.

$$A_e = \text{equivalent area} = (0.30 \times 980 \times 0.01) + (8 \times 8 \times \pi/4 \times 0.10)$$
$$+ (2 \times 4 \times 4 \times \pi/4 \times 0.10) + 1.0$$
$$A_e = 11.47 \text{ ft}^2 = C_{Dp}S; \qquad C_{Dp} \text{ (excluding wing)} = A_e/S$$
$$\text{Total } C_{Dp} = C_{Dp} \text{ (for wing)} + A_e/S = 0.0055 + 11.47/980 = 0.017$$
$$C_{Di} = C_L^2/\pi AR = C_L^2/(3.14 \times 8) = 0.385 \times 0.385/25.12$$
$$C_{D1} = 0.0059$$
$$C_D = C_{Di} + C_{Dp} = 0.0059 + 0.017 = 0.0229$$
$$\text{Drag} = D = C_D\rho v^2 S/2 = 0.0229 \times 0.0019 \times 264 \times 264 \times 980/2$$
$$D = 1486 \text{ lb}$$
$$\text{Unit drag} = D/W = 1486/25{,}000/2000 = 119.0 \text{ lb/ton } (54.3 \text{ kg/ton})$$

Pipeline Resistance   Flow resistance consists principally of (a) internal resistance within the liquid—its viscosity or shearing resistance and the temperature which affects the viscosity; (b) condition or type of flow, that is, whether streamline (laminar) or turbulent; (c) the frictional resistance between the liquid and the inside wall of the pipe, which depends in turn on the roughness factor of the pipe's interior, the pipe diameter, and the number and type of joints and fittings (increasing the diameter of the pipe reduces the percentage of total flow in contact with the walls of the pipe).

These factors are reflected in the basic Fanning equation for flow resistance, $h = f\rho L v^2/2gd$, where $h =$ flow resistance in feet of head, $L =$ length of line in feet, $v =$ flow velocity in feet per second, $d =$ diameter of pipe in feet, $g =$ acceleration due to gravity $= 32.2$ ft/sec^2, $f =$ pipe friction coefficient, and $\rho =$ density in lb/ft$^3 =$ specific gravity $\times 62.4$.

The value of f was determined by Reynolds as a function of the ratio between dv and the viscosity. This dimensionless ratio, called the Reynolds number, is $N = dv\rho/u$, where d, v, and ρ have the meanings given above and u is the viscosity of the fluid. Reynolds found the flow to be laminar or streamlined when $N < 2000$. When $N = 3000$, the flow is likely to be turbulent. See Figure 4.12. Between $N = 3000$ and $N = 2000$ flow may be laminar or turbulent, but conservatively it is considered turbulent. For every value of N there is a corresponding value of f. Further experiments indicated that for streamlined flow $f = 64/N$. If this value for f is substituted in the Fanning equation and head is converted to pounds per square inch of pressure, the resulting equation is the same as Poiseuille's

FIGURE 4.12 STREAMLINED AND TURBULENT
FLOW.

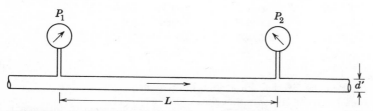

FIGURE 4.13 PRESSURE LOSS IN A PIPELINE.

expression for flow in small tubes,

$$P = P_i - P_f = 2uvL/9gd^2$$

where $P =$ pressure loss in pounds per square inch over the pipe length, L, in feet, and d, g, L, v, and u have the same meaning as in the Fanning equation. P_i and P_f are the initial and final pressures over length L. See Figure 4.13.

A more usable expression for pressure loss in streamlined flow, used by the industry, is expressed in pipeline units.[9]

$$P = 1500\, Bu/D^4 \qquad \text{and} \qquad P = 962{,}000\, Qu/D^4$$

where

$P =$ pressure loss per mile of line in pounds per square inch

$B =$ flow in barrels per hour (1 bbl $=$ 42 gal)

$Q =$ flow in cubic feet per second

$D =$ diameter of pipe in inches

$u =$ absolute viscosity in feet-pounds-seconds, that is, lb-sec/ft²

The Fanning formula for turbulent flow is properly used for N values greater than 2000. The value of f must be obtained experimentally or from experimental values reduced to tables and charts, as in Table 4.1.

[9]W. G. Helzel, Flow and Friction in Pipelines, Pipeline Section, *Oil and Gas Journal*, Tulsa, Oklahoma, June 5, 1930, pp. T-203.T-224.

TABLE 4.1 PIPE FRICTION COEFFICIENT $(f)^a$

$N=dv\rho/u$	f	$N=dv\rho/u$	f
2,500	0.0475	30,000	0.0230
3,000	0.0450	35,000	0.0225
4,000	0.0415	40,000	0.0220
4,500	0.0400	50,000	0.0210
6,000	0.0375	60,000	0.0200
8,000	0.0335	80,000	0.0180
10,000	0.0315	100,000	0.0175
20,000	0.0265		

[a]Based on Figure 13, Pipeline Section Flow and Friction in Pipelines, W. G. Helzel, *Oil and Gas Journal*, Tulsa, Oklahoma, June 5, 1930, p. T-223.

Reduced to practical pipeline units, the Fanning formula becomes

$$P=0.55830 f B^2 \rho / D^2 \quad \text{and} \quad P=229,610 f Q \rho / D^2$$

where

$$P = \text{flow resistence per mile in psi}$$
$$f = \text{pipe friction coefficient—experimental values}$$
$$B, Q, D, \rho = \text{the same as in the preceding equations}$$

Petroleum is usually described by giving the specific gravity in degrees American Petroleum Institute ($°$API) based on an arbitrary scale in which water has a specific gravity of $10°$ API. Standard specific gravity is related to API graivty by the equation spec grav $60°$ F$/60°$ F$= 141.5/(131.5 + °$API).

Viscosity, u, is determined in the petroleum industry by the Saybolt Universal Viscometer and is the time in seconds for 60 cc ($\frac{1}{8}$ pint) to flow through a capillary tube 0.1765 cm ($\frac{1}{16}$ in.) inside diameter and 1.225 cm ($\frac{1}{2}$ in.) long at a specified temperature. Viscosity decreases rapidly as the temperature rises. *Kinematic viscosity* is the ratio of absolute viscosity to density. In English units kinematic viscosity $= u/\rho = 0.00000237t - 0.00194/t$, where u again is absolute viscosity in feet-pounds-seconds, ρ is the density in pounds per cubic foot, and $t =$ time in Saybolt seconds.

This discussion of viscosity is necessary to determine u values for the equation $N = dv\rho/u$.

A convenient expression for the foregoing in pipeline units is $dv\rho/u = N = 0.02381 (B/D)(\rho/u)$, where B is in barrels per hour, and D is the inside diameter in inches.[10] The relation between gravity and viscosity is not clearly defined. Figure 4.14 has been prepared as an aid to students in

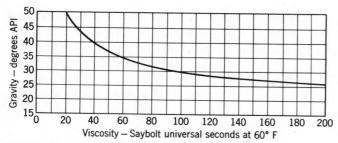

FIGURE 4.14 TYPICAL GRAVITY-VISCOSITY RELA-
TIONS OF CRUDE OIL. (AN AVERAGING OF VALUES
FROM *BUREAU OF MINES BULLETIN 291.*)

solving problems and represents only an averaging of relations from data
contained in the U. S. Bureau of Mines Bulletin No. 291. A problem
example will make the use and significance of these relations more easily
understood.

Problem Example

Given 900 bbl of crude oil per hour to be pumped through a 10-in. line (specific
gravity of the oil is 30° API), what is the loss in pressure in a level line 30 miles in
length?

1. From Figure 4.14, the viscosity of 30° API crude oil is 95 Saybold seconds.
2. $N = dv\rho/u = 0.02381\ B\rho/Du$ in pipeline units
$\rho/u = 1/(u/\rho) = 1/(0.00000237 \times 95 - 0.00194/95)$
$\rho/u = 4884$
$N = (0.02381 \times 900/10) \times 4884$
$N = 10{,}466$
3. Since $N = 10{,}466 > 2000$, flow is turbulent and the Fanning formula in
pipeline units $P = 0.55830 fB^2\rho/D^5$ should be used. D is given, but f and $\rho = 62.4$
$\times$ spec grav must be determined.

$$\rho = 62.4 \times 141.5/(131.5 + 30° \text{ API}) = 62.4 \times 0.88$$

$$\rho = 54.9$$

$$f = 0.0312 \text{ by interpolating in Table 4.1}$$

4. $P = 0.55830 fB^2\rho/D^5 = 0.55830 \times 0.0312 \times 900^2 \times 54.3/10^5$
$P = 7.66$ psi per mile
5. For a 30-mile line, $P_{30} = 30 \times 7.66 = 229.80$ psi

---

[10]W. G. Helzel, Pipeline Section Flow and Friction in Pipelines, *Oil and Gas Journal*, Tulsa,
Oklahoma, June 5, 1930.

Belt Conveyors   The elements of a belt conveyor are shown in Figure 4.15. The belt comprises several plys of a canvas fabric or "carcass" impregnated and coated with rubber. Cord belts, steel wires, and other fibers may be added for strength. The belt-supporting idlers comprise three or more cylindrical rollers of 4 to 7 in. (10.16 to 17.78 cm) diameter, one or more horizontal center rollers and two outer cylinders inclined at about 20 degree angles to form a trough. The idlers are spaced $2\frac{1}{2}$ to $5\frac{1}{2}$ ft (0.76 to 1.68 m) apart depending on the size of the belt and the cargo loading. Return idlers bearing less load are more widely spaced. Other elements include tension-controlling devices, trippers to discharge cargo at inter-mediate points, and, sometimes, belt cleaners bearing or rotating against the belt surface to remove sticking cargo. The drive pulleys or motive power source will be discussed in the next chapter.

Conveyor systems are subject to two principal types of resistance—roller resistance and a form of grade resistance, as the load is raised or lowered through a difference in elevation opposed or aided by the forces of gravity.

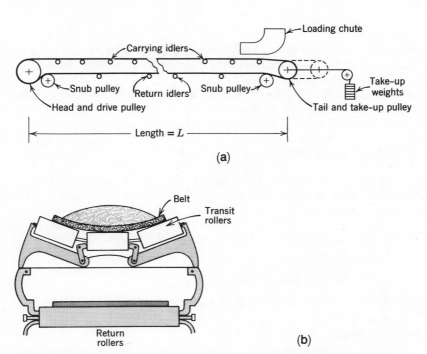

FIGURE 4.15 TYPICAL HORIZONTAL BELT CONVEYOR. (*a*) LONGITU-DINAL SECTION; (*b*) SECTION THROUGH PULLEYS. (COURTESY OF *GOODYEAR HANDBOOK OF BELTING: CONVEYOR AND ELEVATOR*, GOODYEAR TIRE AND RUBBER COMPANY, AKRON, OHIO, P. 9.)

The elevation phases of the problem are considered in a later chapter. An empty horizontal belt encounters resistance in its frictional contact with the rollers and idler pulleys. The amount of frictional resistance will vary with the materials of belt and rollers, the lubrication of the idler pulleys, their spacing, the arc and area of contact, and whether or not the rollers are wet or dry, clean or dirty. An important factor is the amount of tension imposed on the belt by idler pulleys or counterweight or other means to insure slackside tension. There would be no effective or power-producing tension at the driving pulley without some slackside tension because conveyors work by friction drive.

A considerable amount of resistance has to be overcome by the driving pulley even when the belt is empty. The addition of cargo load to the belt imposes additional weight on the pulleys and increases the resistive forces. There is some air resistance, but because of negligible frontal area and low speed, air effects and the variations with speed that accompany air resistance are neglected. Few studies have been made for the resistive forces on belt conveyors in terms of pounds per ton, that is, total resistance divided by the number of tons of cargo on the belt.

In computing the propulsive resistance of belt conveyors, the usual practice is to consider first the resistance of the empty belt, then the additional resistance occasioned by the cargo load, and, finally, the effects of gradient, discussed in the next chapter.

Resistance varies with the length and width of the belt (or flight) and with the load imposed on the rollers over which the belt moves. For a horizontal belt, the length of belt L is taken as the center-to-center distance between the head or driving pulley and the tail and take-up pulley. See Figure 4.15. For inclined belts at the usual angles of inclination, little error is had in using the horizontal projection of the center-to-center contour distances. The greatest error (for a maximum inclination of approximately 23 degrees for the full belt length), is approximately 8.7 percent.[11] It is generally much less because the maximum inclination is usually not involved and the inclination seldom extends over the entire flight. It is significant principally in computing friction loads of empty belts. Even in extreme cases of maximum inclination, it is a much lower percentage with loaded belts, when total effective friction is considered. With empty inclined belts, the actual contour distance should be used when other than approximate solutions are required.

The propulsive resistance or friction force for an empty belt is, therefore, equal to the length of belt × weight of moving parts (belt and rollers) × the

[11]*Handbook of Belting—Conveyor and Elevator*, Section 6, The Goodyear Tire and Rubber Company, Akron, Ohio, 1953, pp. 67–107.

coefficient of frictional resistance. The Goodyear equation expresses this in the form $R_e = CQ(L + L_0)$, where $R_e =$ total resistance of empty belt in pounds and $C =$ frictional factor or the average coefficient of friction.[12] Propulsive resistance comprises approximately 20 to 30 percent idler bearing resistance, 30 percent belt and roller friction, and 50 to 60 percent internal-load friction, that is, shifting and reshaping of the load as it passes over the rollers. The following is the Goodyear evaluation of C:

1. $C = 0.03$ for antifriction installations on temporary, portable, or imperfectly aligned structures.

2. $C = 0.022$ for conveyors with high-grade antifriction idlers on permanent or other well-aligned structures.

3. $C = 0.012$ for conveyors in group 2 but with grades requiring restraint of the belt when loaded to prevent running backward due to gravity.

$Q =$ weight of moving parts per foot, center-to-center distance (of head and tail pulleys), including weight of belt and an average per-belt-foot allocation of the supporting idlers. The value for Q can be computed accurately by the equation

$$Q = 2B + W_1/l_1 + W_2/l_2$$

where $B =$ belt weight in pounds per lineal foot, W_1 and $W_2 =$ weight of revolving parts carrying idlers and return rolls, respectively, and l_1 and $l_2 =$ respective spacing in feet of conveyor idlers and return rollers. For quick computation, the values of Table 4.2 may be used. These data are based on 5-in. (12.9-cm) rollers (except for 42- to 60-in. (106.7 to 152.4 cm) belts, on which 6-in. (15.2-cm) rollers are used), return-roller spacing of 10 ft, (3.05 m) and 3- to 4-ft (0.9- to 1.2-m) spacing for idlers. $L =$ length of horizontal belt in feet, equivalent to the center-to-center spacing of the head and tail pulleys; L also is the horizontal projection of the contour distance for inclined and declined flights. $L_0 =$ an added "distance" to include constant frictional losses, independent of belt length, usually referred to as terminal friction. Corresponding to the three values given for C are L_0 values of 150 to 250, 200 to 1000, and 475. The higher values are used only in empty-belt calculations of tension, where, as in a slightly regenerative decline belt, empty-belt calculations of tension determine the belt design.

The added weight of cargo per foot of belt is $2000T/(60S)$, where $T =$ tons per hour, and $S =$ feet of travel per minute. This equation reduces to $100T/3S$. Total loaded-belt resistance then becomes

$$R_L = C(Q + 100T/3S)(L + L_0)$$

[12]Ibid.

TABLE 4.2 AVERAGE VALUES OF Q FOR PLY-TYPE BELTS[a]

Belt Width (in.)	(cm)	Q (lb)	(kg)	Belt Width (in.)	(cm)	Q (lb)	(kg)
14	35.6	13	5.9	36	91.4	39	17.7
16	40.6	14	6.4	42	106.7	52	23.6
18	45.7	16	7.3	48	126.9	61	27.7
20	50.8	18	8.2	54	137.2	71	32.2
24	61.0	21	9.5	60	152.4	85	38.6
30	76.2	31	14.1				

[a]*Handbook of Belting—Conveyor and Elevator*, Section 6, The Goodyear Tire and Rubber Co., Akron, Ohio, 1953, p.68.

It should be noted that the only variables in this equation are weight and length of belt. Width and speed are assumed constant for any given installation. For a given quantity of load per hour, R_L will decrease as speed increases. This is because the variation in speed affects the load per foot of belt length. If the load per foot of belt length remained fixed, that is, total T became variable, R_L would increase as speed increased. Unit resistance in lb/ft of cargo will equal R_1 divided by $\left(Q + \dfrac{100T}{3S}\right) \times L$.

Problem Example

Given a horizontal 42-in. conveyor belt 1200 ft long, running at a speed of 700 ft/min and moving 1200 tons/hour. What total and what unit resistances are being overcome for (a) the empty belt and (b) the loaded belt?

1. From Table 4.2 for the 42-in. belt, $Q = 52$ lb. The Goodyear frictional constant is $C = 0.022$ (for a permanent structure), and the length equivalent for terminal resistance is $L_0 = 200$.

2. For the empty belt, total resistance $= R_e$.

$$R_e = CQ(L + L_0) = 0.022 \times 52(1200 + 200)$$

$$R_e = 1601.6 \text{ lb } (727.1 \text{ kg})$$

3. Unit resistance $= R_e/$weight of empty belt and rollers. Unit resistance $= 1601.60/(52 \times 1200/2000) = 51.3$ lb/ton.

4. For the loaded belt, total resistance $= R_L$. The load per foot of center-to-center distance $= 100T/3S$.

$$R_L = C\left(Q + \frac{100T}{3S}\right)(L + L_0)$$

$$R_L = 0.022\left(52 + \frac{100 \times 1200}{3 \times 700}\right)(1200 + 200)$$

$$R_L = 3361.6 \text{ lb } (1526.2 \text{ kg})$$

5. Unit resistance $= 3361.6 \div \left(\dfrac{1200}{2000}\right)\left(52 + \dfrac{100 \times 1200}{3 \times 700}\right)$ where the 2000 term changes lb to tons. Unit resistance $= 51.3$ lb/ton (23.3 kg)—the unit resistance has not varied appreciably with the load.

Aerial Tramways   The commonest type of long-distance aerial tramway, ropeway, or cableway consists of a small monorail trolley truck with grooved wheels or sheaves from which the cars or buckets are suspended, running on a fixed cable. Propulsion comes from a parallel, moving cable with a lever or self-locking friction grip that clutches the moving cable. A similar pair of cables is used for the bucket return. See Figure 4.16.

Tractive resistances encountered are the frictional resistance of the grooved wheels on the fixed cable, bearing friction of the grooved wheels in the trolley assembly, air resistance, and the frictional resistance of the traction cable passing over the sheaves at intermediate towers and angle stations and over the driving sheave. The resistance between the trolley wheels and the cable is a function of the coefficient of friction and the weight. It is usually neglected. Resistance of the trolley-wheel bearings is

FIGURE 4.16 AERIAL TRAMWAY, TWO-WIRE SYSTEM. (WILBUR G. HUDSON, *CONVEYORS AND RELATED EQUIPMENT*, 3RD EDITION, WILEY, NEW YORK, 1954, P. 279, FIGURE 11-7.)

also a function of the weight and the bearing-friction coefficient, usually taken as 0.02 for plain bearings and 0.01 for roller bearings. Some resistance arises from side sway when winds are blowing. Air resistance is present but because of the slow speed and small car area it is neglected. The total force to overcome is $T-S$, where T and S are tensions on taut and slack sides, respectively. The cable assumes the form of a catenary, so that the cars are always running more or less on an incline. Grade resistance is therefore usually present in quantities.

Grade Resistance   Overcoming elevation is a major factor affecting operating costs. Elevation effects for pipelines, conveyors, and aircraft will be discussed in Chapter 5; for waterways in Chapter 6. For land-based systems, especially railways and highways, elevation effects appear as grade resistance.

Gradient is defined as the rate of grade, the vertical rise in feet per 100 ft of horizontal distance. The effects of gradients on construction and operating costs are often limiting. Grade resistance of 20 lb per ton of train or vehicle weight per percent of grade must be overcome by the tractive effort of locomotive or vehicle engine.

Assume a car of weight $W=1$ ton (2000 lb) on a unit or 1 percent grade, AB. See Figure 4.17. The weight W, acting vertically downward, can be resolved into two components, V, acting perpendicularly into the plane of the road or rails and F', which tends to roll the car downhill. To overcome grade effects, an equal and opposite force F must be exerted up the grade by the tractive force of the car or truck engine or locomotive. A proportion can be established between force and distance triangles and by assuming that $AB=AC$ (i.e., $\cos\theta=1.000$ approximately). $F/2000=1/100$ from which $F=20$ lb per ton per percent of grade. In metric terms, for a rise of 1 in 100 $F=9.1$ kg/ton; also $F=10$ kg/metric ton. A truck weighing 15 tons loaded on a 4 percent grade would experience total grade resistance of

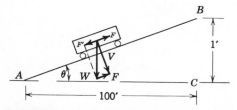

FIGURE 4.17 DERIVATION OF UNIT GRADE RESISTANCE.

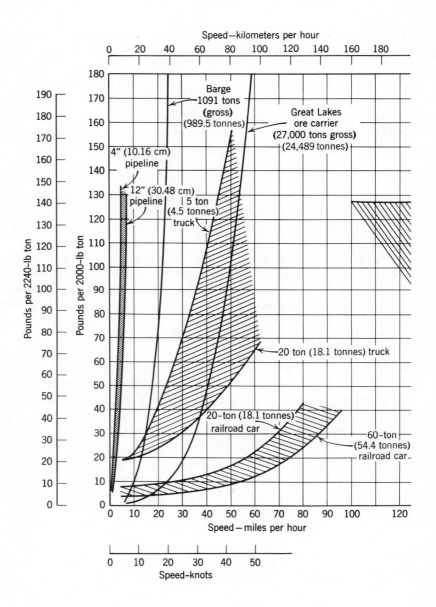

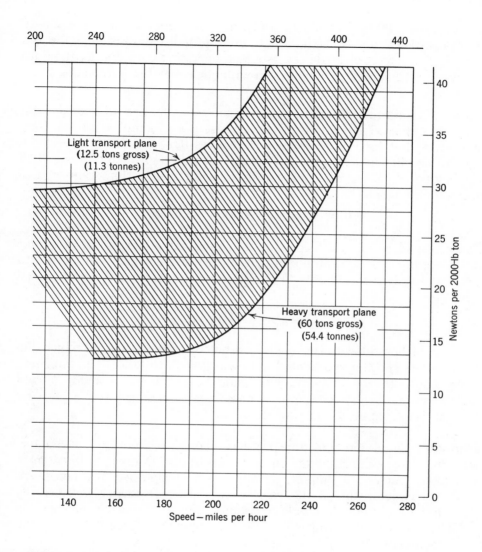

FIGURE 4.18 TYPICAL RESISTANCE-VALUE RANGES.

$15 \times 4 \times 20 = 1200$ lb (544.32 kg), which must be added to tire rim and any other resistance involved to obtain total propulsive resistance for the vehicle. The effects of grade resistance on operating costs and location are discussed in Chapter 5 in conjunction with horsepower requirements.

Curve Resistance   Flanged wheel vehicles encounter additional resistance on curves. The analysis of curve resistance is complex and not fully understood but for rail equipment it involves flange pressure against the rail and lateral slip across the rail head. Test values give 0.80 lb/ton (0.36 kg)/degree of curve. The curve resistance for a 4000 ton train on a 3 degree curve would be

$$4000 \times 0.8 \times 3 = 9600 \text{ lb } (4358 \text{ kg})$$

Curve resistance may be expressed as equivalent grade by dividing unit grade resistance of 20 lb/ton into unit curve resistance and combining into one grade resistance quantity. The equivalent grade for a 5 degree curve would be

$$5 \times \frac{0.8}{20} = 5 \times 0.04 = 0.20 \text{ percent}$$

This would be added to the grade resistance for the grade on which the curve occurs and the two treated simply as grade resistance. The foregoing could apply to monorail systems using conventional railroad flanged wheels. For other systems, curve resistance is minimal and is usually ignored. See Chapter 17 for skid resistance.

Summary   Each mode of transport has inherent technological characteristics that have critical effects on costs and utility. Degrees of freedom in maneuver, guidance, support, stability, and propulsive resistance are primary characteristics of the vehicle. Propulsive resistance is especially operating cost-oriented. Grade resistance can be a major element in propulsive resistance. Ranges of typical resistance or drag values for the more common modes are given in Figure 4.18.

QUESTIONS FOR STUDY
1. A Great Lakes ore carrier has a length of 600 ft, a beam of 65 ft, and a draft with maximum load of 24 ft. The ship with fuel and crew weighs 7000 tons and carries 18,000 tons of cargo. What is its block coefficient?
2. A Great Lakes bulk-cargo carrier is 620 ft long and has a 70-ft beam and a draft of 25 ft when loaded with a cargo of 19,700 tons. If the block coefficient is 0.891, what is the weight of the ship with equipment and fuel but without cargo?

3. What wing area is required to support an airplane flying 200 mph with an angle of attack of 8 degrees at an altitude of 10,000 ft (air density = 0.001756)? Will this same wing area be adequate when landing at a speed of 80 mph, assuming maximum safe angle of attack?

4. Using the Davis equation, compute and plot the unit and total resistance at speeds of 10 to 40 mph of (a) an 8-wheeled empty car weighing 20 tons and (b) an 8-wheeled loaded car weighing 60 tons.

5. Compute and plot comparative curves of resistance for a 10-ton truck with cross-sectional area of 96 ft^2 and a 1500-lb passenger automobile with cross-sectional area of 30 ft^2 at speeds of 10 to 60 mph. At what speed does streamlining become significant for each?

6. What is the unit resistance for a 200-ft towboat (pusher type), 45 ft wide and 12 ft deep, with a draft of 9 ft and block coefficient of 0.87 at speeds of 10 to 40 mph? Assume that the total height of vessel above water line is 30 ft.

7. What is the total resistance of a 16-barge tow at 8 mph when each barge is 230 ft long, 35 ft wide, has a block coefficient of 0.95, weighs 472 tons empty, and has a loaded draft (corresponding to the foregoing block coefficient) of 9 ft and extends 5 ft above the water?

8. A crude-oil pipeline 8 in. in diameter is to carry 800 bbl per hr. The oil has a specific gravity of 32° API. If the pumping pressure of a pump station is 600 psi, how many stations will be required in a level line 200 miles long?

9. Compare the unit resistance of a 42-in. conveyor belt 1600 ft long, fed at a varying rate so that the belt always contains 60 lb of load per foot of length when running at speeds of (a) 600 ft per min and (b) 800 ft per min. What would be the comparison if the belt carried a fixed load per hour of 1080 tons regardless of belt speed?

10. Compute for each of the preceding railroad cars, automobiles, trucks, ships, and towboats (Questions 2, 4, 5, 6, and 7) the resistance at 60 mph and show graphically in a bar or similar type of graph.

SUGGESTED READINGS

1. W. J. Davis, Jr., Tractive Resistance of Electric Locomotive and Cars, *General Electric Review*, Vol. 29, October 1926, pp. 685–708.

2. E. C. Schmidt, Freight Train Resistance, Its Relation to Average Car Weight, *University of Illinois Engineering Experiment Station Bulletin 43*, Urbana, 1910.

3. A. I. Totten, Resistance of Lightweight Passenger Trains, *Railway Age*, July 17, 1937.

4. E. C. Schmidt and F. W. Marquis, The Effects of Cold Weather upon Train Resistance and Tonnage Ratings, *University of Illinois Engineering Experiment Station Bulletin 59*, Urbana, 1912.

5. M. O. Starr, *A Comparative Analysis of Resistance to Motion in Commercial Transportation*, thesis submitted in partial fulfillment of the requirements for the Degree of Master of Science in Mechanical Engineering, University of Illinois, 1945.

6. Train Resistance of Freight Trains Under Various Conditions of Loading and Speed, Report of Committee 16, *Proceedings of the A.R.E.A.*, American Railway Engineering Association, Vol. 43, 1942, pp. 51–71.

7. R. G. Paustian, Tractive Resistance as Related to Roadway Surfaces and Motor Vehicle Operation, *Iowa Engineering Experiment Station Bulletin 119*, Ames, 1934.

8. E. G. McKibben and J. B. Davidson, Effect of Inflation Pressure on the Rolling Resistance of Pneumatic Implement Tires, *Agricultural Engineering*, Vol. 21, No. 1, 1940, pp. 25–26.

9. A. M. Wolf, Practical Tractive Ability Methods, *S.A.E. Journal*, Vol. 27, No. 6, December 1930, pp. 655–664.

10. D. W. Taylor, *The Speed and Power of Ships*, Wiley, New York, 1910.

11. C. D. Perkins and R. E. Hage, *Airplane Performance, Stability, and Control*, Wiley, New York, 1949.

12. *Handbook of Belting—Conveyor and Elevator*, the Goodyear Tire and Rubber Co., Akron, Ohio, 1953, Chapters 2, 4, 5.

13. *Oil Pipe Line Transportation Practices*, E. L. Davis and Charles Cyrus, editors, issued by the University of Texas Division of Extension and the State Board for Vocational Education, Trade, and Industrial Division, 1944, Chapter XIII.

14. W. G. Helzel, Pipeline Section Flow and Friction in Pipelines, *Oil and Gas Journal*, Tulsa, Oklahoma, June 5, 1930, p. T-223.

15. Bernard Etkin, *Dynamics of Atmospheric Flight*, Wiley, New York, 1972.

16. Andrew G. Hammitt, *The Aerodynamics of High Speed Ground Transportation*, Western Periodicals Company, North Hollywood, California, 1973.

17. "Vehicle Operating Characterists—Chapter 2," *Transportation and Traffic Engineering Handbook*, John Baerwald, Editor, Institute of Traffic Engineers, Prentice-Hall, Englewood Cliffs, New Jersey, 1976.

Chapter 5

Propulsive Force, Horsepower, and Elevation

PROPULSIVE FORCE AND HORSEPOWER

Horsepower Propulsive force must be available to overcome resistance —train or tractive resistance, road or tire-rim resistance, drag, skin, wave, and residual resistance of ships and aircraft, roller and belt resistance of conveyors, and flow resistance of pipelines.

Grade resistance or elevation, that is, the effects of gravity on an incline or vertical rise, must also be overcome by the propulsive effort of the power unit.

The propulsive force exerted must be delivered at a certain rate so that all propulsion-resistance eventually resolves into problems of horsepower. Horsepower is the rate of doing work, or, more specifically, it is force times the distance through which the force is exerted in unit time divided by the unit work equivalent per horsepower for the unit of time under consideration. From this, $\mathrm{hp} = F \times v/550$, where $v =$ speed in feet per second, that is, the distance moved in one second, and 550 represents the foot-pounds per second equivalent of one horsepower. Similarly,

$\mathrm{hp} = F \times v'/33{,}000$ where $v' =$ speed in feet per minute

$\mathrm{hp} = F \times V/375$ where $V =$ speed in miles per hour

$\mathrm{hp} = F \times V'/325.6$ where $V' =$ speed in knots

$\mathrm{hp}_m = F \times V/270$ where $V =$ speed in kilometers per hour, F is in kilograms, and hp_m is metric horsepower

The metric horsepower is usually taken as 75 kilogram-meters per second. Converted to the English system, one horsepower (metric) = 542.4 foot-pounds per second and one horsepower (English) = 550/542.4 = 1.014 horsepower (metric). Conversely, one horsepower (metric) = 543.4/550 or 0.986 horsepower (English). The two are approximately equal. Note, however, that the SI system uses watts as a measure of power where 1 hp = 746 watts.

In each of the foregoing equations hp = horsepower, and F = propulsive force, tractive effort, torque, thrust, or pumping force; F may also equal resistive force, tractive resistance, train resistance, road or tire-rim resistance, ship resistance, drag, flow resistance, etc., when the horsepower to overcome any of these is under consideration.

Propulsive force and resistive force are functions of the speed for a given horsepower and vice versa. A prime mover or propulsive unit must be able to exert enough force at a given speed to overcome the resistive forces. At a minimum, the two must be equal if a desired operating speed is to be maintained. In most instances, for acceleration and as a reserve, there should be an excess of tractive force beyond that required to overcome resistance.

Prime Movers   A prime mover is a device that transforms the potential energy of fuel into mechanical energy capable of performing work. Certain types of prime movers and fuels have become associated with one particular mode or another, hence, enter into the modal technoeconomic characteristics. Nonetheless, unconventional power combinations are possible.

Coal has been a common source of fuel and energy for many years, producing steam to drive reciprocating engines and rotative turbines. Gasoline and other petroleum distillates burned in internal combustion engines and in jet chambers have supplemented or even supplanted coal and steam as a source of energy. Water power, usually driving turbo-generator combinations, has also played an active role. Nuclear energy, solar, wind, and thermal energy, and exotic fuels are entering the scene, some already providing a source of power for rocket propulsion and support in a magnitude undreamed of in past decades. The widespread pollution effects from transport exhaust emissions and the increasing shortage of petroleum fuels are intensifying interest and research in developing new types of prime movers and fuels for their use. Coal is experiencing a resurgence as a source of energy.

Steam Engines   Until recent years, steam has been a universal source of force for most transport systems—for ships, railroads, and pumping stations

—offering simplicity in design, construction, and maintenance. The simplest steam engine system must contain one or more cylinders that move a piston connected to driving wheels or a driving shaft, a boiler to generate steam for the cylinders, and a firebox to heat the boiler. A variety of other devices—condensers, feedwater heaters, superheaters—may be added to improve efficient use of fuel. The steam engine is usually heavy and bulky, is often dirty and noisy, and poses problems of air pollution and of water supply and ash disposal for its steam boiler. Most steam engines, conservatively rated, permit a 10 to 25 percent operating overload, even to the point of stalling without injury to equipment. The horsepower curve rises slowly from zero at a standstill to a maximum, (but holds that maximum rather well), thereby limiting acceleration at low speeds. There is little flexibility in control. The transmission of power to the driving wheels or propellers sets up vibrations and dynamic impacts from the reciprocating parts of shafting and rod arrangements. Steam turbines on ships, in pumping stations, and in a few experimental locomotives avoid reiprocating and dynamic impact, but the complex reduction gearing required to bring high initial shaft speeds to practical propeller or axle speed adds to the costs of manufacture and maintenance. The use of the steam turbine in central power plants to generate electric energy for rail and rapid transit systems offers much in flexibility and economy. Light, compact steam turbines to drive generators that provide electrc energy for traction motors in locomotives are still receiving some development effort.

Internal Combustion Engines   Because of its relative simplicity, lightness, flexibility, and ruggedness, the gasoline-fueled, spark-ignited engine has almost universal use in automobiles, trucks, and buses. The internal combustion engine follows a four-stroke cycle of fuel and air intake, compression, ignition and power delivery, and exhaust—taking place within the engine cylinders and delivering through the pistons a reciprocating force to the drive shaft. Compression-ignited Diesel engines achieved greater applicability to trucks, some automobiles, and locomotives when a two cycle engine was developed, reducing bulkiness and weight, by combining intake and compression in the upstroke and ignition-power delivery and exhaust in the downstroke.

Internal combustion engines are high rpm devices. They experience a rise in torque with an increase in shaft rpm until an optimum speed is reached beyond which the torque decreases. See Figure 5.5. Engine speed is usually far greater than the desired axle or propeller speed. Hence some form of speed reduction coupling to the driving wheels or propeller shaft is required through reduction gearing, transmission, or a torque converter that enables the high torque of optimum engine speed to be utilized over a wide range of running speeds. Such mechanical or hydraulic devices have been bulky, complicated and/or weak, and not always dependable (in road locomotives, for example) but have proved more practicable in fixed installations such as pumping stations or on ships.

Electric coupling is used with diesel-electric locomotives. The diesel engine drives a generator or alternator mounted on its shaft. The generator furnishes electric current for dc traction motors. Electric coupling is, in fact, independent of the type of prime mover or power supply. The generator could be driven equally well by steam, gas turbine, or any other device that turns the generator shaft.

Electric Drive Maximum design and operating flexibility accompany electric drive. The power plant, using coal, oil, water, or nuclear energy, can be miles away as with electric locomotives and rapid transit or self-contained as in diesel-electric locomotives and turboelectric ships. Mechanical or hydraulic torque conversion is eliminated, permitting the full horsepower of the prime mover to be utilized over the entire speed range. In ships, the driving motors can be in the stern to permit short propeller shafts, the generating equipment midway in the vessel for better weight distribution and lower bending moments in hull girders. Electric drive creates no air pollution.

The rugged 600-volt, series-wound dc motor has long been standard in suburban and rapid transit equipment and in diesel-electric locomotives. It permits smooth, rapid acceleration and deceleration where frequent starts and stops are necessary as with switching operations or rapid transit and has a high overload capacity useful in starting heavy loads and negotiating steep grades. Motors can be operated well above continuous rated capacity for an established short term period before overheating occurs. Because of heavy line losses direct current motors are seldom used where electric current must be transmitted more than a few miles. Alternating current motors then provide many of the dc machine characteristics but are less compact. Carriers that run at constant speeds such as conveyors, pumps, cableways, and "drag" freight trains find use for constant speed induction motors.

The motor torque, T, in lb-ft that produces propulsive force is

$$T = K\phi I_a$$

where

I_a = armature current in amperes

ϕ = strength of magnetic field in gausses

K = a constant based on motor design,
that is number of poles, paths, and windings

Greater torque and propulsive force are had with a high armature current. But the opposing back electromotive force that is generated during motor action is E_a related to E_t the applied terminal voltage by

$$E_t = E_a + I_a R_a$$

where

$$R_a = \text{armature resistance in ohms}$$

The value of E_a is represented by the equation

$$E_a = K\phi n$$

where

n = revolutions per minute of the armature
K and ϕ are as before

An increase in speed is accompanied by a decrease in armature current, torque, and propulsive force, another example of trading pulling force for speed. Note that the limiting value of E_t across any one of a group of motors can be increased by changing the hookups from series to series-parallel to full parallel or by varying the field current in the pole windings, again with a corresponding loss in tractive force.

Jet Engines   Jet aircraft engines operate on the principle of recoil. A stream of highly compressed gases is ejected at the rear of the engine. Reaction to that backward pressure serves as forward thrust. The jet engine is light in weight, less than one pound per horsepower, burns conventional diesel fuel, requires little or no warmup period, and permits speeds in the sonic and supersonic ranges.

Jet engines follow a cycle of intake, compression, combustion, and exhaust but of air rather than of fuel. These events follow a more or less in-line arrangement of the engine parts and are occurring simultaneously as the engine operates. Air enters through an intake duct or opening at the front of the engine. It must be aerodynamically designed to admit large quantities of air (8 to 10 times that of a piston engine) under a variety of speed and altitude conditions. See Figure 5.1.

The air is drawn into a compressor of one or more stages and of either axial or centrifugal design. Compression ratios of up to 5 to 1 are had with the centrifugal design while the axial arrangement permits ratios up to 24 to 1. The moving elements in the compressor are driven by a central shaft moved by the turbine.

The compressed air, now at a high temperature, enters the combustion chamber where an atomized spray of fuel is injected and ignited by a

(a)

JT9D-7 TURBOFAN ENGINE

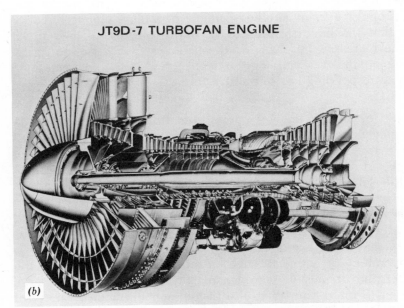

(b)

FIGURE 5.1 PRATT & WHITNEY JT9D-7 TURBOFAN ENGINE. (*a*) OVERALL VIEW. (*b*) CUTAWAY SECTION. THE ENGINE HAS A BYPASS RATIO OF 5.1, A COMPRESSION RATIO OF 22.3, A FAN PRESSURE RATIO 1.5, A TOTAL AIRFLOW OF 1535 LB/SEC, IT WEIGHS 8780 LB AND DELIVERS 40,450 LB OF CONTINUOUS THRUST (47,900 LB AT TAKEOFF WET). (COURTESY OF PRATT & WHITNEY AIRCRAFT DIVISION OF UNITED AIRCRAFT CORPORATION.)

spark. The increased volume arising from combustion following ignition expands through a turbine that is used to drive the compressor unit.

Exhaust gases leaving the turbine are ejected through a nozzle at the rear of the engine and, accelerated from relatively low velocities to sonic or supersonic rates, establish a reaction that drives the aircraft forward. Here too aerodynamic design of the exhaust nozzle is critical for maximum efficiency. One popular type of jet engine has a 17-stage axial flow compressor, a compression ratio of 17 : 1, a 3-stage turbine, and a thrust of 11,000 to 18,000 lb (48,950 to 80,100 N).

Variations of the foregoing include the fanjet or turbofan engine that has an enclosed fan producing an additional thrust without increasing fuel demand by converting the fuel energy into pressure rather than into high velocity energy. It permits a shorter runway takeoff distance and greater speed and fuel efficiency at subsonic speeds. The fanjet engine is exemplified by the Pratt and Whitney Jt9D producing 42,000 to 45,000 lb of thrust (186,900 to 200,250 N). Four of these engines are used by the Boeing 747.

Another variation, the turboprop, uses most of the gas stream energy to turn the compressor and a propellor attached to the turbine shaft. The low velocity, low pressure reaction thrust is only 8 to 12 percent of the total developed.

Gas turbines have been used to drive generators that develop electrical energy for dc series-wound motors on railroad locomotives. Another rail application is found in the recently designed turbotrains that have been placed in passenger service both in the United States, Canada, and Europe. High shaft speed is reduced to a suitable driving wheel speed by a system of reduction gears. The 300-lb engines provide 400 to 500 hp. Two to four engines have been used per power car.

Other Prime-Mover Types   Problems of fuel shortages and pollution are leading to new concepts and designs in automotive propulsion. The Wankel engine that replaces the conventional reciprocating parts with rotary motion and more complete fuel combustion is already in active use with commercial models on the automobile market. Electrically powered vehicles using batteries (to be replenished when not in use by trickle chargers) or fuel cells that derive electric energy from chemical reactions in the cells are among the possibilities being explored. The potential of steam from flash boilers is also being reexamined, as is the more simple Sterling engine.

Atomic energy as a source of transport energy has on-board applications limited thus far to oceangoing vessels such as aircraft carriers and naval submarines. More typically, the generating system is located in a central

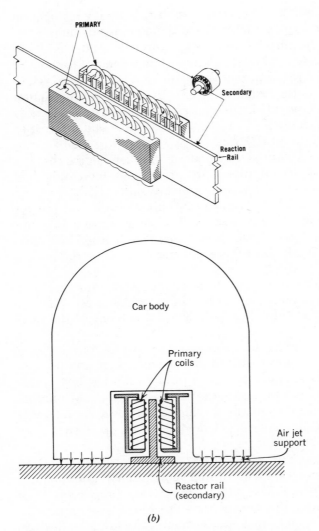

PRIMARY

Secondary

Reaction
Rail

Car body

Primary
coils

Air jet
support

Reactor rail
(secondary)

(b)

FIGURE 5.2 LINEAR-INDUCTION MOTOR WITH FIXED SECONDARY. (a) CONCEPT. (b) APPLICATION. (*FROM FIFTH REPORT OF THE HIGH SPEED GROUND TRANSPORTATION ACT OF 1965* BY THE SECRETARY OF TRANSPORTATION, WASHINGTON, D.C., 1971, P. 37.)

power station and electric energy made available from transmission lines to railroads and rapid transit.

Linear-Induction Motor   The linear-induction motor (LIM) applies electrical energy by using an application of the synchronous induction motor principle. In the Department of Transportation (DOT) test vehicle the primary coils are mounted on the vehicle on either side of a reactor rail or secondary fixed to the center of the track or guideway. Thrust is obtained from magnetic repulsion between localized electric currents in the reaction rail and the fields of the vehicle-mounted windings. See Figure 5.2. The reactor rail used in initial tests at the DOT's Pueblo Test Center is an aluminum extruded section centered between two running rails and standing 21 in. (53.3 cm) above the ties to which it is attached.

The LIM has two air-cooled primary windings 10 in. (25.4 cm) high with a core length of 150 in. (381 cm). Design rating is 2500 hp (1864.3 kW) or 3750 lb (16688 N) of continuous thrust at 250 mph (402.3 kph). The power, 1000-volt 3-phase variable frequency, is supplied by an on-board turbine driving an alternator. The LIM is expected to have application not only to rail systems but more especially to a tracked air cushion research vehicle (TACRV) developed by the DOT's Office of High Speed Ground Transportation.

The choice of making the secondary or primary the stationary element is largely economic. For low density traffic, economy lies in placing the low cost reactor in the guideway, the more expensive primary coils on the vehicle. Where many vehicles are operated, the reactor fin may more economically be placed on the vehicle, the coils on the guideway. Various power sources and pickups are proposed to energize the windings.

Locomotive Tractive Effort   Tractive effort or tractive force is a term for the propulsive force used in railroad and rapid transit engineering. Tractive effort curves for locomotives are prepared by the manufacturer or by the railroad's own test department to show the tractive or propulsive force available at different speeds. A curve for a modern 2400-hp, general-purpose, diesel-electric locomotive appears in Figure 5.3, for a rapid transit car in Figure 5.4. Note that various gear ratios between driving motors and axles give several different speed-pull ranges, that is, they determine what portion of the tractive effort curve can be effectively utilized.

Minimum and maximum speeds can be obtained from the equation:

$$V = \frac{\text{rpm} \times D \times g}{336 \times G}$$

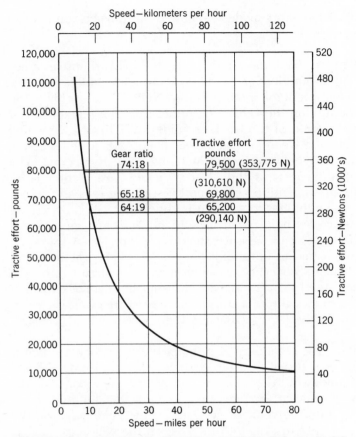

FIGURE 5.3 TRACTIVE EFFORT CURVE OF A ALCO 2400-HP DIESEL ELECTRIC LOCOMOTIVE. (COURTESY OF ALCO PRODUCTS, INC., NEW YORK.)

where

V = maximum or minimum speed of locomotive

rpm = safe maximum or minimum allowable motor speed (500-600 ± rpm and 2000-3000 ± rpm)

D = driving wheel diameter in inches

g and G = number of pinion and driving gear teeth, that is, g/G = gear ratio

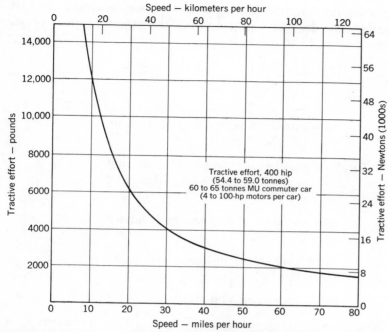

FIGURE 5.4 TRACTIVE EFFORT CURVE OF A RAPID TRANSIT CAR.

An approximate tractive effort curve for diesel-electric locomotives follows from the horsepower equation; it takes into account mechanical and electrical losses and losses to auxiliary units:

$$TE = (\text{hp}_r\text{-hp}_a) \times 375 \times e/V$$

where

hp_r = rated output of the diesel engine

hp_a = horsepower used by auxiliaries

e = efficiency factor, mechanical-electrical, taken as 82.2 percent

V = velocity in miles per hour

When these factors are given numerical values from conventional design and operation, the equation reduces to

$$TE = 308 \times \text{hp}_r/V$$

The tractive effort range established by gearing is obtained by multiplying the foregoing expressions by the gear ratio.

By equating work done by the driving wheel in one revolution to that done at the same time by the motor torque, the tractive effort of an electric locomotive or rapid transit car

$$TE = T \times 24 \times G \times e \times N/(D \times g)$$

where

$T =$ torque in pound-feet at a 12-in. (30.48cm) radius

G and $g =$ number of teeth in gear and pinion, respectively; the gear ratio

$N =$ number of motors, each having torque T

$e =$ mechanical efficiency of gears, 95 to 97 percent

$D =$ driving wheel diameter in inches

Similar expressions can be determined for gas-turbine locomotives or other types having self-contained power plants generating electric energy for the driving motors.

Rated tractive effort for a steam locomotive is assumed constant to about 15 mph and is therefore used in most ruling grade studies in which the train speed is reduced by elevation effects to within that range. By equating work performed in the cylinders to work at the rim of the drivers during one revolution of the driving wheel, the rated or starting tractive effort is found to be

$$TE = 0.85 Pd^2 s/D$$

where

$TE =$ tractive effort in pounds

$P =$ boiler (gage) pressure in pounds per square inch

$d =$ cylinder diameter in inches

$s =$ length of stroke in inches

$D =$ diameter of driving wheel in inches

$0.85 =$ factor to account for pressure drop between boiler and cylinder

Regardless of engine capacity there must be sufficient weight giving adhesion to the rails to prevent slippage. Diesel and electric locomotives with steady rotative drive delivered to the wheels can operate on an average coefficient of 0.25 to 0.30, but experience indicates 0.18 to 0.20 as practical values to cover starting resistance, slippery rails, and other unfavorable conditions; 0.25 is the accepted maximum for steam locomotives. A locomotive weighing 240,000 lb (108,862 kg) on drivers thus has a working potential of 43,200 lb (192,240 N) and a practical maximum of 60,000 lb (267,000 N) of tractive effort. On a very clean, dry rail slightly higher coefficients of adhesion can obtain.

The tonnage rating is the number of tons a locomotive can pull under stated speed and grade conditions. It is obtained by dividing the drawbar pull of the locomotive (net tractive effort after deducting the locomotive's own propulsive resistance) by the unit resistance of the trailing load.

Problem Example

A 240-ton locomotive with tractive effort of 30,800 lb of tractive effort at 30 mph has a unit tractive resistance of 3.76 lb/ton (by the Davis equation) and a total resistance of 902 lb. The unit resistance of a 40-ton car at 30 mph (Davis equation) is 6.5 lb/ton. The locomotive can haul a trailing load, on level tangent track, of $(30,800-902) \div 260$ gross tons or a train of 115 cars.

Automotive Tractive Effort   The tractive effort for a highway vehicle is usually expressed as torque, the rotative force in lb-ft exerted at the rim of a fly wheel of unit (1-ft) radius. Brake horsepower, frequently used in rating highway-vehicle motors, follows from the general horsepower equation where the turning force is exerted at the rim of a 1-ft (0.3-m) radius flywheel and the distance turned by the flywheel in one minute is the flywheel perimeter $\times$ the number of revolutions per minute.

$$\text{hp} = 2\pi R \times F \times N/33,000$$

where

$$F = \text{tractive force in lb, that is,}$$
$$\text{hp} = 2\pi R \times TE \times N/33,000$$

But

$$TE \times R = \text{torque} = T \,(\text{for}\, R = 1\text{-ft})$$
$$N = \text{revolutions per minute}$$

So

$$\text{hp} = 2\pi \times T \times N/33,000 = 0.00019\,TN$$

Torque and horsepower curves are prepared by the manufacturer or obtained by purchaser's tests. See Figure 5.5. Torque curves of gasoline, spark-ignited engines show low values at low engine speeds, rise to a peak as engine speed increases, and then fall off rapidly. Diesel, compression-ignited engines show a flat curve with good torque at low speeds and little falling off at high engine speeds, making these particularly suitable for truck service. The most economical fuel operation comes within that speed range in which maximum torque is developed. For passenger vehicles in direct drive, this speed is about 45 mph. Maximum torque or tractive force does not occur at maximum horsepower. Higher vehicle speeds depend on higher rpm at the expense of pulling (tractive) force.

Tractive effort curves can be computed from torque-hp-rpm curves by taking into account the effective gear ratio between engine shaft and rear wheels (or front wheels, as the case may be) because of the transmission and differential gear ratios. The several possible gear combinations give a series of tire-rim torque curves. Tractive effort at the rear wheel tire rim is

$$TE = TG_t G_d e / r$$

where

TE = tractive effort in pounds

T = engine flywheel torque in pound-feet at a given rpm

G_t = transmission gear ratio

G_d = differential gear ratio

e = factor to cover mechanical transmission losses between engine shaft and tire rim; taken to be 0.85 to 0.90

r = loaded radius in feet of rear tires, varying with load on the vehicle and the degree of inflation

Problem Example

The tractor of a highway tractor-trailer combination weighs 16.0 tons gross vehicle weight (GVW) and has an engine torque of 758 lb-ft at 1600 rpm; maximum brake horsepower is 289 at 725 rpm. What would be the *maximum hauling capacity* and at what speed in each of four gear combinations, ratios being: 6:10 to 1, 2.90 to 1, 1.60 to 1, and 1.0 to 1: the differential ratio is 5.80 to 1? The vehicle is assumed to have a tire diameter of 42 in. with a 1-in. radial deformation under load.

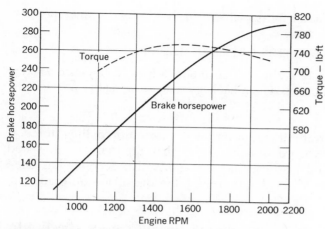

FIGURE 5.5 TORQUE—RPM—HORSEPOWER FOR A V-8 DIESEL TRACTOR.

Maximum torque and pulling force occur at 1600 rpm. The corresponding horsepower is

$$hp = 0.00019 \times 758 \times 1600 = 230.4$$

In low gear the tractor will have a tractive effort of

$$TE = 758 \times 6.1 \times 5.8 \times 0.85 \times (21-1)/12$$

$$TE = 37,992 \text{ lb}$$

If we use the speed-tractive effort for horsepower, the speed is

$$V = \frac{230.4 \times 375}{37,992}$$

$$V = 2.3 \text{ mph}$$

By using the same procedures, we find that the tractive effort in second gear is 18.062 lb at a speed of 4.8 mph; in third gear the tractive effort becomes 9965 lb at a speed of 8.7 mph; in high gear the speed is 13.9 mph and the tractive effort is 6228 lb.

Assuming a tractive resistance of 20 lb/ton, the total resistance of the 16-ton tractor will be 320 lb. This leaves a net hauling capability at 13.9 mph of 6228 lb − 320 lb = 5908 lb. If tractive resistance for the trailer is also 20 lb/ton, the total gross trailer weight that can be moved by the tractor would be 5908 lb divided by 20 = 295 tons.

Assuming the same horsepower but with a combined weight of load and vehicle of 40 tons, the combination would have on level grade a total resistance of 20×40 lb or 800 lb. If we use the horsepower equation and allow for a mechanical efficiency of 0.85, the maximum speed would be $V = 375 \times$ horsepower $\times$ efficiency $\div$ resistance.

$$V = 375 \times 289 \times 0.85 \div 800 = 115 \text{ mph}$$

Here, the maximum speed for a given set of conditions is being determined whereas in the proceeding example the maximum hauling capacity, no matter how low the speed, was the goal.

These examples illustrate the relation that maximum pulling force is obtained only at the expense of speed and that an increase in speed reduces tractive effort and load-carrying capability. No account has been taken of grade effects or of acceleration. (See the discussion on grades, p. 127.)

Thrust   The propulsive power of a ship or aircraft is made effective by the propeller thrust expressed in pounds. It must at least be equal to the resistance of the ship or drag of the aircraft. Forward motion is imparted by frictional contact or thrust between propeller blade and water. The wake or tendency of water to follow behind a ship contributes to the thrust, but there are numerous losses. The blades do not bear at right angles to the water in comparison with the direction of the ship's motion. The formation of air pockets, around the propeller blades, called cavitation, reduces the thrust. Shape of blade, diameter of screw and hub, number of blades, amount of screw out of water, position of propeller with relation to the ship's hull—these and other factors have a determining effect on the actual horsepower delivered by the propeller and its design and efficiency. The exact determination of these losses and comparative design effects involves procedures too complex and extended to be presented here. No great error will result for purposes of this text in assuming that 50 percent of the indicated horsepower of the ship's engines is lost in bearing and thrust-collar friction and in propeller inefficiency. Since the speed of a vessel is usually expressed in knots, the horsepower-thrust equation becomes $hp_i = $ thrust $\times$ velocity $\times 2 \times e/326$, where $hp_i = $ indicated horsepower of the ship's engines, velocity is in knots, 2 is a factor to cover the 50 percent loss of indicated horsepower in the propeller and shafting, and e is the mechanical efficiency of the engine, representing a 6 to 8 percent friction loss. For diesel engines the frictional loss would be about five percent or less.

Propulsion was initially provided by steam-powered multiple-expansion engines driving directly on the propeller shaft. The marine power plant of today may be one or more coal- or oil-fired steam turbines turning electric generators. These deliver energy to electric motors geared to the shaft. A diesel engine may be coupled to the shaft through reduction gearing. With diesel-electric drive the diesel engine replaces the steam turbine.

Screw propellers found their first field of usefulness in deep-draft shipping. Early river boats were propelled by paddle wheels, first side mounted, and later placed at the stern. Modern pusher boats are built with screw propellers. The propellers are located in grooves or tunnels in the hull of the vessel, permitting shallow draft required for river boats, and protecting the propellers from excessive damage when the draft is too great for the water's depth. Flow of water to the propellers is concentrated by Kort nozzles that add 20 to 25 percent to propeller thrust. See Figure 5.6.

Grade Effects   Inland waterways—canals and rivers—impose elevation problems. In free-flowing rivers the current flow aids or opposes tow movements. For downstream movements the current speed adds to the

FIGURE 5.6 SCREW PROPELLER ON A RIVER BOAT. UNDERSIDE VIEW OF THE TOWBOAT *A.D. HAYNES II*, SHOWING PROPELLERS, RUDDERS, AND KORT NOZZLES. (COURTESY OF THE DRAVO CORPORATION, PITTSBURGH, PENNSYLVANIA.)

speed attained by the propellers. In computing resistance the relative speed between craft and water should be used in determining the skin friction and residual resistance, but water-to-land speed is required for the air resistance term. For upstream movements the river current opposes the vessel, subtracts from propeller speed but adds to the velocity for skin and residual resistance.

In old meandering rivers, the currents are slower and less of an obstacle. When the upstream reaches flow too rapidly or are too shallow, dams are constructed to form slack-water pools of required depth. Canal grades are almost level.

Elevation is overcome by locks that raise (or lower) the vessels vertically in one or more stages through the lock chambers. The water level in the lock chamber is then raised or lowered by gravity, as explained in a later chapter. The engineer must balance the costs of crossing a divide at a low level with few locks against the cheaper route (but higher cost in locks) of locating over a higher elevation. He or she also must decide whether to make the lift in several short stages or a few big stages.

Aircraft vs. Elevation   A solution to the propulsive force problems for aircraft derives from the thrust-horsepower equation

$$hp_t = \text{thrust in pounds} \times \text{speed in feet per minute}/33,000$$

Similarly $hp_d = \text{drag in pounds} \times \text{speed in feet per minute}/33,000$. In uniform level flight the thrust and drag must be at least equal to maintain a given constant speed.

An airplane engine may be nominally rated in brake horsepower (Bhp) at a given number of revolutions per minute, but the effective or thrust horsepower is what can be delivered by the propeller at a given altitude. $Thp = Bhp \times e_p$ where $Thp = $ thrust horsepower and $e_p = $ propeller efficiency. Engine and propeller efficiency vary with speed and with air density at different altitudes. Engine efficiency, as in diesel-electric locomotives, can be improved at high altitudes with supercharging (forcing air to the fuel mixture under pressure). Curves prepared for each manufacturer's model show the brake horsepower at different revolutions per minute for various altitudes. See Figure 5.7.

In climbing, an airplane must lift its total weight upward at a certain rate. The rate of climb $V_c = $ change in altitude in feet per minute $= da/dt = V \sin \theta$, where $\theta = $ the angle made by the path of a steady climb with the horizontal. From the horsepower equation,

$$V_c = hp_e \times 33,000/W$$

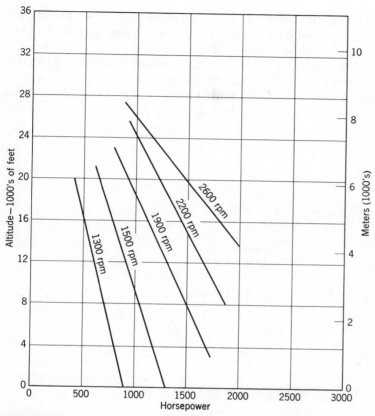

FIGURE 5.7 ENGINE PERFORMANCE AT VARIOUS ALTITUDES AND RPM.

where

V_c = rate of climb in feet per minute

hp_e = excess thrust horsepower available at a given altitude (excess over that necessary to maintain horizontal flight)

W = weight in pounds of the plane and its contents

At any given altitude, the maximum speed is determined by the maximum horsepower available. At any speed less than the maximum, there is an excess of horsepower available for climb at any given altitude. See

Figure 5.8. By inserting the value of excess horsepower in the equation for V_c, the rate of climb at a given altitude is obtained. Figure 5.9 shows a series of such points plotted for various altitudes.

The point at which the maximum thrust horsepower available equals the maximum thrust horsepower required establishes the ceiling of the airplane. The point at which rate of climb becomes less than 100 ft/min usually sets the service or cruising ceiling of the craft.

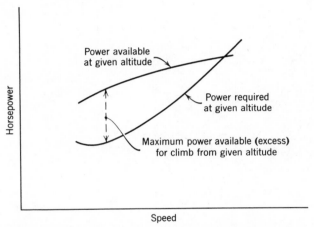

FIGURE 5.8 EXCESS HORSEPOWER AVAILABLE FOR CLIMB.

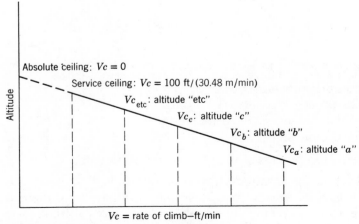

FIGURE 5.9 CEILING OF AN AIRPLANE.

When the airplane is on the ground in takeoff or landing, the more familiar rules of tractive resistance apply. The total ground resistance is equal to the sum of air resistance and rolling resistance as for a land vehicle, or $R_g = D + R_r$. For an airplane, the air resistance is the drag, or $D = C_{Dp}\rho v^2 S/2$ and $R_r = f(W-L)$, where F varies from 0.02 for concrete runways to 0.10 for an average unpaved field with long grass, depending on the smoothness, and $L =$ lift $= C_L v^2 S/2$. As the run starts, the effective angle of attack is minimal, giving a maximum L/D ratio. The accelerating force available to give the plane sufficient speed for takeoff is the excess thrust $T_e = T_v - (D + R_r)$, where $T_v =$ ground speed thrust, and D and R_r are as given above. As the speed increases, the drag becomes greater because of the higher speed, tending thereby to reduce T_e. At the same time, however, the lift increases with speed and lessens the load on the wheels as more and more of the load is borne by the wings. As the angle of attack is changed abruptly to make the craft airborne, the wing drag becomes much greater thereby reducing the speed. Takeoff speed must accordingly be greater than the stalling speed to counteract the sudden drop in velocity. A takeoff speed corresponding to 70 to 90 percent of the maximum C_L is used, whereby (using 90 percent$_{stall}$ v (at takeoff)$= v_{stall} W/C_{Lmax})S/2$ $= 1.054 v_{stall}$.

Problem Example

Referring to the problem example of Chapter 4, the propulsive force available for acceleration at the beginning of the takeoff run (when, for example, the airplane has attained a speed of 20 mph) is $T_e = T_{v20} - (D + R_r)$. Assuming an 85-percent propeller efficiency, $T_{v20} = 2 \times 1600 \times 0.85 \times 375/20 = 51,000$ lb. $T_e = 51,000 - [C_{Dp} v^2 S\rho/2 + 0.02(25,000 - C_{L min} v^2 S\rho/2)]$. From Figure 4.6, $C_L = 0.43$, for an effective angle of attack of 2 degrees, and $C_{Dp} = 0.0056$. Inserting these and values for the wing area, speed of 20 mph, and ground-level density of 0.0024 in the foregoing equation, $T_e = 50,503$ at 20 mph.

As the plane is about to be airborne, practically all the wheel load is taken by the wing so that $R_r =$ zero and $T_e = T_v - D$. Assuming a stalling speed of 80 mph, v (at takeoff)$= T_{84.32} - 0.013 \times 0.0024 \times (84.32 \times 1.47)^2 \times 980/2 = 102$ lb. $(T_v = 2 \times 1600 \times 0.85 \times 375/84.32 = 12,000)$. $T_e = 12,000 - 102 = 11,898$ lb. This is a somewhat simplified approach to the problem.

Gradients for airport runways should be as level as conditions permit. Only enough gradient to produce adequate drainage should be allowed in a commercial airport. As long-distance flights are normally made in the stratosphere to benefit by prevailing winds to reduce air drag, and to avoid traffic at lower altitudes, variations in terrain have no meaning except in the presence of very high mountain peaks or ridges. Navigation charts must indicate these points, and routes will follow passes through the barriers or circumvent the peaks as a ship sails around a reef or island. Planes may also go to still higher altitudes and pass over these obstructions. For lower-flying craft, the passes must be clearly charted. The design of radio beacons and markers for established routes is a matter of navigational control and is discussed in a later chapter.

In regard to airport location, altitude has an important effect in determining the lengths of runways and therefore the space requirements. At high altitudes longer runways are required than for the same plane and load at lower altitudes because the density of the atmosphere is less and a greater initial speed is required to secure the necessary lift for takeoff. A runway requirement of 4700 ft (1433 M) at sea level becomes 5700 ft (1737 M) at a 4000 ft (1219 M) elevation and 6700 ft (2042 M) at 8000 ft (2438 M).

As in other modes of transport, an airplane may descend from a high to a lower altitude aided by the force of gravity, with reduced fuel consumption during the time of descent. If an initial terminal is higher than the destination terminal, there will be some economy in the downward flight, as contrasted to one in the opposite direction. However, as a significant percent of the airplane's power and fuel are exerted in producing lift, the relations between thrust, gravity, and drag are not so important as for land-based vehicles.

Pumping Pressures   Oil is forced through a pipeline by a pumping pressure P, expressed either in pounds per square inch or in feet of head, which overcomes the flow resistance expressed in the same units. The total force to move the liquid and the corresponding horsepower required by the pumping equipment is the power actually delivered to the liquid, that is, the weight of the liquid $\times$ the head:

$$\text{hp}_f = W_f \times h / 33{,}000$$

where

$\text{hp}_f =$ horsepower delivered to the fluid

$W_f =$ weight in pounds of the fluid flow per minute

$h =$ total head of liquid in feet and is equal to pumping pressure-/62.4 $\times$ specific gravity of the fluid

Pumping pressure is provided by pumps in initial and booster stations along the line of flow. Reciprocating pumps were commonly used in early stations and have been installed in some more recent ones, but the multistage centrifugal pump is widely used today. Steam and internal-combustion engines (usually diesels) and electric motors drive the pumps. The diesel engine has an advantage in pumping some products in that the fuel can be drawn from the cargo being pumped. See Figure 5.10.

The horsepower delivered by a reciprocating pump is the product of the force (pumping pressure P in pounds per square inch $\times$ area of the piston in square inches) and the distance in feet through which that force is applied in unit time:

$$\text{hp} = P\,(\pi d^2 s / 4) \times n \times N / 33{,}000$$

where d is the piston diameter in inches, s is the stroke in feet, n is the number of strokes per minute, and N is the number of pump barrels or pistons.

FIGURE 5.10 LONGVIEW STATION, MID-VALLEY PIPE LINE COMPANY—TWO 800-HP DIESEL ENGINES PUMPING CRUDE OIL. (COURTESY OF R.B. WARD AND THE OIL AND GAS JOURNAL, TULSA, OKLAHOMA.)

In practical units the flow B is usually stated in barrels per hour. The volume of oil in barrels displaced per hour by the pump is $A \times s \times n \times 12 \times 60/9702$, where A is the piston area in square inches, s and n are as before, and 9702 is the number of cubic inches in a 42-gal barrel. The horsepower, then, is $hp = 0.0004PB$, where B is the volume pumped in barrels per hour. For a multicylindered pump, the foregoing is multiplied by N, the number of pistons, and, for a double-acting pump, the number of strokes must be multiplied by two. The actual volume and therefore the horsepower delivered to the liquid may be 3 to 20 percent less than that computed from pump dimensions because of slip. Mechanical efficiency is the ratio of the horsepower delivered to the fluid hp_f to that delivered to the pump shaft hp_s by the prime mover, or $e = hp_f/hp_s$. Working pressures for piston pumps may be as much as 1400 psi (9,652 MPa).

Centrifugal pumps have different characteristics. The quantity of discharge varies directly with the speed, so that horsepower characteristics are usually expressed at a certain rpm and curves are prepared to show the relation between that speed and head, capacity, horsepower, and efficiency. The head or pumping pressure varies as the square of the speed and the horsepower as the cube of the speed. At a given speed, the capacity decreases as the pressure increases. Pressures of 1200 to 1400 psi (8,273 to 9,652 MPa) are possible with these pumps. Two or more may be connected in series to develop the combined pressure of each. When combined in parallel, the capacities are correspondingly increased.

Elevation in Pipelines   The problem of grades as such does not arise in pipeline location except as it affects the difficulties of actual construction. Differences in elevation initial and final terminals or between any two pumping stations on the line are, however, important and add to or subtract from the pumping effort, depending on whether the initial point is higher or lower than the terminal point.

The high point between two stations is determined, and a hydraulic gradient is prepared. The hydraulic gradient is the loci of equivalent heights of columns of liquid if piezometer tubes were to be placed at intervals along the line. It shows the pressure change between any two points. In Figure 5.11a CB is the hydraulic gradient for the line AB. The only resistance to the 600 psi(4,137 MPa) pumping pressure is flow resistance, which is sufficient to reduce the pressure to approximately zero at B; in practice it is usually limited to 50 to 100 psi (345 to 689 MPa). AC is the pressure in equivalent feet of head obtained as shown in the following problem example.

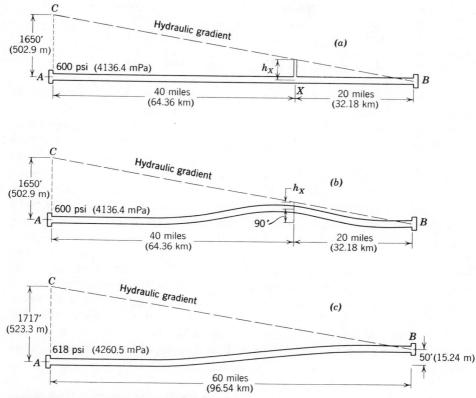

FIGURE 5.11 HYDRAULIC GRADIENTS.

Problem Example

The unit pressure at the base of a 1-ft column of water is $62.4/144 = 0.434$ psi. Oil with a specific gravity of 0.82 weighs 0.82×62.4, or 51.2 lb/ft³ density, and unit pressure is $51.2/144 = 0.360$ psi. The head AC corresponding to 600 psi is $h_a = 600/0.36$, or 1667 ft. At X, 40 miles from A, the head h_x by similar triangles is 550 ft and pressure is $550 \times 0.36 = 198$ psi. As a check, the average pressure loss of $600/60 = 10$ lb/mile. In 40 miles the pressure loss is 40×10, or 400 psi, leaving 200 psi pressure at X. A piezometer tube placed in the pipe would rise to the height $n_x = 550$ ft.

If elevation is introduced between A and B, as in Figure 5.11b, the head at X is no longer $h_x = 550$ ft.

Assume that the elevation of the line at X is 90 ft. The head required to overcome the 90 ft of rise would be 90 ft, leaving $h_x = 460$ ft and pressure at $X = 165.6$ psi. Since A and B are still at the same level, the loss of pressure is raising the oil over the crest of X is regained in going back to the lower elevation at B. If, however, B is at higher elevation than A, sufficient initial head must be added at A to overcome the resistance of elevation. If the 600 psi is just enough to overcome flow resistance on an effective level grade, there must be additional pressure and head to lift the oil 50 ft higher to the elevation at B. Pressure at A must be $600 + (50 \times 0.36) = 618$ psi and $h_{AC} = 618/0.36 = 1717$ ft.

If B had been lower than A, gravity would have aided the flow. In general, the difference in heads between "upstream" and "downstream" ends of the line must be added to or subtracted from the pressure losses caused by resistance alone. If an intermediate elevation were higher than the hydraulic gradient, the line would not continue to flow unless there were suction (vacuum) applied at the downstream end. Thus the locating engineer's principal concern with intermediate grades and elevation is to stay below the hydraulic gradient (or increase the initial pressure, thereby changing the hydraulic gradient) and select the easiest route for construction. The overall differences in elevation should be kept to a minimum to avoid the need for excess pumping pressures. A balance must be secured between costs of building to a low elevation and costs of building and operating higher-pressured pumping units. This problem must be resolved at the time the hydraulic gradient is prepared. Abrupt changes in elevation, even within the hydraulic gradient, should be avoided, especially close to a pumping station. It is desirable to have the initial terminal at a lower level than the gathering lines so that gravity will aid in feeding oil to the point of concentration.

Design Features   The gathering lines, (there can be a hundred or more), feed an initial collecting, storage, and pumping area. The design and frequency of an intermediate or booster station location are functions of the elevation of initial and terminal nodes, initial pumping pressures, hydraulic gradient, pipe diameter, and desired flow capacity. Flow capacity varies approximately as the 2.6 power of the pipe diameter. Thus one large pipe is the equivalent of several smaller pipes. For a given pipe diameter, the throughput or flow capacity will be a function of the horsepower and pumping pressure; the greater the distance between stations the greater must be the initial pressure. Usually there is allowed a residual 50 to 100 psi of pumping pressure remaining at the downstream node of a station interspace. Thus there are possible economic tradeoffs between the number of stations (spacing), pipe size, and pumping capability (horsepower and pressure). The monitoring of flow and the computerized central control of the system are discussed in Chapter 11. More than one type of fluid can be shipped simultaneously. A product line may have batches of various grades of gasoline adjacent to diesel fuels and heating oils. Rubber spheres or batch separators can be used between different grades, but tariffs permit a limited amount of comingling when separators are not used.

"Solids" Pipelines   Pipelines to transport solids have been found useful where the cost of alternative methods was too high and where a high load factor (large volumes of traffic) is available. As with other pipelines, the problem of returning empty containers does not exist, but neither does the opportunity for moving other types of traffic. Flow is one-directional. In this system there is no flexibility in deciding location and types of traffic.

Two methods are available for transporting solids by pipelines: (1) slurry flow (the one in common use) and (2) in capsules, in which the cargo is kept separate from the conducting liquid. Capsule flow is only in the experimental stage. Slurry methods utilize a suspension of finely ground particles (of coal or ore, for example) in a liquid, usually water. The suspension generally follows the laws of liquid flow. High liquid velocities slide large diameter particles slowly along the pipe invert; in a nonhomogeneous flow nonuniform particles may move more slowly than the liquid. The third and most commonly used method transports particles 25 to 30 microns or smaller in size, with turbulent flow in homogeneous suspension. Turbulence is needed for uniform suspension, but the particles will reslurry if the flow is halted. If the flow is too slow, the particles will settle and wear the pipe; particles will also settle if the particle grind is too large.

Technological requirements are that (a) the solids not undergo any harmful change or reaction from contact with the suspension liquid, (b) the solids should go into suspension readily and be easily separated from the liquid at destination, (c) the suspension be not unduly corrosive or abrasive to the pipe, (d) the suspension should be handled readily with available preparation, pumping, and retrieval equipment, and (e) large quantities of flow liquid, usually water, be available. This last requirement can be a serious handicap because of the increasing demand for water, especially unpolluted water (a possible problem in the disposal of flow liquid following separation at the destination terminal). Slurry storage tanks must be continuously agitated to prevent settlement of the solids.

The Southern Pacific's Black Mesa Coal Slurry Pipeline, serving the 1.5 million kW Mohave Generating Station, pumps 600 tons (544.3 tonnes) per hour of finely ground coal particles that form 50 percent of a water solution over approximately 200 miles (322-km) in an 18-in. (45.7-cm) line. Three days are required to travel the total route, which includes 1600 ft (487.7 m) of rise in the first 25 miles (40.2km) and a 300 ft (914.4 m) drop near the end (where the diameter reduces to 12 in. (30.48 cm). Four large piston displacement pumps delivering pressures of 1000-psi (6894 MPa) are located at the originating pump station and at three booster stations. The coal is recovered by a centrifuge dryer, forming a damp cake that is then pulverized to a dry powder for burning. Water recovered from the centrifuge is used for cooling. The system has demonstrated the feasibility of automating the entire operation and of shutting down and restarting the pumping action. Furthermore, it has demonstrated that terrain need not be a limitation. A 1000-mile (1609-km) line is now under study by the Southern Pacific. Selection of a solids line over other transport modes must be based on the economics of a particular location and demand situation.

Aerial Tramways   Aerial tramways are examples of band drive (Figure 5.12) that find most frequent use over rugged terrain and inclines where no other type of transport can penetrate economically. A complete discussion of the design and horsepower requirements is contained in the first edition of this text, pp. 210-214 and in the section by Edward B. Dunham, "Aerial Tramways and Cableways" in Peele's *Mining Engineer's Handbook*, 3rd edition, Wiley, New York, 1941, pp. 6-7, 24-25.

Conveyors   The belt conveyor, like the cableway, is an example of band or belt drive. Resistance of rollers, load, and elevation are usually overcome by an electric motor mounted on, coupled to, or geared to the drive-pulley shaft. As with other modes frictional adhesion is required between the driving-pulley surface and the belt in order for the belt to move. An effective or tractive tension is needed equal to the difference in tight and slack side tension, $E = T - S$, where $E =$ effective tension in pounds and T and S are, respectively, the tight and slack side tensions in pounds.

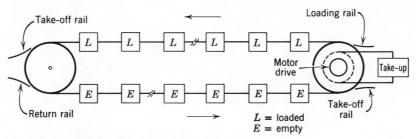

FIGURE 5.12 AERIAL-TRAMWAY LAYOUT.

The horsepower to drive the belt is

$$\text{hp} = \frac{E \times v}{33,000} = \frac{(T-S) \times v}{33,000}$$

where v is the belt speed in feet per minute. There must be tractive force (effective tension) and horsepower to (a) move the empty belt on the incline, (b) move the load horizontally, (c) raise the load vertically.

When a load is to be raised H feet above the horizontal on an inclined belt (Figures 5.13 and 5.14), the length of belt on the incline is

$$L_i = \frac{H}{\sin A}$$

The cargo load per foot of belt length, from Chapter 4, is

$$= \frac{100 T}{3 S}$$

FIGURE 5.13 CONVEYOR BELT ON AN INCLINE. RAISES 1200 TONS OF COAL PER HOUR A HEIGHT OF 868 FT. (COURTESY OF THE GOODYEAR TIRE AND RUBBER COMPANY, AKRON, OHIO.)

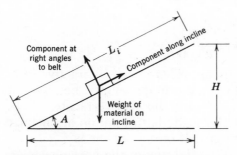

FIGURE 5.14 CONVEYOR BELT ON AN INCLINE—THEORY OF VERTICAL LIFT.

The total weight of cargo, W, acting downward on the belt rollers is

$$W = \frac{100T}{3S} \times \frac{H}{\sin A} = \frac{100TH}{(3S)\sin A}$$

A component of gravity R tends to move the belt downward. It must be restrained by an equal and opposite force F.

$$F = W \sin A$$

Substituting the value for W, we get

$$F = \frac{100TH}{(3S)\sin A} \times \sin A = \frac{100TH}{3S}$$

The horsepower to raise the load (or generated by a descending load) is force×distance moved in one minute divided by the horsepower work equivalent.

$$\text{hp}_v = \frac{100TH}{3S} \times \frac{S}{33{,}000}$$

$$\text{hp}_v = \frac{TH}{990}$$

Using values from Chapter 4, the horsepower to move a level empty belt will be hp_e where

$$\text{hp}_e = CQ(L + L_0) \times \frac{S}{33{,}000}$$

Rearranging terms and multiplying numerator and denominator by 0.03

$$\text{hp}_e = \frac{C(L + L_0)(\text{speed})}{33{,}000} \qquad \text{where } W = \frac{100T}{3S}$$

whereby

$$hp_1 = C(L + L_0) \frac{(100T)}{3S} \frac{(S)}{33,000}$$

and

$$hp_1 = C(L + L_0) \frac{(T)}{990}$$

The total horsepower on the incline will be the sum of the horsepower for the empty belt plus the horsepower needed to move the load horizontally plus or minus the horsepower to raise or lower the load H feet:

$$hp = hp_e + hp_1 \pm hp_v$$

From which

$$hp = \frac{C(L + L_0)(0.03QS)}{990} + \frac{C(L + L_0)(T)}{990} \pm \frac{TH}{990}$$

Note that $(TH/990)$ is added if the point of discharge is higher than the point of origin, and subtracted if it is lower.[1]

The maximum or tight side tension is a function of adhesion to the driving pulley. The value of n, the number of half-turns around the driving pulley is the arc of contact in half-turn or 180-degree multiples. The expression, $T = Se^{\pi f n}$ for band drive may be converted to the logarithmic base 10, and, putting the equation in its more useful form, $T - S = S(10^{0.00758 fa} - 1)$, where $a = $ arc of contact and f is the coefficient of friction. A lower value of S, and therefore of the maximum T, is obtained by increasing the coefficient of friction either by lagging the pulley or by snubbing the belt to increase the area of contact or both. The coefficient of friction between rubber (covered) belting and pulley steel is 0.25. If the pulley has a rubber lagging, f is increased to 0.35. Theoretically, $f = 0.55$ to 0.75 for clean rubber surfaces in contact. These ideal conditions do not exist in practice, where belt, pulley, or more likely both become greasy, dirty, wet, or dusty. The arc of contact can be further increased by using two drive pulleys interlocked or dual pulleys driven by separate motors.

In the design of the tractive requirements for a conveyor, the necessary horsepower is determined first, followed by the effective tension E, maximum tension T, and slack side tension S. The horsepower to move the

[1]*Goodyear Handbook of Belting: Conveyor and Elevator*, The Goodyear Tire and Rubber Company, Akron, Ohio, 1953, pp. 70-71.

empty belt is a function of its weight in pounds, the length of belt (per flight) in feet, and the belt speed in feet per minute. The weight of the belt is established by the belt materials, the number of plies, method of manufacture, and the width. Once a capacity in tons per hour has been determined, the belt width and speed to handle that tonnage are selected. These depend in turn on the weight per cubic foot of the materials being handled. The horsepower to move horizontally and raise the load is next determined by the methods earlier described or by use of manufacturer's or handbook values. An approximate 10 percent increase in speed requirements is desirable to compensate for belt slip.

The vertical rise in the load is a function of the length and gradient of the belt. The gradient on the belt is limited by the angle of repose of the materials being conveyed. The angle varies between 10 degrees for briquettes, 12 degrees for washed gravel, 15 degrees for grain and dry sand, 18 degrees for pit-run gravel and mine-run coal, and 20 degrees for bituminous slack coal, crushed ore, and damp sand.[2] The horsepower to overcome elevations usually exceeds that for moving the load and belt horizontally. A brake is needed to prevent the belt from running backward when stopped.

The conveyor belt is a constant-speed operation with an overload requirement at starting. A favorite type of conveyor drive is the double-wound rotor or slip-ring type of squirrel-cage induction motor with high torque but low starting current requirements.[3] Variation in the starting torque is desirable because of the higher torque required to put the belt in motion, especially when the bearings are cold. Other types of transport experience the same difficulty.

GRADES AND ELEVATION

The effects of elevation on aircraft and conveyors have already been considered. There remains the effect of grades and elevation on land-based vehicles and craft, especially railroads and highways.

Gradients   The addition of 20 lb (9.07 kg)/ton/percent of grade increases markedly the resistance to be overcome by tractive effort and adds to operating costs in other ways. The locating engineer must therefore keep his gradients as light as possible. In rugged, difficult terrain light grades

[2]Wilbur G Hudson, *Conveyors and Related Equipment*, 3rd edition, Wiley, New York, 1954, p. 229.
[3]*Ibid.*, pp. 236-237.

usually require expensive construction by heavy excavation and fill, bridging, or tunneling. Development, the introduction of additional distance to reduce the rate of grade, lowers the gradient but may be difficult to secure. Additional distance and curvature thus require increased construction and operating costs.

The selective utilization of grade in rapid transit design may, on the contrary, prove beneficial. A slightly ascending grade toward the station at each end aids deceleration for platform stops and contributes to acceleration on departing the platform.

Ruling Grade   The ruling grade of a railroad is that which establishes the maximum train load to be hauled at a given speed by a given horsepower. Steeper grades may appear on the profile, but must be operated with helper engines, by reducing train loads or, for very short grades, by momentum. A ruling grade on a highway defines the maximum load a truck or tractor can haul at a given speed with a stated gear combination, usually low gear.

In the following problem examples both tractive and grade resistance are included. The conditions of the earlier problems on tractive resistance are the same (page 145) except that the train is operating on an 0.80 percent grade, the truck on a 4.0 percent grade.

Locomotive resistance becomes R_L where

$$R_L = 240(3.76 + 0.80 \times 20) = 4742 \text{ lb}$$

Net tractive resistance or drawbar pull $= 30,800 - 4742 = 26,058$ lb. Unit resistance for the cars on the grade becomes R_c where

$$R_c = 6.5 + 0.80 \times 20 = 22.5 \text{ lb/ton}$$

The tonnage rating of the locomotive under these conditions at 30 mph will then be $26,058 \div 22.5 = 1158$ tons. The train will now consist of $1158 \div 40 = 29$ cars. Introducing the 0.80 percent grade has reduced the train load and rating from 115 cars and 4600 gross tons to 29 cars and 1158 gross tons, a reduction in tonnage rating of 75 percent. If 20,000 gross tons are to be moved in a day, 5 trains will be required on level track, and 17 trains on the 0.80 percent grade (assuming no reduction in the stated speed of 30 mph).

Using the earlier tractor-trailer problem data but assuming a 4-percent grade, high gear speed of 13.9 mph with a tractive effort of 6228 lb and rolling and air resistance still 20 lb per ton, the tractive resistance now becomes R_g where

$$R_g = 16 \times (20 + 4 \times 20) = 1600 \text{ lb}$$

The net tractive effort or drawbar pull for the tractor will be $6228 - 1600 =$ 4628 lb. The total gross load that can be pulled by the tractor at 13.9 mph is $4628 \div (20 + 4 \times 20) = 46.3$ tons. If the trailer weighs 14 tons, then the payload will be $46.3 - 14$ or 32.3 tons.

Limiting Grades; Weight/Horsepower Ratios The ruling grade concept is not widely used in highway design and operation because other factors are usually more restrictive. An analogous feature for highways is the limiting grade, the grade that is so steep it reduces vehicle speed and thereby limits the number of vehicles that can traverse the grade in a given time. Maximum flow occurs at a speed of 30 mph; hence congestion will begin on a crowded highway when speeds drop below that critical rate. The lack of ability to maintain speed on an incline can have a significant impact on highway capacity. See page 289.

Grade resistance reduces vehicle speed. If a vehicle approaches the foot of a grade at a speed of 40 to 60 mph (64.4 to 96.5 kph) the speed will decrease rather rapidly corresponding to the maximum for a particular gradient, the balancing speed. At this speed, total resistance and total tractive effort are equal.

Ignoring the constraints of gearing and using the 289-hp example vehicle of page 146 with 40 tons of total vehicle weight, the effects of gradient on vehicle speed can be determined. On a one-percent grade, a grade resistance of 20 lb/ton must be added to the tire rim resistance. If an average value for tire rim resistance, also of 20 lb/ton, is taken, total resistance becomes $40 \times (20 + 20) = 1600$ lb. The maximum velocity on a one-percent grade is then:

$$V = 375 \times 289 \times 0.82 \div 1600 = 55.6 \text{ mph}$$

Table 5.1 gives the total resistance and the maximum speed attainable by the foregoing vehicle on various gradients. The fourth column gives the distance from the bottom of the grade at which the critical speed of 30 mph occurs for a vehicle having a 277 lb/hp ratio and approaching the foot of grade at 40 mph. Longer grades lead to a reduction in lane capacity. For a higher weight/hp ratio, that is, more weight per horsepower, a shorter decelerating distance ensues; if the ratio is lower the critical distance is less.

Climbing lanes or passing bays can be added on long slow climbs to permit faster-moving passenger and other light vehicles with a lower weight/hp ratio to continue without appreciable reduction in speed. See the paragraphs on momentum, page 167.

TABLE 5.1 GRADE-SPEED CHARACTERISTICS FOR A 40-TON, 289-HP HIGH-WAY VEHICLE

Grade (%)	Resistance (lb)	(Newtons)	Maximum speed (mph)	(kph)	Approximate Distance to Reduce from 40 to 30 mph[a] (ft)
1	1,600	7,120	55.6	89.5	Can maintain 40 mph (64.4 kph)
2	2,400	10,680	37.0	59.5	Can maintain 40 mph (64.4 kph)
3	3,200	14,240	27.8	44.7	2965 ft (904 m)
4	4,000	17,800	22.1	35.6	1342 ft (409 m)
5	4,800	21,360	18.5	29.8	867 ft (264 m)
6	5,600	24,920	15.9	25.6	640 ft (195 m)
7	6,400	28,480	13.9	22.4	508 ft (155 m)

[a] By use of acceleration-deceleration equations of Chapter 8.

Momentum   The reader is probably aware that a hill can be negotiated with more ease, to a greater elevation, and at a faster speed if approached at a high rate of speed by "taking a run for it." The aid thus rendered by speed is called momentum. From mechanics, a body, by virtue of its elevation, possesses potential energy and the ability to acquire speed, because of the acceleration of gravity, as it falls from that elevation. Expressed mathematically, $h = v^2/2g$, where h = the elevation or height of fall in feet, and v is the final speed in feet per second that the body has acquired. Conversely, a body moving at a velocity of v feet per second can expend that kinetic energy in rising to an elevation of h feet. If the velocity is expressed in miles per hour and approximately 5 percent is added to include rotative energy stored in moving wheels, $h = 0.035 V^2$. The energy of momentum is additive to the energy being exerted by the prime mover of the locomotive or highway tractor.

Such additional energy might be used to overcome a grade at a high average speed or to cross a summit at a higher elevation in the same horizontal distance, that is, by a steeper grade. Thus, if by virtue of the motive power alone, a grade of G_t can be overcome in the distance L (in 100-ft engineering stations), the maximum grade G_m that can be attained by motive power and momentum together would be

$$G_m = G_t + 0.035(V_i^2 - V_f^2)/L$$

where V_i and V_f are the initial and final velocities, respectively.

For example, a train or truck can just maintain motion on a 2400-ft, 1.2 percent grade at a speed of 10 mph by virtue of motive power alone. If the grade is approached at an initial speed of 30 mph, the 10-mph minimum could be maintained over a steeper grade of the same length of $G_m = 1.2 + 0.035(30^2 - 10^2)/24$ or a 2.4 percent grade.

Momentum should never be considered, except as a factor of safety, in the design of ruling grades. A tonnage train stopped on the ruling grade would lose all its momentum and be unable to start. A truck could shift gears (unless already in low gear) but at a sacrifice of speed.

Short grades can be negotiated by momentum both on railroads and highways. The loss of speed on the ascending slope can be regained on the downward slope. On a rolling profile it is thus possible to operate with no throttle change. The demand on the engine is constant, as if the train or truck were on a level grade. There is some loss in time but a saving in fuel. This is, in a sense, a rewording of the floating-and drifting-grade concept.

OTHER POWER-RELATED FACTORS

Thermal Efficiency and Fuel Consumption   A pound of coal or a gallon of fuel oil contains a certain number of heating units, expressed in British Thermal Units (BTU), from which work-performing energy is obtained. One BTU is the amount of heat or energy necessary to perform 778 ft-lb of work or to raise the temperature of one pound of water $1/180$ of $32°$ F to $212°$ F; that is, a rise of $1°$ F at atmospheric pressure. A pound of coal contains approximately 8500 (8967.5×10^3 joules) to 14,500 BTUs ($15,297.5 \times 10^3$ joules). A gallon of fuel oil contains 130,000 to 136,000 BTUs ($137,150 \times 10^3$ to $143,480 \times 10^3$ joules) or about 18,000 to 20,000 BTU/lb. The equivalent of one kilowatt hour may be taken as 3412 BTU (3600×10^3 joules). Thus one pound of coal has the potential for doing 6,613,000 to 10,892,000 ft-lb of work.

Only a small part of this potential energy is realized as power and propulsive force. Thermal efficiency represents the percent of energy in the fuel that is available at the tire rim, driving wheel, or at the propellor blade. A modern stationary steam power plant can convert 18 to 32 percent of the heat in coal into steam energy. When a steam plant is crowded into the frame of a locomotive, given a fire-tube boiler, and radiation and internal friction losses are included, the net available energy for pulling loads drops to less than 10 percent of that originally present in the coal and to little more than 5 percent after the locomotive has overcome its own inertia and tractive resistance. If an efficient fixed steam plant drives an electric generator feeding electric current via a power transmission line to the motors of an electric locomotive, an improved efficiency is achieved. With a steam plant efficiency of 30 percent, an electric generator efficiency of 90 percent, a power transmission efficiency of 90 percent, and traction-motor and gearing efficiencies of 85 percent, an overall efficiency of 20 percent (20 percent of the energy in the initial pound of coal) is available in propulsive force at the driving wheel of the locomotive, trolley bus, or rapid transit car. In a ship such an efficiency is further reduced about 50 percent by losses in the shaft and propellor. The initial potential for a water-power plant depends on the hydraulic and velocity heads of the water, which are reduced by internal flow losses of the water, conduit friction, etc. (See the section on flow resistance, p. 118), and on the mechanical losses and utilization factor of the turbine wheels.

An internal-combustion engine can convert about 30 to 35 percent of the potential heat in a pound of liquid fuel into useful energy. Deductions must be made for incomplete combustion, moisture in the fuel and atmosphere, and radiation losses. A diesel-electric locomotive, assuming 92 percent generator and 85 percent motor-gear efficiencies, would possess an overall average thermal efficiency of 23.5 percent, the range is 18 to 26 percent. If the engine is designed to drive a truck with 15 percent mechanical losses, the final recovery at the tire rim is 25 percent; if the engine is driving an airplane propeller with an efficiency of 75 percent, the recovery drops to 18.1 percent; and if the engine is on a ship with 50 percent shaft and propellor losses, the efficiency is about 12.8 percent.

Because diesel fuels contain approximately 6000 more BTU per pound than coal, the internal-combustion engine offers fuel economy. There is also a BTU difference between gasoline, kerosene, and other petroleum distillates. Questions arise about the relative abundance of various types of fuel and the effect on the economy and the conservation policies of a country, as well as on the individual carrier and of using one fuel in preference to another. A related problem is the corresponding effect on the economy and conservation of resources of using one carrier mode as opposed to another because of the degree and kind of fuel utilization for each. Of rapidly increasing importance is the quantity of air pollutants put forth by each type of motive power (see page 340). These are questions that call for hard decisions to insure the country's future energy supply and the health of its citizens.

Thermal efficiency has a very simple and direct effect on fuel costs. A prime mover that has a thermal efficiency of 25 percent will require four times the BTUs it would require with 100 percent efficiency. One with only 20 percent efficiency would require five times as many BTUs. What this means in pounds of fuel and dollars depends on whether coal, kerosene, gasoline, or other fuel is used as the source of BTUs; however, for a given type of fuel, the vehicle propelled by the prime mover with the highest thermal efficiency will have the lowest fuel cost.

The foregoing represent average theoretical relations. In practice a number of other factors determine the actual amount of fuel consumed—quality of the fuel, engine adjustment, skill of the operator, idling time, and weather conditions, for example. Additional material on energy use is in Chapter 9.

The next several pages discuss additional factors affecting the problem of fuel consumption.

Dead-Load-to-Pay-Load Ratio   A relation has been established between the resistance of a vehicle or craft, the horsepower required of the prime mover to overcome that resistance, and the fuel consumed in the process. One other factor is important in this vehicle-motive power system, the relation between vehicle weight and cargo weight as expressed in the empty-or deadload weight-to-payload ratio. In general $R_{P/e}$ = payload capacity/empty or dead load. Total vehicle weight is the sum of pay load and empty weight.

Thus a tractor-trailer combination weighing 20 tons when empty (but with a full fuel complement) and with a cargo capacity of 40 tons would have a payload-to-empty-load ratio of 2.

With passenger vehicles the ratio is more often expressed as empty load or weight per seat. In Figure 5.15, for example, the six-passenger automobile has 812 lb (368.6 kg) (total vehicle weight of 2.4 tons) (1.09 tonnes) of engine, body, tires, frame, and accessories for each of the six seat spaces. The UMTAs "State-of-the-Art" rapid transit car has a ratio of 890 to 1, that is, 890 lb (404 kg) of car weight for each of 100 passengers seated and standing. If the full capacity of 300 persons, mostly standees is considered, the ratio is 297 lb (134.8 kg) of car weight per passenger.

It is evident there are several ways in which the vehicle weight–cargo capacity relation can be expressed. The values in Table 5.2 may, for example, be shown in reverse by computing the tons of empty weight per ton of payload.

The lighter the tare or dead weight of the vehicle as contrasted to pay load, the more efficient from all standpoints is the operation. Designers strive to reduce dead load by the use of light weight materials and the elimination of all nonessential equipment and gadgets. Limits to this effort are imposed by considerations of structural strength for safety, passenger comfort, and the relative costs of maintenance. Sometimes a radical change in design may be introduced to decrease dead weight, for example, tubular

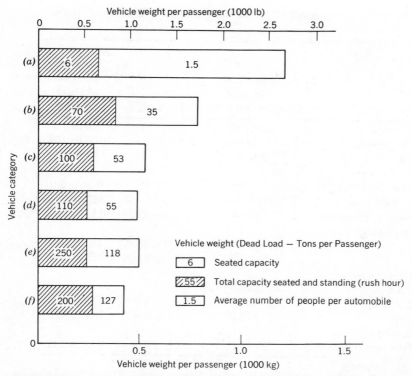

FIGURE 5.15 VEHICLE WEIGHT (DEADLOAD) PER PASSENGER SEAT IN TYPICAL URBAN AREA TRANSIT VEHICLES.

(*a*) 2-TON (1.18 TONNES) AUTOMOBILE—6 AND 1.5* PERSONS.
(*b*) 30-TON (27.21 TONNES) BUS—35 AND 70 PERSONS.
(*c*) 30-TON (27.21 TONNES) BUS—53 AND 100 PERSONS.
(*d*) 30-TON (27.21 TONNES) SUBWAY CAR—55 AND 110 PERSONS.
(*e*) 65-TON (59 TONNES) MU SUBURBAN CAR—118 AND 250 PERSONS.
(*f*) 60-TON (54.42 TONNES) MU SUBURBAN CAR—127 AND 200 PERSONS.

framing as used in aircraft and Turbotrains, and by the use of light weight materials. The problem is further complicated when fuel constitutes a large part of the load. Much of the gross lifting force of an airplane is lost in fuel carriage, especially on long-distance and overseas flights. Similarly, there is a significant relation between the cruising and cargo-carrying capacity of a ship. The greater the distance between ports, the more fuel required—and the less space available for cargo. Faster speeds consume fuel more rapidly for a given distance of travel and require more fuel space in relation to

TABLE 5.2 TONS OF PAY LOAD PER TON[a] OF EMPTY WEIGHT—TYPICAL VALUES

Carrier	Tons of Light, Tare, or Empty weight	Tons of Pay Load	Tons of Pay Load per Ton of Dead Load
Railroad freight cars	20 to 30	50 to 100	2.5 to 4.2
Railroad passenger cars	40 to 60	2.3 to 4.5	0.06 to 0.08
Highway trucks	2 to 8	4 to 20	2.1 to 2.5
Highway tractor-trailer combinations	11 to 20	10 to 40	0.91 to 2.0
Automobiles—6 persons	1.2 to 2.6	0.32 to 0.48	0.27 to 0.18
Motor buses—28 to 45 persons	5 to 16	2.10 to 3.18	0.21 to 0.42
Great Lakes bulk-cargo carriers	6,000 to 14,000	9,000 to 26,000	1.5 to 1.96
River barges	160 to 550	1,000 to 3,000	6.25 to 5.45
Towboat with 10 barges (pusher-boat weight = 500 to 800 tons	2100 to 8800	10,000 to 30,000	4.76 to 3.41
Airplane—freight	12 to 240	3.6 to 100	0.3 to 0.42
Airplane—passenger (28 to 450 passengers and cargo	12 to 240	2.1 to 100	0.18 to 0.42

[a] One ton = 0.907 tonne).

cargo or passenger space. Railroad trains and highway vehicles carry little fuel in proportion to their pay load because their runs are relatively short and it is convenient ot stop from time to time for refueling. Also, their unit propulsive resistance is low in contrast to that of ships and airplanes. Pipelines, conveyors, cableways—and electric railways—do not face this problem. All vehicle space is available for cargo. The power plant is stationary and in all cases the fuel or its equivalent electrical power is brought to the source of propulsive force rather than being carried along with the cargo. An exception to this is the situation where pipeline pumps may be operated with diesel fuel drawn from the cargo being pumped through the line. Typical ratios are given in Table 5.2.

Horsepower per Net Ton of Pay Load   The horsepower required to move 1 ton of pay load 1 mile or one passenger one mile is basic in determining capacity and cost. Horsepower costs money. A high ratio of horsepower to gross tons indicates that only a small percentage of the power goes to providing transportation, that is in moving pay load. The major percentage may go to overcoming the resistive forces of the power plant, fuel, and cargo containing space. The figures cited for this unit in Table 5.3 are based on conventional conditions, equipment, and practice.

Truck capacities (pay load or net tons) usually range from 0.50 (453.6 kg) to 50.0 tons (45.36 tonnes). About 45 percent have a gross vehicle rating of 5000 lb (2270 kg); 4 percent over 26,000 lb (11,804 kg). Horsepower ratings vary from 100 to 356. The conventional over-the-road semitrailer and tractor combination can haul about 20 to 40 net tons (18.1 to 36.2 tonnes) in one or more trailers containing approximately 1900 to 2500 ft³ of capacity per trailer. It will be pulled by a 180-to 280-hp tractor furnishing 6.4 to 7.07 hp per net ton. The usual load in a semitrailer, however, is more likely to be 10 to 15 tons (9.1 to 13.6 tonnes) giving a horsepower-per-ton ratio of 11.33. These ratios, higher than for railroads, reflect the higher propulsive resistance of highway vehicles. Heavy duty vehicles used for

TABLE 5.3 TYPICAL HORSEPOWER-PER-NET-TON RATIOS

Carrier	Horsepower[a] per Net Ton	Horsepower per Passenger
Railroads—Freight	1.25 to 2.54	
Railroads—Passenger		5 to 30
BART rapid transit car		8.3 (seated)
UMTA State-of-the-Art car		6.0 (seated)
		2.3 (with standees)
Bilevel commuter cars		1.9 (seated)
Chicago Transit Authority cars		8.0 (seated)
		2.6 (with standees)
Passenger automobile	24.0 to 92.0	6.0 to 43.0
Highway trucks and semis	6.0 to 7.0	
River tows	0.12 to 0.20	
Bulk Cargo ships	0.35 to 0.26	
Airplanes—Freight	6.67 to 240	
Airplanes—Passenger		230 to 140
Pipelines (petroleum)	2.0 to 3.0	
Belt conveyors	10.0 to 20.0	
Aerial tramways (cableways)	0.27 to 2.0	

[a] 1 horsepower = 745.6 watts.

hauling coal and ore may have a capacity of 100 or more tons (90.7 tonnes) with horsepowers of 480 to give a ratio of 4.8 hp per ton of pay load.

Passenger cars possess to a greater degree than trucks the limitation of horsepower to payload, 200 hp per ton or 15 hp per passenger for a 90-hp, six-passenger vehicle. The ratio is even greater for modern high-horse-power automobiles.

A modern freight train may consist of 5000 gross tons (4535.5 tonnes) including 2270 net tons (2059 tonnes). Assuming 5 lb/ton (2.3 kg) of train resistance at 20 mph (32.2 km/hr) and 12 lb/ton (5.5 kg/ton) grade resistance on an 0.60 percent grade, there is a unit resistance of 17 lb/ton (7.7 kg/ton) and a total resistance of 85,000 lb (38590 kg). Evaluating the horsepower equation

$$\text{hp} = \frac{R \times V}{308} = \frac{85,000 \times 20}{308} = 5519 \text{ hp}$$

Three 2000-hp units would be able to haul this train on an 0.40- to 0.60-percent grade at 20 mph, 2.64 hp per net ton. Even better perfor-mance, 1.25 hp per net ton, can be obtained with a lower speed, 15 mph, on an 0.50-percent grade when unit trains of 8000 net tons (7257 tonnes) (13,000 ± gross tons) (11,792 ± tonnes) are operated with five 2000-hp locomotive units.

The Urban Mass Transit Administration (UMTA) State-of-the-Art rapid transit car, weighing 89,000 lb (40,406 kg) empty is designed for a normal load of 100 persons. With that loading it weighs 104,000 lb (47,216 kg) and is powered by four 175 hp motors. This gives a hp per passenger ratio of 6.0. The high horsepower meets requirements for 80 mph top running speed and for a rapid acceleration rate of 3.0 mph/sec. With its full capacity of 300, seated and standing, the ratio drops to 2.00 hp per person and to 12 hp per ton.

Modern lake freighters carry as much as 40,000 tons (36,284 tonnes) of cargo (ore, coal, grain, etc), and, powered by 10,500-hp engines, have a horsepower-per-net-ton ratio of 0.26 one of the lowest. The largest oc-eangoing tankers and bulk-cargo carriers built to circumnavigate the Suez Canal haul 40,000 to 60,000 cargo tons with engines of 12,500 to 16,000 horsepower-per-net-ton ratio of 0.31 to 0.27.

River transport can combine a 2400-hp towboat with as many as 10 or more 2000-ton barges, a total of 20,000 tons (18,142 tonnes) of payload. The ensuing ratio of 0.12 hp per net ton gives an even lower relation than do the lake and oceanic carriers.

However, this heavy loading is not customary, and less favorable ratios are the rule. The slow rate of speed is also a factor. Nevertheless, even with a six-barge tow, a common loading, the combination gives 12,000 net tons for a ratio of 0.20 hp per net ton.

Fuel Economy   For a given speed, fuel consumption will be a function of horsepower, thermal efficiency, total vehicle weight, and unit propulsive resistance. Grade resistance enters not only into the costs of operating one vehicle but, more important, determines for surface vehicles the number of those required to move a stated volume of traffic. The more traffic moved by a given horsepower, quantity of fuel, and deadload the greater the fuel and dollar economy. For a given horsepower, more fuel and cost will be incurred if more speed is required (propulsive resistance increases as the square of the speed). For very high speeds larger prime movers with higher capital and maintenance costs are required. Vehicles supported by air jets or cushions (ACVs) will need additional horsepower and fuel for that support. These relations are shown diagramatically in Figure 5.16.

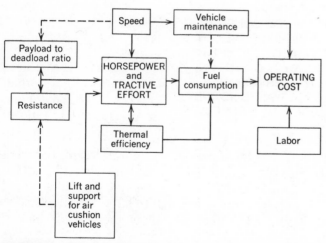

FIGURE 5.16 TECHNOLOGICAL FACTORS IN VEHICLE OPERATING COSTS.

Reducing vehicle weight increases fuel economy. Figure 5.17 shows fuel consumption as a function of vehicle weight. According to this chart, a reduction in vehicle weight from 3000 to 2000 lb will produce a BTU energy saving of about 14 percent. Thus the role of small, compact automobiles in energy conservation is indicated.

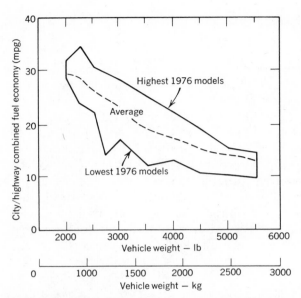

FIGURE 5.17 AUTOMOBILE FUEL CONSUMP-
TION AS A FUNCTION OF VEHICLE WEIGHT
(FROM "PASSENGER CAR FUEL ECONOMY
TRENDS THROUGH 1976" BY T.C. AUSTIN,
R.B. MICHAEL, AND G.R. SERVICE, PRE-
SENTED TO THE SOCIETY OF AUTOMOTIVE
ENGINEERS, OCTOBER 1976, P.11, FIGURE 2.)

Vehicle weight can be reduced by improved structural design, reducing capacity, and overall size, and by use of light weight metals and plastics. Tests conducted by the Environmental Protection Agency showed a spread of 28.7 mpg to less than 8 mpg for a range of small to large automobiles.[4] A second study estimated a 30-percent fuel saving by the use of smaller, lighter cars.

Resistance increases approximately with the square of the speed with corresponding increased demand for fuel. Low speeds waste fuel through incomplete combustion. The most economical speed range for automobiles is said to be in the 50 to 55 mph (80.5 to 88.5 kph) range. A reduction in speed of 10 percent would reduce fuel requirements by 20 percent. If a car were obtaining 20 mpg while cruising at 60 mph (96.5kph), its miles per gallon could be increased to 24 mpg by cruising at 54 mph (86.9 kph).

[4]"The Potential for Energy Preparedness," U.S. Office of Energy Preparedness, Washington, D.C., October 1972.

Service conditions also affect fuel economy. An automobile or bus driven in an urban area experiences idling time at stop signs and traffic lights. Excess fuel is consumed in frequent decelerating and accelerating at signs and lights or with the changes in traffic flow. Rapid transit trains experience a high energy demand because of frequent station stops and speed reductions followed by rapid acceleration. In distinguishing between fuel consumption for in-town and country driving, the EPA fuel consumption test data show a 20- to 40-percent decrease in mpg for vehicles driven in town.

Other factors enter into fuel consumption and operating costs—route design, capacity, schedules, and state of maintenance. The use of air conditioners reduces fuel economy by 10 percent and automatic transmission by the same amount.[5] Radial tires and streamlining reduce fuel demand.

Fuel saving per unit of traffic can be more significant than per vehicle or per mile. The fuel consumed per passenger by a normal commuter-use automobile with an average occupancy of 1.5 persons is four times what it would be if the full capacity of six passengers were utilized; hence the virtue in car-pooling. The average fuel consumption per person in a crowded rapid transit train can be very low indeed during rush hours, but can rise sharply when the train runs lightly loaded during off-peak hours. The proper mix of modes to combine adequate service with necessary fuel conservation is a sharply debated problem. The reader is referred to *Suggested Readings*, especially 14, 15, and 16 at the end of this chapter for more extended discussions. The further significance of operating cost is discussed in Chapter 12.

QUESTIONS FOR STUDY

1. Using thermal and mechanical efficiency values as given in this chapter, develop the overall thermal efficiencies of pipeline, a conveyor, and a cableway.
2. A 2400-hp diesel-electric locomotive weighs 130 tons and has a unit tractive resistance of 5 lb/ton at 20 mph on tangent level track. How many 60-ton cars can this locomotive haul on tangent level track? Compare its level track performances with what it could do on an 0.80-percent grade.
3. Using the locomotive of question 2, plot a curve showing the relation between horsepower per ton and gradient on $\frac{1}{2}$, 1, 2, and 3 percent grades.
4. A motor truck, weighing 5 tons empty, develops a horsepower of 180 at 1440 rpm. Gear ratios are low, 6.00 to 1, second, 3.00 to 1, third, 1.70 to 1, fourth, 1 to 1, and the differential ratio is 5.90 to 1. Tires are 32×6 in., depressing an

[5]"The Potential for Energy Preparedness," U. S. Office of Preparedness, Washington, D. C., October 1972.

average of 0.40 in. Cross-sectional area of the truck is 80 ft. The truck is on a smooth concrete raodway. Determine (a) the available net tractive effort for hauling pay load and the speed in high gear when empty, (b) the pay load at that speed, and (c) the effect on pay load of a 5-percent grade.

5. Refer to the ship in question 2, Chapter 4. What propeller horsepower and what shaft horsepower are required to drive the ship?

6. Refer to Question 8 of Chapter 4. What pump horsepower and what prime-mover horsepower are required if the pump is a three-pistoned, double-acting reciprocating machine?

7. Refer to Question 9 of Chapter 4. What horsepower will be required for the two belt speeds if the belt is (a) level throughout, and (b) on a 12-degree slope throughout?

8. Compute the theoretical fuel consumption per mile en toto and per ton for
   (a) The train of Problem 2 on level track and on the 0.80-percent grade.
   (b) The truck in Problem 4b.
   What practical factors will make actual consumption differ from these theoretical values?

9. Refer to Question 3 of Chapter 4 and plot the shaft horsepower requirements for the airplane flying at 10,000 ft altitude at speeds of 200, 250, and 300 mph, assuming a propeller efficiency of 80 percent.

10. A train approaching a 1.0 percent grade 4000 ft long is able to surmount the summit at 10 mph if it first stops at the foot of the grade. At what speed can the train cross the summit if it is traveling 50 mph at the foot of the grade?

11. What are the possible effects on fuel conservation and overall operating costs of a 55 mph speed limit on the nation's motor freight carriers?

SUGGESTED READINGS

1. R. P. Johnson, *The Steam Locomotive*, Simmons-Boardman, New York, 1945.
2. Charles F. Fowll and M. E. Thompson, *Diesel-Electric Locomotive*, Diesel Publications, New York, 1946.
3. W. W. Hay, *Railroad Engineering*, Vol. I, Wiley, New York, 1953.
4. A. M. Wellington, *The Economic Theory of the Location of Railways*, Sixth Edition, Wiley, New York, 1887, Chapters IX, X, XIV, XX.
5. Wilbur G. Hudson, *Conveyors and Related Equipment*, 3rd edition, Wiley, New York, 1954, Chapters 10 and 119
6. F. V. Hetzel and Russel K. Albright, *Belt Conveyors and Belt Elevators*, 3rd edition, Section I, Belt Conveyors, Wiley, New York, 1941
7. John Walker Barriger, *Super-Railroads*, Simmons-Boardman, New York, 1656, Chapter II.
8. Noël Mosert, *Supership*, Knopf, New York, 1974.
9. *Vehicle Characteristics*, REC 344, Transportation Research Board, Washington, D. C., 1971.
10. Berry, Blomme, Shuldiner, and Jones, *Technology of Urban Transportation*, Northwestern University Press, Evanston, Illnois, 1963.

11. A. S. Lang and R. M. Soberman, *Urban Rail Transit*, Joint Center for Urban Studies, The M.I.T. Press, Cambridge, Massachusetts, 1964.
12. *Urban Rapid Transit: Concepts and Evaluation*, Carnegie-Mellon University, Transportation Research Institure, Research Report No. 1, Pittsburgh, Pennsylvania, 1968.
13. *Big Load Afloat*; U. S. Domestic Water Transportation Resources, publishes by The American Waterways Operators, Inc., Washington, D. C., 1973.
14. *The Potential for Energy Conservation*, United States Office of Emergency Preparedness, Washington, D. C., October 1972.
15. *The Role of the U. S. Railroads in Meeting the Nation's Energy Requirements*, Proceedings of a Conference sponsored by the U. S. Department of Transportation, the Federal Railroad Adminstration, and the Wisconsin Department of Transportation, published by the Graduate School of Business, University of Wisconsin, Madison, Wisconsin, October 1974.
16. A. C Malliaris and R. L. Strombotne, *Demand for Energy by the Transportation Sector and Opportunities for Energy Conservation*, presented at a "Conference on Energy—Demand, Conservation, and Institutional Problems," Massachusetts Institute of Technology, Cambridge, Massachusetts, 12.14 February 1970.
17. E. Hirst, *Energy Intensiveness of Passenger and Freight Transport Modes*, 1950-1970, Oak Ridge National Laboratory Report ORNL-NSF-EP-44, April 1973.

Chapter 6

Roadway

The roadway is one of the five major elements in transportation technological systems. Roadways are related to vehicle and motive power design through vehicle loads and by the grades and curvatures that place constraints on vehicle size, speed, and tractive effort. Roadways are also related to operational control through route capacity, guidance, and vehicular separation.

ROADWAY FUNCTIONS

The land area devoted to the route and facilities of railroads, highways, canals, pipelines, and conveyors is the right of way. Natural waterways have their own basins and river beds. Pipelines are laid within the earth itself, preferably below the frost line. Airways are related to the terrain only by visual or radio-range route markers and at airport runways and taxiways.

Roadways give all-weather support to the vehicles, facilitate drainage, provide frictional adhesion for acceleration, deceleration, and cornering and, through geometric design of widths, intersections, side slopes, drainage, sight distances, etc., provide for movement and passing with safety at established levels of service.

Guidance Railway tracks, pipelines, conveyors, and center or side beams give direct vehicle or cargo guidance. The term *guideway* is sometimes substituted for roadway to indicate these functions.

180

Support A principal function of pavement and track structures is to bear and distribute wheel loads to within the bearing capacity of the subgrade soil. Some deformation or deflection, nevertheless, will occur. The success of the load-distributing action and the probable life of the pavement or track structure are functions in part of the amount of deflection of the superstructure and of the deformation or distress of the subgrade. In some design systems subgrade failure is indicated by the deflection exceeding a maximum allowable amount. Track distress can arise from deflections of 0.10 to 0.40 in. (0.3 to 1.0 cm). Pavement distress can occur with deflections 0.01 in. (0.03 cm); limiting values of 0.10 to 0.20 in. (0.3 to 0.6 cm) are frequently used.

The pavement-track cannot be analyzed in isolation from the vehicle. The two form a roadway–vehicle system with interaction between the vehicle and its support. This concept includes the simple relations of vehicle wheel arrangement vs. load distribution, dynamic impact effects, and the more elaborate studies now being conducted in track–train dynamics. Random vibration concepts and optimization of construction and maintenance of pavements take account of pavement roughness, speed, shock and vibratory frequency of the vehicle. Track and pavement loads are usually much greater than the bearing capacity of the subgrade material. See Table 6.1.

Subgrade A subgrade serves (a) to bear and distribute the imposed loads with diminished unit pressures, (b) to facilitate drainage, and (c) to provide a smooth platform conforming to the established grades on which the running structure can be laid. Wheel loads must be supported with a minimum of elastic and plastic deformation that shortens the life of the roadway structures and gives a rough ride to vehicles and their contents. Few soils possess such a degree of resistance and permanence. The imposed wheel loads will usually exceed the 5- to 50-psi (34.475- to 344.750-MPa) bearing capacity of most soils. Because the passage of wheels would abrade or rut the subgrade surface, a bearing course or cover is laid over the subgrade. Track and ballast fulfill this function for railroads; for highways and airport runways and taxiways, some form of pavement is used. Discussion of track and ballast are deferred until later.

The load-bearing capacity or stability of a subgrade is a function of the properties and characteristics of constituent soils, adequacy of drainage, loads to be imposed, depth and intensity of load distribution, and type of subgrade cover. Load intensity can be reduced through design of the subgrade cover, but the generally lower costs of subgrade construction warrant that considerable effort be expended there to increase its load-bearing capabilities.

TABLE 6.1. TYPICAL WHEEL LOADINGS FOR VARIOUS MODES OF TRANSPORT

Transport Vehicle		Total Wheel Load (lb)	Area of Contact (in.²)	Unit Load on Rail or Pavement (psi)
Locomotives				
Diesel-electric				
Road engines		25,000	0.3 ± in.²	83,333
		30,000		100,000
Switchers		10,000	0.3 ± in.²	33,333
		20,000		66,667
Railroad cars (loaded)		26,000	0.3 ± in.²	86,667
		38,000		125,000
Automobiles (loaded)		1,000	20 ± in.²	50
Trucks		4,000	60 ± in.²	67
Heavy pneumatic tire loads[a]		8,000	100 ± in.²	80
Single tires				
Gross load per wheel		60,000	527 in.²	114
		100,000	798 in.²	125
		140,000	1080 in.²	130
Dual tires				
Gross load	60,000	30,000	325 in.²	92
	100,000	50,000	464 in.²	108
	140,000	70,000	612 in.²	114

[a]H. O. Sharp, G. R. Shaw, and J. A. Dunlop, *Airport Engineering*, Wiley, New York, 1944, p. 71.

THEORIES OF LOAD DISTRIBUTION

If a soil mass is loaded with a tire, railroad tie, or a pavement slab there is a tendency for the soil to be depressed and even forced upward around the loaded area. If the depressed soil (subgrade) rebounds after the load is removed, the soil has experienced an elastic deformation, but if it stays depressed and rutted, plastic deformation (failure) has occurred (Figure 6.1).

Distribution of Pressure   The intensity of pressure from a load can be shown to decrease with depth and to be distributed over a plane at a given

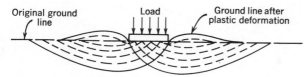

FIGURE 6.1 SUBGRADE FAILURE CONCEPT;
PLASTIC CONDITION.

depth with an intensity variation having a normal or Gaussian bell-shaped frequency distribution. Pavement deflection can be directly related to Hooke's law that says stress is proportional strain, that is, $S = E\sigma$ where S is the stress, σ the strain, and E is the modulus of elasticity of the material. The Talbot analysis of track expresses the relation initially as $p = uy$ where p is the load, y is the deflection, and u is the modulus of track elasticity or track support stiffness.

Flexible pavements are usually multilayered containing at least a bearing surface, a base course, and, if subgrade soils are weak, a subbase above the subgrade. A similar situation exists for track in the rail-tie-ballast-subballast system. One must either settle for the inaccuracies of an average stiffness modulus E (a lumped parameter) or attempt to develop individual moduli for each layer of the system. This problem has brought forth the elastic layered system approach of Burmister, the shear layer theory of Barenberg, finite element analyses, and others.

With Boussinesq's application of Hooke's law, the deflection at a given depth z can be calculated for a given load. If only a point load is considered, the Boussinesq relation would be

$$\sigma = K\frac{P}{z^2}$$

where σ_z is the vertical stress in psi, P is the point load in lb and z is the depth below the point of load application. In this equation

$$K = \frac{3}{\pi}\frac{1}{\left[1+(r/z)^2\right]^{5/2}}$$

where r is the radial distance from the point of load application. See Figure 6.2.

For stress on a vertical plane passing through the center of a loaded plate,

$$\sigma_z = p\left[1 - \frac{z^3}{\left(r^2+z^2\right)^{5/2}}\right]$$

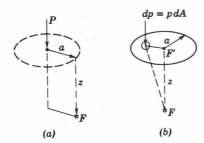

FIGURE 6.2 SUBGRADE PRESSURE UNDER A LOADED AREA ACCORDING TO BOSSENESQ. (*a*) POINT LOAD. (*b*) LOADED CIRCULAR PLATE.

where p is the unit load on a circular plate of radius r (or of a tire of known contact area and pressure).

Poisson's ratio μ is the ratio of strain normal to the applied stress to the strain parallel to that stress. It is usually taken as 0.50 for soil. If we combine equations for vertical stress on a vertical plane through the center of the plate and radial strains dependent on the 0.5 value for Poisson's ratio, the Boussinesq equation for deflection at the center of a circular plate becomes

$$\Delta = \frac{3p\,(r^2)}{2E\,(r^2 + z^2)^{1/2}}$$

which may be rewritten $\Delta = (p(a)/E)F$ where

$$F = \left(\frac{3}{2}\right) \frac{1}{\left[1 + (r/z)^2\right]^{1/2}}$$

F is a term that reflects the depth-radius ratio. With a flexible plate F has a value of 1.5 when taken at the contact surface where $z=0$ whereby

$$\Delta = 1.5 \frac{pr}{E}$$

For a rigid plate $F = 1.18(pr/E)$. Values of F for various z/r combinations may be computed to facilitate the computation of deflection or modulus for a given plate loading. By measuring the deflection under a known load and contact pattern, the modulus of elasticity of a soil or a pavement layer can be computed. The classic method assumes the subgrade to be a heavy liquid in which the subgrade reaction is a linear function of the deformation. A plate with a radius of 30 in. is usually used for these determinations. The application of the rigid plate deflection procedure to rigid-type pavements by Westergaard is discussed in the rigid pavement section. The load (tire pressure) and radius of a tire contact area may also be used to determine direct effects.

Talbot's Equation   Dr. A. N. Talbot and his Committee on Stresses in Track developed an empirical relation

$$p_c = \frac{16.8 p_a}{h^{1.25}}$$

where p_c is the pressure in psi at any depth h in inches below the center of a railroad cross tie under the rail and p_a is the average unit load over that area of the tie face in compacted contact with the ballast. For the pressure at any point x inches to the right or left of the center of tie under the rail,

$$p_x = \frac{16.8 p_a}{h^{1.25}} (10)^m$$

where

$$m = -6.05 \left(\frac{x^2}{h^{2.5}} \right)$$

These expressions are reasonably accurate for depths of 4 to 30 in. (10.2 to

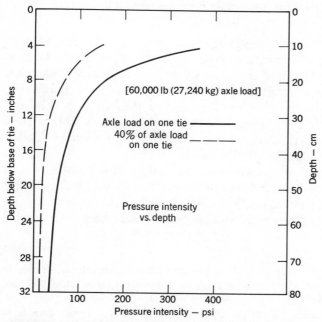

FIGURE 6.3 DISTRIBUTION OF LOAD WITH VARIA-TIONS IN DEPTH USING TALBOT'S EQUATIONS.

76.2 cm) below the tie.[1] Figure 6.3 shows the reduction in pressures beneath the track under static loads of a two-axle railroad car truck with a load of 60 kips (27,240 kg) per axle.

Newmark charts permit analysis at greater depth using a graphical application of Boussinesq.[2]

THE ROADWAY STRUCTURE

Because of the heavy wheel loads of modern vehicles and the relatively low bearing capacity of subgrade soils, there must be an intermediate element between the actual load and the subgrade.

Pavements   For highways and airport runways and taxiways, a pavement acts as subgrade cover, performing several functions.

(a) It bears and distributes the load with sufficiently diminished unit pressure to be within the bearing capacity of the subgrade soil, thereby reducing rutting tendencies.
(b) The pavement waterproofs the roadway surface by draining moisture away from the load-bearing areas and the subgrade.
(c) Wheel abrasion on subgrade materials is reduced or eliminated.

Pavement Types   Pavement types range from nearly rigid concrete slabs laid directly on the subgrade to various types of single or multilayer flexible pavements to the mere placing of the most select (sand or gravel) materials in the upper levels of a subgrade where load intensity is maximum. Highway pavements are often classified as *rigid* or *flexible* with sub- and intermediate groupings. The difference between rigid and flexible is only one of degree. The most rigid has some flexibility and many so-called flexible pavements approach the rigidity of concrete.

The low flexibility of rigid pavements distributes wheel loads over a broad subgrade area. Thus small nonuniformities in the subgrade bearing capacity are not unduly significant. The slabs may be laid directly on the subgrade, but in heavy modern construction one or more base courses will probably be underneath.

[1]A. N. Talbot, Second Progress Report of Special Committee to Report on Stresses in Railroad Track, *Proceedings of the American Railway Engineering Association,* Vol. 21, 1920, Chicago, Illinois, pp. 645–814.
[2]Nathan M. Newmark, *Influence Charts for Computation of Vertical Displacements in Elastic Foundations,* University of Illinois Engineering Experiment Station Bulletin 367, Urbana, Illinois, 1947.

Flexible pavements use a relatively thin, wearing surface that in some low grade roads is placed over a thin base of gravel or broken stone laid on the subgrade or, for higher grade roads, with one or more subbase courses. Where an "open" type of base or subbase material is used, a filter blanket of graded material (or one of the newly developed fabric covers) may be interposed between the lower course and the subgrade to reduce the capillary rise of moisture and the interpenetration of subgrade and base course materials.

The well-known *macadam* road, a flexible type, is essentially one or more base courses of broken aggregate with a top dressing of finer particles, such as limestone screenings, formed into a hard-wearing surface by binding with water, limestone dust, or bitumens.

FLEXIBLE PAVEMENT DESIGN

Design procedures range from empirical methods that relate thickness to a few index properties of the support system materials to mathematical analyses that require great detail about the complex nature of the materials and the environment in which they are used. The simpler methods seem to prevail, partly because of their simplicity and partly because of the difficulties in securing reliable numbers for the more complex evaluations.

Pavement Thickness   A primary problem is determining the required thickness of base and subbase courses for a given combination of materials, loading, and environment to provide needed bearing strength. The amount of deflection a pavement experiences is a measure of its probable life and load-bearing capacity. It is a function of the load, of the support capacity of the subgrade, and of the load-distributing capability of the pavement. A greatly simplified illustration of the factor of pavement thickness in relation to load distribution assumes the load from each wheel to be distributed in the form of a cone with a slope of approximately 45 degrees. A uniform subgrade reaction (bearing capacity) has a value of p psi, Figure 6.4. Flattening of the tire spreads the load over a small area, assumed to be a circle, for passenger cars and light trucks, having a radius r. A conservative value for r is the nominal tire width divided by 4. Equating the imposed load W to the subgrade support at the base of the cone, $W = \pi(t+r)^2 p$ and $t = 0.546 W/p$.[3] For the heavier wheel loads of large trucks and airplanes, the area of contact is assumed to be an ellipse with a width approximately equal to the nominal width of a tire. The length of the

[3]A. H. Oakley Sharp, G. Reed Shaw, and John A. Dunlop, *Airport Engineering*, Wiley, New York, 1944, p. 67.

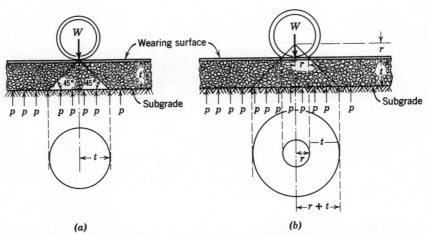

(a)   (b)

FIGURE 6.4 CONE THEORY OF WHEEL LOAD DISTRIBUTION (*a*) TIRE ASSUMED UNFLATTENED; (*b*) TIRE ASSUMED FLATTENED.

ellipse may be computed by assuming the actual load = *inflation pressure* × *area of the ellipse* = *inflation pressure* × πab where a is the long radius of the ellipse of contact, b the short radius. As before, $W = \pi(a+t)(b+t)p$ and

$$t = \sqrt{\frac{W}{\pi p} - ab + \left(\frac{a+b}{2}\right)^2} - \left(\frac{a+b}{2}\right)$$

California Bearing Ratio   A necessary factor in the design procedure for flexible pavements, especially for the AASHTO method, is the load-bearing capability of a soil or soil layer. The CBR test is frequently used for that purpose. In the test, the penetration of a standard sized 3 in.^{2} (19.38 cm^2) piston into a prepared sample is observed. The ratio of the loading causing an 0.10-in. (0.254-cm) penetration to the load required for the same penetration into a sample of high quality crushed stone (CBR = 100) is the CBR for the material being evaluated.

AASHTO Method   The AASHTO procedure is applicable to the design of a flexible pavement with surface, base, and subbase construction. It is derived from experience obtained from a series of full-scale tests conducted near Ottawa, Illinois on a road loop containing various types of flexible pavements. The tests ended in 1960 and were reported in 1962.

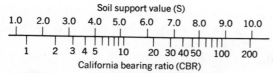

Soil support value (S)

California bearing ratio (CBR)

FIGURE 6.5 CORRELATION BETWEEN SOIL SUPPORT VALUE, *S*, AND CALIFORNIA BEARING RATIO (*CBR*), UTAH DEPARTMENT OF HIGHWAYS. (COURTESY OF ASSOCIATION OF AMERICAN STATE HIGHWAY AND TRANSPORTATION OFFICIALS, WASHINGTON, D.C., *INTERIM GUIDE FOR DESIGN OF CONCRETE PAVEMENTS*, FIGURE C.3-1, P. 68, 1972.)

Out of that report has come an "Interim Guide" that adapts the test findings to pavement design.[4]

Because the load concentration is greatest near the pavement surface where the load is applied, the highest quality layers are placed near that surface. Strength arises not from bending strength in the pavement slab (as with rigid pavements) but rather by building up layers to distribute load over the subgrade.

The procedure involves determining the total thickness of the pavement structure and the thickness of base, subbase, and cover courses. The design is made according to a selected level of serviceability expressed as a *serviceability index*. The level of serviceability represents the amount of wear and deterioration to be permitted in a pavement before surfacing or reconstruction must be performed. It is based on the smoothness of ride vs. rutting, cracking, and other surface irregularities. The index has values ranging from 0 to 1 (very poor) to 4 to 5 (very good). A value of 2.5 (fair) is usually taken for major highways and 2.0 (lower limit of the 2 to 3 range) for lesser highways.

A measure of soil strength is required. It is convenient to convert CBR values into *soil support values* using a correlation chart similar to that of Figure 6.5. The daily traffic in equivalent 18 kip single axle loads is developed as explained earlier. Other needed data include the *structural number SN* and the *regional factor*. The dimensionless structural number *SN* expresses pavement strength in terms of the soil support value, daily

[4]*AASHTO Interim Guide for Design of Concrete Pavements.* American Association of State Highway and Transportation Officials, Washington, D.C., 1972.

equivalent 18-kip single axle loads serviceability index, and the regional factor. Appropriate coefficients convert the SN value into the actual thicknesses of the surface, base, and subbase courses.

The *regional factor* relates the foregoing structural number to local conditions of climate and other environmental conditions such as gradient rainfall, frost penetration, temperatures, groundwater table. The selection of a suitable regional factor is largely based on judgment. As a guide, 0.2 to 1.0 is used where roadbed materials freeze to depths of 5 in. (12.7 cm) or more, 0.3 to 1.5 for materials in dry summer and fall, and 4.0 to 5.0 for roadbed materials in spring thaw.

In practice the foregoing are conveniently related by a nomograph such as those of Figures 6.6 and 6.7. Using a straight edge one enters the graph with the soil support value and daily equivalent 18-kip axle load applications to obtain the unweighted structural number, SN. With the un-

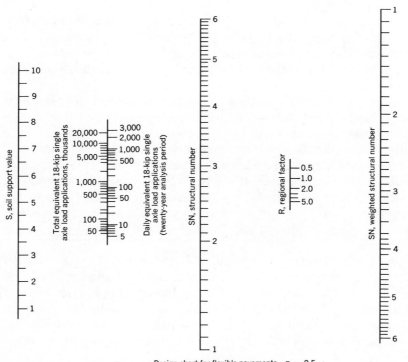

Design chart for flexible pavements, $p_t = 2.5$

FIGURE 6.6 DESIGN CHART FOR FLEXIBLE PAVEMENTS, $P_T = 2.5$. (COURTESY OF ASSOCIATION OF AMERICAN STATE HIGHWAY AND TRANSPORTATION OFFICIALS, WASHINGTON, D.C., *INTERIM GUIDE FOR DESIGN OF CONCRETE PAVEMENTS,* FIGURE II-1, P. 19, 1972.)

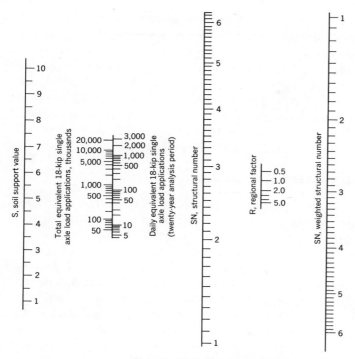

Design chart for flexible pavements, $p_t = 2.0$

FIGURE 6.7 DESIGN CHART FOR FLEXIBLE PAVEMENTS, $P_T=2.0$. (COURTESY OF ASSOCIATION OF AMERICAN STATE HIGHWAY AND TRANSPORTATION OFFICIALS, WASHINGTON, D.C., *INTERIM GUIDE FOR DESIGN OF CONCRETE PAVEMENTS*, FIGURE II-2, P. 20, 1972.)

weighted structural number and a selected regional factor, a second application of the straight edge gives the weighted structural number.

The *SN* value for the entire pavement is related to the individual layers through the equation

$$SN = a_1 D_d + a_2 D_2 + a_3 D_3$$

where

$SN =$ the structural number

$a_1, a_2, a_3 =$ relative strength coefficients assigned respectively to the surface, base, and subbase courses

$D_1, D_2, D_3 =$ thickness in inches of courses 1, 2, and 3

The foregoing equation must be satisfied by combinations of thicknesses and characteristics of materials. The selection is based on availability of materials and the relative economics of different combinations. Recommended practical minimums have been established as 2 in. (5.08 cm) for surface courses, 4 in. (10.16 cm) for base courses, and 4 in. (10.16 cm) for subbase courses when used. Average suggested values of a are 0.20 to 0.44 for asphaltic concrete surface courses, 0.14 for crushed stone base courses, 0.07 to 0.11 for sandy gravel, and 0.05 to 0.10 for sandy clay base and subbase courses. Actual materials may vary from these rather widely.

RIGID-SURFACED PAVEMENTS

Heavy-duty rigid pavements include asphaltic concretes and Portland cement concretes. The asphaltic concretes consist of well-graded aggregates mixed either before, during, or after laying with bituminous oils. Strength is obtained by control of aggregate quality and number and thickness of base courses. Portland cement concrete may be laid directly on the rolled and compacted subgrade surface or placed as a wearing surface over one or more base courses. Asphaltic concrete may be similarly placed. Portland cement base courses are sometimes combined with asphaltic concrete wearing surfaces. See Figure 6.8a and 6.8b.

Portland cement concrete pavement is subject to a variety of stresses caused by the nature of concrete as a material. Concrete is high in compressive strength and low in tensile strength, with a resultant low beam or flexural strength. Concrete expands and contracts as it is wet or dry. Thus in curing or setting contraction occurs. It expands as temperature increases, contracts as temperature decreases. Often the stresses of temperature increases are offset by an opposite stress resulting from the drying effect.

Abrasive Stress   Abrasive stress is caused by rolling wheels on the wearing surface. Although there is no reliable measure of abrasive stress, experience indicates a relation to compressive strength. Design usually calls for a compressive strength of 4000 to 4500 psi (27576 to 31023 MPa) in 28 days, using a water-cement ratio of 6 gal of water to 1 bag of cement. Abrasive stress is not considered a problem with modern pneumatic ties.

Direct Compression and Shear   These conditions result from wheel loads. Concrete pavement is resistive to relatively high compressive loads of 4000 to 8000 psi (27576 to 55152 MPa). Wheel loads are limited in many states to a maximum of 9000 lb (4086 Kg), although a few states in the East permit as much as 11,200 lb (5085 Kg). An average impact factor of 1.5 is

Standard design 24'(7.32 m) portland cement concrete pavement with 10' shoulders

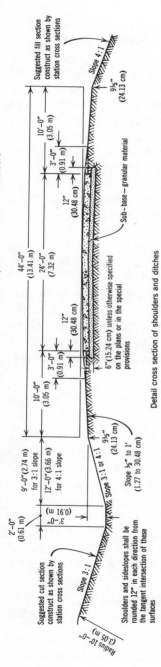

Detail cross section of shoulders and ditches

FIGURE 6.8a TYPICAL PAVEMENT CROSS SECTION, TWO-LANE. (COURTESY OF ILLINOIS STATE HIGHWAY DEPARTMENT.)

Standard design for dual portland cement concrete pavement with 40' (12.19 m) depressed median (Interstate highways)

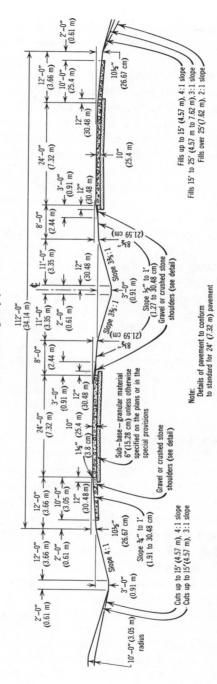

Note:
Details of pavement to conform to standard for 24' (7.32 m) pavement

Fills up to 15' (4.57 m), 4:1 slope
Fills 15' to 25' (4.57 m to 7.62 m), 3:1 slope
Fills over 25'(7.62 m), 2:1 slope

FIGURE 6.8b TYPICAL PAVEMENT CROSS SECTIONS; DIVIDED LANES. (COURTESY OF ILLINOIS STATE HIGHWAY DEPARTMENT.)

commonly used in design, although the range is 1.25 to 2.00. Failures of highway slabs in direct shear and compression have been relatively few.

Bending Stresses   These stresses are caused by *flexure* of the pavement under wheel loads and are far more significant than the foregoing. The term "rigid pavement" implies a resistance to deflection or bending where subgrade support is inadequate. Actually, bending and deflection do occur. In 1925 the late H. M. Westergaard published the results of theoretical studies in which he assumed that the slab acted as an elastic plate, elastically and continuously supported by the subgrade.[5] He also assumed that vertical subgrade reactions were directly proportional to the slab deflections and were related to them by the modulus of subgrade reaction, k, expressed in pounds per square inch per inch of deflection. (Note that Westergaard's k modulus differs from Talbot's modulus of track elasticity u, in that u is expressed in pounds per inch of track per inch of deflection, a linear rather than an area index.) The subgrade modulus thus reflects both the stiffness of the subgrade and stiffness of the slab.

Westergaard considered the effects of loads placed at three critical positions on slabs of uniform thickness: the interior, the edge, and the corner of the slab. For these slabs he found that the maximum unit stress occurred at the corners or edges rather than in the interior. He established, empirically, a measure of the relative stiffness of the slab in relation to that of the subgrade

$$l = \sqrt[4]{Et^3 / \left[12(1 - u^2)k \right]}$$

where l = radius of relative stiffness in inches, a measure of stiffness of slab in relation to subgrade stiffness, t = slab thickness in inches, E = modulus of elasticity of concrete in pounds per square inch, conservatively taken as 5×10^6 psi, u = the Poisson ratio for concrete, varying between 0.10 and 0.20 but usually taken in design as 0.15, and k = subgrade modulus in pounds per square inch per inch of deflection. The subgrade modulus can be determined by test-loading a circular plate of 30-in. (76.2-cm) diameter. Values of k vary from 50 psi for poor subgrades to 700 psi (4825.8 MPa) for very stiff subgrades. A value of 100 psi (689.4 MPa) for general use has been recommended by E. F. Kelly.[6]

Empirical formulas, modifying the Westergaard equations, have been developed by the Bureau of Public Roads. Typical of these is the following:

[5]*Public Roads,* April 1926, and *Proceedings of the Highway Research Board,* Part I, 1925.
[6]E. F. Kelly, Application of the Results of Research to the Structural Design of Concrete Pavements, *Public Roads,* July and August 1939.

$\sigma = (3P/t^3)[1-(a\sqrt{2/l})^{1.2}]$, where σ = maximum tensile stress in pounds per square inch produced by a load P at the corner of the slab, P = load in pounds including an allowance for impact, t = slab thickness in inches, l = radius of relative stiffness in inches, and a = radius of area of load (tire deformation) in square inches. Westergaard's studies stand in the same relation to stresses in pavements as Talbot's to stresses in railroad track.

Compressive Stresses   Compressive stresses arise from contraction in the cooling and drying processes as concrete sets. Contraction in the slab is resisted by frictional contact with the subgrade, and cracks occur. Goldbeck equated these forces in the equation $L/2 \times W_s \times b \times f = 12btS + SaEs/E_c$ (Figure 6.9) where W_s = weight of slab in pounds per square foot of slab area, S = allowable tensile stress in concrete in pounds per square inch, usually taken as 30 psi, f = coefficient of friction between slab and subgrade with an average value of 2.0, b = breadth of slab in feet, L = slab length in feet, E_s = modulus of elasticity of reinforcing steel = 30×10^6, E_c = modulus of elasticity of concrete = 5×10^6, and E_s/E_c = 6 (approximately). Also, a = cross-sectional area of reinforcing steel in square inches, and S = working stress in reinforcing steel, 20,000 psi being an average value. Use of the Goldbeck formula indicates that cracks will occur on the average about every 30 ft (9.1 m) of pavement length. Such cracks can be anticipated by introducing contraction joints at 20- to 35-ft (6.1 to 10.7 m) intervals. Some designers permit the cracks to occur naturally, thus forming approximately 30-ft (9.1-m) slabs and then filling the cracks as they

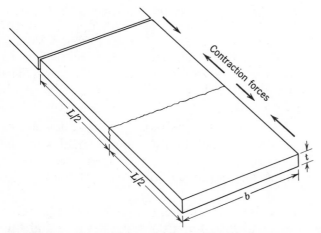

FIGURE 6.9 RESISTANCE TO MOVEMENT OF A CONCRETE SLAB.

open with asphaltic compounds. However, one body of opinion holds that contraction joints are necessary to prevent blowups caused by direct compressive stress.

Thermal, Tensile, and Compressive Stresses   These stresses, due to changes in temperature, cause variations in slab length and longitudinal movement of the slab relative to the subgrade. Cracks will ensue unless provision is made for expansion and contraction joints. Such thermal joints are usually spaced every 90 to 100 ft, a multiple of contraction-joint spacing, replacing the contraction joints at those locations. While low temperatures may cause tensile stresses that contribute to transverse cracking, high temperatures contribute to expansion and compressive stresses. On very hot days pavement failure or blowups can occur.

Studies made with continuous steel reinforcing have shown that only 500 to 1000 ft at each end of a long slab move over the subgrade with changes in temperature and with no movement of intervening slab. The end portions have thus developed enough frictional resistance with the subgrade to restrain the central portion. This is similar to the way continuous welded rail is restrained against temperature stresses by anchoring the rail through cross ties to the ballast.

More serious than longitudinal expansion and contraction is the warping or curling about a longitudinal axis as the upper and lower surfaces of the slab warm and cool. As the sun heats the upper surface, there is a tendency for the outer edges to curl downward because of the cooler shrinking effect of the underside of the slab working in conjunction with the warm and expanding upper surface. The opposite effect occurs at night when the base courses and underside may be warmer than the upper surface. Temperature differentials caused by changing seasons produce similar effects in which the underside of the slab changes its temperature more slowly than the upper. Thermal stresses may be as high as 200 psi (1379 MPa). Compensation is afforded by thickened edges on the slab, which react less to temperature stresses, thus giving rise to 9-7$\frac{1}{2}$-9-, 10-7$\frac{1}{2}$-10-, 8-6$\frac{1}{2}$-8-in., etc., thickness of interiors and edges of slabs for various classes of highways. Longitudinal joints also aid in relieving thermal stresses. Westergaard developed empirical formulas, not reproduced here, setting forth temperature-warping stresses at edges and interiors of slabs.

Axle Loads   Design factors include not only soil characteristics but also the wheel loads and the number of load repetitions. Most states and the federal government have limited single axle loads to 18,000 lb (18 kips)

TABLE 6.2 CONVERSION FACTORS: SINGLE- AND
TANDEM-AXLE LOADS TO EQUIVALENT 18-KIP
(8172-KG) LOADS[a]

Loads			
Kips	Kilograms	Single Axle	Tandem Axle
2	908	0.0004	
6	2,724	0.01	
10	4,540	0.09	0.01
18	8,172	1.00	0.08
22	9,988	2.21	0.20
26	11,804	4.34	0.38
30	13,620	7.87	0.67
34	15,436	13.46	1.10
38	17,252	21.91	1.70
40	18,160	27.53	2.08

[a]Based on AASHTO Test Road findings. The equivalent 18-kip
(8172-kg) single-axle load is had by multiplying the actual load
by the corresponding conversion factor.

(8172 kg) and to an equivalent 32,000 lb (32 kips) (14528 kg) for tandem
axles.[7] From AASHTO tests the single axle load equivalent is taken as
$0.57 \times$ the tandem axle load. From traffic surveys or estimates, the number
of daily one-direction axle loads of various weights are obtained. These are
multiplied by the appropriate equivalency factor to obtain the equivalent
number of 18-kip axle loads per day for the design lane, that is,

$$W_e = (W_a \times T_e)$$

where

W_e = equivalent daily 18-kip single axle loads

W_a = actual axle load

T_e = 18-kip traffic equivalence factor

Various equivalency charts and tables have been prepared for T_e. Table 6.2
gives the equivalence factors for single- and dual-axle vehicles based on
AASHTO test results.

[7]The Congress in early 1975 passed legislation that permits single axle loads to be increased
to 20 kips (9080 kg) and tandem axle loads to 30 kips (13620 kg) on federally supported
highways. This is permission only. Individual states must act to incorporate these higher
limitations into their state requirements.

Geometric Design   Roadway design must include, of course, factors in addition to stability. Geometric standards for curvature, sight distance, profile, and safety must be considered. The reader is referred to Chapter 17 where geometric design is discussed and to Chapter 8 for capacity design factors.

AIRPORT PAVEMENTS

The foregoing materials apply to airport runways and taxiways but with additional factors involved. One major difference is in width. Landing strips range in width from 250 to 500 ft (76.2 to 152.4 m) depending on airport classification and size of the aircraft. The paved runway portion is normally 75 to 150 ft (22.9 to 45.7 m) wide. This great width necessitates crowning of the pavement for drainage in contrast to highway pavements that can be sloped to assist runoff.

The total weights and wheel loads are greater for aircraft than for trucks. A truck will have 18-kip (8172-kg) axle loading or 9 kips (4086 kg) on a set of dual wheels, whereas the larger aircraft will have wheel loads of 100 kips or more. Tire pressures for trucks are in the 60 to 90 psi (414 to 620 MPa) range but up to 200 psi (1379 MPa) for aircraft. Runways for small aircraft will of course have a much lighter total weight and wheel loads. All but the busiest airport runways will experience fewer load applications than will a normally busy highway with equivalent wheel loads.

The wheel arrangement and loading patterns are different. Trucks possess a conventional inside-outside in-line pattern that places the load within two to four feet of the outside edge of the pavement. Flexible pavements especially show a high distress rate at the edge and may be given added thickness at the edge as a relief measure. Aircraft generally have a tricycle landing gear arrangement with a steerable lead wheel or wheel set. Loads are channelized on the middle portion of the runway with 80 percent of the loads falling within about 8 percent of the pavement area. Runway distress is thus concentrated within the center third of the pavement.

Because of higher gross weights of commercial aircraft, pavement thickness for runways will usually be greater than for highways. Runways may have a tapered thickness because the load is concentrated on the initial third or more of the runway length. The lifting action at takeoff lessens the load, and when landing a plane does not impose load until actual contact has been made.

Runways must also withstand warmup vibrations, the exhaust from jet engines, and landing impacts. See the Suggested Readings for references giving a fuller discussion of details on runway design. Data on runway length will be found in Chapter 9 Terminals.

SOILS

Modern subgrade design requires a determination of *load-bearing capacity* for safe and economic proportioning of design. Load bearing characteristics vary greatly with different soils, and the nonuniformity of soils is a frequent cause of uncertainty. Bearing capacity may be determined by laboratory tests on soils or from less refined field tests. The latter, usually some form of loading or penetration test, are more in favor for transportation subgrades. The California Bearing Ratio as a measure of subgrade strength in highway design was explained in an earlier section.

Soils possess *index properties* of grain size, internal friction, cohesion, shearing strength, capillarity, permeability, compressibility, liquid and plastic limits and mineral content that determine their bearing capacity and stability characteristics. Attempts have been made to classify soils according to their index properties or characteristics and bearing capacities. These range from relatively simple grain size classifications to the elaborate and complex classifications used in the construction of highways and airport runways. A classification developed by the AASHTO based on soil properties include particle index, grain size distribution, liquid limit, and plasticity index (AASHTO Designation: *M* 145-73) is shown in Table 6.3. One enters the required test data into the table and proceeds from left to right. The first group from the left into which the test data fit is the correct classification.

Materials within the Groups A-1-a, A-1-b, A-2-4, A-2-5, and A-3 are suitable for well-drained and compacted subgrades under pavements of moderate thickness. Groups A-2-6 and A-2-7 and the silt-clay Groups A-4 through A-7 range from the approximate equivalent of good A-2-4 and A-2-5 to fair and poor subgrades that require a subbase layer or an increased thickness of base course.

A Group Index procedure for evaluating subgrade materials has the formula:

$$\text{Group Index} = (F\text{-}35)\left[0.2 + 0.005(LL\text{-}40)\right] + 0.01(F\text{-}15)(PI\text{-}10)$$

F = percent passing 0.074-mm sieve, expressed as a whole number, LL = liquid limit, and PI = plasticity index. The equation is based on a liquid limit of 40 or above and plasticity indexes of 10 or above are considered critical. The Group Index is considered as 0 for nonplastic soils or where the liquid

TABLE 6.3 AMERICAN ASSOCIATION OF STATE HIGHWAY AND TRANSPORTATION OFFICIALS CLASSIFICATIONS OF SOILS AND SOIL-AGGREGATE MIXTURES.[a] AASHTO DESIGNATION: M 145-73

General Classification	Granular Materials (35% or Less Passing 0.075 mm)							Silt-Clay Materials (More Than 35% Passing 0.075 mm)			
	A-1		A-3	A-2				A-4	A-5	A-6	A-7 A-7-5. A-7-6
Group Classification	A-1-a	A-1-b		A-2-4	A-2-5	A-2-6	A-2-7				
Sieve analysis, percent passing:											
2.00 mm (No. 10)	50 max	—	—	—	—	—	—	—	—	—	—
0.425 mm (No. 40)	30 max	50 max	51 min	—	—	—	—	—	—	—	—
0.075 mm (No. 200)	15 max	25 max	10 max	35 max	35 max	35 max	35 max	36 min	36 min	36 min	36 min
Characteristics of fraction passing 0.425 mm (no. 40)											
Liquid limit			—	40 max	41 min	40 max	41 min	40 max	41 min	40 max	41 min
Plasticity index	6 max		N.P.	10 max	10 max	11 min	11 min	10 max	10 max	11 min	11 min
Usual types of significant constituent materials	Stone fragments, gravel and sand		Fine sand	Silty or clayey gravel and sand				Silty soils		Clayey soils	
General rating as subgrade	Excellent to Good							Fair to poor			

[a]From *Specifications for Materials*, Table 2, p. 223 (courtesy of the American Association of State Highway and Transportation Officials, Washington, D.C.).

[b]Plasticity index of A-7-5 subgroup is equal to or less than LL minus 30. Plasticity index of A-7-6 subgroup is greater than LL minus 30.

limit cannot be determined. A group index of 0 indicates a good subgrade material, but one of 20 or above represents a "very poor" material. The Group Index equation is found on page 222 of the AASHTO's *Specifications for Materials*, pp. 218–224.

Design for Bearing Capacity and Stability   The requirements for constructing a stable subgrade include the following:

1. A soil survey to determine the characteristics of the natural ground and its suitability for use as a fill material.
2. Locations that avoid such troublesome ground as swelling clays, varved clays, false shales, swamp muck and muskeg, and unstable side hills.
3. Adapting the geometric features of the subgrade—width, depth, and side slope—to the characteristics of the soil to be used. The cross section must include adequate drainage.
4. Placing the soil in well-compacted thin layers with the moisture content controlled to give maximum dry density. The least stable soils should be placed where they will do the least harm—on the shoulder and side slope or under the weight of more stable materials. The upper reaches of the fill should contain select materials that promote drainage and discourage the capillary rise of moisture.
5. Protecting the slopes by planting root-producing vegetation, riprapping, sodding or by similar means.
6. Performing the foregoing under the supervision of someone trained in soils engineering principles and applications.

Moisture Effects   An essential condition for soil and subgrade stability is freedom from excess moisture. A change in moisture content can quickly convert a stable material into one that is highly unstable. Nevertheless, the addition of moisture during the compaction process decreases the surface tension between soil grains, permitting the particles to be more closely consolidated into a denser, more stable mass with greater shearing strength and less space for moisture. As more moisture is added, however, a point is reached where the particles are widely separated by moisture and a less dense, less stable mass ensues. The point of maximum dry weight density is termed the *optimum* and the amount of moisture is the *optimum water content*. The optimum water content and density are obtained through a standard compacting process in the laboratory but must be modified for actual field conditions and the type of compacting equipment being used (Figure 6.10). A practical contract requirement may be for compaction to within 95 percent of optimum. Care must be exercised with certain soils

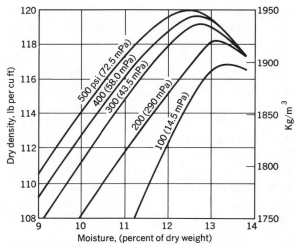

FIGURE 6.10 EFFECT OF COMPACTION PRES-
SURES ON DRY DENSITY.

that have swelling tendencies. If compacted to the full optimum, these may absorb additional moisture and experience volume change when in place.

When excess moisture enters the soil mass, the particles are no longer in close contact, the soils may experience volume change and, since water has a shearing strength approaching zero, become instable. Excess moisture enters from many sources—from surface flow, rainfall and snow melt, from capillary rise accelerated by the repetitive pumping action of passing wheel loads, and from subsurface seepage and flow. Pockets of moisture may occur in lenses of dirt or fine-grained soils in the upper reaches of a subgrade, in pavement base or subbase courses, or in ballast sections. Such pockets freeze and undergo volume change (as much as 10 percent) distorting the surface with frost heaves. When the frozen lenses thaw during a warm period or in the spring, there can be a disastrous loss of support and a breakup and rutting of paving surfaces or distortion of track geometry. The necessity for good drainage is evident.

DRAINAGE

Relation to the Subgrade   Drainage is undoubtedly the most important single factor contributing to stability. The first prerequisite is to keep water away from the subgrade structure. This involves a system of ditches and culverts. Roadside and track ditches border the ballast section and pavement shoulder through cuts and level country to give immediate drainage

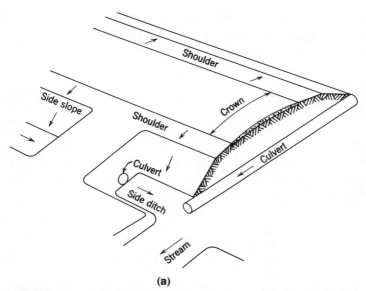

(a)

FIGURE 6.11a ROADWAY DRAINAGE—SURFACE WATER. (COURTESY OF EUGENE Y. HUANG, *MANUAL OF CURRENT PRACTICE FOR THE DESIGN, CONSTRUCTION, AND MAINTENANCE OF SOIL-AGGREGATE ROADS,* ENGINEERING EXPERIMENT STATION, UNIVERSITY OF ILLINOIS, URBANA, JUNE 1959, P. 57, FIGURE 3.8.)

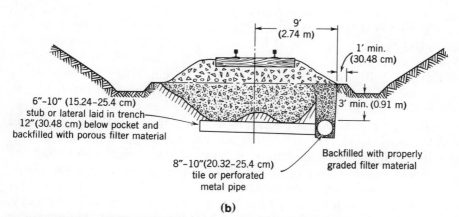

(b)

FIGURE 6.11b ROADWAY DRAINAGE—SUBSURFACE WATER.

to track or pavement. Intercepting ditches collect runoff before it reaches the subgrade. Culvert openings are necessary at intervals in the subgrade to conduct runoff and drainage channels to the other side. Figure 6.11 shows schematically these several drainage features.

Drainage Design   A design problem, basic and common to culverts and ditches, is to determine the cross-sectional area of ditch or culvert that will provide sufficient capacity for the quantity of flow that must be handled. Expressed mathematically, $Q_c = A \times v = Q_r$, where Q_c = capacity of channel or opening in cubic feet per second, A = cross-sectional area of opening or channel in square feet, and v = rate of flow in feet per second. From hydraulics, the Manning formula gives a value for flow velocity as $v = (1.486/n)R^{2/3}S^{1/2}$, where R = the hydraulic radius = area of the cross section divided by the wetted perimeter, S = slope in feet per foot, and n = coefficient of roughness varying from 0.02 for ordinary earth smoothly graded and for corrugated metal pipe to 0.016 for concrete paved and smooth tiled channels. An average value of 0.04 may be used for stream channels when other data are lacking and 0.06, for weed-grown ditches. The value of v, the velocity, should not exceed 10 fps in culvert pipes to prevent scour at the outlet and should preferably be no more than 4 to 6 fps. Concepts of critical flow from hydraulics lead to the equation of $S = 2.04/D^{1/3}$, where D is again the diameter of the pipe in inches and S = slope in feet per foot = that critical slope at which a pipe must be placed so that the water may be taken away without any backwater effect, the condition of maximum flow.[7] The pipe or channel capacity Q_c must, of course, be equal to the runoff Q_r, the amount of water coming from the drainage area. By assuming a value for A and determining corresponding values for S and v, the pipe capacity Q_c is determined and compared with Q_r. If the first comparison is not in close agreement, a second value for A is chosen, the choice being guided by the error in the first assumption.

The rational method of determining Q_r is based on the hydrology formula $Q_r = AIR$, where Q_r = rate of runoff in cubic feet per second = 1 acre-in. per hour. A = area of watershed or drainage area in acres, I = intensity of rainfall in inches per hour for a selected storm of given duration and frequency (the maximum design storm, see Figure 6.12) and R = the runoff factor. R is a difficult value to determine accurately, depending as it does on topography, vegetation, permeability and other soil characteristics, and extent of paved and built-up areas. It will vary from 0.10 to 0.15 for flat, vegetated, or gently rolling country, from 0.3 to 0.5 for built-up sections, from 0.8 to 0.9 for completely built-up sections or rocky, hilly, or mountainous areas, and will be 1.00 (or even more when snow is

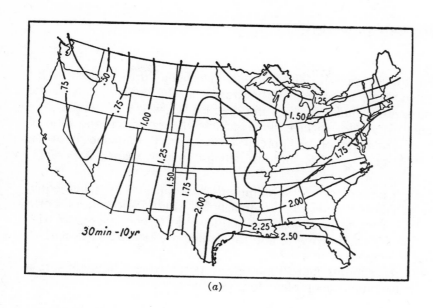

(a)

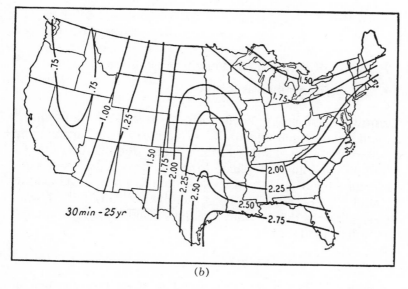

(b)

(1 in. = 2.54 cm)

FIGURE 6.12 RAINFALL INTENSITY IN THE UNITED STATES (TYPI-
CAL EXAMPLES OF DEPTHS OF PRECIPITATION OF VARIOUS
DURATIONS). (a) 30-MINUTE RAINFALL, IN INCHES, TO BE EX-
PECTED ONCE IN 10 YEARS; (b) 30-MINUTE RAINFALL, IN INCHES,
TO BE EXPECTED ONCE IN 25 YEARS. (COURTESY OF C. O.
WISLER AND E. F. BRATER, *HYDROLOGY*, 2ND EDITION, WILEY,
NEW YORK, 1959, P. 90, FIGURES 33 AND 34.)

melting) for frozen ground. This formula has been further modified to account for the time of rainfall concentration at the culvert, that is, the time for maximum runoff to reach the opening. $Q_r = AIR/f$, where $f =$ a factor to compensate for surface slope, which in turn affects the time of concentration. For slopes of 0.50 percent or less, $f = 3.0$; for slopes between 0.5 and 1.0 percent, $f = 2.5$, and for slopes greater than 1.0 percent, $f = 2.0$.[7]

In selecting a design storm, the minimum duration may be taken equal to the time of concentration for the area under study. Selection of storm frequency and intensity will also be influenced by the degree of risk involved for persons and property and the economics of low risk design vs. costs of possible washouts or floodings.

Where an area exhibits more than one slope or cover characteristic, a weighted average runoff coefficient R_{ave} may be used.

$$R_{ave} = \frac{R_1 A_1 + R_2 A_2 + \cdots + R_n A_n}{A_1 + A_2 \cdots + A_n}$$

where

$$R_1, R_2, \ldots, R_n = \text{runoff coefficients for subareas}$$

$$A_1, A_2, \ldots, A_n = \text{subareas in acres}$$

Empirical Formulas   Simple empirical formulas have been determined for local areas and given attempted universal application by introducing coefficients to compensate for changes in topography. Rainfall and time of concentration are usually ignored. Such formulas should be used with caution and understanding of limitations. The Talbot formula, developed in central Illinois by Dr. A. N. Talbot, has been used by railroads and highway departments for small areas that require openings of 60 in. (152.40 cm) or less in diameter. It has a value of

$$A = c\sqrt[4]{A_d^3} \, ,$$

where $A =$ area in square feet of required opening, $A_d =$ drainage area in acres, and $c =$ a coefficient having a value of 1 for steep rocky ground with abrupt slopes, $\frac{1}{2}$ for uneven valleys that are long in comparison with their width, $\frac{1}{3}$ for rolling agricultural country with a valley length three to four times the width, and $\frac{1}{5}$ for flat, level land. The required pipe diameter given by this equation is on the high side for small openings and thereby

[7]*Handbook of Drainage and Construction Products*, W. H. Spindler, Editor, Armco Drainage and Metal Products, Middletown, Ohio, 1955, p. 215.

contributes to the safety of its use. Because the openings are small, the excess size does not constitute a severe economic handicap. More accurate methods are employed when openings exceeding 60 inches in diameter are contemplated.

Subdrainage   Soils that are kept excessively wet by subsurface flow, seepage, or percolation may require some form of subdrainage to lower the water table and dry out the soil. Large, flat areas, such as airfields, railroad yards, and parking lots, also benefit from drainage. A subdrain usually consists of tile laid with open joints or of perforated corrugated metal pipe laid in a ditch and backfilled with a porous, well-graded material (usually sand or gravel). The drain may be a short stub laid transversely under a railroad or highway embankment and discharging down the slope of the subgrade, a similar pipe in a cut draining into a longitudinal header, or a long header and series of stubs through a large, flat area. The success of the drain will depend on its location with respect to the water table, source of percolation or underground flow, or bottom of water or ballast pocket. The type of soil is also important. Impervious clays will not drain readily, and subdrainage is then useful primarily to drain pockets and intercept underground seepage. There must be sufficient fine material in the backfill grading to prevent fine-grained silts and clays from entering and clogging the drain. Soils engineering texts contain specifications for such filter design.

THE TRACK STRUCTURE

Railroad Ballast   A ballast section of coarse, granular material is interposed between the rail-tie assembly and the subgrade. Its function is (1) to bear and distribute tie loads, (2) anchor the track against lateral and longitudinal movement, (3) provide immediate drainage for the track, (4) discourage vegetation, and (5) facilitate maintenance (adjustments to surface and alignment). The ballast section is, in effect, an extension of the subgrade. The most select materials—crushed granite, basalt, trap rock, slag, gravel, or other coarse materials, because of greater load-bearing capability—are used. Particle sizes vary from $1/2$ to $3\frac{1}{2}$ in. (1.27 to 8.89 cm) with $3/4$ to $1\frac{1}{2}$ in. (1.91 to 3.81 cm) having frequent use.

Pressure distribution would be no problem with a bed rock subgrade (there would still be need for ballast to anchor the track), but subgrades with load-bearing capacities of 5 to 30 psi (34.5 to 206.8 MPa) are encountered. The wheel loads transmitted by the ties must be further reduced in magnitude until equal to or less than the support offered by the subgrade to provide vertical stability.

Talbot's pressure distribution equations on pages 185 and 186 give reasonably accurate pressures at depths of 4 to 30 in. (10.2 to 76.2 cm) below the tie. Greater accuracies can be had with Newmark's charts. Pressure distributions vary across the face of the tie from the point of load application (Figures 6.13 and 6.14).

From these diagrams and later experiments, uniform distribution of pressure occurs at a depth equal approximately to the center-to-center spacing of the cross ties. Type of material has little or no effect on pressure distribution, whereby the lower half of a ballast section can use a lesser grade of material. The actual pressures in Figure 6.14 are expressed as percentages of the average pressure on the bearing surface of the tie in pounds per square inch. That average pressure, p_a in Talbot's pressure equation, is

$$p_a = \frac{2P}{(2/3)b \times l} = \frac{3P}{b \times l}$$

where

$P =$ wheel load on ties, often taken as 40 percent of actual wheel load

b and $l =$ width and length of cross tie, usually 8×102 in. $(20.32 \times 259.08$ cm$)$

$(2/3)$ $=$ area of tie that receives compaction and distributes load

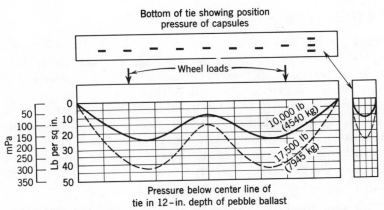

FIGURE 6.13 PRESSURE DISTRIBUTION ACROSS THE FACE OF TIE. (A. N. TALBOT, "SECOND PROGRESS REPORT OF SPECIAL COMMITTEE TO REPORT ON STRESSES IN RAILROAD TRACK," A.R.E.A. PROCEEDINGS, VOL. 21, AMERICAN RAILWAY ENGINEERING ASSOCIATION, CHICAGO, 1920, PP. 645–814.)

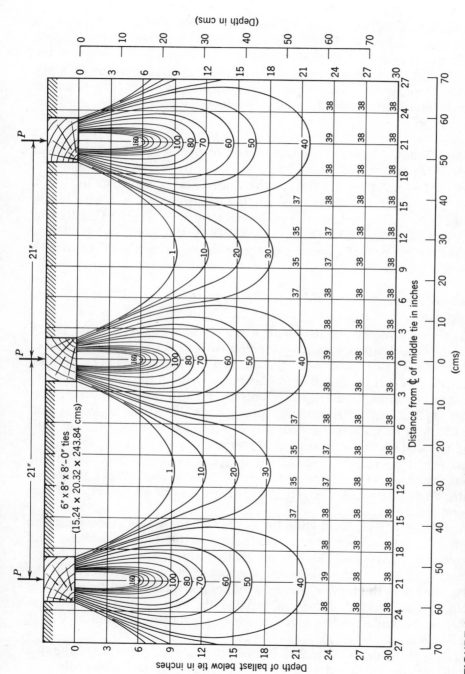

FIGURE 6.14 PRESSURE DISTRIBUTION FROM THREE TIES (IN PERCENT OF UNIT PRESSURE ON FACE OF TIES). (A. N. TALBOT, "SECOND PROGRESS REPORT OF SPECIAL COMMITTEE TO REPORT ON STRESSES IN RAILROAD TRACK," *A.R.E.A. PROCEEDINGS*, VOL. 21, 1920, FIGURE 97, P. 807.)

The design of ballast section is seen to be closely related to the strength and design of subgrade. When the subgrade is composed of fine-grained materials, it is advisable to place a well-graded filter blanket between the subgrade and the ballast section, as in highway construction for example, to reduce capillary rise and prevent intermingling of subgrade and ballast materials. Standard textbooks on soils engineering contain filter specifications.

Railroad Track   Railroad track is formed of parallel steel rails that bear and guide the flanged wheels of cars and locomotives. The rails are held to proper gage, 4 ft, $8\frac{1}{2}$ in. (1.435 m) in the United States (meter, 42 in., 5 ft, 0 in. 5 ft, 3 in., and 5 ft, 6 in. in various other countries), on timber cross ties. The rail is manufactured from open-hearth steel in 39-ft (11.88-m) lengths (which just fit inside a 40-ft (12.19-m) gondola for shipment) and is referred to by the weight per yard and the design of the cross section. Thus 132 RE rail weighs 132 lb (59.9 kg) per yard of rail length and was designed by the AREA.

Ties vary from 7×8 in. (17.78×20.32 cm) to 8×9 in. (20.32×22.86 cm) in cross section and from 8 to 9 ft (2.44 to 2.74 m) in length (for 4 ft, $8\frac{1}{2}$ in. gage). Ties are usually treated with a creosote-oil preservative to retard decay and insect attack and are protected with steel tie plates and rubber or fiber tie pads placed between the ties and the base of rail. The rails are secured end to end by bolted joint bars or by welding the ends together to form long continuous welded rails. Rails are held to the ties and tie plates by nail spikes, screw spikes, or spring clips. Thermal expansion and contraction of continuous welded rails is restrained by anchoring the rails to the ties and by the compression of spring clips.

Concrete ties, widely used in Europe, have only limited use in the United States where their design and economic viability is still a subject of debate. Their future use is largely dependent upon the availability and cost of wood ties.

The ballast section, in addition to distributing the wheel loads, must also anchor the track. The interlocking between the ballast particles and between the ballast and the ties serves to resist longitudinal and lateral movement of the track. Additional anchorage is obtained by filling the ballast section to the top, or nearly to the top, of the ties and by extending it 6 to 12 in. (15.34 to 30.48 cm) beyond the ends of the ties. A tie in crushed stone ballast offers a resistance to movement per rail of 800 to 1000 lb (363.2 to 454.0 kg).

Because of the close 19- to 24-in. (48.26- to 60.95-cm) spacing of the ties, the rail may be analyzed as a flexible continuously and elastically sup-

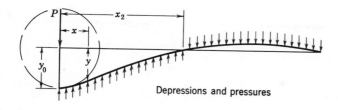

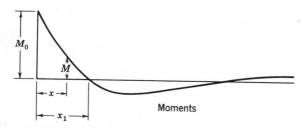

FIGURE 6.15 CONTINUOUS BEAM THEORY; DE-FLECTION AND BENDING.

ported beam. The primary vertical deflections and bending moment relations are indicated in Figure 6.15. A wheel load applied at any point in the track structure causes a deflection of the rail and an upward reaction accompanied by a downward movement and compression of the ties and ballast immediately under and adjacent to the load. Beyond the point of load application, a reverse bending occurs in the rail, with downward components of force acting in the rail. In effect, the wheel continually runs uphill. The energy expended in compressing the track and pushing the reverse bending wave ahead of the wheel reduces by a small amount the energy available to accelerate the train and to haul payload.

With a constant proportionality factor u and the assumption of elastic, continuous support, the track depressions and upward pressures resulting from the applied wheel load P are proportional to each other, so that $p = uy$. See Figure 6.15. The maximum deflection $P = uY$ occurs under the point of load application. The proportionality factor u is the modulus of track elasticity that depends upon the stiffness of rails, ties, ballast, and subgrade. It is the load-per-unit length of rail required to depress that unit length an equal unit distance. The value of u must be determined by test, assumed by comparison, or computed when other related factors are known. Typical u values as determined by the Special Committee on Stresses in Track are given in Table 6.4. The larger the value of u, the

TABLE 6.4 MODULUS OF TRACK ELASTICITY (u)[a]

Rail	Ties	Track and Ballast	u
85 lb	7×9 in.×8 ft-6 in. spaced 22 in. c to c	6-in. fine cinder ballast, in poor condition, on loam subgrade.	530
85 lb	6×8 in.×8 ft-0 in. spaced 22 in. c to c	6-in. limestone on loam and clay roadbed. Good before tamping.	970
85 lb	7×9 in.×8 ft-0 in.	12-in. limestone on loam and clay roadbed. After tamping.	1090
130 lb RE	7×9 in.×8 ft-6 in. spaced 22 in. c to c	24-in. gravel ballast plus 8 in. of heavy limestone on well-compacted roadbed.	2900,3000
110 lb RE	7×9 in.×8 ft-0 in. spaced 22 in. c to c G.E.O. fastenings	Flint gravel ballast on wide, stable roadbed.	2500, 2600, 3600 average 2900
110 lb RE	7×9 in.×8 ft-0 in. spaced 22 in. c to c G.E.O. fastenings	Limestone ballast on wide, stable roadbed.	3700, 5500, 6200 average 5100
Concrete roadbed			6000,7000

(1 inch = 2.54 cm)

[a]First and Sixth Progress Reports of the Special Committee on Stresses in Railroad Track, *Proceedings of the American Railway Engineering Association*, Vol. 19, 1918, and Vol. 35, 1934, A.R.E.A., Chicago, Ill.

stiffer the track will be, with a corresponding reduction in track deflection, bending moment, and stress.

In developing this concept of an elastically supported continuous beam, the Special Committee followed the mathematical solution in Foppi's *Technische Mechanik*, Vol. III, pp. 254–266, 1900 edition[8] The following development summarizes portions of the solution. The first, second, third, and fourth derivatives of the elastic curve for a continuously supported beam are respectively proportional to (1) the slope of the elastic curve, (2) the bending moment, (3) the shear, and (4) the load intensity. The differen-

[8]Report of the Special Committee for Study of Track Stresses, A. N. Talbot, Chairman, First Progress Report, *Proceedings of the A.R.E.A.*, Vol. 19, 1918.

tial equation of equilibrium, from the fundamental condition is

$$EI\,(d^4\,y/dx^4) = uy$$

This differential equation is satisfied by the following equation:

$$y = \left(-\frac{P}{\sqrt[4]{64EIu^3}}\right)e^{-\lambda x}(\cos\lambda x + \sin\lambda x) \qquad \text{where } \lambda = \sqrt[4]{\frac{u}{4EI}}$$

Special values developed from the several derivatives of the foregoing are $x_1 = \pi/4\sqrt[4]{4EI/u}$, where $x_1 =$ distance from the wheel load to the point of zero bending moment in the rail $(M=0)$, $M_0 = P\sqrt[4]{EI/64u} = 0.318Px_1$, where $M_0 =$ maximum bending moment (at the wheel load where $x=0$), $Y_0 = -P/\sqrt[4]{64EIu^3}$, where $Y_0 =$ maximum deflection (also at the wheel load where $x=0$), $P_0 = P\sqrt[4]{u/64EI} = -uY_0$, where $P_0 =$ the maximum intensity of upward pressure (at the wheel load, $x=0$), and finally the distance $x_2 = 3x_1$, where $x_2 =$ the distance from the wheel load to the point of zero upward pressure on the rail $(p=0)$. $E=$ modulus of elasticity of steel, 30,000,000 psi; $I=$ moment of intertia of the rail.

Dr. Talbot and his committee developed a useful tool for determining the effects of a wheel load by means of a master diagram. See Figure 6.16. From this diagram, values of x, y, and M can be computed at any distance from the wheel in terms of percent of maximum value. The diagram clearly illustrates how the effects of a single wheel load are distributed over several ties, both behind and in front of the wheel (a simplifying rule of thumb has assumed that a wheel load is distributed equally over three ties, the one under the load and the two adjacent). The combined effect of two or more adjacent wheels is determined by superposition, that is, by taking the alegebraic sum of moments and/or deflections of each wheel at any one x position.

A stiff rail contributes to the stiffness and stability of the total track structure. From the properties of steel beams, rail weight varies with the cross-sectional area; stiffness varies with the area also and therefore with the square of the weight. Also from beam properties, stiffness varies with the cube of the height. This has tended to develop a high, girder type of rail for maximum stiffness with minimum steel. The measure of stiffness is the usual section modulus I/c, where I is the moment of inertia and c is the distance from the outer fiber (base of rail) to the neutral axis. The greater the value of I/c, the stiffer and stronger the rail. Dimensions and pertinent properties of the most frequently used rails are given in Table 6.5.

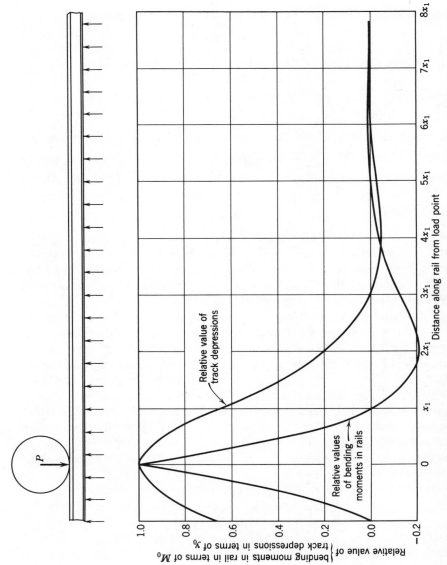

FIGURE 6.16 MASTER DIAGRAM FOR TRACK ANALYSIS. (A. N. TALBOT, "REPORT OF THE SPECIAL COMMITTEE FOR THE STUDY OF STRESSES IN RAILROAD TRACK, FIRST PROGRESS REPORT," *A.R.E.A. PROCEEDINGS*, VOL. 19, 1918, P. 886, FIGURE 5.)

TABLE 6.5 PROPERTIES OF TYPICAL RAIL SECTIONS

Section	Weight per Yard (lb)	Moment of Inertia	Base to neutral Axis (in.)	section Modulus (base)	Area (sq in.)	Height (in.)	Base Width (in.)	Head Width (in.)
133 RE[a]	133.4	86.0	3.20	27.0	13.08	$7\frac{1}{16}$	6	3
132 RE	132.1	88.2	3.20	27.6	12.95	$7\frac{1}{8}$	6	3
131 RE	131.2	89.0	3.20	27.0	12.90	$7\frac{1}{8}$	6	3
130 RE	130.0					$6\frac{3}{4}$	6	$2\frac{15}{16}$
120 RE[b]	120.9	67.6	2.92	23.1	11.85	$6\frac{1}{2}$	$5\frac{3}{4}$	$2\frac{7}{8}$
115 RE	114.7	65.6	2.98	22.0	11.25	$6\frac{5}{8}$	$5\frac{1}{2}$	$2\frac{23}{32}$
112 RE	112.3	65.5	3.00	21.8	11.01	$6\frac{5}{8}$	$5\frac{1}{2}$	$2\frac{23}{32}$
110 RE[b]	110.4	57.0	2.83	20.1	10.82	$6\frac{1}{4}$	$5\frac{1}{2}$	$2\frac{25}{32}$
100 RE	101.5	49.0	2.75	17.8	9.95	6	$5\frac{3}{8}$	$2\frac{11}{16}$
90 RA-A[c]	90.0	38.7	2.54	15.23	8.82	$5\frac{5}{8}$	$5\frac{1}{8}$	$2\frac{9}{16}$
75 AS[f]	75.2	23.0	2.36	10.0	7.38	$4\frac{13}{16}$	$4\frac{13}{16}$	$2\frac{15}{32}$
106 CF&I	106.6	53.6	2.85	18.8	10.45	$6\frac{3}{16}$	$5\frac{1}{2}$	$2\frac{21}{32}$
119 CF&I[d]	118.8	71.4	3.12	22.9	11.65	$6\frac{13}{16}$	$5\frac{1}{2}$	$2\frac{21}{32}$
136 CF&I	136.2	94.9	3.35	28.3	13.35	$7\frac{5}{16}$	6	$2\frac{15}{16}$
155 PS[e]	155.0	129.0	3.38	36.7	15.20	8	$6\frac{3}{4}$	3
152 PS	152.0	130.0	3.50	37.0	14.90	8	$6\frac{3}{4}$	3
140 PS	140.6	97.0	3.37	29.0	13.80	$7\frac{5}{16}$	6	3

[a]RE signifies American Railway Engineering Association design.
[b]Signifies not presently being rolled.
[c]RA signifies American Railway Association (forerunner of AAR) design.
[d]CF&I signifies Colorado Fuel & Iron Steel Corp. design.
[e]PS signifies Pennsylvania Railroad design.
[f]AS signifies American Society of Civil Engineers design.

An initial step in rail design is to determine the allowable stress S in rail steel in pounds per square inch from the beam formula $S = M/(I/c)$, where M is the bending moment in inch-pounds, and c is in inches. The bending moment for the contemplated wheel load is computed on the basis of a design of rail and the stress is determined and compared with the allowable value. Allowable stress is based on the yield point of rail steel, usually taken as 70,000 psi (482,580 MPa). This initial allowable stress is generally reduced to 32,000 to 35,000 psi (220608 to 241290 MPa) by allowing for lateral bending and such external factors as temperature stress in the rail, unbalanced superelevation, rail wear (especially on curves), and variations from the wheel-loading diagram. The proposed wheel loading, rail design, and stiffness of ties and ballast are then considered with regard to the maximum allowable working stress. The final selection of rail and ballast is essentially an economic matter. Several conventional rail weights, for example, may be adequate for load support, but under conditions of dense, heavy traffic the heavier rail section, having a minimum of deflection, usually requires less track maintenance and affords a longer rail and tie life.

Shearing Stresses   Rail sections in main line use today, when properly assigned, are capable of withstanding the bending stresses normally in effect, although a continuation of the trend toward heavier wheel loads may impose more exacting requirements.

Shearing stresses pose a different problem, one of rail quality and toughness. The problem arises from the contact stresses created directly under the point of wheel-rail contact. From such stresses arise a variety of railhead defects—head checking, flaking, spalling, corrugating, and shelling. Whereas rail end batter (a function of wheel load and impact) was often a limiting factor for rail life, shelly rail is likely to limit the life of continuous welded rail (CWR) where rail joints have been vastly reduced or eliminated. Shells are more or less horizontal cracks or separations occuring anywhere from $\frac{3}{8}$ to $\frac{5}{8}$ in. (0.95 to 1.59 cm) below the gauge corner of the rail. Shells will grow under traffic and may break out. Shells are not in themselves a serious problem but are very serious in that they can develop a downward transverse component that leads to a detail fracture, a type of transverse defect that cannot be visually detected prior to failure of the rail; hence hazardous. Shells will grow and develop under traffic. Some originate in usual types of mill defects—porosity and inclusions. For most shells, however, the originating source is still obscure. A concentration of residual stresses in the gauge cover is a likely cause of growth.[9] Heavy traffic loadings will cause shells and other defects to grow to a condition of failure.

Contact stresses are evaluated by an adaptation of the Hertzian solution for two cylinders in rolling contact—in this instance the wheel and the curved surface of the rail head. The solution becomes

$$\tau = \frac{23,500 P^{1/3}}{\left(\dfrac{R_1}{R_2}\right)^{0.271} \times \left(R_2\right)^{2/3}}$$

[9]G. C. Martin, *The Influence of Wheel-Rail Contact Forces on the Formation of Rail Shells,* Ph.D. thesis, University of Illinois, Champaign-Urbana, Illinois, 1971.

where

τ = the contact stress in psi; it should not exceed 50,000 psi.

P = the wheel load in lb; it should be given an increase factor to account for dynamic effects of moving loads of 1 percent per mile per hour over 5 mph

R_1 = radius of the larger cylinder (usually the wheel)

R_2 = radius in inches of the smaller cylinder (the rail)

Wheels tend to develop a hollow wear in the tread with a constant radius of 17 in (43.18 cm). Rails tend to wear to a constant railhead radius of 11.5 in. (29.21 cm). To account for the hollow tread wear, the rail head is, in effect, extended to fill that hollow by use of the relation:

$$R_1 = R' = \frac{R_h \times R_r}{R_h - R_r}$$

where

R_h = worn wheel hollow wear radius

R_r = worn railhead radius

Thus R' becomes 35.5 in. (90.17 cm), R_2 becomes 18 in. (45.72 cm) for a 36-in. (91.44 cm) wheel.[10]

The Hertzian theory assumes an isotropic, homogeneous material and takes no account of orthogonal (lateral and longitudinal) forces, nor of the load being concentrated on the gauge corners. The solution is therefore only approximate for rail that has been cold rolled by traffic and subject to lateral and longitudinal as well as to vertical wheel loads usually offset toward the gauge corner of the rail. The allowable value of τ is usually taken as 50,000 psi (344700 MPa), a value that is frequently exceeded when wheel loads exceed 810 lb (368 kg) per in. of wheel nominal diameter for 36-in. (91.44-cm) wheels.

[10]H. R. Thomas, *Proceedings of the American Railway Engineering Association,* Vol. 39, 1938, pp. 835–840.

Geometric Excellence Safety and smooth riding require that rails be held to true alignment, gage, and surface. Irregularities in any of these factors produce shock, sway, and vibration that cause discomfort to passengers, damage to lading and equipment, and even derailment. Irregularities in any one of these three usually lead to irregularities in the others. Of the three, however, surface is the most important. Exact cross level must be maintained between the rails by good initial design and construction and by appropriately timed track-surfacing operations. The basic surfacing operation is tamping or packing ballast materials under the ties. It is now largely carried out with powered mechanical tamping equipment.

Track Safety Standards were promulgated in 1972 by the Office of Safety, Federal Railroad Administration. Track is classified according to speed. Track segments failing to meet requirements for their intended class are reclassified to the next lower classification for which it meets requirements. No operation is permitted over track failing to meet Class 1 requirements. The rules further establish general conditions for drainage, vegetation, ballast, and stability. The rules become specific in details of track geometry—gauge, cross level, warp, superelevation, alignment, and profile. Minimum requirements for use and soundness of ties, spikes, tieplates, rail anchors, turnouts, and rails are also set forth. Measurements taken must reflect the loaded condition of the track. Inspection requirements and penalties for violations are a part of the rules.

Class	Maximum Allowable Operating Speed for Freight Trains	Maximum Allowable Operating Speed for Passenger Trains
Class 1 track	10 mph (16.09 kph)	15 mph (24.14 kph)
Class 2 track	25 mph (40.23 kph)	30 mph (48.27 kph)
Class 3 track	40 mph (64.36 kph)	60 mph (96.54 kph)
Class 4 track	60 mph (96.54 kph)	80 mph (128.72 kph)
Class 5 track	80 mph (128.72 kph)	90 mph (144.81 kph)
Class 6 track	110 mph (176.99 kph)	110 mph (176.99 kph)

Minimum safety standards should not be confused with recommended practice for new construction such as that recommended by the American Railway Engineering Association nor do safety standards necessarily relate to optimum economic maintenance practices.

The widespread use of unit trains that have a uniform consist and roller-bearing trucks with limited lateral play is another reason for excellence in maintaining the geometric qualities. Each car in a unit train will respond to any track irregularity in the same way as the preceding car and offer a repetitive impact. This could be the originating cause of many shelly rails and corrugations.

When adequate maintenance has been deferred, the track is likely to loose its correct alignment, surface, and stability. Ties become decayed, plate cut, and split, losing their spike and gauge-holding capability. Rail experiences excessive end batter and abrasive wear, and develops a high incidence of shells, corrugations, and other defects. Until these conditions are corrected, the speed must be reduced according to the foregoing FRA Safety Rules.

In very recent years a major maintenance effort on some railroads has been the rehabilitation of deferred maintenance. In other instances, track is upgraded to permit higher speeds, heavier wheel loads, or a greater volume of traffic. Subgrades are stabilized, ballast may be plowed out and new ballast applied, or the old ballast is cleaned and returned to track, ties and rail are renewed, the latter usually continuously welded into long 1440 ft+ strings, and the track surfaced and lined. Powerful and expensive machines have been developed to perform many of the operations involved. Such maintenance work, in addition to meeting the wide variety of emergency situations that arise in normal operation, poses an interesting and demanding challenge to those engaged in railway engineering.

Pipelines The problems of stability and support for pipelines are less critical than for other carrier types. Usually the natural ground gives sufficient support in the trench, and the rigidity of the pipe will carry it over soft spots a few feet in length. However, consideration should be given to soil conditions for access roads, which must often be constructed to bring in equipment and materials along the route.

Location reconnaissance and surveys seek to avoid troublesome ground. A unique soil characteristic in regard to pipelines is the corrosive quality of some soils. A combination of ground water and soil chemicals is likely to corrode the pipe and calls for expensive protective coverings and cathodic devices to prevent electrolysis. Sulfur, for example, may combine with water to form sulfuric acid. Not only naturally corrosive soils but those made corrosive by seepage of industrial wastes, which often contain corrosive substances, must be avoided if possible.

In swamps and small streams there should be a minimum depth of cover of 4 to 5 ft (1.22 to 1.52 m) and a 10- to 20-ft (3.04- to 6.1-m) minimum coverage under large streams and rivers. In addition, the pipe may have to be weighted with a concrete casing to aid in resisting forces of stream flow and scour.

Proper compaction of the backfill is especially desirable when a surcharge might be placed over the line, a likelihood in built-up areas or areas of potential development. The first 6 to 8 in. of backfill should be moderately fine material free of large lumps and rocks to prevent damage to the pipe and its protective covering as coarser, heavier backfill is added and compacted.

Problems arise when a pipeline must cross permafrost. In order to prevent settlement, breakage, and leakage, the adopted design will place the pipe on a trestlelike structure above ground supported on a subsidence-absorbing foundation. Openings between the supports are intended to permit migrations of wildlife common to permafrost country. The effect of the proposed design on such migrations is as yet undetermined.

WATERWAYS

Natural Waterways   Natural waterways—lakes, seas, and rivers—sometimes serve as established by nature. Usually, however, such natural waterways, especially rivers, cannot be used consistently without considerable development and maintenance effort. To provide and maintain navigable depths and widths, the engineer must resort to direct blasting and excavation, to dredging with bucket or suction dredges, or to confining and concentrating the flow to provide higher velocities, thereby making a channel self-flushing or self-scouring. Rivers young in geologic time have steep gradients and swift currents. Navigation depth is secured by means of dams impounding slack-water pools. The older, meandering streams flow slowly but are subject to silting and sandbar formation as well as change of channel.

Primary channel requirements for barge operations were set forth by Act of Congress, May 15, 1928, as 300 ft (91.4 m) of width and 9 ft (2.74 m) of depth. The Act of December 22, 1944, revised the depth to 12 ft (3.66 m) for the Mississippi River. The characteristic profile of erosive, slow-flowing, nontidal rivers (the best for navigation purposes) is that of a series of pools and bars. The deep pools usually occur at bends where crossing bars have accumulated and act as small dams. See Figure 6.17.

In periods of high water, flow across the bars of the open stream is usually sufficient for the required depth for navigation. At times of low water only the pools furnish the necessary depth. Channels can be dredged through the crossing bars, but the operation must be repeated frequently because the bars are soon reformed by silting. The practice of channel regulation in the United States is based on the theory of making the flow of

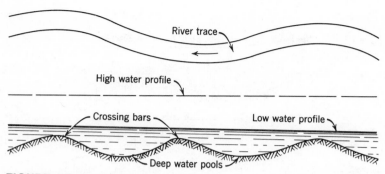

FIGURE 6.17 PROFILE AND TRACE OF A SLOW-MOVING RIVER.

the rivers accomplish as much of the regulation as possible. In this instance the design aims at confining most of the flow to a single, narrowed channel. This increases the flow velocity and the scour to a level that is sufficient to keep the channel open without disturbing the low-level, pool-crossing bar (which provides minimum low-water navigation depth) or the cross section of channel at high water. Within these limits, no more than 3 ft of additional depth are usually obtained at the bars. Contraction of the silted channel is usually accomplished with spur dikes. See Figure 6.18. The velocity increase must not be great enough to interfere with navigation. The current should not exceed 4 mph for uncanalized rivers, and the increase in flow velocity should stay within this range. Such rates of flow will possibly be exceeded for short periods in flood stages and require a suspension of river traffic.

Another problem of regulation is the guiding of the current to provide a sinuous rather than a straight trace or alignment over long reaches. The straight alignment, in the absence of training works, may afford an unstable location. Stability of trace is obtained by the use of training dikes and jetties. Side channels and sloughs are closed by pike dikes that cause sedimentation and damming of the unwanted channels by silt carried in the river's flow. Channel widening may be required at bends of sharp curvature because of the difficulty of steering the rigid type of United States' tow around it. Such curvature must permit maximum discharge during flood conditions as well as navigation during normal and low-water periods. Both requirements are met by placing spirals at the ends of the circular curves that form the main trace of the curve. Spirals based on the cubic parabola (similar to railroad spirals) are used. These spirals are only approximately accurate from the standpoint of geometry. Design is fre-

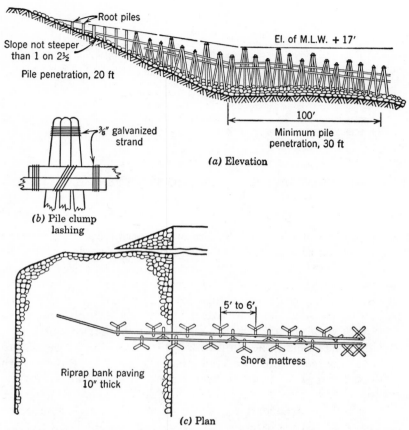

Root piles

Slope not steeper
than 1 on 2½

Pile penetration, 20 ft

El. of M.L.W. + 17′

100′

Minimum pile
penetration, 30 ft

(a) Elevation

⅜″ galvanized
strand

(b) Pile clump
lashing

5′ to 6′

Shore mattress

Riprap bank paving
10″ thick

(c) Plan

FIGURE 6.18 SPUR DIKES FOR SELF-REGULATION OF RIVERS.
(COURTESY OF ROBERT W. ABBETT, EDITOR, *AMERICAN CIVIL
ENGINEERING PRACTICE*, VOL. II, WILEY, NEW YORK, 1956, P.
15–105, FIGURE 44.)

quently accomplished by drawing a curve section to scale and maneuvering scale models of tows on them in passing positions.

Other regulatory and maintenance works include revetments for bank
protection, levees to protect shoreline communities in times of flood,
dredging, and the removal of snags.

Canalized Rivers   When sufficient depth of flow is not otherwise obtainable throughout the year, resort can be made to storage reservoirs to
provide artificial pools and slack-water navigation. Such canalization in-

creases silting and a loss of valley storage, interferes with local land drainage, and causes increased and heavy evaporation from the larger surfaces of slack-water pools. The design of navigation dams is outside the scope of this book, but basic requirements and principal navigation features should be noted.

There are two significant requirements for a lock and dam design: (a) there must be provision for the passage of floods without increasing flood height by impounded back water, and (b) the ability to transport sediment delivered by tributaries must not be impaired; otherwise silting, dredging, or both will become excessive.[11]

The requirements for an economical warrant are rivers that are natural routes for water-borne commerce and that possess permanent beds and banks. Rivers with a heavy bedload movement, as indicated by meandering, bank caving, and bar-building activity, and rivers with steep gradients are usually not suited physically or economically for conversion into canalized waterways.

Currents should not exceed 4 mph (6.4 kph) for efficient water transport; otherwise there is difficulty in maintaining upstream operation. Canalized rivers offer a solution to this problem whereby a series of pools impounded behind dams can limit current to 3 mph (4.83 kph) or less except during periods of very high water. The resultant efficient back-haul ability aids in developing a balanced upstream-downstream traffic.

The dams used in canalized waterways are the weir type that permits flow over crest or sill. These dams are further classified as navigable and nonnavigable. In times of highwater the navigable dam can be lowered to permit movement over its crest or sills. See Figure 6.19. It is used where flow is of sufficient volume to permit open-river navigation for long periods each year. Nonnavigable dams may have either a fixed crest or a movable crest of the taintor or roller type. See Figure 6.20. The nonnavigable, movable-crest dam has gained preference in the United States because with it a nearly constant pool level can be maintained. This is advantageous to real estate, industrial, and other development along the shore. Taintor gates are simple and effective and are most frequently used today for nonnavigable dams. The roller gate can discharge water either underneath or over the top. It is useful in skimming ice and debris with a minimum of water loss.

A necessary requisite of dams is a stable foundation. Protection against piping caused by erosion must be provided, and a cutoff wall or baffle,

[11]James H. Stratton, Canalized Rivers and Lock Canals, Section 15, River Engineering, Abbett's *American Civil Engineering Practice*, Vol. II, Wiley, New York, 1956, pp. 15–148.

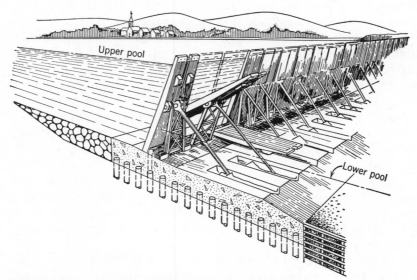

FIGURE 6.19 NAVIGABLE-TYPE DAM. (COURTESY OF ROBERT W. ABBETT, EDITOR, *AMERICAN CIVIL ENGINEERING PRACTICE*, VOL. II, WILEY, NEW YORK, 1956, P. 15–150, FIGURE 68.)

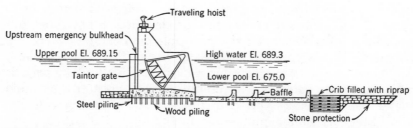

FIGURE 6.20 MOVABLE-CREST DAM. (COURTESY OF ROBERT W. ABBETT, EDITOR, *AMERICAN CIVIL ENGINEERING PRACTICE*, VOL. II, WILEY, NEW YORK, 1956, P. 15–150, FIGURE 71.)

combined with a filter on the downstream side, are required to prevent seepage flow and piping under the dam. Very extensive subsoil explorations are warranted.

Navigation Locks Locks are used to overcome elevation, whether in artificial or manmade canals, around falls and rapids in open rivers, or to change from one pool to another in combination with nonnavigable dams.

Shipping is raised or lowered over the obstructing crest and elevation in a rectangular compartment having a movable gate at each end. For upstream locking, a vessel enters the chamber through the open downstream gates. The downstream gates are then closed behind it, and water from the upper pool is allowed to flow into the lock chamber. When the chamber water level has risen to the upper pool level, the upstream gates are opened, and the vessel proceeds on its way. A downstream movement occurs in reverse order of the upstream move. Normal locking time is from 20 to 30 minutes, but may require one to two hours if the tow is so large that it has to be broken and reassembled with two or more lockings required for the entire tow.

Locks are constructed of concrete, bedrock, or steel sheet piling, with earth or rock backfill. Unless underlain with bedrock, a floor is necessary to prevent erosion from the turbulence of emptying and filling the chamber. The problem of chamber-wall design is similar to that of dams and retaining walls, and is outside the scope of this book. Lock gates are of timber, steel, or concrete and of miter, rolling, sector, or vertical-lift design. See Figure 6.21.

The lock chambers are filled and emptied through culverts in the walls with valve control over the culvert flow. Ports must be placed both for upstream and downstream movement. Taintor, cylindrical, and butterfly valves are used for control.

Rapid emptying and filling of the lock chamber is desirable to speed movement through the locks. However, a flow that is too rapid, especially at the start of the operation, causes unfavorable turbulence both in chamber and pool. If the chamber is filled too rapidly for an upstream movement, it may cause sufficient drawdown of the upper pool level to allow shipping to touch bottom. Too heavy a downstream surge may seriously reduce vertical clearances under bridges and other overhead obstructions. Either situation may also contribute to bank erosion.

The chamber design must include means of securing the vessels by cables to minimize their movement in the swirling turbulence and reduce damage to vessel and chamber. The pull on river-barge mooring cables may vary between 2000 lb (908 kg) for small vessels and 5500 lb (2497 kg) or more for large ones. Many variations in culvert design, location, and control have been devised to reduce chamber turbulence. Some designs place ports in the base of the walls, others in the floors, and still others use a combination. The 1200-ft (365.75-m) Chain of Rocks locks in the Mississippi River above St. Louis can be filled safely in $7\frac{1}{2}$ minutes.

Lock-chamber dimensions are determined largely by the size of vessels to be locked but are also partly determined by the trace of the stream. The

FIGURE 6.21 FEATURES OF A LOCK CHAMBER, SHOWING MITER GATES AND FLOODING PORTS OF LOCK CULVERT LATERAL SYSTEM DURING CONSTRUCTION OF MARKLAND LOCKS, WARSAW, KENTUCKY. (COURTESY OF DRAVO CORPORATION, PITTSBURGH, PENNSYLVANIA.)

lock chamber and land-side guide wall should be in a straight line. (The river wall is extended only on the upstream side to prevent vessels from going over the dam or through the rough water being circumnavigated.) Long locks and land walls are difficult to place when the river has considerable sharp curvature.

The usual size of Mississippi River system locks has been 110×600 ft. These dimensions are no longer acceptable for modern design of floating equipment. Tows may consist of 6 to 20 barges, varying in size from 25×175 ft, with 10 ft of draft, to 48×280 ft, with 11 ft of draft. The 1200-ft lock is becoming the standard wherever conditions permit. It enables 18 units 195×35 ft (59.4×10.7 m) 17 barges and one towboat—six units long and three units abreast—to be in the lock at one time. The 600-ft (182.9-m) lock permits only 12 such units, 4 in line and 3 abreast.

On the Great Lakes the size of bulk-cargo freighters governs. These ships may be as long as 730 ft (225.5 m). At the Soo the MacArthur Lock, largest and most recently built (1949), is 800 ft (243.8 m) long, 80 ft (24.4 m) wide, 31 ft (9.45 m) deep, and has a lift of 21 ft (6.40 m). Its cost was $14 million. The Davis and Sabin locks are both 1350 ft (411.46 m) long and 80 ft (24.4 m) wide but have depths of 23.1 ft (7.04 m). The new Poe lock, designed to be 1200 ft (365.8 m) long and 110 ft (33.5 m) wide, will accommodate vessels of 45,000 tons (40,815 tonnes), 1000 ft (304.8 m) long, and with 100 (30.5 m) to 105 ft (32 m) of beam.

Channel Design   Efficient channel design requires among other factors, adequate width and depth. Width is a function of terrain, of traffic volume and density, and whether or not two-way movement (or one-way with occasional passing points) is required. Most inland channels are 200 ft (60.96 m) or more wide. With more narrow channels, the crowding of water between channel walls and ship hulls increases ship resistance and may reduce channel depth.

There must be sufficient depth to permit intended draft plus 4 to 6 ft (1.83 m) additional clearance (depending on the height of waves and the intended speed) to accommodate the squat or downward drag of the propellers. At minimum depth the speed of the vessel may be reduced to lessen the squat factor. Adequate and uniform depth are also factors in reducing ship resistance.

The St. Lawrence Seaway-Great Lakes System provides a 27-ft (8.23-m) minimum depth throughout. Most ocean shipping can thus enter the Great Lakes in full draft. Larger vessels may also enter in light draft by off-loading at Montreal and later add to their outbound cargo at Montreal.

This route has been further improved by a straightening (opened in 1973) of the southern end of the Welland Canal and removal of several bridges over the canal.

Lock-Dam Economics   The design of a canalized river system offers problems in economics involving the choice of a few high-lift dams versus more numerous low-lift dams. The trend is toward the high lifts. The Ohio River, for example, was canalized in 1929 with a system of 52 locks and dams. Starting 1955 these are being replaced with 19 new high-lift locks. The locks are increased in length from 600 ft (182.88 m) to 1200 ft (365.76 m), the width of 110 ft (33.53 m) remaining the same.[12] The combination of high lifts and larger locks will save, it is estimated, two days in locking time alone for tows moving between Pittsburgh and Cairo, Illinois. Similar improvements are planned for other parts of the inland waterways system.

Manmade Canals   Manmade canals are two general types: (a) those connecting bodies of water at approximately the same elevation throughout and (b) those which must overcome elevation in their course and for which locks must be used as described above. Vessels move through canals under their own power or are pulled by a horse, mule, or electric locomotive on a towpath or track alongside the canal. Electric locomotives supplement ships' power in the Panama Canal. Ships use their own power exclusively at the Soo.

Embankment materials must not permit excessive leakage or percolation of water from the channel. Part of the slope may be the levee type, composed of soil excavated from the canal bed. The embankment or levee for a canal, as for a river, is built much as an earth dam with an impervious core of fine-grained soils topped on the slopes with coarser materials to resist wave wash.

A principal problem in canal design and location is that of obtaining and holding water to float the vessels. In a constant-level design water may flow in from both directions from the bodies of water being joined. Water may also be obtained for any type of canal by tapping parallel rivers or lakes with connecting channels (as has been done for the New York State Barge Canal), by drawing on streams flowing transversely to the canal, or, if sources of this kind are not available, by building storage reservoirs from which the canal can be supplied. Gatun Dam in the Panama Canal system fulfills this function. Pumping, except for very small channels, is usually an uneconomical method of supplying canals. Economy in initial construction

[12]"Big Load Afloat," the American Waterways Operators, Washington, D.C., 1973, p. 40.

and water supply can sometimes be obtained by using existing stream channels for part of the canal. This economy may prove illusory, for floods or low water may sometimes render the system unnavigable. Contact with adjacent streams might better be avoided except as water is drawn under control to fill the artificial channel. Natural watercourses may be bypassed by damming and diverting their flow, by crossing their courses on bridges, or for small streams, by siphoning the flow under the canal bed.

QUESTIONS FOR STUDY

1. Using the cone theory of wheel-load distribution, determine the bearing capacity of a soil to support a wheel load of 6000 lb, transmitted by a 32×6 in. pneumatic tire through a 6-in. pavement.
2. Discuss the basic errors in the cone theory of wheel-load distribution and show how these errors have been compensated for in the equations developed for heavy truck and airplane wheel loadings.
3. What thickness of flexible pavement would be necessary for an airplane runway designed for a gross weight of 60,000 lb, using 17.00×18 in. tires, inflation pressure of 65 psi, laid on a subgrade with a bearing capacity of 45 psi?
4. A survey of daily traffic flow shows 50 tandem-axle vehicles carrying approximately 22 kips on each axle pair, 20 single-axle trucks with axle loads of 2 kips, 35 with loads of 6 kips, and 25 with loads of 15 kips. What is the daily equivalent 18-kip single-axle load carried by this pavement lane?
5. Find the weighted structural number, total thickness, and thickness of surface cover, base course, and subbase course for a proposed pavement that will carry about 200 dual-axle loads of 28,000 lb (1270 kg) each. The subgrade has a CRB value of 5.0. Roadbed materials are likely to freeze to a depth of 4 in. or more and become wet during spring thaws and extended periods of rainfall. Available materials permit the cover to be of a highly stable asphaltic concrete, with crushed stone and sand available as choices for base and subbase materials.
6. Given an axle load of 60,000 lb and 21-in. spacing of 7×8 in $\times 8$ ft, 6 in. ties, what depth of ballast would be required to give a uniform distribution of pressure to the subgrade?
7. (a) Describe the characteristics of an ideal soil for subgrade purposes. (b) To what extent does such an ideal soil occur in nature?
8. What can be done to compensate for the deficiencies of a soil in using it for subgrade construction?
9. In what way are the problems of subgrade and canalway design and construction similar? Dissimilar?
10. What are the difficulties and possible sources of error in determining the size of drainage openings by use of the rational method? What are the advantages and the dangers of using an empirical equation for the same purpose?

11. Find the percent of maximum allowable bending stress developed at 50 and at 90 mph in a 115-lb RE rail, laid in track having a modulus of track elasticity of 2800. *Note:* a rule of thumb for dynamic loading is to increase static load one percent per mile per hour.

12. A railroad track is laid with 136 lb CF&I rail on 8×9 in. $\times 9$ ft 0 in. cross ties with 21-in. centers. It is supported by rock ballast laid on a subgrade that has a normal supporting strength of 8 psi. What minimum depth of ballast would be required under the ties for stable track carrying wheel loads of 30,000 lb?

13. Using the data of Problem 12 and assuming a modulus of track elasticity of 2500, compute the maximum deflection, bending moment, and bending stress in the rail.

14. Using a 132-lb RE rail and 60,000-lb axle load, compute contact stresses for:
    (*a*) New wheel on new rail
    (*b*) New wheel on old rail
    (*c*) Old wheel on new rail
    (*d*) Old wheel on old rail
    Tabulate the results and comment upon the relation of these to allowable contact stresses.

15. A drainage area, more or less elliptical in shape, located in Central Iowa, has two unlike portions. The first consists of 80 acres that covers a densely settled residential area having a slope of 2 ft in 1000 ft. The second is 100 acres of wooded parkland and is practically level. Concentration time at a highway culvert at the lower end of the long, narrow area is 30 minutes. The highway is an important, heavily traveled two-lane connection between a large urban area and a satellite community housing commuter residents. What size of drainage opening is recommended?

16. What is the likely spacing of transverse cracks in a concrete pavement laid on the subgrade when the pavement is 12 ft wide, 8 in. thick, and the concrete weighs 150 lb/ft^3?

17. What is the *approximate* time to lock the following through a 600 ft lock on the Mississippi?
    Towboat and 6 barges
    Towboat and 11 barges
    Towboat and 20 barges

SUGGESTED READINGS

1. Karl Terzaghi and R. B. Peck, *Soils Mechanics in Engineering Practice*, Wiley, New York, 1948, pp. 372–406.
2. H. O. Sharp, G. R. Shaw, and J. A. Dunlop, *Airport Engineering*, Wiley, New York, 1948, Chapters VIII and IX, pp. 63–110.
3. L. I. Hewes and C. H. Oglesby, *Highway Engineering*, Wiley, New York, 1963, Chapters 13 to 19.
4. W. W. Hay, *Railroad Engineering*, Vol. I, Part Two, Wiley, New York, 1953.
5. *Handbook of Drainage and Construction Products*, Armco Drainage and Metal Products, Middletown, Ohio, 1958, Sections VI, VII, and VIII, pp. 195–379.

6. *Concrete Pavement Design,* Portland Cement Association, Chicago, Illinois, 1951.
7. Eugene Y. Huang, *Manual of Current Practice for the Design, Construction, and Maintenance of Soil-Aggregate Roads,* Engineering Experiment Station, University of Illinois, Division of Highways, State of Illinois, and Bureau of Public Roads, U.S. Department of Commerce, Urbana, 1958.
8. N. M. Newmark, *Influence Charts for Computation of Stresses in Elastic Foundations,* Engineering Experiment Station, Bulletin Series No. 338, University of Illinois, Urbana, May 1951.
9. Reports of the Special Committee for the Study of Track Stresses, *A.R.E.A. Proceedings, First Progress Report,* Vol. 19, 1918, *Second Progress Report,* Vol. 21, 1920.
10. H. M. Westergaard, Stresses in Concrete Pavements Computed by Theoretical Analysis, *Public Roads,* Vol. 7, No. 2, April 1926.
11. H. M. Westergaard, Analytical Tools for Judging Results of Structural Tests of Concrete Pavements, *Public Roads,* Vol. 14, No. 10, December 1933.
12. Nai C. Yang, *Design of Functional Pavements,* McGraw-Hill, New York, 1972.
13. E. J. Barenberg, *A Structural Design Classification of Pavements Based on an Analysis of Pavement Behavior, Material Properties, and Modes of Failure,* PhD thesis, University of Illinois, Urbana, Illinois, 1965.
14. *Full Depth Pavements for Air Carrier Airports,* The Asphalt Institute, Manual Series No. 11 (MS-11), College Park, Maryland, January 1973 edition.
15. *Flexible Pavement Design Guide for Highways,* National Crushed Stone Association, Washington, D.C., 2nd edition, October 1972.
16. *AASHTO Interim Guide for Design of Concrete Pavements,* American Association of State Highway Officials, Washington, D.C., 1972.
17. M. T. Salem and W. W. Hay, *Vertical Pressure Distribution in the Ballast Section and on the Subgrade Beneath Statically Loaded Ties,* Civil Engineering Studies, Transportation Series No. 1, University of Illinois, Urbana, Illinois, July 1966.
18. *Handbook of Steel Drainage and Highway Construction Products,* American Iron and Steel Institute, New York, 1971.
19. *Handbook of Concrete Culvert Pipe Hydraulics,* Portland Cement Association, Skokie, Illinois, 1964.
20. Horonjeff, R. *Planning and Design of Airports,* McGraw-Hill, New York, 1962.
21. E. J. Yoder, *Principles of Pavement Design,* Wiley, New York, 1967.
22. *Soils and Base Characteristics, Classification, and Planning,* Highway Research No. 405, Highway Research Board, Washington, D.C., 1972.
23. *Performance of Composite Pavement, Overlays, and Shoulders,* Highway Research Record No. 434, Highway Research Board, Washington D.C., 1973.
24. *Asphalt Concrete Pavement Design and Evaluation,* Transportation Research Record No. 521, Transportation Research Board, Washington, D.C., 1974.
25. *Bituminous Mixtures, Aggregates and Pavements,* Transportation Research Record No. 549, Transportation Research Board, Washington, D.C., 1975.

Chapter 7

Systems for the Future

RESEARCH AND DEVELOPMENT LOGIC

Need for Improved Systems Human beings rise continually to the challenge of bettering the various hardware and software elements of their living system. Transportation presents critically challenging problems that cry for solution before the system breaks down or loses its capacity to serve.

Population growth places increasing demands for transport capacity to move its numbers and to make available the food and other essentials for their livelihood. A major effort is directed toward increasing speed, that is, reducing travel time, especially in moving people. Reduced travel time is sought as cities expand outward engulfing suburban and intermediate communities that become mere names in long corridorlike concentrations of urbanized land use. Commuter rides get longer in miles and congestion adds to travel time. The inner city, too, loses its mobility through congestion and lack of accessibility to its many neighborhoods.

The high speed of aircraft is negated by congestion and delays at air terminals of both planes and passengers, and by difficulties attendant on ground travel to and from the airport. The much-vaunted speed and safety of railroads have been lost in a seeming inability of the industry to continue to provide adequate maintenance and dependable operation.

Automobiles and trucks are in universal use but contribute in a major way to air pollution. Significant air pollution comes also from aircraft. Water transport adds to water pollution. All modes create unwanted noise and unsightliness. Since all modes are intensive users of fuel, energy-efficient transport is an urgent goal for future development. The ominous inroads of inflation establish economy as another urgent need. Changes in the location of population centers, markets, sources of raw materials, and production centers require the development of new or little-used routes with the opportunity to start afresh with new but proven technologies.

There is an unevenness in the transportation system. Aircraft, for example, have experienced technological and financial progress, but form only one link in the total portal-to-portal transport requirement. All links in the door-to-door trip must be equally adequate, each mode must be used in its area of greatest utility. A balanced transport system still represents an unsolved problem.

Basis for Design-Analysis The thinking, planning, research, and development required to meet the needs of today and of the future must be projected from several bases. One of these is time, immediacy. Each of the problem areas named above cries for immediate solution. We cannot depend today on the exotic systems that are proposed for a later time. It is likely that many amazing innovations will become available at some future date. Several of these have already shown technological feasibility in demonstration models. But technological feasibility is far removed from operational practicality. Technological design must be developed into a fully workable debugged system that gives safe, dependable, economical service from day to day, regardless of the weather or other demands placed upon it.

Such a stage of development is many years in the future for most exotic systems. The high investment in present facilities tends to make changes incremental. To the time required for technological research and development must be added a period of debate over which systems are to be selected for a particular use, the funding of research and development, the time required to sell the idea to decision-making groups, the actual design, the land acquisition, the legal arguments, and the final construction and debugging. All this can add to 15 or 20 years of delay or lag time between the initial serious conception of an idea and its functioning as an operating reality. The still unfinished system of Interstate Highways entered initial planning stages in the mid-1930s; the enabling act was passed in 1956. The Toronto Subway's Yonge Street line experienced 15 years of lag time, the Montreal system 16 years. The BART system of rapid transit officially entered the planning stage in 1951. As of 1976 all lines are operating including the Bay tunnel. Problems with the automatic train control system and equipment failures that have limited frequency of service and dependability are being corrected. Ridership, below anticipated levels, is rising, and the public response is considered favorable, but financial problems continue.

On 29 March 1976, seven years after construction started and 14 years after the National Capital Planning Act of 1952, trains began operation in Washington, D. C. over a five-station, 4.6 mile segment of the system. The final plan calls for 98 miles of line serving 86 stations.

These systems are using "state of the art" equipment and methods that have required no extensive period of research and development. Excessive power requirements, relatively low vehicle capacity, and high roadway costs for the exotic systems indicate that economic factors may increase the lag time by several years. Alternatively, the pressure from energy shortages and pollution may require extraordinary efforts to meet critical demands in a lesser time.

If a 10 to 20 year lag time must be anticipated before a new transport concept can be brought to productive fruition, it is necessary that research and development now be underway for innovative systems to meet the needs of the next three decades.

Development Goals   What goals should govern research and development for the future? There are several possible areas for improvement.

(a) Speed   High speed, usually oriented toward moving people, has always featured the Sunday-supplement type of modal forecasting. This is despite an obvious incompatability between high speed and the needs of urban areas for which it is forecast. High speed, nevertheless, is a definite need for intercity and corridor travel of the future. If the extension of suburban areas and commuting travel time continue, higher speeds will be needed to bring people and goods to and from the decentralized zones within urban complexes.

A major thrust is the reducing of door-to-door travel time by high speed ground transportation (HSGT) to a point that it is competitive with the air plus ground time of air travel. Distances of 300 to 600 miles are involved at the present state of technology. Whether society has a real need for rapid movement need not influence technological design, but it poses philosophical questions that should not be overlooked in the planning process.

Commercial jet aircraft already fly at speeds of 600+ mph (965.4 kph), military craft attain more than a thousand miles per hour. It is probable that conventional land modes can be improved to move at speeds of 125 to 200 mph (201 to 322 kph). The costs of developing new land modes at faster speeds demand that speeds must be significantly higher to justify research and development costs. Hence 300 to 500 mph (483 to 805 kph) may be considered the probable range for such vehicles with some of them being predicted as capable of 1000 mph (1609 kph). Such speeds would permit one-day business round trips between major city pairs.[1]

(b) Accessibility   For urban transport, route flexibility is more significant than speed. Accessibility to expanding land uses will be required. Multi-mode devices or an automobile compatible with energy and environmental resources or a combination of the two may be the answer.

(c) Intensive Land Use   An entirely different approach is based on the type of land use intended. A decentralized sprawl requires an acceptable design of individualized vehicle. The intensive use of land, identified with high-rise living units, congregated in small land areas and served by electronic high-speed elevators would best be served by a mass-transit type of system.

(d) Transport Alternatives   Future development might well combine intensive land use with a greater reliance on transport alternatives. Television is an example. It provides entertainment in its many forms and educational opportunities. Additional progress in development of the use of facsimile transmission and the photophone could greatly reduce the need for personal and shopping trips. A TV display of goods in a shop could eliminate the need to go there for a selection. The possibilities of combining living quarters, work opportunities, and shopping and entertainment in one high-rise building have already been demonstrated in a few instances such as the Hancock Building in Chicago. Small, self-contained cities and self-contained neighborhoods within large cities are other possible alternatives to the problems and costs of transport. Many possibilities lie in imaginative patterns of land use.

SHORT TERM IMPROVEMENTS

As noted earlier, reliance must be placed almost entirely on conventional modes of transport for the next 5 to 10 years. These include the flanged wheel on steel rail technologies, the automobile, bus, truck, and aircraft. Their basic principles and configurations will not change greatly, but improvements in the state of the art are constantly being made to give these conventional modes greater effectiveness in their performance.

[1]*Research and Development of High Speed Ground Transportation,* report of the Panel on HSGT for the Commerce Technical Advisory Board, U. S. Department of Commerce, Washington, D. C., March 1967, p. 10.

Walking   As a ready and available solution to urban transport problems, walking is often overlooked. Walking can be encouraged by the construction of malls in the central business districts and other activity areas, by the contiguous grouping of business and other activity enterprises as in a shopping center, by providing over- or underpasses for the safe crossing of busy streets, and by covered walkways to protect pedestrians from the elements. For certain situations, walking may be supplemented by moving sidewalks. Landscaping and art displays that add interest to the walk provide additional incentives.

Bicycles   A growing dependence on the economy and convenience of bicycle transport over distances of one to three miles will assist in solving problems of traffic congestion and human health. The greatest handicap to bicycle use is the lack of safe and adequate bikeways where riders can have reasonably direct access to destinations that is safe from the obvious hazards of street traffic but without endangering pedestrians on sidewalks. Special bikeways have been built in activity areas such as college campuses, and streets have been designated as bikeways on which auto traffic is prohibited or, at least, must move with caution looking out for cyclists. These of course give no protection from inclement weather, another cycling handicap.

Rail Improvements   The problem with rail transport for people is not so much improving the technology as in recapturing the speed, comfort, and dependability that characterized rail travel prior to World War II. This step requires new, properly maintained equipment, well-maintained track, and a willingness on the part of all concerned to operate passenger trains with priority and dispatch. Amtrak, which originally started with worn and outmoded equipment purchased two to three decades ago by the member railroads, is acquiring new modern cars and locomotives. Amtrak has also been given responsibility for the electric Metroliners, deluxe trains in the Northeast Corridor designed for speeds of 125 to 165 mph (201 to 266 kph).

To these have been added gas turbine trains, both domestic and foreign designs, for use in nonelectrified corridor territory and on short runs in the Midwest such as Chicago to St. Louis. The United Aircraft version is designed for high speeds on conventional track and sharp curves. Use is made of air support springs that carry the car weight suspended by struts from the ceiling. Rotating arms enable a car to maintain a vertical position even though the car trucks and wheels are tilting outward in response to centrifugal force and of superelevation on curves. The Japanese National

Railways have developed a pendulum-type car support that performs essentially the same function by tilting the car body toward the center of the curve. The British APT (Advance Passenger Train) and similar concepts are bringing into production trains with improved speed and comfort and with relative economy.

State of the Art rapid transit cars combine new concepts in automated control with improved springing and suspension, coupling devices, heating, air conditioning, and interior decor and comfort. The Bay Area Rapid Transit (BART) contains a variety of features that make rail suburban service attractive. Features of innovative interest are the systems of automatic train control and dispatching, low noise levels, high speed (up to 80 mph or (129 kph)), automatic ticketing, electronic destination signs, and pleasing decor and seating.

For conventional rapid transit, larger cars 60 to 75 ft (18.3 to 22.9 m) long, made possible by wider clearances in new tunnel construction and longer platforms, provide large passenger capacities. Motors of 115 hp (vs. older 100-hp designs) permit smoother, more rapid acceleration and higher speeds. A new application of an old principle, the storage of energy in a flywheel during braking and deceleration, permits the use of that energy for acceleration, traction, and fuel economy.

Light rail systems with lower construction and operating costs, high capacity, and updated for speed and comfort are offering transport solutions for cities too small (less than 500,000 population) for full-scale rapid transit. Single rail cars can carry 10,000 to 12,000 an hour and can be coupled into 2- to 3-car trains. The ride is smooth and quiet. The major objection to street cars as interfering with and being obstructed by street traffic is eliminated by the use of private rights of way. Such systems can establish the location and roadway for future development into full-scale rapid transit systems as volume of demand warrants.

Recent studies have shown economic viability for railroad electrification in other than mountainous territory and high density passenger corridors. Electrification of selected intercity rail lines is a likely prospect for the near future, especially as liquid fuel shortages and demand for less air pollution increase.

There is also a likely return to the good operating and track and equipment maintenance practices of earlier decades. The extended use of welded rail adds smoothness to the ride, life to the rail and ties, and economy in maintenance. Concrete offers an alternative for a dwindling supply of tie wood. The problems of financing a return to standards of safety and excellence established by the Federal Railraod Administration's Bureau of Safety for high speed operation is commanding the attention of the Congress and of state legislatures.

In freight movement the operation of long trains and high capacity cars has already become commonplace, but more is in prospect. A program of research in track/train dynamics has been undertaken; it gives promise of early improvements in the safety and efficiency of such operation. Automatic car identification (ACI) and computerized control of car movements, both loaded and empty, are improving the efficiency of car usage. Revised work rules permit smaller train crews and shorter, more frequent trains. Automated classification yards and improved scheduling speed traffic between and through terminals. Studies conducted by the AAR and FRA are designed to introduce more efficiency and reduce congestion and delay in complex terminal areas, St. Louis, for example.

Coordination between highway and rail is secured with trailer-on-flatcar (TOFC) and containerization (COFC) with much package freight being shifted from LCL to express via air and highway. Auto-carrying trains operating on passenger schedules carry both passengers and their cars to a terminal destination.

Highways   A major concern with the automotive mode has been the high accident rate. The mandatory use of seat belt and shoulder harness has, however, been relaxed because of public outcry over the inconvenience of their use. Inflatable air cushions to reduce injury from collision impacts have also been proposed, but again a public protest against the mandatory use of this device may reduce the incidence of its use. There is hope that an adequate bumper will reappear. Dependable, errorfree manufacture is always a desirable safety device along with protective bracing and padding in the car itself.

Air pollution from automobiles and trucks is a major concern. The catalytic device described earlier may be on all 1978 vehicles, but questions have arisen as to its overall effectiveness and the possibility of its contributing to a different type of air pollution, pollution by sulfuric acid vapors.

The rotary type of internal combustion engine is already in commercial use. A possible competitor is the Stirling engine powered by heat from an external source. One design would use the expansion of hydrogen gas to move a sliding piston. The gas would be heated by a burner that could use almost any kind of fuel. Fuel economy, less noise, and less air pollution are suggested advantages.

The possibilities in electric energy are being explored and hold promise, at least for small vehicles making short trips in urban areas. A zinc chloride battery for power has been proposed. Small electric vehicles are in limited use today with operating ranges on one battery charge of 50 to 150 miles (80.5 to 241 km) at speeds of 30 to 54 mph (48.3 to 86.9 kph).

There seems to be a definite trend in the United States for smaller cars that will use less energy, a trend long established in Europe.

A second look is being given to electrically powered trolley buses running as conventional rubber-tired vehicles on city streets, but drawing energy from an overhead contact conductor. The use of a flywheel to recapture lost energy earlier mentioned is also proposed for these units.

Another recent development is the articulated bus capable of moving its two hinged sections with over 100 passengers through the sharp turns of urban streets and traffic.

Automated and computerized traffic control to permit maximum flow can be anticipated. Entry to freeways is often controlled by ramp signals and detector surveillance, sometimes supplemented with closed-circuit TV. Completion of the Federal Interstate Highway System and construction of spurs to it will expand the speed, safety, and accessibility offered by good design.

Within urban areas, the designating of express lanes for buses and car pools can reduce the burden on other traffic lanes. At intersections equipped with traffic lights, bus sensors have been installed on an experimental basis to give priority of movement to buses. Minibus systems facilitate economy and flexibility in congested areas. Fleets of minisized rental cars on a pick-up-here, deliver-there basis are a possibility.

So-called contra lanes for bus operation reserve one lane of a multilane facility for bus movement in a direction opposite to the normal use of that lane. The direction of bus flow can be changed to accommodate morning and evening peaks. An example is the 8.4 miles (13.6 km) inside lane of the Boston Southeast Expressway where speed and capacity have been increased.

An example of express bus lanes is the San Bernadino Busway of 11 miles between Los Angeles and City of El Monte. Paralleling an existing expressway and a set of railroad tracks, the busway has 4 miles (6.44 km) of a 54-ft (16.45-m) wide section with two-way operation, and seven miles where the freeway has been widened to provide bus lanes in each direction in the median with 10 ft (3.05 m) of common shoulder between the busway and the freeway lanes. There is a bus station and 700-car parking lot in El Monte. Sound walls have been provided to deaden the sound of the buses. Exclusive bus lanes have been established elsewhere too: between the New Jersey Turnpike and the Lincoln Tunnel; also in San Francisco over the Golden Gate Bridge, and in Washington, D. C.

Types of automatic highway control are noted in Chapter 10. More innovative is the automated highway concept on which vehicles would be equipped with detecting devices that respond to signals of a given intensity or frequency transmitted from a cable laid in the middle of the traffic lane.

If the vehicle tended to veer from its correct path, there would be a change of signal intensity in one or other of the two sensors and the vehicle would be adjusted back to the point where signals of equal intensity are being received. This would be a "no hands" system while on the highway so equipped. Signals for brake and throttle control would also be transmitted through electrical impulses from the cable. Means of entry and exit to these roadways and the braking and spacing of vehicles are aspects of the system that require further refinement.

Airways   Aircraft development seems caught for the moment in economic constraints. The Boeing 747 and similar craft have apparently reached an economic capacity plateau. There is some difficulty in filling enough seats to offset the high costs of these large capacity craft. There are also unsolved terminal problems in the efficient handling of the concentrated numbers of passengers and luggage posed by high capacity craft.

The present speeds of 600±mph (965 kph) are not likely soon to be markedly increased for commercial jets. That speed satisfies current schedule requirements. Higher speeds cause marked increases in energy consumption. The supersonic transport (SST) has met opposition because of its adverse effects on the environment. It is, however, its adverse economic effects and a limited market for the craft and for its services that have delayed its acceptance and development, perhaps for years to come.

There is no technological reason why helicopters should not see increasing use as feeder service and to give accessibility to sites with limited landing areas. Vertical takeoff and landing craft (VTOL) also perform on limited ground space but have not yet satisfied economic requirements. Both types present a noise and pollution problem for use in populated areas.

The biggest improvement in air travel may occur on the ground through the use of available technology to give greater capacity and faster accessibility to airports and more speed and comfort in moving through the terminal to and from the aircraft. Improved building layouts, use of moving sidewalks, intraterminal transport between buildings and departure gates, improved luggage handling, and service to the airport by rapid transit are means to this end. The extension of Instrument Landing Systems will aid safety and dependability.

The Hovercraft, an air cushion supported system midway between air and watercraft, has the ability to hover at zero speed and to travel over land or water without the presence of a prepared roadway. It gives scheduled services carrying as many as 200 passengers and 75 autos at speeds of 60 to 80 mph (96.5 to 128.7 kph) across the English Channel and across shallow expanses of water in Western England and Wales.

Hydrofoils   The second of two types of ground effect machines is closely allied to the Hovercraft design in which the craft is supported by a cushion of pressurized air constrained beneath the vehicle. The second type is supported by the aerodynamic lift as the vehicle moves at a high rate of speed. Lift may be augmented by a cushion of air created by lift fans and constrained within catamaran-style hulls. One design of jetfoil being tested for commercial use makes use of fully submerged foils. As the craft gains speed it rises above the water until the hull is clear in the same way in which an aircraft rises above the ground at takeoff. Added support is gained from the submerged foils. Various means of propulsion may be used. The 100-ton jetfoil, tested at 85 mph, has a waterjet for that purpose using gas turbines to drive a system of pumps. The possibility of applying aerodynamic lift principles to ships of 2000-ton size at 100 knots has received serious consideration.

SLOW SPEED INNOVATIONS

Long-term innovative concepts can be grouped into speed and areal hierarchies. Usually the slower speed systems are also projected for extensive route flexibility and responsiveness to varying land use patterns and densities. Trips in the one to ten mile range are generally contemplated.

Personalized Rapid Transit (PRT)   PRT at various levels of speed and capacity are adaptive to serving areas of activity such as CBDs, airports, college campuses, research and development centers, and shopping centers, or to meet the travel needs of any centralized complex. Larger models might serve small cities or low density areas.

A variety of concepts has been projected to give a personalized transport service. These usually consist of small rubber-tired vehicles holding four to six passengers, traveling in a fixed guideway. Similar systems with capacities up to $30\pm$ are frequently included in discussions, as is done here. Because of the low capacity, human operators are not economically viable, so the vehicles are automated under computerized control.

The guideways could be rails or, more likely, concrete troughs or running pads with center rail or side-of-trough guidance. Support would be had from steel wheels on rails, rubber tires on concrete pads (as is the design at the Morgantown, West Virginia pilot model and the Dallas-Fort Worth Airport System), or by air cushions or magnetic levitation. The specific features of PRT include an exclusive right of way forming a loop or, for urban areas, a network of lines spaced 1 to 2 miles (1.61 to 2.32 km) apart to serve all but suburban areas. Stations, usually off-line, would be closely spaced with empty vehicles available at each station. Controls would be automated, directed by a central computer. In loop systems the vehicles would simply circulate with stops at each station (as in an airport complex). For urban grids, the vehicle would move in response to the designated destination indicated by the passenger upon entering the system. The rider would be taken to that destination at speeds variously proposed of 10 to 60 mph (16.1 to 96.5 kph) with no stopping. The lower speeds are more probable.

A *bimodal* variation of this system permits the rider to be taken automatically over the line to an entry-exit point from which he takes control and drives the vehicle from the guideway over city streets to his ultimate destination. The vehicle is returned to the system on a later trip from that destination and left at some terminal or intermediate point for the next user.

A different type of dual mode system uses a light rubber-tired vehicle operating in trains on an automated guideway. At designated points individual vehicles are cut from the train and are driven over local streets for service that is almost door-to-door. Another system places retractable flanged wheels on rubber-tired buses. The vehicle can thus run both on railway tracks and on pavements. A more fanciful concept carries conventional vehicles suspended from an aerial cableway by attachment on the car roof. In another scheme a vehicle is powered by a battery that can be recharged from house current. It is guided by inboard flanged wheels when on the system's guideway but uses outboard rubber-tired wheels for driving to final destinations over city streets.

Several working systems are undergoing demonstration and test. The PRT at Morgantown, West Virginia utilizes rubber-tired, eight-passenger (and 13 standees) vehicles with a top speed of 30 mph (48.3 kph). The troughlike guideway is supported on concrete piers with steel framing in the superstructure. The riders will have push-button selection of their destinations. Guidance comes from wheels that bear laterally against side panels on the walls of the trough. Propulsion is from a 60 hp dc motor with an onboard ac-dc converter. Conductor contact shoes are mounted on the sides of the vehicle. The body is made of fiberglass, but the chassis and frame are put together from standard automotive components. Cars are expected to arrive within two minutes of pushing a selector button. Completion of this pilot model has encountered difficulties because of costs that are greater than anticipated.

Another rubber-tired system known as Airtrans, similar in many respects to the Morgantown model, is in operation at the Dallas-Fort Worth airport. It seats 16 (with 24 standees), uses electrical energy, and receives guidance from wheels bearing laterally against guide rails on the sides of the troughlike guideway.

The Transit Expressway or Skybus uses rubber-tired, electrically driven wheels on longitudinal concrete pads atop a steel trestle. Laterally suspended guide wheels bear against a center rail for guidance.[2] The concrete roadway and guide rail are supported on steel I-beams and longitudinal and cross members supported by rectangular steel plates welded into box columns placed 50 to 60 ft (15.2 to 18.3 m) apart. All vehicles in the system are under computerized control with no operators in the individual cars. A top speed of 50 mph (80.5 kph) is proposed but 30 mph (48.3 kph) is probably a more realistic figure. Vehicle seating capacity is 28 persons and total capacity with standees is 70 persons. The weight per seat is 696 lb (316.0 kg) and per person 279 lb (126 kg). The problem of switching has not been entirely solved in practice. This system is presently in operation at the Miami and Seattle airports and has been proposed on an area-wide basis for the city of Pittsburgh.

PRT systems offer hourly capacities of 1000 to 10,000 persons. Costs of guideways and equipment are said to be low enough to make such systems attractive to smaller cities and in low to medium density corridors. The technology is fairly well developed, but the economics of these systems are still in controversy.

Dial-A-Bus   Midway between conventional and completely innovative systems is the dial-a-bus, demand-responsive concept. Conventional taxis, minibuses, or limousine-type vehicles may be used. The innovative feature is the ability to give door-to-door service by a communication and monitoring system that correlates calls for service with the nearest vehicle and directs that vehicle to answer the call. Vehicle selection based on the location of caller, the direction of bus movement, the destination called for, and the number and destination of passengers already aboard is optimized by the monitoring computer control. Local vehicles could also rendezvous with express vehicles and regular bus lines for longer distances. The system is especially adaptable to low density areas with greatest efficiency expected for demands of 100 trips per square mile per hour. Many dial-a-bus systems are now in operation.

Freight Handling Systems   For the foreseeable future, the motor truck is the likely distributor of freight within urban areas. The routing of these vehicles below ground levels is a possible solution that will require changes not only in street layouts but in building construction to provide access to

[2]"Transit Expressway Concept and Accomplishments," Westinghouse Electric Corporation, Pittsburgh, Pennsylvania, 1 April 1967, pp. 16, 17.

underground receiving and shipping rooms and platforms. Containerization and trailer-on-flatcar techniques are giving greater flexibility in the location of light industry and commercial establishments. Detailed access to any location is made available by highway in combination with the long-haul advantages of rail intercity service.

HIGH-SPEED SYSTEMS

Very high-speed ground transport refers primarily to speeds of 300 mph (483 kph) or higher. It also refers primarily to passenger movement. Speed in freight movement is more dependent on a reduction in terminal time than an increase in road speed (The Santa Fe's former Super C TOFC train that attained speeds of 90 mph is a notable exception). Obviously there is a gray area in which a speed category may extend from one hierarchy to another. A variety of problems are encountered with high-speed systems, both of a human and of a technological character. The severity of these problems varies with the concept offered for solution.

Location   An important initial feature is the location of the guideway with respect to the ground surface. For obvious reasons of safety, a ground level or at-grade location of a high-speed system would be unwise. Such a location would have to be free of grade crossings. The scarcity of land in the urban corridors where such systems are proposed work against ground level locations. Elevated structures also use considerable land space, cast shadows, and may distribute noise pollution over wide areas. An underground tunnel location therefore is usually considered for very high-speed systems. Such locations have many problems.

Human Problems   High speed itself, but especially combined with underground location, constitutes a hostile environment for people. Certain specific problem areas should be noted.[3]

(a) *Acceleration and deceleration rates* must be limited to comfortable human endurance yet kept as high as possible to reduce delays from speed reductions and at station stops. The well-trained astronauts in pressurized suits and strapped to reclining padded couches experience brief accelerative forces of several units of gravity (g), but the ordinary

[3]*High Speed Rail Systems*, report by TRW Systems Transportation for Office of High Speed Ground Transportation, U. S. Department of Transportation, Contract No. 353-66, Washington, D. C., February 1970, pp. 5.4-1 to 5.4-26.

traveler, especially the very young, the aged, or the infirm, are not conditioned for such rates. Aircraft takeoffs, aided by seat belts may reach 0.50 g or more, but a longitudinal acceleration rate of 0.10 to 0.15 is about the limit for normal situations; 0.15 to 0.50 for emergency situations.[4] Lateral acceleration should not exceed 0.10 to 0.13 for comfort. An actual rate of acceleration of 3 mph per second (0.28 g) is considered the maximum desirable for the normal conditions of high speed rapid transit. At this rate about 8 miles (12.9 km) would be required to attain a speed of 300 mph (483 kph, which means a limit of 16 miles (25.7 km) of station spacing for the system. Faster rates, such as might be felt in airplane takeoffs where rates up to 0.50 g are experienced, require seat belts and complete avoidance of standees. Intermediate stops and speed restrictions must be kept to a minimum.

(b) *Claustrophobia*, a fear of confining places, presents a likely hazard for those subject to its effects. Long, high-speed journeys underground can create apprehension and fear. As with aircraft, the problem can be partially overcome by the use of pleasing decor, lighting, music, or any other means of en-route activity that diverts the traveler's mind from the actual situation.

(c) *Motion sickness*, as used here, is caused by the effects on some individuals of seeing a succession of objects passing at high speed, the ribs in a tunnel lining for example, or a series of catenary poles. Riders can experience nausea and disorientation similar to what airplane pilots experience when light passes through an airplane propellor. This effect is further evidence for placing high-speed systems underground, possibly with curtained windows or no windows at all. The comfortable decor and ambience suggested for claustrophobia would be effective here as well.

(d) *Air pressure* on vehicle movement in a tunnel of restricted cross section will be discussed later. Pressure can also build up inside the vehicle, causing physical discomfort to the passengers. It is probable that vehicles at the highest speeds will require a pressurized interior similar to that of high altitude aircraft. The rate of pressure change should not exceed 0.10 psi per minute.[5]

(e) *Noise levels* will depend in part on the method of support and propulsion. Air-cushioned support or contact support and any form of jet or combustion propulsion will create noise. Electric motors and magnetic levitation should prove relatively quiet. There is evident need

---

[4]Ibid.
[5]Ibid., p. 5.4-17.

for sound insulation, perhaps combined with an overlaying sound of a different pitch or frequency to make unpleasant sounds less noticeable. Music may fill the need. The conversational level of 55 dB should be the goal. Sound must be kept below the 80-85 dB level at which injury to the ear may develop. Subway cars presently sometimes have an interior level of 98 dB, auto interiors 90 dB.

(f) *Ventilation* in long tunnels is always a problem. Aircraft experience indicates that a solution is feasible. In addition, heat from motors and compacted air must be dissipated and fresh air produced. The problem becomes most severe when any form of combustion engine is used to propel the vehicle.

(g) *Vibration*, both vertical and lateral, contributes to motion sickness and fatigue. The human body tends to tense its muscles to support the head and internal organs against the vibratory movements. At vibratory rates of 20 to 30 cycles per second, the head tends to resonate and vision is disturbed. Lower rates, especially in the 6 cycle per second range, can also be disturbing. Problems in support and suspension systems are posed and require designs that keep such vibrations within tolerable levels.[6]

(h) *Human control*. Design for high speed requires answers to the problem of how much control will rest with the operator and how much with an automated or semiautomated control system. At very high speeds, situations can arise too quickly for human decision and response. The operator can have full-decision responsibility or an override capability. With complete automation, the operator, if any, serves only as a monitor of the vehicle's functions. The experience of 600 mph (965 kph) commercial jets and 1000+ mph (1609+ kph) military aircraft indicate that considerable decision-making latitude can be left to the operator. A vehicle with roadway guidance offers an even greater capability for full automation.

TECHNOLOGICAL PROBLEMS

Many technological problems arise in the effort to develop high speed ground transportation systems.

Propulsion   In high-speed performance it is essential to have sufficient propulsive force and horsepower to accelerate to and maintain a high speed. The use of electrical energy from a central station will provide the

[6]Ibid., pp. 5.4-12 and 5.4-13.

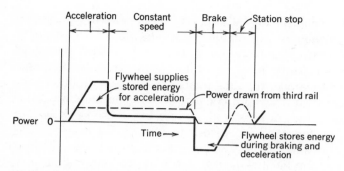

FIGURE 7.1 FLYWHEEL ENERGY STORAGE SYSTEM.
(COURTESY OF AIRESEARCH MANUFACTURING COM-
PANY OF CALIFORNIA, TORRANCE, CALIFORNIA, A
DIVISION OF THE GARRETT CORPORATION.)

rapid acceleration and ease of speed control usually associated with electric operation. There will be minimum noise, and in a tunnel location the
problem of exhaust gases from an internal combustion engine is eliminated.
The fuel supply will not take up vehicle space. Power requirements,
varying with the square of the speed or higher, mean heavy and expensive
demands for horsepower and the energy to produce it—raising again the
question of whether high speed is worth its cost. Present efforts are
directed toward the use of linear-induction motors (Chapter 5) as the
preferred mode of propulsion.[7] For nontunnel locations, the gas turbine
engine offers an efficient, lightweight on-board power source. The turbine
may develop propulsion through geared mechanical drive, by jet or propellor action, or through coupling to a motor-generator system.

One concept undergoing recent development for use by rapid transit
cars and buses is the flywheel storage principle in which a 150-lb (68.1-kg),
20-in. diameter (50.8-cm) flywheel (turning in a vacuum) is first energized
by an electric motor at a station stop or during a braking operation and
given a speed of 9000 to 14,000 rpm. The energy stored in the flywheel is
then used to accelerate and propel the vehicle toward the next station. The
system saves fuel. See Figure 7.1 for the flywheel power cycle.

Power Pickup   There is general agreement that electrical energy will be
required on board either for direct propulsion, for energizing elements in

[7]The Department of Transportation's linear-induction motored flanged rail vehicle reached
a speed of 255 mph (410.3 kph) at the Pueblo test track.

an induction principle system, or for lighting and other auxiliary purposes. Size and ventilation problems, especially in tunnels, tend to rule out on-board generation. The usual pickup system is through a sliding contact —shoes on a third rail or pantograph or trolley on a contact conductor cable. Experience has indicated problems in maintaining continuous contact at speeds above 150 mph (241 kph) because of wave action, sway, and vibration in the conductor and/or rock, roll, and nosing of the vehicle. The method can probably be improved for speeds of 200 to 250 mph (322 to 402 kph) but may prove limiting for higher speeds. Noncontact pickup using microwave transmission has been suggested and proven in laboratory conditions. Its technological feasibility for application to HSGT awaits further development.[8] Capacitive and inductive coupling as transmission media have also been suggested. Power efficiency and safety to passengers remain unsolved problems with these methods.

Speed Control   High speed can be attempted safely only when there is sufficient corresponding capability to control and reduce speed. While decelerating rates of 0.2 g will do for service application of vehicles brakes, a greater effort, up to 0.50 g, is required for emergencies.[9] Friction brakes in common use today, including rim, rib, disk, and rail brakes may be supplemented by regenerative braking (commonplace with electrically propelled locomotives and multiple unit (MU) cars), by aerodynamic braking, by which flaps and drag components are extended to use the potential in air resistance, or by hydraulic brakes in which the greater resistance of a liquid in a roadway trough provides brake action on a vane or scoop dipped into it; the depth of vane extension regulates the amount of braking.

Guidance   At high speeds, completely safe positive guidance is a necessity. Positive guidance is desirable at low speeds as well. Two general systems (with numerous variations) are emerging as alternatives to flanged wheels on rails. The first is the center rail guidance in which auxiliary wheels suspended laterally beneath the vehicle bear against an I-shaped girder or guide beam placed in the center of the roadway. See Figure 4.1a. A variation in use by certain subways (and proposed systems) has the lateral bearing wheels moved outward to bear against guide rails set

---

[8]*Research and Development for High Speed Ground Transportation*, Report of the Panel on HSGT to the Commerce Technical Advisory Board, U. S. Department of Commerce, March 1967, p. 22.
[9]Ibid.

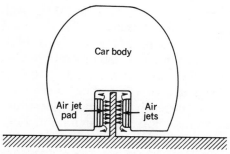

FIGURE 7.2 CENTER RAIL GUIDANCE
WITH AIR JETS.

alongside the track or roadway or against the high curb of a roadway. See
Figures 4.1*b* and 4.1*d*. The second method replaces the laterally hung
wheels with pads through which air is ejected against the guide beam or
against the walls of a confining trough or tunnel. If the center guide beam
type is used with air jets, the vehicle will probably straddle it as is the
design for the Aerotrain and the (wheel-guided) Alweg monorail. See
Figure 7.2.

Support and Levitation   The type of support or levitation is fundamental in high speed systems. Contact with the roadway is made conventionally by flanged steel wheels on rails or by rubber tires on pavements.

For contact support, that is, flanged wheels on rail or rubber tires on
pavement, a high degree of precision in profile, cross level, and alignment
must be maintained to provide smooth riding qualities, reduce shock and
impact, and keep the vehicle on the roadway. Present FRA Safety Standards for Class 6 railroad track permit no greater variation in cross level
than $\frac{5}{8}$ in. (1.59 cm) in 62 ft (18.9 m).[10] The White Sands Proving Grounds
guideway for tests of high speed "sleds" up to 2900 fps (1977 mps)
permitted no greater alignment deviation than ±0.036 in. (0.09 cm) in
20,000 ft (6096 m). Obviously a high degree of precision would be required
for speeds in the 300-600 mph (483-965 kph) range. Such high precision is
costly to build and maintain.

There is some evidence that steel wheels on rails lose adhesion at speeds
above 200 mph (322 kph), but the actual performance at higher speeds has

[10]Federal Railroad Administration Track Safety Standards as amended 22 December 1972,
para. 213.63 Track Surface.

not been investigated.[11] This, and the effects of centrifugal forces on wheel metallurgy await further research. Rubber tires have withstood $200 \pm$ mph speeds on race tracks and on single runs (on the Salt Lake Flats) at over 500 mph (804.5 kph), but the viability at speeds over 200 mph of rubber tires on conventional highway surfaces awaits further demonstration. The general acceptance of difficulties both in wheel and in roadway design for speeds above 200 mph (322 kph) thus leads to a concentration of effort in noncontact systems. Two such systems are presently undergoing development: (a) fluid (air) support and (b) magnetic support or levitation. Both systems create a demand for power to provide the support beyond that required for propulsion.

The fluid support systems require induced power that varies in magnitude with the vehicle speed, lift, lift area, vehicle configuration, and clearance between the vehicle and the roadway surface. The problem is primarily one of clearance vs. power demand. Very close clearance systems make use of air "lubrication" exemplified by shoes sliding on a guideway rail with air pressure provided through holes in the shoe. An air film of 0.001 in. (0.00254 cm), more or less, is established.

Close clearance can also be had with plenum chambers or pads that maintain a cushion of air within the chamber. The ground effects machine (GEM) design uses an 0.01- to 0.05-in. (0.025- to 0.127-cm) air cushion with an open flexible plenum skirt about 4 in. (10 cm) long. A reduction in total power is possible if air jets can be directed backward in such a manner as to provide both lift and propulsion. The confining space of a close clearance tunnel would be helpful in this respect, although adding to parasitic drag on the vehicle.

A rather high degree of precision in roadway construction and maintenance is required. The surface itself should be smooth. An order of smoothness of 1 cm in 10 meters is desirable. Power ratios for lift vs. propulsion are in the order of 1 to 3. See also Chapter 4.

Magnetic Levitation   Magnetic support systems (MAGLEV) utilize the fundamental attraction-repulsion properties of magnetized materials. Two systems are undergoing study: (a) the magnetic repulsion system (in the United States) by which vehicle repulsion (levitation) occurs as the vehicle is "repulsed" upward from a metallic portion of the roadway, and (b) a magnetic attraction system by which the vehicle is levitated by the attraction between shoes on the vehicle and an armature fixed above the edge of the roadway. See Figure 7.3.

---

[11]*Research and Development for High Speed Ground Transportation*, Report of the Panel on High Speed Ground Transportation convened by the Commerce Technical Advisory Board, U. S. Department of Commerce, Washington, D. C., 1967, pp. 10–13.

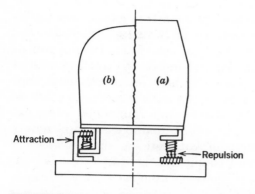

FIGURE 7.3 MAGNETIC LEVITATION CONFIGURATIONS: (*A*) ATTRACTION, (*B*) REPULSION.

In the repulsion system, a nonmagnetic conductor such as aluminum is used for the track. Superconducting magnets are in the underside of the train. As speed increases, eddy currents generated by magnetic interaction lift the vehicle 8 to 10 in. (20.3 to 25.4 cm) from the track. The system gains lift with speed.

The attraction system uses electromagnets placed both in the shoe of the vehicle and in a groove in an inverted rail. A gap of one inch is maintained by controlling the strength of the magnets. Lift is not dependent on speed.

For noncontact support, that is, air cushion or magnetic levitation, the degree of precision in roadway construction and maintenance varies with the clearance required between vehicle and roadway. Close clearances in the order of 0.001 to 0.05 in. (0.00254 to 0.127 cm) for the GEM system, up to 1 in. (2.54 cm) for MAGLEV, require a high degree of precision. The greater the clearance, however, the more horsepower is needed to maintain the necessary volume of air or magnetic field strength for support. Such tradeoffs must be evaluated economically.

Earth Movements   In addition to the distortions in cross level and alignment that arise from vehicle impact and dynamics or from local support deviations, the problem of earth movements must be considered for deep tunnel locations where subsurface adjustments of a minor nature frequently occur. Special design precautions were introduced into the BART system Trans-Bay tunnel between San Francisco and Oakland to protect against the shock of earthquakes. More frequent minor shocks and tremors can be a disturbing influence where roadway surfaces requiring high precision are at more than nominal tunnel depth.

Tunnel Construction   Long distance tunnel construction presents problems in construction procedures, cost, effects of earth tremors and movements, and disposal of muck or waste materials excavated. The support of unstable ground is always a problem that has led to the development of tunneling machines with supporting shields from which permanent supporting linings can be installed. Giant earth augers have been designed for drilling through stiff clays. The use of a laser beam, flame jets, or plasma arcs to "melt" or otherwise disintegrate rock materials is undergoing research to supplement or supplant the conventional jumbo with its time-consuming drill, blast, and muck sequence. Drainage, ventilation, exhaust resistant tunnel linings, and escapeways in the event of accident or system failure pose additional problems.

Air Compression   The narrow confines of a tunnel tend to cause a buildup of hot air masses ahead of a rapidly moving vehicle that fills most of the tunnel cross section, the so-called plunger effect. Not only are propulsive resistance and pressure within vehicles increased, but the heat so created must be dissipated. A 10,000-hp vehicle of normal size passing through a tunnel of 200-ft^2 cross section at 300 mph (483 kph) can raise the temperature 4° F with electric propulsion, but with turbine propulsion it can rise 16° F.[12] A vehicle entering at high speed where the ratio of tunnel area to vehicle area approaches one experiences shock and air compression that can be harmful both to the vehicle and to its occupants.

Ventilation, forced or through air shafts, will be necessary to overcome air resistance and compression and heating and to allow for gas escape where chemical fuels are used. One method proposes to solve the problem with a series of air locks by which air is withdrawn from a route section ahead of a vehicle to reduce air resistance; it is then introduced behind the vehicle to aid in propulsion. Admission of air ahead of the vehicle would assist in speed control and braking. Another system would draw air from the front of the vehicle, give it further compression with an on-board compressor, and eject it at the rear as a means of propulsion. At the very least, adequate air escape vents and fans are required.

SPECIFIC HIGH-SPEED CONCEPTS

A listing of HSGT concepts could go on for pages. A few of those showing feasibility and with some record of successful testing are summarized below.

[12]Ibid., p. 22.

Automatic Highway Possible improvements in highway transport have already been described. In addition an innovative feature is the automatic highway on which vehicles are guided and controlled by inductive coupling to an electric cable laid in the center of each lane. Despite foreseeable improvements in roadway, engine, and vehicle design, it is unlikely that speeds greatly exceeding 100 mph (160.9 kph) will become commonplace.

Auto Carriers Supplementing the improvement of conventional high technology is the use of auto carriers to move automobiles (and trucks) long distances with a predicted saving in energy, pollution, time, highway construction, safety, and convenience.

One proposal calls for new rail tracks with rails spaced about 18 ft (5.49 m) apart, carrying vehicles with a capacity of 10 to 12 automobiles in slots *transverse* to the track. In another system towers are constructed to support aerial tramway cables from which cars are hung by attachments on the car roofs.

These systems are perhaps solving the wrong problem. There is still ample capacity on the major highways for intercity movement. The congestion occurs in and at the approaches to cities. Auto carriers thus provide capacity where it is not needed, then dump a high concentration of vehicles in the areas where congestion is greatest. Nevertheless, an increase in capacity demand and a reduction in highway uses of energy and pollution may indicate a need for such systems at a later time. Currently, autos are being carried over conventional railroads, positioned longitudinally in railroad cars, while the car driver relaxes in coach-lounge cars. The Auto-Train Corporation provides such service between northern points and Florida. Its chief attractions are the convenience of having one's own car at the desired destination without the tiresome driving and expense of an intervening highway journey. A similar service is under consideration for trucks.

Monorail There is no such thing as a practical, true monorail, that is, a vehicle supported on one rail by a row of wheels in line, stabilized by a gyroscope. The nearest approach to this system is an operation in Wuppertal, Germany in which the car hangs suspended from a single rail and travels at speeds between 20 and 30 mph (32 and 48 kph). A few low capacity variations of this system have been used at expositions, parks, and airports. Other so-called monorail systems use many wheels on axles and/or more than one supporting surface.

Monorails are of two types—suspended and supported. The suspended type may use a split box girder guideway with the vehicle suspended from an electric power truck with closely spaced but conventional flanged wheels running on two parallel rails. The steel wheels may be replaced with rubber tires on concrete pads and the flanges by side rails or center rail guidance. High vertical clearance, side sway, especially approaching station platforms, and longitudinal thrust on the supporting roadway structure are problems with this design. Switching, too, presents a problem. The system presently works best with a pendulum or loop roadway configuration. Speeds of 150+ mph (241+ kph) have been predicted, but pilot models have seldom exceeded speeds of 70 to 80 mph (113 to 129 kph). Several systems are in successful commuting operation in the Metro area of Tokyo.[13]

Supported types, exemplified by the Disneyland and Seattle models, follow the Alweg design and straddle a heavy concrete "beam" used as a rail and with several sets of rubber-tired wheels bearing on the surface for support and traction and on the sides for guidance. It, too, requires high vertical clearance, the concrete rail surface becomes corrugated and has to be ground smooth from time to time. Riding qualities are not of the best. A major problem is switching. A section of the entire beam has to be moved, a process requiring approximately 45 seconds. Speeds of over 130 mph (209 kph) have been attained but 20 to 30 mph seem to characterize present models. Figure 4.1 shows monorail configurations.

Monorails have been proposed to connect CBDs with airports, for inter- and intra-airport connections, for suburban commuter service, and even for intercity travel. There seems little reason for their use, however. The known technology of the dual rail, supported systems can accomplish everything claimed for monorail and, according to various studies, with less cost and effort.

Air Support Systems   The basic principles of air support systems have already received mention. In application, the high clearance Hovercraft, noted earlier, represents an operating system, most notably its use to cross the English Channel from Dover to Boulogne in approximately 45 minutes.

A second system is represented by the French-designed Aero-Train and by the tracked air cushion vehicle (TACV) undergoing development and test by the DOT's Office of High Speed Ground Transportation at their HSGT Test Center near Pueblo, Colorado. The Aero-Train is supported by an air cushion of only 0.01 in. (0.0254 cm) or so above a concrete roadway made in the form of an inverted Tee. The vehicle straddles the inverted stem of the T

[13]*The Japan Times*, 14 May 1973, p. 3.

FIGURE 7.4 DEPARTMENT OF TRANSPORTATION TEST VEHICLES: (*a*) LINEAR-INDUCTION MOTOR VEHICLE (LIMV) AND (*b*) TRACKED AIR CUSHION VEHICLE (TACV). (COURTESY OF THE U.S. DEPARTMENT OF TRANSPORTATION, HIGH SPEED GROUND TRANSPORTATION TEST CENTER, PUEBLO, COLORADO.)

(a)

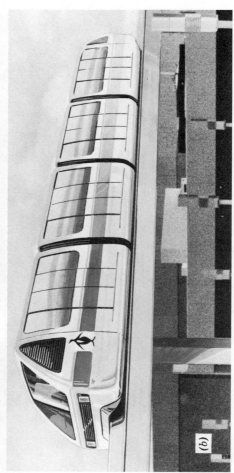

(b)

FIGURE 7.5 FUTURE AND DEVELOPING MODES. (a) TRANSIT EXPRESSWAY THREE-CAR TRAIN (COURTESY OF WESTINGHOUSE ELECTRIC CORPORATION, PITTSBURGH, PENNSYLVANIA.) (b) HAWAIIAN MONOTRAIN INSTALLATION. (c) SUPPORT AND GUIDE BEAM FOR ROHR MONOCAB. (d) MAGNETIC LEVITAVED VEHICLE. [VEHICLES (a), (b), AND (c) ARE COURTESY OF ROHR INDUSTRIES, CHULA VISTA, CALIFORNIA.]

and secures guidance by air jets acting against that stem. Propulsive force has come from an outboard jet engine but linear-induction motors are contemplated for the future.

The DOT's TACV travels in a troughlike roadway. Air jets against the sides of the trough provide guidance. Again, on-board jet engines provide propulsion, but linear induction motors will be used in future designs. See Figure 7.4*a*. The rail-mounted version with a linear-induction motor (LIMV) has achieved a speed of 230 mph (370.1 kph). The TACV is designed to carry 200 passengers at speeds of approximately 150 mph (240 kph). See Figure 7.4*b*. Air turbulence would be a problem in operating these vehicles in trains or with very close headway. For slow speed terminal and switching movements, conventional-tired wheels can be provided.

Tube Systems   Vehicles in underground tubes vary from the conventional dual-rail rapid transit systems to exotic aerodynamic support and propulsion systems.

One design, the *Edwards Tube*, proposed a series of air lock sections. Air is evacuated ahead of rail-mounted vehicles to reduce air resistance then introduced behind the vehicle to aid propulsion. This system with deeply buried tunnels depends largely on gravity both for propulsion and for deceleration as the vehicles descend then rise to stations near ground elevation.

The *Foa Tube* uses the ram jet principle. Air is drawn in at the front of the vehicle, thereby reducing air resistance. The air is compressed and ejected in a whirling jet to form a propulsive vortex at the rear of the vehicle. The vehicle is supported and guided by air jets directed from pads against the tunnel walls. Speeds of 500 to 1000 mph have been projected.

Summary   There are obviously many innovative concepts, close to 100 or more, available for development and use. Of these many, both systems of monorails, the Transit Expressway (modified PRT), Dial-A-Bus, and the Hovercraft are in commercial operation. These systems have reached a state of technological development where their selection and use are governed primarily by economic factors.

The linear-induction motor, discussed in more detail in Chapter 6, and the close clearance air support systems are undergoing tests on pilot models in Europe and by the DOT. Additional technological development is required by these systems. A pilot system of magnetic levitation continues to be under study. Figure 7.5 shows various proposed new modes.

While technological considerations still pose serious problems, the biggest factor is economic. Can these new systems provide a technologically and economically viable alternative to known and functioning systems? An affirmative answer is yet to be established.

QUESTIONS FOR STUDY

1. Explain the need for placing considerable reliance on making short-term improvements in existing modes of transport.
2. Account for the lag time that usually exists between the development of a transport concept and the time it is actually available in a viable day-to-day state of operation.
3. Describe some of the "state-of-the-art" improvements that are available to make existing modes of transport more useful.
4. Distinguish between services that require low-speed innovations and those requiring high-speed ones and explain the reasons for those different needs.
5. Define high-speed ground transportation and explain why the effort toward higher speed must be directed toward substantial increases in speed capability.
6. What is personalized rapid transit? Explain its proper area of usefulness or utility.
7. What technological problems do tunnel locations for high speed transportation present?
8. Explain the statement that high-speed represents a hostile environment to people and define the human problems that arise from it.
9. Outline the areas of utility represented by the several types of high-speed transport proposed for the future.
10. Give specific reasons to indicate which of the proposed innovations have the greatest likelihood of succeeding.

SUGGESTED READINGS

1. W. W. Siefert et al., *Survey of Technology for High Speed Ground Transportation*, Massachusetts Institute of Technology report for the United States Department of Commerce, Northeast Corridor Transportation Project, PB 168 648, Washington, D. C., June 1965.
2. *The Glideway System*, an interdisciplinary design project by students of the Massachusetts Institute of Technology for the United States Department of Commerce, M. I. T. Press, Cambridge, Massachusetts, 1965.
3. Howard R. Ross, "New Transportation Technology," *International Science and Technology*, November 1966.
4. *Research and Development for High Speed Ground Transportation*, Panel on High Speed Ground Transportation convened by the Commerce Technical Advisory Board, U. S. Department of Commerce, March 1967.
5. William S. Beller and Frank Leary, "Megalopolis Transportation," *Space/Aeronautics*, New York, September 1967.
6. *Tomorrow's Transportation: New Systems for Urban Development*, Office of

Metropolitan Development, Urban Transportation Administration, U. S. Department of Housing and Urban Development, Washington, D. C., 1968.

7. *Report and Recommendations to Governor Kerner and the 75th General Assembly* by the High Speed Rail Transit Commission, State of Illinois, Springfield, Illinois, March 1967.

8. *Proceedings of the Midwest High Speed Rail Transit Conference*, sponsored by the Illinois High Speed Rail Transit Commission and the Chicago Association of Commerce and Industry, Chicago, Illinois, 12 January 1967.

9. Robert B. Meyersburg, *Commercial V/STOL and the California Corridor*, paper presented at the National Aeronautic meeting, Society of Automotive Engineers, New York, 25–28 April 1966.

10. *Transit Expressway: Concept and Accomplishment*, Westinghouse Electric Corporation, Pittsburgh, Pennsylvania, April 1967.

11. *R Rollway*, General American Transportation Corporation, Chicago, Illinois, May 1966.

12. Joseph V. Foa, *High Speed Ground Transportation in Non-Evacuated Tubes*, TR AE 604, Rensselaer Polytechnic Institute, Troy, New York, 1966.

13. Robert A. Wolf, *Elements of A Future Integrated Highway System Concept*, Transportation Research Department, Cornell Aeronautical Laboratory, Buffalo, New York, March 1965.

14. D. L. Atherton, *Study of Magnetic Levitation and Linear Synchronous Motor Propulsion*, Annual Report for 1972 by the Canadian Maglev Group, Department of Physics, Canadian Institute of Guided Ground Transport, Queen's University, Kingston, Ontario, Canada, December 1972.

15. *Surface Effects Ships for Ocean Commerce*, Final report on a study of the technological problems by the SESOC Advisory Committee convened by the Commerce Technical Advisory Board, U. S. Department of Commerce, February 1966, Washington, D. C.

16. *A Systems Analysis of Short Haul Air Transportation*, prepared for the U. S. Department of Commerce by the Massachusetts Institute of Technology, Cambridge, Massachusetts, August 1965, Part III, Contract C-88-65, Technical Report 65-1, PB 16 95 21.

17. *High Speed Rail Systems*, by TRW Systems Group, Contract No. C-353-66(Neg), for the Office of High Speed Ground Transportation, U. S. Department of Transportation, Washington, D. C., February 1970.

18. *Personal Rapid Transit II*, a report of the 1973 International Conference on Personal Rapid Transit, edited by J. Edward Anderson and Sherry H. Romig, published at the University of Minnesota, Minneapolis, Minnesota, February 1974.

19. R. F. Kirby, K. U. Bhatt, M. A. Kemp, R. G. McGillivray, and Martin Wohl, *Para-Transit: An Assessment of Experience and Potential*, Final Report Volume II, Para-Transit Program Design, The Urban Institute, Washington, D. C., June 1974.

20. *Demand Responsive Transportation Systems*, Special Report No. 136, Highway Research Board, Washington, D. C., 1973.

21. R. F. Kirby et al., *Para-Transit*: *Neglected Options for Urban Mobility*, Para-Transit Program Design, The Urban Institute, Highway Research Board, National Research Council, Washington, D. C., 1974.
22. *Light Rail Transit*, Special Report 161, Transportation Research Board, National Research Council-National Academy of Sciences, Washington, D. C., June 1975.

Part 3

Factors in Operation

Chapter 8

Level of Service Factors – Performance Criteria

In Chapter 4 and 5 the technological characteristics of buoyancy, stability, guidability, resistance, propulsion, and the effects of elevation and gradients were briefly reviewed. In this and the following chapter certain characteristics more closely related to the way in which a carrier is used and operated are surveyed. These characteristics relate to the *level of service* required to meet the volume of demand: capacity, speed, accessibility, flexibility, frequency; it also relates to the *quality of service*: safety, dependability, speed, acceleration, door-to-door travel time, comfort and amenities, effects on environment, pollution, energy use, land use, and effects on the community.[1] These factors, within which there must be some obvious overlap, combine with the characteristics of Chapters 4 and 5 to establish modal utility. Technological considerations establish the absolute limits for this group of characteristics. It is only within these limits that other considerations, such as the economic, operate to determine the utility of these characteristics as known and used in practice.

CAPACITY

A fundamental requirement for any transport system is its ability to meet the volume of demand. A system's traffic capacity is measured by the quantity of freight or number of passengers that can be moved per hour or

---

[1]Note that later in the chapter a more restricted use of the term "level of service" is used in discussing highway capacity.

per day between two points by a given combination of fixed plant and equipment. Traffic capacity is a function of vehicle capacity, speed, and the number of vehicles, trains, or craft that can be on a roadway at any one time, that is, the route capacity. The next section reviews the combined effects of speed and vehicle capacity.

Vehicle Capacity   The capacity of a given craft or vehicle is largely a problem in wheel loading, lift or buoyancy, horsepower, and space utilization. As earlier discussed, the highest possible payload to dead load ratio is generally sought, but when moving persons comfort factors may limit the designed capacity. Widths, heights, and wheel loadings of vehicles and craft are restricted by law, pavement width, roadway strength, and operational economy. Highway vehicles are limited in most states to a width of 96 in. (243.8 cm). Heights are limited by clearance under overhead structures, generally 10 to 12 ft (3.05 to 3.66 m) on older roads, 16 ft on Federal Interstate Highways. Railroad equipment is limited by bridge and tunnel portals and by overhead clearances of 22 to 24 ft (6.71 to 7.32 m) between the top of rail and the underside of obstructions. The usual widths of rail equipment are 10 to 11 ft (3.05 to 3.35 m), but subway cars may be no more than 8 to 9 ft (2.44 to 2.74 m) wide. Barges, tows, and Great Lakes carriers are limited by channel depth and lock widths and lengths. Barge and towboat widths seldom exceed 40 ft (12.19 m). Lock widths varying from 66 to 110 ft (20.12 to 33.53 m) and lengths of 360 to 1200 ft (109.7 to 365.75 m) establish how many barges can be taken through at one time. Locks on the Great Lakes-St. Lawrence Seaway System have a maximum width of 105 ft (32 m). Aircraft and oceangoing vessels capacities are generally limited only by the availability of traffic as related to economic design but also find limitations in harbor and channel depths and locks in the Panama Canal. Roadway-bearing capacity can limit the weights of trucks and railroad cars.

Freight stowage creates special problems for handling irregularly shaped items. Bulk cargo carriers such as barges, hopper cars, tank cars, and corresponding highway vehicles make maximum utilization of space. In railroad cars and airplanes, passenger comfort vs. economy is the problem as exemplified by the two- vs. three-abreast seating arrangements. Rapid transit cars provide only nominal seating capacity, relying instead on standees.[2] Railroad commuting cars may seat 70 to 80 people on one level and 160 or more in double-decked cars.

---

[2]The 150 light rail cars being built by Boeing Vertol for the Massachusetts Bay Mass Transportation Authority and 80 similar cars for the San Francisco Municipal Lines seat 52 but including standees can accomodate as many as 210 people.

Net Ton Miles per Vehicle Hour   Transport agencies exist to produce transportation. It is their only product. It cannot be stored but must be consumed as it is produced. The rate at which transportation can be produced is one measure of transport productivity and capacity as measured by the *net ton mile per vehicle hour = net tons per vehicle × the speed* in miles per hour. Operationally, the weight of the vehicle must also be moved, so that gross ton miles per vehicle hour, a figure including both vehicle and cargo weight, is of concern. For moving persons, the unit becomes *passenger-miles per vehicle hour = passengers per vehicle × the speed*.

This productivity measure can be maximized in several ways:

1.   *By placing as much tonnage as possible in one carrying unit*, such as a train or ship. Skillful loading and a full utilization of the vehicle's propulsive force and horsepower are involved. If traffic is not abundant, the vehicle may be held until a full tonnage becomes available. This practice makes the statistical unit look good but downgrades service by delaying traffic movement and causing dissatisfied customers. Pipelines and conveyors, although holding only nominal amounts of cargo per unit of length, move very large amounts in a day by the continuous, full utilization of propulsive force and horsepower.

2.   *By making the vehicle itself large enough to hold many tons of freight (or many passengers)*. The airlines' 747 is an effort in this direction, but the potential for improvement in this area for airlines and trucks is rather restricted. Increasing vehicle size and/or allowable wheel loads will increase damage to roads and increase the accident hazard for automobiles. Less restriction exists for bulk cargo ships, towboats, and trains although the lock size, harbor depth, and track gauge impose restrictions. The costs of increased track wear from heavier railroad cars is creating concern in that industry.

3.   *By increasing vehicle speed*. This requires light loads combined with high speed—or the utilization of a relatively enormous power plant. Airplanes possess high speeds to the greatest degree; railroads and highways next, and ships, tows, pipelines, and conveyors the least. Carriers tend to utilize all three methods to the limit of their technological capabilities.

Using the statistically average train of 1973, 1844 net tons (1672.5 tonnes), and an average train speed of 20 mph (32.2 kph) transportation was produced at the rate of 36,880 net ton miles per freight trains hour.[3]

[3]*Yearbook of Railroad Facts*, 1973 edition, Economics and Finance Department, Association of American Railroads, Washington, D.C., p. 44.

Gross ton miles per freight train hour for the same period were 76,726.

A 40-ton (36.3-tonne) tractor-trailer combination at 55 mph (88.5 kph) produces 2200 ton-miles per truck hour. Single truck units will carry $1\frac{1}{2}$ to 10 tons (1.4 to 9.1 tonnes) or more; tractor single-trailer combinations as much as 20 tons (18.2 tonnes). Iron ore is being hauled in Labrador from the mine to rail loading sites in 100- and 125-ton (90.7- and 113.4-tonne) vehicles. At speeds of 20 mph (32.2 kph) these produce 2000 and 2500 net ton miles per vehicle.

The most recent design of Great Lakes bulk cargo carrier will carry 42,000 tons (38,094 tonnes) at a speed of 16 mph (25.7 kph) (knots are seldom used in Great Lakes terminology) producing 672,000 net ton miles per ship hour. Oceangoing ships holding 40,000 to 140,000 cargo tons (36,280 to 126,980 tonnes) and moving 15 to 20 mph (24.1 to 32.2 kph) produce 600,000 to 2,800,000 net ton miles per ship hour.[4] Even larger vessels are in use, primarily in the petroleum trade.

Net ton miles per towboat hour, based on a conventional speed of 8 mph (12.9 kph) and ten 2000-ton (1814 tonne) barges, amounts to 160,000. Tows of even greater size and speed, up to 22,500 tons (20,408 tonnes) at 12 mph (19.3 kph) pool speed, give approximate maximums of 270,000 net ton miles per towboat hour. Six or eight-barge tows are more common, producing 96,000 to 128,000 net ton miles per towboat hour at 8 mph (12.9 kph).

Pipelines vary considerably but two typical operations may be noted. An 8-in. line at 1 mph will produce 1000 ton-miles per pumping-station hour. The same line at a speed of 4 mph (more powerful pumps being required) produces 4000 net ton-miles per pumping-station hour. A 24-in. line at 1 mph will produce 9300 net ton-miles per pumping-station hour. The same line at a speed of 4 mph produces 37,000 net ton-miles per pumping-station hour.

Airplanes have a low unit load capacity but fly that load at high speeds. The D-6A flying at 300 mph (482.7 kph) with 15 tons (13.6 tonnes) of payload produces 4500 net ton-miles per airplane hour and the Globe-master flying 25 tons (22.7 tonnes) at the same speed produces 7500 net ton-miles per airplane hour. The Douglas DC-8 transport, powered by four turbojet engines offers a maximum speed of 598 mph (962 kph) with a payload of 35,000 lb (15890 kg) for 10,665 net ton-miles per airplane hour. Other modern jet transports afford similar capabilities. Modern passenger jets fly at speeds of 500 to 600 mph (804.5 to 965.4 kph). The Boeing 747

[4]On 24 November 1973 the Japanese ship *Kohnan Maru* loaded 139,300 long tons of iron ore at Sept Iles, Quebec. *Iron Ore*, Vol. 1, No. 3, 24 December 1973.

TABLE 8.1 NET TON-MILES PER VEHICLE HOUR

Carrier	Unit	Quantity (Normal to Maximum)
Railroads	Net ton miles per train hour	25,000 to 360,000+
Trucks and semi-trailers	Net ton miles per truck hour	1260 to 2,400
Ships—Great Lakes	Net ton miles per ship hour	200,000 to 300,000+
Ships—deep-water	Net ton miles per ship hour	100,000 to 2,500,000+
Towboats	Net ton miles per towboat hour	100,000 to 270,000
Airplanes—propeller-driven	Net ton miles per airplane hour	4,500 to 7,500
Airplanes—jets	Net ton miles per airplane hour	8,000 to 10,600
Helicopters	Net ton miles per helicopter hour	200 to 1200
Pipelines	Net ton miles per pumping-station hour	1,000 to 37,100+
Conveyors	Net ton miles per flight hour	600 to 13,500+
Cableways	Net ton miles per mile of cable-way per hour	50 to 600

designed to carry 320 to 490 passengers can thus produce from 160,000 to 294,000 passenger miles per airplane hour. Table 8.1 gives typical values for both freight and passenger carriage.

These figures relate only to actual road-haul time. Terminal delays and stops en route are not included. Continuous-flow carriers and ships have a low hourly ton-mile output but compensate for that by a 24-hour day operation, thus giving continuous delivery. A greater output tonnage may then be delivered than by assembled-unit carriers with a high level of productivity between terminals but with high terminal delays at the originating, interchanging, and terminating terminals. A railroad freight car, for example, can achieve its average daily mileage in one hour. From 60 to 90 percent of the remaining 23 hours is usually spent in some form of terminal activity or delay.

Vehicle Performance   For most types of transport, vehicle performance involves three basic factors: vehicle miles, gross ton miles, and road time. Road time may be expressed in terms of hours per trip or hours per 100 vehicle miles. The time T in hours to move a cargo over a distance of S miles with a given horsepower at a speed of V mph is $T = S/V$. Earlier chapters on propulsive and resistive forces have shown that as the load is increased, speed decreases and the minimum time per trip or per 100 vehicles increases.

In the following examples, a somewhat-hypothetical wheeled land vehicle is assumed, weighing 30 tons (27.2 tonnes) empty, having a cross-sectional area of 144 ft,[2] and a 250-hp engine, moving upon a smooth concrete pavement of level grade over a 100-mile (161-km) trip distance. Rolling resistance is assumed to be 20 lb per ton[5] (9.1 kg per ton). Figure 8.1 shows the simple relation between gross vehicle load and speed.

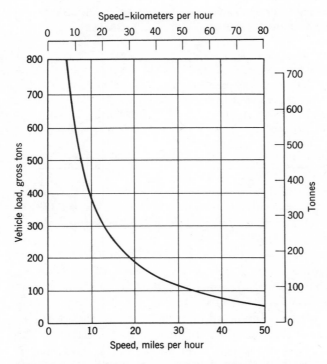

FIGURE 8.1 MAXIMUM GROSS VEHICLE LOADS WITH VARIOUS SPEEDS AT CONSTANT HORSE-POWER.

Figure 8.2 shows variations in gross ton-miles per vehicle hour for different weights and corresponding speeds. The product of the *gross weight per vehicle × length of trip ÷ hours per trip = gross ton-miles per*

[5]A tractor-trailer combination weighing 30 tons (27.2 tonnes) empty, accelerating to 45 mph (72.4 kph) when empty and to about 30 mph (48.3 kph) when carrying 80 tons (72.6 tonnes) of payload is in use near Peoria, Illinois, hauling coal from a strip-mining operation to the railroad.

vehicle hour, or $GTM / T_s = (W_e + W_c)S \div T_s$, where W_e = vehicle weight in tons, W_c = cargo weight in tons, S = length of trip, and T_s = time for one trip. As the loading increases, the gross ton miles per vehicle hour increase, up to a certain speed, at which point the decrease in speed resulting from the heavier load *may* bring about a decrease in total gross ton-miles per vehicle hour. Figure 8.2 has been plotted for the same equipment and conditions as Figure 8.1. Here the increase in time per trip due to lower speed has been compensated by the increase in propulsive force at the lower speed, which permits heavier loads. This is usually the case.

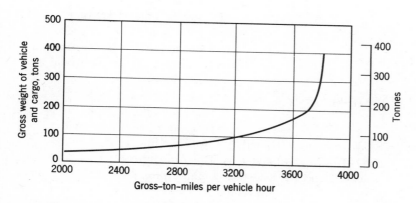

FIGURE 8.2 VARIATIONS IN GROSS TON MILES PER VEHICLE HOUR WITH VARIOUS VEHICLE LOADINGS.

The number of vehicles to move a given tonnage between two points including return trip time and terminal or loading and unloading time may be computed in the following manner. The number of trains, planes, ships, or trucks for other time periods can be obtained by making appropriate changes in the time factors.

VEHICLES TO MOVE A GIVEN NET TONNAGE

A. Let W = total number of net tons to move per day

W_g = gross tons equivalent, that is, net tonnage plus vehicle weight

$W_g = (W + 2W/R_p)$ where R_p = payload to empty weight ratio and the factor 2 represents return trip empty

B. W_d = gross tons moved per *vehicle day*

$W_d = (W_n + 2W_e)N_t$ where W_n = net tons in vehicle, that is, cargo

$\quad W_e$ = empty weight of vehicle

$\quad N_t$ = number of round trips per day per vehicle

$$N_t = \frac{24}{T_c + T_e + T_t}$$

where (for a 24-hour day):

$\quad T_c$ = time to travel with cargo, that is, loaded travel time

$\quad T_e$ = time in return direction, empty

$\quad T_t$ = terminal and delay time

(*Note*: the foregoing can also be based on an 8-hour day or shift, in which case the numerator would be 8 instead of 24.)

C. But: $T_c = S/V_1$ and $T_e = S/V_e$ where V_1 = speed loaded in mph

$\quad V_e$ = speed empty in mph

$\quad S$ = distance one way in miles

Then: $T_c = S/(375 \times \text{hp} \times e/R_1)$ and $T_e = S/(375 \times \text{hp} \times e/R_e$

$T_c = R_1 S/(375 \times \text{hp} \times e)$ and $T_e = R_e S/(375 \times \text{hp} \times e)$

where R_1 and R_e = loaded and empty propulsive resistance of the vehicle,

$\quad \text{hp}$ = horsepower available per vehicle

$\quad e$ = mechanical efficiency of vehicle

And $W_d = (W_n + 2W_e)\left[\dfrac{24}{\dfrac{(R_1 + R_e)S}{375 \times \text{hp} \times e} + T_t}\right]$

D. Number of vehicles required = $N = W_g \div W_d$

Then $N = (W + 2W/R_p) \div (W_n + 2W_e)\left[\dfrac{24}{\dfrac{(R_1 + R_e)S}{375 \times \text{hp} \times e} + T_t}\right]$

Note that the foregoing takes into account most of what has been studied thus far about the vehicle—propulsive resistance, propulsive power, and the payload to empty weight ratio. These are further related to speed and the practical factor of terminal and road delays.

ANOTHER APPROACH

E. Total number of vehicles in terms of loading time:

$Nt = T + \dfrac{S}{V_1} + \dfrac{S}{V_2}$ where t = time to load one vehicle and

N = number of vehicles to balance loading operation

Nt = time for one vehicle to make a round trip.

T = time factor or constant loading, unloading, tuning, regular delays

S = distance one way

V_1 = speed going, loaded

V_2 = speed returning, empty

Therefore, $N = \dfrac{T}{t} + \dfrac{S}{V_1 t} + \dfrac{S}{V_2 t}$

The foregoing procedures assume a fixed type of motive power and horsepower and a fixed route and system of operation with relatively constant speed and no excessive traffic or other interference, that is, adequate route capacity.

Route Capacity The discussion of vehicle capacity assumes sufficient route capacity to handle all needed vehicles. Route capacity can, however, be a limiting factor in system capacity. It might seem that the maximum capacity would obtain in a saturated condition with vehicles placed "bumper to bumper" as in Figure 8.3a. Because of the slow speed with such an arrangement, the output is generally not as great as with some spacing or headway at a higher speed (Figure 8.3b). In that sketch, the *headway* is

$$H = L_v + S_d$$

where

H = distance from the front of one vehicle to the front of the following vehicle (or from rear to rear), that is, the headway. See evaluation of H for rapid transit lines on p. 281.

L_v = vehicle length in feet, assumed to be uniform

S_d = stopping distance at v mph including operator's reaction time, braking distance, and a factor of safety

By determining the number of feet of vehicles and headway that can pass a given point in an hour at V mph and dividing that length of line by the headway, the capacity in vehicles per hour (vph) is obtained. That is,

$$C_v = \frac{5280 \, V}{H}$$

$$C_v = \frac{5280 \, V}{L_v + S_d}$$

where

$$C_v = \text{capacity in vehicles per hour}$$

$$V = \text{speed in miles per hour}$$

Traffic capacity is then obtained in terms of tons or persons per hour by multiplying C_v by the vehicle capacity in tons or persons, that is, $C_t = C_v \times L_c$ where $L_c =$ vehicle capacity. L_c is a function of vehicle size and loading. The holding capacity of a system would be the number of route miles $\times$ the traffic capacity in persons or tons for one mile.

Spacing constraints are somewhat different for a rapid transit line or railroad where block systems (usually automatic) are in effect. Figure 8.3c shows the normal signal sequence of "Stop," "Approach," and "Clear-Proceed," a so-called two-block system. In order to maintain maximum authorized speed with minimum headway, at least two full blocks are taken by any one vehicle—that occupied by the vehicle and the block to the rear, that is, the block between the "Stop" and the "Approach" signals. Headway now equals $2L_b$, where L_b is the minimum length of block and must at least equal the maximum stopping distance at maximum authorized speed. For other than two-block systems, the problem involves the speed desired to be maintained, the number of blocks between "Stop" and "Proceed" and the length L_b based on the maximum decelerating distance for the several speed reductions involved. These sequences apply both to single and multiple track.

In practice, many nonuniformities are encountered. Speeds are not uniform, stopping distances vary, and vehicles are of different lengths. One can obtain an approximate value for C_v by using average values for terms in the equation. The following sections present specific approaches to the problem for several common modes.

Theoretical Track Capacity   Track capacity procedures have direct application to railroads but may also be applied to other modes. Theoretical track capacity, C_t, can be expressed for a 24-hour period in train hours

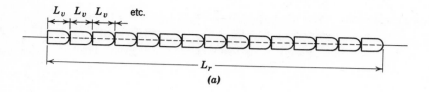

(a)

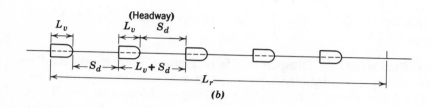

(b)

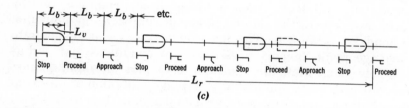

(c)

FIGURE 8.3 THEORETICAL ROUTE CAPACITY: (a) ROUTE OC-
CUPANCY, COMPLETE SATURATION; (b) ROUTE OCCUPANCY,
HEADWAY BETWEEN VEHICLES; (c) ROUTE OCCUPANCY, TWO-
BLOCK SYSTEM.

and is equal to $24n$ where n is the number of spaces available for a train.
On single track, train spaces are between sidings so that $n=$ number of
sidings equally spaced by distance or time. For multiple track, the situation
of Figure 7.3b pertains with headway assumed to equal $2L_t$, that is, L_t is
assumed to give adequate stopping distance where $L_t=$ a train length in
feet. Then

$$C_t = \frac{L_r + L_0}{2L_t} \times 5280 \times 24$$

where

$$L_r = \text{miles of road}$$
$$L_0 = \text{miles of other main track}$$
$$L_t = \text{train length in feet}$$

When block systems are used as in Figure 7.3c

$$C_t = \frac{L_r + L_0}{2L_b} \times 5280 \times 24$$

where

$$L_b = \text{the block length}$$

The denominator would become $3L_b$ or $4L_b$ for a three- or four-block system of signaling.[7]

Train-Hour Performance Diagram   Many factors unite to determine the time to cover a given distance such as a 100-mile subdivision—weather, train tonnage, horsepower, station stops, handling by the engineman. If one takes a large sample of train performance, from a train sheet for example, some trains will have made the run in a minimum time. Other trips under less ideal conditions will have taken a longer time depending on the foregoing variables. Some of the elapsed time will approach the minimum, some will extend far beyond it. The majority of times will concentrate around an average point. Since the variations in conditions and performance are largely accidental, the similarity of a plot of running times against number of vehicles to one half of a normal distribution or probability curve has led to the train-hour performance diagram, (Figure 8.4).

The vehicle-hour performance diagram has a sampling of trips, 100 trips for example arranged and plotted in order of elapsed times as shown in Figure 8.4[8] (*Note*: the use of the probability curve makes for simple computations. Dr. Mostafa K. K. Mostafa has shown that a more precise representation is given by the first integration of the probability curve.[9]) However, plotting actual data from test results and dispatchers' records is often as easy as processing the same data in order to plot a curve mathematically.

[7]*A.R.E.A. Bulletins*, No. 462, November 1946 and *A.R.E.A. Proceedings*, Vol. 48, 1947, pp. 125-144.

[8]Mr. E. E. Kimball, Consulting Engineer, General Electric Company, developed this concept as a train-hour diagram in studying train performance. Its application to any transport operation is obvious. Mr. Kimball's work is found in the *A.R.E.A. Bulletin*, Vol. 47, No. 462, November 1947, pp. 125-144 (and ensuing *Proceedings*); in *Track Capacity and Train Performance*, a report of a subcommittee, Mr. E. E. Kimball, Chairman of the A.R.E.A. Committee 16 on Economics of Railway Location and Operation; and in earlier studies and reports noted herein.

[9]Mostafa K. K. Mostafa, Actual Track Capacity of a Railroad Division, Ph.D. thesis, University of Illinois, Urbana, Illinois, 1951.

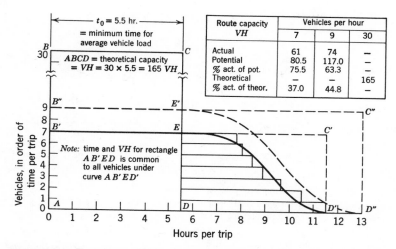

The following table appears in the figure:

| Route capacity | Vehicles per hour | | |
VH	7	9	30
Actual	61	74	—
Potential	80.5	117.0	—
% act. of pot.	75.5	63.3	—
Theoretical	—	—	165
% act. of theor.	37.0	44.8	—

FIGURE 8.4 TRAIN-HOUR PERFORMANCE DIAGRAM. (AFTER E.E. KIMBALL AND THE A.R.E.A.)

When traffic becomes heavy, especially as a route's maximum capacity is approached, the problem of traffic interference becomes very important in creating delays and increasing minimum and average times per trip. It should be recalled that the fewer the tons per vehicle, the more vehicles will be used (in moving a given tonnage) to cause traffic interference. Also, the slower speeds of heavily moving trains may be an added source of interference and delay. Traffic interference is an important factor in railroad operation, especially on a single-track lines and on rapid transit lines.

Actual Track Capacity In Figure 8.4, the rectangle $ABCD$ represents a theoretical track capacity of 165 train hours developed by foregoing procedures. The minimum time to cover the route is 5.5 hours. Dividing 165 by 5.5 gives 30 trains as the maximum theoretical traffic capacity of the line.

Because of train interference and other causes, not all trains cover the distance in 5.5 hours. If seven trains are operated, one train may cover the distance in 8 hours, another in 8.2 hours, etc. and one train will require 11.5 hours. The area of the figure under the curve $AB'ED' = 61$ train hours, the actual capacity utilized. There is a "potential" capacity given by rectangle $AB'C'D' = 80.5$ hours. If nine trains are operated, train interference increases the times for most trains, giving 74 actual train hours for the figure $AB''E''D''$ and potential train hours of 117 for the rectangle

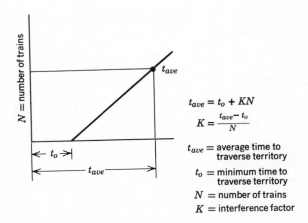

$$t_{ave} = t_o + KN$$

$$K = \frac{t_{ave} - t_o}{N}$$

t_{ave} = average time to traverse territory

t_o = minimum time to traverse territory

N = number of trains

K = interference factor

FIGURE 8.5 EFFECT OF TRAIN INTERFERENCE ON AVERAGE RUNNING TIME (AFTER E.E. KIMBALL AND THE A.R.E.A.)

$AB''C''D''$. If 10 trains were operated, the longest train time would become 15 hours and potential train hours 150. Should 11 trains be attempted, however, the maximum time train would be 176. Since 176 train hours are greater than the theoretical maximum of 163, it follows that 10 is the maximum number of trains that can be operated under the given conditions.

E. E. Kimball, in analyzing the problem for train operation, presents the theory, supported by statistical data, that the average interference time for a given set of conditions is proportional to the number of trains operated in a given time. From this theory he developed the straight-line equation, $T_{av} = T_0 + kN$, where T_{av} = average road time, T_0 = minimum road time (without traffic interference in ideal conditions), N = number of trains in a given time, and k = an interference factor depending on track capacity and miscellaneous conditions. See Figure 8.5

A. S. Lang suggests a curvalinear rather than a straight line relation as being more nearly correct, a relation not incompatible with the scatter of Kimball's plots.[10] Thus according to Lang,

$$T_{ave} = t_0 + K\,(N/2)(T_{ave}/24)$$

or

$$T_{ave} = t_0 + \frac{KNT_{ave}}{48}$$

[10]Letter to the author dated August 4, 1958, from A. S. Lang, then Professor of Transportation Engineering, Massachusetts Institute of Technology, Cambridge, Massachusetts.

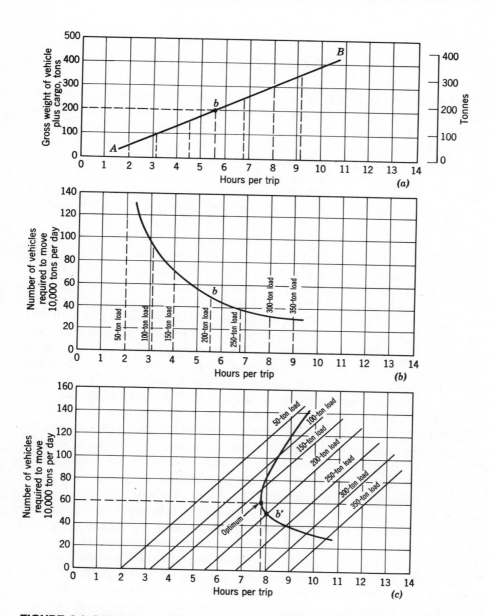

FIGURE 8.6 OPTIMUM VEHICLE LOAD: (*a*) GROSS LOAD VS. HOURS PER TRIP; (*b*) NUMBER OF VEHICLES VS. HOURS PER TRIP; (*c*) OPTIMUM VEHICLE LOAD FOR MINIMUM HOURS PER TRIP. (AFTER E.E. KIMBALL AND THE A.R.E.A.)

whereby

$$T_{ave} = \frac{t_0}{(1 - KN/48)}$$

Optimum Train (Vehicle) Size   The following presentation is based on Kimball's more simple if less accurate procedure, a procedure that can be extended to other modes of carriage. The factors governing t_0 have already been considered. Kimball's t_0 is the T_s of preceding pages where theoretically perfect operation was considered. When the increased number of lightly loaded trains or the slower speeds of heavily loaded trains or vehicles are traffic interference factors, the train load and speed that will give the minimum time per trip (or per 100 train miles) must be determined.

Figure 8.6a has, as a straight-line representation, the hours per trip for various train or vehicle loadings—as already explained in the section "Vehicle Performance". The average load of 2000 tons (1814 tonnes) with a computed road time of 5.5 hours represents theoretically perfect performance ($t_0 = 5.5$ hours) and is indicated by point b.

In Figure 8.6b, the number of trains required to handle a given 10,000 tons (9070 tonnes) per day is presented, using the hours per trip and train loadings of Figure 8.6a.

In Figure 8.6c the average road time *from test results* or *dispatchers' records* is found to be 8 hours rather than 5.5 hours. This is a 2.5 hour increase because of traffic interference. Point b' is plotted and extended to $t_0 = 5.5$ hours and a performance line drawn through the two points. Performance under traffic-interference conditions for other vehicle weights of Figure 8.6a is obtained by plotting their minimum road times from that figure on Figure 8.6c and drawing performance lines through those points parallel to the average performance line passing through point b'.

To determine how the average road time will vary with different train or vehicle weights and traffic densities for a given volume of traffic, project from the number of vehicles (Figure 8.6b) to the corresponding performance line on Figure 8.5c and locate on it the point corresponding to the number of vehicles. The curve EF drawn through these points indicates that a minimum time of 7.75 hours per trip will be obtained with approximately 5.5 trains of 1750 tons each. This represents the optimum train load for minimum trip time. If total train hours are to be minimized, the low point F is chosen on the basis of the maximum load per train.

Rapid Transit Capacity   A fair degree of uniformity prevails in rapid transit operations. Uniform train consists and headways are the rule, at least for major blocks of time throughout the day. The number of persons carried per hour can be expressed as

$$Q = \frac{60 \, KnL_c}{H}$$

where

Q = capacity in persons per hour per track

K = loading coefficient, that is, passengers per foot of train length = car capacity divided by car length; it has values of 2 to 4, the latter with a high standee incidence or with double-decked commuter cars

L_c = car length in feet

n = number of cars in the train

H = headway in minutes[6]

Headway, as earlier explained, is a function of train length, speed, and stopping distance. It is also affected by station stop time and frequency and the time taken in acceleration and deceleration, which leads to a nonuniform speed. When station stops are involved, headway in seconds becomes

$$h = T + L/v + \frac{v}{2a} + 5.05 \frac{v}{2d}$$

where

T = train halt time at stations in seconds

T = will vary with volume of traffic to load and unload with the station platform and its operation

v = maximum running speed in fps

L = train length = $n \times L_c$

a and d = rates of acceleration and deceleration in fps

L/v = time for train to travel its own length.

Where there are no stations in a particular track section,

$$h = t + \frac{L}{v} + 2.03 \frac{v}{d}$$

t = motorman's reaction time and brake application time in seconds, usually 1 to 3 seconds

[6] After Lang and Soberman in *Urban Rail Transit*, MIT Press, Cambridge, Massachusetts, 1964, pp. 61-63.

Capacity related design alternatives include a choice between slow short trains at frequent intervals or longer faster trains at less frequent intervals. Chapter 8 details the relation between station spacing, station stop time, and train speed. See Table 7.9 for capacities per track (one-directional movement) at various speeds and train lengths. These high capacities of 44,000 to 82,000 passengers per hour can only be achieved when there is a high demand and at high capital and operating costs.

Theoretical and mathematical solutions should be tempered by the criteria of observation and experience. With train orders alone or with manual block, no more than 25 to 30 trains should be operated in 24 hours on a single-track railroad. When one attempts to run more, excessive interference delays are likely to occur. When automatic block signaling is used, the number may increase to 40 or 50 trains. L. F. Loree told of putting 60 trains a day over a single-track line without specifying the type of operation or signaling.[11] The South African Railway, using a station-to-station system of operation, regularly handles 30 to 40 trains on single track, but train interference causes frequent delay. The addition of Centralized Traffic Control (CTC) will increase capacity 50 to 100 percent and may permit as many as 60+ trains to run over a single-track line in 24 hours under favorable conditions. Double-track main lines can handle as many as 90 to 100 trains daily with manual or automatic block and up to 200 with CTC. Four tracks are desirable when as many as 150 trains are being operated, even with automatic block (but without CTC). The four-track New York Division of the Pennsylvania (more than four tracks in some places) has regularly handled 340 to 360 trains daily, and a four-track New York subway system operates over 400 trains per day. One passenger train requires about the same track capacity as two freight trains when the two are run on the same track. The capacity of any line is being exceeded when traffic interference causes such delays that trains continually fail to maintain schedules.

Highway Capacity   Highway capacity may be crudely approximated by the previously discussed formula, $C = 5280 \times V/(L_v + S_d)$. An important constraint with this equation is the driver's inherent recognition of what is a safe stopping distance. Highway capacity as expressed by the number of cars that pass a given point in an hour will vary with the speed and headway. The driver of an automobile tends to increase the headway as his or her speed increases. A slow speed gives close headway but, because of

[11]L. F. Loree, *Railroad Freight Transportation*, 1931 edition, D. Appleton and Company, New York, p. 25.

the slow speed, few cars. High speed also gives few cars because of the greater spacing. There must therefore be an intermediate or optimum speed that will allow the maximum number to pass a given point in an hour. Studies by the Highway Research Board have shown the maximum theoretical capacity of a single traffic lane to be about 2000 vehicles at 30 mph under ideal conditions.[12] See Figure 8.7. It should be noted that maximum capacity is not synonymous with maximum density, which usually leads to reduced capacity (congestion). The ultimate in density is where there are so many vehicles that movement is completely halted. See Figure 8.3a. The few instances of sustained capacities in excess of 2000 vph probably represent unsafe operation, that is, too small a headway and stopping distances that can lead to the chain reaction accidents involving scores of vehicles. A four-lane divided highway can move 8000 vph and a three-lane highway 4000 vph, all under ideal conditions. The 2000 vph ideal capacity figure also applies to the combined two lanes of a two-lane highway with a preponderance of traffic in one direction to provide slots in the opposing lane for overtaking and passing.[13] These values are for conditions of uninterrupted flow, a situation that seldom fully exists because of "fleeting" and queuing of cars in the lines of flow.

As traffic volume increases, other conditions being equal, speed will decrease. In practical operation, all drivers do not drive at the same speed. Unless the slowest-moving vehicle is to establish the rate for a group, there must be an opportunity for faster-moving vehicles to overtake and pass those of slower speed. This opportunity is afforded by a second lane in the same direction as in four-lane highways or in the opposing lane of a two-way, two-lane highway when opposing traffic is light. However, when the traffic density increases to a point where there is no room for safe overtaking and passing, all traffic must move at approximately the same speed and the relative speed between vehicles will approach zero. The critical or optimum density has been reached and any increase in density will lead to a reduction in speed and in traffic volume.

Ideal conditions for maximum uninterrupted flow include 12-ft (3.66-m) lane widths, 6-ft (1.83-m) shoulder widths free of lateral obstructions, no commercial vehicles, only passenger cars, no restricting sight distances, vertical and horizontal alignments suitable for speeds of 70 mph (112.6

[12]O. K. Normann and W. P. Walker, *Highway Capacity Manual*, Bureau of Public Roads, U. S. Department of Commerce, Washington, D. C., 1950, p. 27.

[13]*Highway Capacity Manual 1965*, Highway Research Board Special Report 87, National Academy of Sciences, National Research Council, Publication 1328, Washington, D. C. 1965, p. 76, Table 4.1.

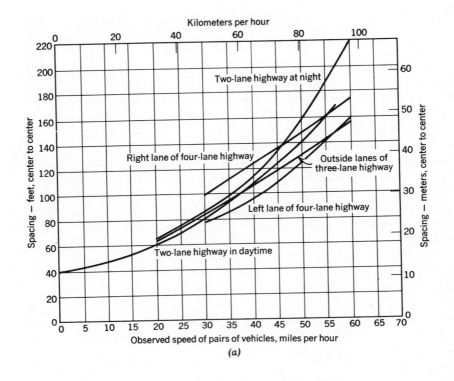

(a)

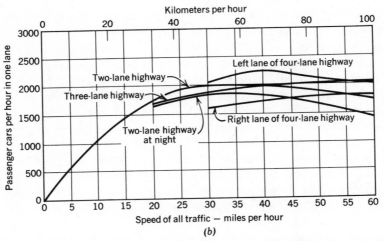

(b)

FIGURE 8.7 MAXIMUM CAPACITY OF A TRAFFIC LANE. (*TRAFFIC CAPACITY MANUAL*, BUREAU OF PUBLIC ROADS, DEPARTMENT OF COMMERCE, WASHINGTON, DC. COURTESY OF L.I. HEWES AND C.H. OGLESBY, *HIGHWAY ENGINEERING*, WILEY, NEW YORK, 1954, P. 144, FIGURES 3 AND 4.)

kph), and no side or lateral interference from vehicles or pedestrians; obviously, no cross traffic, stop signs, or stop lights.

There must be adequate roadway capacity both upstream and downstream and with favorable weather conditions.[14]

Unless the opportunity for free overtaking and passing is present, the maximum theoretical capacities will seldom be reached. Various types of traffic interference operate to reduce the maximum to a more easily attained value. Cross interference comes from two streams of traffic crossing at grade. Stop signs and stop lights may here have a controlling effect upon speed and capacity, especially on city streets. Marginal interference occurs between moving persons and vehicles or objects along the roadway edges. Internal interference occurs between cars moving in the same direction (the overtaking and passing problem), and medial interference occurs between cars moving in opposing directions with no separation between lanes. All these tend to reduce the maximum capacities to the practical design capacities for uninterrupted flow given in Table 8.3. These capacities, based on 12-ft (3.65-m) lanes, are further reduced by narrow lanes (marginal interference). An 11-ft (3.35-m) lane has only 86 percent of the capacity of a 12-ft (3.65-m) lane on two-lane roads, and 97 percent on multilaned roads. Correspondingly, a 9-ft lane has only 70 and 81 percent, respectively.

Lateral obstructions above curb height such as retaining walls, signs, fences, light poles, or a row of parked vehicles cause a certain often unrealized apprehension, a shying away from the "obstruction," and a reduction in speed. Regardless, adequate shoulders are desirable in providing a space for disabled vehicles and, if paved, can add to the effective width of lanes less than 12 ft wide. The combined effect of narrow lanes and shoulders on capacity is shown in Table 8.2.

Commercial vehicles are larger than passenger cars and often move at slower speeds, especially on ascending grades. In a lane carrying the maximum number of passenger cars at 20 mph, two passenger cars would have to be removed for each truck or bus placed in that lane. At 60 mph the ratio is eight passenger cars to one truck or bus. Observed performance shows that one truck will have the same effect on capacity as two passenger cars on level grade, four passenger cars on rolling profiles, and eight passenger cars in mountain territory. Similar values for buses are 1.6, 3, and 5 where volumes of bus traffic are significant.[15] Additional slow-speed lanes on ascending grades aid in keeping the traffic moving.

[14]Ibid, chapter five, pp. 88–100.
[15]Ibid., Table 9.3a, p. 257.

TABLE 8.2 COMBINED EFFECT OF LANE WIDTH AND RESTRICTED LATERAL CLEARANCE ON CAPACITY AND SERVICE VOLUMES OF DIVIDED FREEWAYS AND EXPRESSWAYS WITH UNINTERRUPTED FLOW[a]

Distance from Traffic Lane Edge to Obstruction (ft)	Adjustment Factor,[b] W, for Lane Width and Lateral Clearance							
	Obstruction on One Side of One-Direction Roadway				Obstructions on Both Sides of One-Direction Roadway			
	12-ft (3.66-m) Lanes	11-ft (3.35-m) Lanes	10-ft (3.05) Lanes	9-ft (2.74-m) Lanes	12-ft (3.66-m) Lanes	11-ft (3.35-m) Lanes	10-ft (3.05-m) Lanes	9-ft (2.74-m) Lanes
(a) Four-Lane Divided Freeway, One Direction of Travel								
6	1.00	0.97	0.91	0.81	1.00	0.97	0.91	0.81
4	0.99	0.96	0.90	0.80	0.98	0.95	0.89	0.79
2	0.97	0.94	0.88	0.79	0.94	0.91	0.86	0.76
0	0.90	0.87	0.82	0.73	0.81	0.79	0.74	0.66
(b) Six- and eight-Lane Divided Freeway, one Direction of Travel								
6	1.00	0.96	0.89	0.78	1.00	0.96	0.89	0.78
4	0.99	0.95	0.88	0.77	0.98	0.94	0.87	0.77
2	0.97	0.93	0.87	0.76	0.96	0.92	0.85	0.75
0	0.94	0.91	0.85	0.74	0.91	0.87	0.81	0.70

[a]*Highway Capacity Manual*, Special Report 87, Highway Research Board, National Academy of Sciences, Washington, D.C., 1966, Table 9.2, p. 256.
[b]Same adjustments for capacity and all levels of service.

The foregoing have general application, but where truck or bus traffic is heavy and grades are long, steep, and frequent, a grade by grade analysis using more refined methods and tables of the Highway Capacity Manual is in order. See Tables 8.4 and 8.5.

Level of Service   It has been noted that the volume of traffic moving past a point is a function of vehicle spacing or headway, that is, of the operating speed. The old concepts of "possible" and "practical" capacities, or capacities under prevailing conditions other than "ideal," have been replaced in the Highway Research Board's[16] Capacity Manual by a single "capacity" and a series of speed-oriented levels of service that reflect adjustments to that capacity. The "capacity" is that which obtains under ideal conditions, 2000 vph per lane, occuring at a speed of 30 to 35 mph. If a stream of traffic is to move more rapidly, there must be a corresponding increase in the headway between vehicles and therefore a reduction in

[16]Now known as the Transportation Research Board.

TABLE 8.3 LEVELS OF SERVICE AND MAXIMUM SERVICE VOLUMES FOR FREEWAYS AND EXPRESSWAYS UNDER UNINTERRUPTED FLOW CONDITIONS[a]

Level of Service		Traffic Flow Condition Operating Speed (mph)	Service Volume/Capacity (v/c) Ratio[b] — Basic Limiting Value for Average Highway Speed (abs) of 70 mph, for:			Approximate Working Value for Any Number of Lanes for Restricted Average Highway Speed of		Maximum Service Volume under Ideal Conditions, Including 70-mph Average Highway Speed (Total Passenger Cars per Hour, One Direction)															
Level	Description		Four-Lane Freeway (Two Lanes/Direction)	Six-Lane Freeway (Three Lanes/Direction)	Eight-Lane Freeway (Four Lanes/Direction)	60 mph	50 mph	Four-Lane Freeway (Two Lanes) One Direction				Six-Lane Freeway (Three Lanes) One Direction				Eight-Lane Freeway (Four Lanes) One Direction				Each Additional Lane above Four in One Direction			
Peak-hour Factor (PHF)[f] →								0.77	0.83	0.91	1.00[d]	0.77	0.83	0.91	1.00[e]	0.77	0.83	0.91	1.00[e]	0.77	0.83	0.91	1.00[d]
A	Free flow	>60	<.35	<.40	<0.43	—[c]	—[c]	1400				2400				3400				1000			
B	Stable flow (upper speed range)	>55	<0.50	<0.58	<0.63	<0.25	—[c]	2000				3500				5000				1500			
C	Stable flow	>50	<0.75 (PHF)	<0.80 (PHF)	<0.83 (PHF)	<0.45 (PHF)	—[c]	2300	2500	2750	3000	3700	4000	4350	4800	5100	5500	6000	6600	1400	1500	1650	1800
D	Approaching unstable flow	>40	<0.90 (PHF)	<0.90 (PHF)	<0.90 (PHF)	<0.80 (PHF)	<0.45 (PHF)	2800	3000	3300	3600	4150	4500	4900	5400	5600	6000	6600	7200	1400	1500	1650	1800
E[f]	Unstable flow	30-35[f]	<1.00					4000[f]				6000[f]				8000[f]				2000[f]			
F	Forced flow	<30[f]	Not meaningful					Widely variable (0 to capacity)																

(1 mph = 1.609 kph)

[a] *Highway Capacity Manual*, Special Report 87, Highway Research Board, National Academy of Sciences, Washington, D.C., 1966, Table 9.1, p. 56

[b] Operating speed and basic v/c ratio are independent measures of level of service; both limits must be satisfied in any determination of level.

[c] Operating speed required for this level is not attainable even at low volumes.

[d] Peak-hour factor for freeways is the ratio of the whole-hour volume to the highest rate of flow occuring during a five-minute interval within the peak hour.

[e] A peak-hour factor of 1.00 is seldom attained; the values listed here should be considered as maximum average flow rates likely to be obtained during the peak five-minute interval within the peak hour.

[f] Approximately.

287

capacity volume. Six levels of service are identified by the letters A through F (fastest to slowest speed) and are defined primarily in terms of operating speed for rural highways (and average overall travel speed for urban areas) and volume as measured by the v/c or volume/capacity ratio. Table 8.3 gives the levels of service and the maximum service volumes for freeways and expressways with divided lanes under uninterrupted flow conditions with full access control. The level of service, E, at a speed of 30 to 35 mph (48.3 to 56.3 kph), represents maximum capacity under ideal conditions, that is, the v/c ratio is 1. It represents the optimum balance between density and speed.

Using the four-lane divided expressway of Table 8.3, service level C represents stable flow at a speed of 50 mph (80.45 kph). The observed service volume under ideal conditions and a peak hour factor of 1 is 3000 vph. Since capacity under ideal conditions has been defined as 4000 vph for two lanes in one direction at a speed of 30 mph (48.3 kph), the volume capacity or v/c ratio is $3000/4000 = 0.75$. Different service volumes can be selected depending on the level of service desired, that is, operating speed. Level of service tabulations similar to those of Table 8.3 are also available in the *Highway Traffic Manual* for multilane undivided highways, for two-lane, two-directional highways, and for urban-suburban arterial streets.

Capacity is also a function of the prevailing conditions of lane and shoulder width, incidence of commercial vehicles, gradient, sight distance, and other factors peculiar to the type of roadway under consideration. A given length of roadway may contain stretches that offer different prevailing conditions and hence different levels of service and capacities.

For conditions of uninterrupted flow, capacity is determined by multiplying the capacity under ideal conditions by factors for prevailing conditions, that is, by use of the equation

$$c = 2000 \times N \times W \times T_c \times B_c$$

where

c = capacity under prevailing conditions in vehicles per hour, vph

N = number of lanes in one direction

W = adjustment for lane and shoulder width (lateral clearance); Table 8.2, for divided highways

T_c = truck factor at capacity from Tables 8.4 and 8.5

B_c = bus factor from Table 8.5; used only where buses are being analyzed separately, otherwise omitted or combined in a more general way with trucks.

By using the methods of the *Highway Capacity Manual*, the Service Volume may be computed directly from capacity under ideal conditions.

$$SV = 2000 \times N \times (v/c) \times W \times T_L$$

where

SV = service volume in mixed vehicles per hour, total for one direction

v/c = volume-capacity ratio obtained from Table 8.3

N = number of lanes in one direction

W = adjustment for lane width and lateral clearance, Table 8.2

T_L = truck factor at a given level of service from Table 8.4 (over extended sections of highway) or Table 8.5

Problem Example

A four-lane divided freeway in rolling terrain has an operating speed of 55 mph (88.5 kph). Lane widths are 10 ft (3.05 m) with shoulder widths of 6 ft (1.83 m).

TABLE 8.4 AVERAGE GENERALIZED ADJUSTMENT FOR TRUCKS[a] ON FREEWAYS AND EXPRESSWAYS, OVER EXTENDED SECTION LENGTHS[b]

Percentage of Trucks, P_T	Factor, T, for All Levels of Service		
	Level Terrain	Rolling Terrain	Mountainous Terrain
1	0.99	0.97	0.93
2	0.98	0.94	0.88
3	0.97	0.92	0.83
4	0.96	0.89	0.78
5	0.95	0.87	0.74
6	0.94	0.85	0.70
7	0.93	0.83	0.67
8	0.93	0.81	0.64
9	0.92	0.79	0.61
10	0.91	0.77	0.59
12	0.89	0.74	0.54
14	0.88	0.70	0.51
16	0.86	0.68	0.47
18	0.85	0.65	0.44
20	0.83	0.63	0.42

[a]Not applicable to buses where they are given separate specific consideration; use instead Table 8.5.

[b]*Highway Capacity Manual*, Special Report 87, Highway Research Board, National Academy of Sciences, Washington, D.C., 1966, Table 9.3b, p. 257.

TABLE 8.5 ADJUSTMENT FACTORS[a] FOR TRUCKS AND BUSES ON INDIVIDUAL ROADWAY SUBSECTIONS OR GRADES ON FREEWAYS AND EXPRESSWAYS (INCORPORATING PASSENGER CAR EQUIVALENT AND PERCENTAGE OF TRUCKS OR BUSES)[b]

| Passenger Car Equivalent, E_T or E_B^d | Truck Adjustment Factor T_c or T_L (B_c or B_L for Buses)[c] | | | | | | | | | | | | | | |
| | Percentage of Trucks, P_T (or of Buses, P_B) of: | | | | | | | | | | | | | | |
	1	2	3	4	5	6	7	8	9	10	12	14	16	18	20
2	0.99	0.98	0.97	0.96	0.95	0.94	0.93	0.93	0.92	0.91	0.89	0.88	0.86	0.85	0.83
3	0.98	0.96	0.94	0.93	0.91	0.89	0.88	0.86	0.85	0.83	0.81	0.78	0.76	0.74	0.71
4	0.97	0.94	0.92	0.89	0.87	0.85	0.83	0.81	0.79	0.77	0.74	0.70	0.68	0.65	0.63
5	0.96	0.93	0.89	0.86	0.83	0.81	0.78	0.76	0.74	0.71	0.68	0.64	0.61	0.58	0.56
6	0.95	0.91	0.87	0.83	0.80	0.77	0.74	0.71	0.69	0.67	0.63	0.59	0.56	0.53	0.50
7	0.94	0.89	0.85	0.81	0.77	0.74	0.70	0.68	0.65	0.63	0.58	0.54	0.51	0.48	0.45
8	0.93	0.88	0.83	0.78	0.74	0.70	0.67	0.64	0.61	0.59	0.54	0.51	0.47	0.44	0.42
9	0.93	0.86	0.81	0.76	0.71	0.68	0.64	0.61	0.58	0.56	0.51	0.47	0.44	0.41	0.38
10	0.92	0.85	0.79	0.74	0.69	0.65	0.61	0.58	0.55	0.53	0.48	0.44	0.41	0.38	0.36
11	0.91	0.83	0.77	0.71	0.67	0.63	0.59	0.56	0.53	0.50	0.45	0.42	0.38	0.36	0.33
12	0.90	0.82	0.75	0.69	0.65	0.60	0.57	0.53	0.50	0.48	0.43	0.39	0.36	0.34	0.31
13	0.89	0.81	0.74	0.68	0.63	0.58	0.54	0.51	0.48	0.45	0.41	0.37	0.34	0.32	0.29
14	0.88	0.79	0.72	0.66	0.61	0.56	0.52	0.49	0.46	0.43	0.39	0.35	0.32	0.30	0.28
15	0.88	0.78	0.70	0.64	0.59	0.54	0.51	0.47	0.44	0.42	0.37	0.34	0.31	0.28	0.26
16	0.87	0.77	0.69	0.63	0.57	0.53	0.49	0.45	0.43	0.40	0.36	0.32	0.29	0.27	0.25
17	0.86	0.76	0.68	0.61	0.56	0.51	0.47	0.44	0.41	0.38	0.34	0.31	0.28	0.26	0.24
18	0.85	0.75	0.66	0.60	0.54	0.49	0.46	0.42	0.40	0.37	0.33	0.30	0.27	0.25	0.23
19	0.85	0.74	0.65	0.58	0.53	0.48	0.44	0.41	0.38	0.36	0.32	0.28	0.26	0.24	0.22
20	0.84	0.72	0.64	0.57	0.51	0.47	0.42	0.40	0.37	0.34	0.30	0.27	0.25	0.23	0.21
21	0.83	0.71	0.63	0.56	0.50	0.45	0.41	0.38	0.36	0.33	0.29	0.26	0.24	0.22	0.20
22	0.83	0.70	0.61	0.54	0.49	0.44	0.40	0.37	0.35	0.32	0.28	0.25	0.23	0.21	0.19
23	0.82	0.69	0.60	0.53	0.48	0.43	0.39	0.36	0.34	0.31	0.27	0.25	0.22	0.20	0.19
24	0.81	0.68	0.59	0.52	0.47	0.42	0.38	0.35	0.33	0.30	0.27	0.24	0.21	0.19	0.18
25	0.80	0.67	0.58	0.51	0.46	0.41	0.37	0.34	0.32	0.29	0.26	0.23	0.20	0.18	0.17

[a]*Highway Capcaity Manual*, Special Report 87, Highway Research Board, National Academy of Sciences, Washington, D. C., 1966, Table 9.6, p. 261.

[b]Used to convert equivalent passenger car volumes to actual mixed traffic; use the reciprocal of these values to convert mixed traffic to equivalent passenger cars.

[c]Trucks and buses should not be combined in entering this table where separate consideration of buses has been established as required, because passenger car equivalents differ.

Truck traffic is approximately 10 percent of the total. What is the service volume for this section of the freeway?

From Table 8.3, a v/c ratio of 0.50 for a speed of 55 mph (88.5 kph) is found and a lane width adjustment factor of $W = 0.91$ is taken from Table 8.2. The truck factor, T_L, obtained from Table 8.4, is 0.77. Applying these to the service volume equation, we get

$$SV = 2000 \times 2 \times 0.50 \times 0.91 \times 0.77 = 1402 \text{ vph.}$$

Adjustments must be made for the appropriate level of service where different from those of capacity and the appropriate v/c ratio applied for the level of service desired. It is important to check the results to confirm that both the volume and operating speed criteria for the desired level of service are met, giving due consideration to the prevailing average highway speed.

The foregoing procedures are presented to demonstrate the principles involved and the general character of computational procedures. Variations in procedures and in tabular values for divided multilane and for two-lane highways are found in the *Highway Capacity Manual*.

Signalized Intersection Capacity   Thus far, highway capacity has been discussed in terms of uninterrupted flow. A prime source of flow interruption is the intersection, especially a signalized intersection. Unsignalized intersections either have a relatively small volume of traffic or are protected over two directions of approach by stop signs on the minor approach giving, in effect, uninterrupted flow on the major approaches. Four-way stops are used where prevailing volumes of traffic are normally light or as a temporary expedient, prior to installation of lights, for example.

Levels of service have been established also for signalized intersections. These are identified as A through F where level A represents free flow with a load factor of 0.0, no delays at all. B and C have respective load factors of 0.1 and 0.3 that represent stable flow. D approaches stable flow, with a load factor of 0.7 (some waits and delays). And E is unstable flow representing capacity with a load factor of 1 (but more often 0.85), involving queues, waits, and delays. F represents jammed conditions including backups from other intersections.[17]

The capacity and level of service of a signalized intersection are functions of:[18]

(a) *Physical and operating conditions* such as approach widths—curb-to-curb distance and lane widths, whether one- or two-way operation, and whether or not parking is permitted, especially within 250 ft of the intersection.

(b) *Environment conditions* that includes the *load factor* or incidence of vehicles available to utilize the green portion of the light cycle. The load factor represents the degree to which the green cycle has vehicles available and utilizing the green light through its duration. It is the ratio of number of green phases fully loaded or utilized to the number

[17]Ibid., pp. 130-131.
[18]Ibid., pp. 112-129.

available during the same period. Its value may range from 0.0 (no cycle loaded) to 1.0 (all cycles loaded). With a factor of 0.3, vehicles may be held occassionally for more than one red signal, while 0.7 represents substantial delay.

The *peak hour factor* for each approach measures the consistency of demand and is defined as the ratio between the number of vehicles counted during the peak hour and four times the number of vehicles during the highest consecutive 15 minutes. A factor of 1.0 has been used for very heavy demand, but a more frequently used value is 0.85 for approaches with high loads most of an hour; 0.6 and 0.7 can be used for high flow over short periods.

Intersections are found to have larger capacities in large metropolitan areas than in smaller cities. Capacities tend to increase with distance from the central business district, in part, because of the lower incidence of pedestrians.

(c) *Traffic characteristics* that include the percent of turning movements, especially the left turns; trucks and buses that have slower acceleration and greater size than passenger cars; incidence of local transit buses that stop at intersections to pick up and discharge passengers (far side curb stops have less adverse effect than near side stops, except when right turn movements are numerous).

(d) *Control measures* that include utilization of traffic signals—location, cycle length (especially the green time to cycle length ratio, G/C), and approach lane markings that establish lane width and provide a type of channelization.

Capacity and service volumes are expressed as "vehicles per hour of green light." Knowing the percent of total time cycle that the light is green, the G/C ratio, one can compute the number of vehicles that can pass through the intersection in an hour. Charts have been prepared for one- and two-way streets with and without parking that take account of the foregoing factors. Figures 8.8a and 8.8b for one-way and two-way streets are typical of such charts.

The average conditions of traffic for which these charts are developed are indicated in the lower right-hand corner of each chart. Additional adjustments must be made for a particular intersection to account for percents of right and left turns and of truck and bus traffic that differ from those on which the charts are based. See Table 8.6. The computation of an hourly capacity is best illustrated by an example.

Assume a two-way street intersection, parking prohibited, located in the residential area of a city of 100,000 population. Width of curb to the division line is 24 ft. A level of service, C, giving stable flow and a load factor of 0.3 is

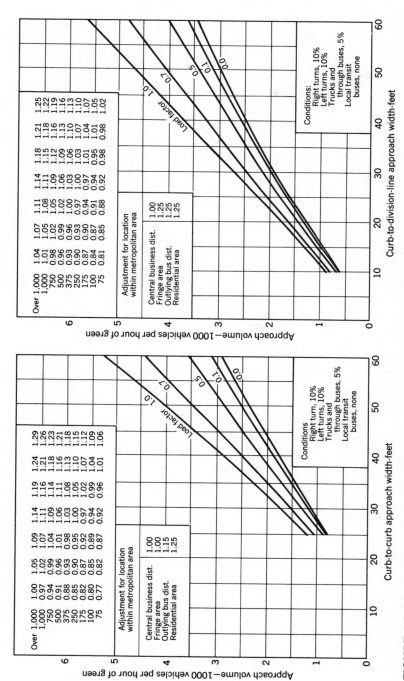

FIGURE 8.8 (*a*) URBAN INTERSECTION SERVICE VOLUME, IN VEHICLES PER HOUR OF GREEN SIGNAL TIME, FOR ONE-WAY STREETS WITH PARKING BOTH SIDES. (*b*) URBAN INTERSECTION APPROACH SERVICE VOLUME, IN VEHICLES PER HOUR OF GREEN SIGNAL TIME, FOR TWO-WAY STREETS WITH NO PARKING. (*HIGHWAY CAPACITY MANUAL*, HIGHWAY RESEARCH BOARD SPECIAL REPORT 87, NATIONAL ACADEMY OF SCIENCES–NATIONAL RESEARCH COUNCIL PUBLICATION 1328, WASHINGTON, DC., 1965.)

TABLE 8.6 ADJUSTMENT FACTORS: DETERMINING EQUALIZED INTER-SECTION CAPACITY

Part A   Right turns on two-way streets, right turns on one-way streets, and left turns on one-way streets[a]

Turns	No Parking Approach Widths			With Parking Approach Widths		
(%)	≤ 15 ft	16–24 ft[b]	25–34 ft	≤ 20 ft	21–29 ft	30–39 ft
0	1.20	1.050	1.025	1.20	1.050	1.025
10	1.00	1.000	1.000	1.00	1.000	1.000
20	0.90	0.950	0.975	0.90	0.950	0.975
30	0.85	0.901	1.000	0.85	0.900	1.000

Part B   Left turns on two-way streets[c]

Turns	No Parking Approach Widths			With Parking Approach Widths		
(%)	≤ 15 ft	16–34 ft	≥ 35 ft	20 ft	21–39 ft	≥ 40 ft
0	1.30	1.10	1.050	1.30	1.10	1.050
10	1.00	1.00	1.000	1.00	1.00	1.000
15	0.90	0.95	0.975	0.90	0.95	0.975
20	0.85	0.90	0.950	0.85	0.90	0.950
30	0.80	0.85	0.900	0.80	0.85	0.900

Part C   Truck and through bus adjustments

Percent	Adjustment Factors	Percent	Adjustment Factors
0	1.08	15	0.90
5	1.00	20	0.85
10	0.95		

Part D   Local bus factors—CBD (near side bus stops, no parking)

Lanes and Width	Buses per Hour					
	0	20	40	60	80	90+
2 lanes (24 ft)	0.59	0.92	0.82	0.73	0.63	0.58
3 lanes (36 ft)	0.67	0.92	0.86	0.77	0.71	0.67
4 lanes (48 ft)	0.74	0.95	0.90	0.84	0.77	0.74
—Outlying and Fringe Areas[c]						
2 lanes (24 ft)	0.58	0.95	0.87	0.79	0.72	0.67
3 lanes (36 ft)	0.67	0.96	0.90	0.84	0.78	0.74
4 lanes (48 ft)	0.74	0.97	0.92	0.85	0.84	0.81

[a]*Highway Capacity Manual*, Special Report 87, Highway Research Board, National Research Council, Washington, D.C., Table 6.4, p. 140.
1 ft = 0.3048 m.
[b]Ibid., Table 6.5, p. 141. Ibid., Table 6.6, p. 142.
[c]Ibid., Table 6.11, p. 143.

required. Truck traffic is three percent of the total and there are 10 buses an hour making near side stops. Right turns are made by 20 percent of the traffic; left turns by 12 percent.

This type of street requires the use of Figure 8.8b. For a 24-ft (7.3-m) curb-to-division line width and a load factor of 0.3, the approach volume per hour of green time is found to be 1800 vehicles. Adjustments must be made for peak hour conditions using a value of 0.85 to represent approaches with high loads most of the time. Combining 0.85 with a population of 100,000 gives an adjustment factor of 0.94. A second adjustment of 1.25 is found for intersections in residential areas. The adjusted vehicles per hour of green light becomes

$$1800 \times 0.94 \times 1.25 = 2115 \text{ vehicles}$$

Further adjustments must be made for right turns and left turns. Using Table 8.6 and foregoing turning percentages, adjustment factors of 0.950 for right turns and 0.980 for left turns are obtained. An adjustment for the three-percent truck traffic is taken as 1.02. The local buses, 10 per hour stopping on the near side require a correction factor of 0.558 taken from Table 8.6. The fully adjusted vehicles per hour of green light now becomes:

$$2115 \times 0.950 \times 0.980 \times 1.02 \times 0.558 = 1121 \text{ vehicles}$$

But this is vehicles per hour of green light. The number of vehicles per hour is $1121 \times$ the G/C ratio, that is, $1121 \times 30/50 = 673$ vehicles per hour.

Waterway Capacity   Traffic is seldom heavy enough on open waterways to impose any serious route-capacity restrictions. However, narrow channels, locks, and harbor entrances may impose rather severe restrictions. Traffic is then limited to that which can move through the restricting section in a given time according to the general principles just set forth. In open channels, a headway limited by stopping distance must be observed. In passing through locks, a headway equal to the time it takes to lock through a vessel or tow is automatically imposed. The time may vary from 20 minutes to an hour or more depending on the size and flow capacity of the lock and on the number of units in the tow. See Chapter 6.

Airway Capacity   Federal airways obtain maximum safe air capacity by dividing the skies into 1000- and 500-ft (304.8- and 152.4-m) levels and assigning each kind of craft by direction and class to its proper level. The ability of a landing field to receive and discharge planes has traditionally imposed limitations upon route capacity. Landing capacity is especially restricted when bad weather forces instrument landings, more time-consuming than contact landings under some conditions, and incoming airplanes have to "stack up" over the field, that is, fly a prescribed waiting pattern bringing them lower and lower over the field until their turn comes to land. Nevertheless, increased numbers of planes in the air and the increasing speeds of flight, with a corresponding increase in the air space required for safe flight and maneuverability, are already imposing air-route-capacity limitations.

Runways can handle peak loads of 40 to 60 landings and takeoffs per hour under visual conditions and 25 to 35 landings and takeoffs per hour under instrument conditions. Airport capacity and, in a roundabout way, route capacity, is then a function of the number of runways and of airport capacity.

A new problem has accompanied the introduction of big jets—air turbulence created by the wake of these giant planes. Violent air currents caused by the passing of such craft have been held responsibile for the crashes of more than a score of small planes. The Federal Aviation Agency has expanded permissible takeoff and landing distance between large jets and small craft from three to a current five miles.

All planes flying above 10,000 ft (3048 m) which includes most domestic trunk-line flights, now fly on instrument-flight rules.[19] In addition to the 1000-ft (304.8 m) vertical separations noted earlier, air-traffic-control regulations for instrument flight have required a 10-minute headway between airplanes flying the same route and elevation (plus a 10-mile or 16.1 km lateral spacing). At a speed of 180 mph (289.6 kph), a headway of 30 miles is thereby imposed; at 300 mph (483 kph), a headway of 50 miles (80.5 km). For the $600 \pm$ mph ($965 \pm$ kph) speeds of jet transports, an equivalent headway would be 100 miles (161 km). According to this calculation, no more than three such airplanes could be in the air at any one time on a given route and elevation between New York and Buffalo. Jet transport speeds are becoming too fast for safe contact navigation and the more stringent regulations and space requirements of instrument flight must be observed, including landing approaches. A conventional propeller-driven transport plane requires from 12 to 15 square miles for approach maneuvering at airports. Jet transports require over 1000 square miles. Cross-country superskyways connecting New York and Washington with San Francisco and Los Angeles have been established by the CAB at altitudes between 17,000 and 22,000 ft (5182 and 6706 m). No airplane can use or cross these routes without specific permission from the traffic-control center involved. Radar separation is being provided for aircraft operating above 24,000 ft (7315 m).

Other Modes   The route capacities of pipelines are a function of the capacity of one line, already described as depending on pumping pressure

---

[19]*Air Transport Facts and Figures 1959*, Air Transport Association of America, Washington, D. C., p. 9.

and flow resistance, etc., and the number of lines. Many systems are composed of two, three, or even four or more lines of pipe. The capacity is then a multiple of the capacity through one line.

Capacities for conveyors are based on the load per foot of belt (a function, in turn, of the width and strength of the belt) and its speed. These matters too have been considered in earlier chapters.

Cableways have a relatively fixed capacity as there is little speed variation for a given installation. Capacity depends on the load per carriage, the number of carriages, and the speed of the cable. These are all fixed within rather narrow limits for most installations. More cars may be added to a cable but not many more because the design of the cable, strength of its supports, and allowable sag between towers are based on a given number of loaded cars, which cannot be exceeded with safety.

Summary   Typical route capacity values are given in Table 8.7.

TABLE 8.7 ROUTE CAPACITIES

Route	Average Speed	Train or Vehicle Hours	Number of Trains, Vehicles, etc.
Railroads: (100-mile district)			
Single-track (theoretical)			
10 sidings	10 mph[a]	240	24 per day
	20 mph	240	48 per day
	40 mph	240	96 per day
20 sidings	10 mph	480	48 per day
	20 mph	480	96 per day
	40 mph	480	192 per day
Single-track (practical)			20–30 per day
(CTC-practical)			45–60 per day
(CTC-potential)			80–100 per day
Double-track (theoretical)			
1-mile block (1 mile of train)			
2-block headway	20 mph	2400	480 per day
	40 mph	2400	960 per day
1-mile block (1 mile of train)			
3-block headway	20 mph	1600	320 per day
	40 mph	1600	640 per day
Double-track (practical)			
Manual or automatic block			
(potential)			60–80 per day
Manual or automatic block			
(potential)			80–200 per day

TABLE 8.7 (Continued)

Centralized Traffic Control	160–200 per day
Four-track (practical)	300–360 per day
(potential)	360–460 per day

Rapid transit, one-track, one-direction; $T = 40$ sec, $K = 3.1$.

Running Speed	Average Acceleration	Passengers per Hour, $(-)$ = Headway in Sec		
		$L = 400$ ft[b]	$L = 500$ ft	$L = 600$ ft
20	3.0 mph/s	60,000 (83)	72,400 (87)	83,200 (90)
30	3.0 mph/s	56,200 (94)	68,300 (96)	79,700 (98)
40	2.65 mph/s	52,100 (106)	63,300 (109)	73,200 (111)
50	2.0 mph/s	44,600 (126)	55,100 (127)	65,000 (129)

(*Note*: rapid transit capacity data are from *Urban Rail Rail Transport* by Lang and Soberman, Joint Center for Urban Studies, the M.I.T. Press, Cambridge, Massachusetts, 1964, p. 65.

Highways: (level of service A, B, C, and E):
   A two-lane, two-way: free flow > 60 mph and 400 vph
   B two-lane, two-way: stable flow > 50 mph and 900 vph
   C two-lane, two-way: stable flow > 40 mph and 1400 vph
   D two-lane, two-way: approaching unstable flow > 35 mph and 1700 vph
   E one-lane, one-way: maximum flow $30 \pm$ mph and 2000 vph
(*Note*: the foregoing highway capacities are from the *Highway Capacity Manual*, Special Report 87, Highway Research Board, National Academy of Sciences, Washington, D.C., 1966, Table 9, p. 66.)

Signalized intersection capacity examples: for level of service C, load factor 0.30, $G/C = 0.50$, peak-hour factor of 0.85, 10% right turns, 10% left turns, trucks and through buses 5%, 20 local transit buses per hour making near side stops.

	Capacity—Vehicles per Hour					
Population	One-way, No Parking, Curb-to-Curb Width (Feet)			Two-way, No Parking, Curb-to-Dividing Line (Feet)		
	24	36	48	24	36	48
100,000	846	1340	1850	685	1058	1396
250,000	900	1426	1968	728	1125	1485
500,000	954	1512	2086	772	1193	1575

(*Note*: based on Figures 6.5, 6.8, and 6.11, pp. 124, 135, and 143 of the *Highway Capacity Manual*.)

TABLE 8.7 (Continued)

Aircraft Runways: practical hourly capacities:	IFR	VFR
Single runway (arrivals = departures)	42–53	45–99
Two parallel, 5000 ft or more apart	84–106	90–198
Open V, dependent operations away		
from intersection	50–57	53–108

(*Note*: runway capacities from *Airport Capacity Criteria Used
in Long-Range Planning*, Federal Aviation Administration, U.S.
Department of Transportation, Washington, D.C., 1969, Table 1, pp. 6-7.

Pipelines: (For 40-A.P.I.-gravity, 0.825-specific gravity petroleum with a viscosity
of 40 Saybolt universal seconds)

Pipe diameter:	2 in.	4–150 bbl per hour
	4 in.	10–400 bbl per hour
	6 in.	50–2000 bbl per hour
	8 in.	100–4000 bbl per hour
	10 in.	100–4000 bbl per hour
	12 in.	400–5600 bbl per hour

Belt conveyors: (actual, carrying slack coal, etc.—Capacities vary with
types and densities of materials being handled)

Belt width—24 in.	200 fpm	65 tons per hour
	400 fpm	130 tons per hour
	600 fpm	196 tons per hour
Belt width—48 in.	200 fpm	325 tons per hour
	400 fpm	649 tons per hour
	600 fpm	974 tons per hour
	800 fpm	1298 tons per hour
Aerial tramways:		6–100 tons per hour

(*Note*: data for belt conveyors is from the *Handbook of Belting: Conveyor and Elevator*, 1953 edition,
The Goodyear Tire and Rubber Company, Akron, Ohio, Section 6.)

[a]1 mph = 1.609 kph.
[b]1 ft. = 0.3048 m.

ACCESSIBILITY AND FREQUENCY

Not only must a transport system have adequate carrying capacity, but that
capability must be placed within a reasonable distance of the intended
user. Otherwise it is as if the service didn't exist. Accessibility is thus a
function of route location and network design. Accessibility is also related
to the route flexibility of an individual mode. A bicycle or an automobile
places the means of transport close at hand with a wide choice of routes

over which it can be operated. The ready availability and route flexibility of the automobile accounts in large part for its popularity. Piggybacking gives highway accessibility to intercity rail lines.

Location   Even with highway modes the degree of proximity of through routes—arterials, expressways, freeways—may determine their usefulness to a particular area. Communities strive today to be on or near an Interstate Highway just as earlier communities strove to be on or near a railroad. Not only is proximity of the route significant but also the means of access to it, the frequency of station locations and stops on a rail or bus line, or the frequency and spacing of access and exit ramps for an expressway. Industry may be able to ship products by truck or piggyback but many companies still need a rail location to receive raw materials and fuel.

Airport locations can support or negate the speed advantages of air travel. It is often true, unfortunately, that the time spent going to and from an airport is as great or greater than the time spent on the actual air journey. The in-town location of railroad stations gives rail transportation a location-time advantage over air transport in competing for core city traffic. Rail transport has been slow to capitalize on this advantage. Rapid transit networks are generally oriented toward the central business district with maximum utilization occurring during the commuter hour peaks in the morning (7 to 9 a.m.) and afternoon (4 to 6 p.m.).

Lack of accessibility can be overcome, in part, by means of feeder and collector-distributor systems. Bus lines carry passengers between ultimate origin or destination and rapid transit, commuter railroad, or express bus stations. Trucks give access between outlying points and freight houses for rail, highway, and airway. Parking lots at public transit stops permit the patron to provide his own feeder service.

Within the City, where dependence is placed on walking to a public transit line, bus and rapid transit routes should be located sufficiently close together to permit walking times of no more than five to ten minutes. This requires route spacings of $\frac{1}{4}$ to $\frac{1}{2}$-mile (402 to 805 m), an ideal not always realized in practice. These generalized spacings are adjusted in terms of origination-destination studies and traffic assignments. There is no general agreement as to the proper spacing of expressways and circumferentials around large urban areas, but 6 to 10 miles (9.7 to 16.1 km) seems to represent a common pattern. Any fixed guideway whether it be commuter rail line, rapid transit system, PRT, or even an expressway introduces a degree of inflexibility and inaccessibility for those remotely situated.

[20]A.M. Wellington, "The Economic Theory of the Location of Railways" Wiley, New York, 1916 edition, pp. 712–713.

Wellington stated long ago that traffic availability varied inversely as the distance of the carrier from the traffic source.[20] The general concept is still correct. Freight traffic facilities are located ideally, close to traffic sources, but having those facilities close to intercity rail and highway routes avoids long movements through congested city areas. Motor freight terminals transfer between intercity and local vehicles and make interline transfers. Double bottoms (two trailers) are not permitted on city streets. Local drivers may be required by union contracts. These factors combine to modify the role of accessibility, suggesting a location at the city's edge with short, easy access to intercity freeways. This, in turn, may be modified by the penetration of freeways into the urban core area.

Frequency of Service   All the transport capacity needed in a day could, in some instances, be combined in one large ship, train, or plane, or in a series of these moving in close succession. The demand requirements would not, however, have been met. Capacity must be available *as needed.* This is possible only when there is a suitable frequency of movement. Adequate frequency involves problems of scheduling and vehicle availability.

For the owner-driver of an automobile or truck, frequency is not a problem. The vehicle is there and available when needed. Frequency is a problem, however, for public or for-hire transport systems—bus lines, rapid transit and commuter railroads, and intercity services offered by all modes.

For scheduled operation there is a question of how frequently and at what hours buses, trains, planes, or trucks should operate. Commuter service, for example, requires a concentration of capacity and schedules during the morning and afternoon rush hours (7–9 a.m. $\pm$ and 4–6 p.m. $\pm$).

There is a nice problem here of matching frequency with demand. There is waste when more service is provided than is needed or utilized. Insufficient service means an inadequate level of service. The problem involves the ability to finance sufficient equipment to provide the requisite number of vehicles. Speed, then, becomes a factor, not only running speed, but also the speed in getting a plane, train, car, or bus or truck through the terminal processes of loading, unloading, transferring, interchanging, and servicing so it will be ready for another trip—that is, the rapidity of turnaround. Slow turnaround time can mean a poor utilization of equipment, too few trips per vehicle, and a greater capital outlay to provide needed equipment. The problems of terminal design for efficient operation are discussed in Chapter 10.

Because certain types of equipment involve large capital investment and operating expenses, there is a temptation to reduce service frequency in order to obtain intensive utilization of high capacity equipment. A 400-seat aircraft can be a good service investment when there is sufficient demand to fill it on a reasonably frequent schedule. When, however, the number of flights must be reduced in order to accumulate enough passengers to fill those seats, there has been a level of service deterioration. Smaller planes flying more often might be better. The problem involves a question of the economic load factor required for profitable operation vs. service. Energy shortages may induce less frequent service.

In rail transport there is a tendency to delay the forwarding of cars in order to accumulate a large number of cars in a train. Operating costs per train mile are reduced and more intensive use is made of the motive power and train and engine crews. A high 1000 gross ton mile per freight train hour indicates the "efficiency" thus secured. What is lost, however, is the investment in rolling stock and service to the shipper and consignee, along with their good will and patronage. They want their shipments to move with the utmost dispatch.

The needed frequency of service must be determined during the analysis of demand study data. See Chapter 15. No generalized rules can be set forth. A suitable frequency must be obtained by a study of the individual community or situation. Some railroads, for example, have found an overall time saving by holding cars until enough have accumulated to run a trainload to a distant point without stopping at intermediate terminals for fill-outs. A long distance truck haul might similarly avoid stops at intermediate freight stations. Each situation has to be evaluated on its own merits.

QUESTIONS FOR STUDY

1. Using typical or average values for vehicular capacities, determine how many units for each mode of transport would be required to move 10,000 tons 50 miles in 24 hours. *Note*: vehicles must make the return trip empty and travel at speeds appropriate to each vehicle.
2. How much horsepower is being used to move dead weight in a train of 8000 gross tons at 30 mph on a 0.5% grade assuming empty cars to weigh 32 tons, 60 tons of payload to be in each car, and the train to be pulled by an 8000-hp diesel-electric locomotive (4-2000-hp, 120-ton units)?
3. What reduction in horsepower, still giving the same ton-mile performance, would be possible if a truck with a 240-hp engine and gross vehicle weight of 40,000 lb had the weight of the 15-ton vehicle reduced by 10 percent? What added cargo weight could be had with the same 240-hp engine?
4. Determine, for the several modes of transport, typical values for net ton-miles per vehicle mile and explain the significance of this figure.
5. Determine the theoretical track capacities and number of trains per day for (*a*) A single-track-railroad district 120 miles long, with sidings spaced 12 miles apart, and trains running at an average speed of 24 mph. (*b*) A double-track-

railroad district 120 miles long, blocks 1.5 miles in length, each train commanding three blocks, and trains running at an average speed of 40 mph.

6. Given: the following data for a railroad operation in which 40,000 gross tons per day must be moved over a 100-mile district: minimum road time for a train of 4000 gross tons average weight is 2.35 hours. Average road time is 4 hours. Road times for trains of 2500, 3000, 5000, 6000, 7000, and 8000 tons are, respectively, 1 hour, 55 minutes; 2 hours, 5 minutes; 2 hours, 38 minutes; 2 hours, 56 minutes; 3 hours, 14 minutes; and 3 hours, 32 minutes. Determine the optimum road time and the interference time for the optimum among these several trains. (Based on data used by E.E. Kimball.)

7. Using the times for the trains of Question 6, prepare a train-hour diagram and determine therefrom the track capacity in train hours actually utilized by these trains.

8. A mining company desires to move 32,000 tons of ore per eight-hour day to a lakefront site 40 miles away. Assuming an average speed of 20 mph, how many trucks and how many lanes of highway would be necessary for this movement? If a railroad were used instead, how many tracks and how many trains would be required? Could a conveyor be used for this movement? How many ships per eight-hour day would be needed to move the ore down the lake?

9. What would be the service capacity and what level of service would be represented by a six-lane divided highway at 40 mph under ideal conditions and uninterrupted flow with a peak hour factor of 0.91? What is the v/c ratio?

10. A four-lane divided highway in rolling terrain (grade of 1 to 2 percent) has 11-ft lanes and 4-ft shoulders. It carries 8-percent truck traffic and averages two buses per hour. If traffic moves at 60 mph in uninterrupted flow, what is the total capacity of the highway?

11. It is desired to move a minimum of 6000 passenger/vph over a highway at a speed of 70 mph. What must be the design specifications for such a highway and what level of service would be represented?

12. An airport is to be designed for a minimum capacity of 60 landings and takeoffs per hour under all conditions. What is the minimum number of runways required for this specification?

13. What must be the speed of a 36-in. belt conveyor designed to carry 1200 tons/hour up an 0.10 percent incline? The belt flight is 1600 ft long.

14. A signalized intersection (light cycle of 50 sec green, 3 sec yellow, and 37 sec red) in a one-way street, located in the residential area of a city with a 250,000 population has 20 percent of the traffic making right turns, 10 percent making left turns, and includes 5 percent through bus and truck traffic. The curb-to-curb distance is 36 ft.

    (a) What is the hourly capacity of the intersection for a level of service C with a load factor of 0.30 and a peak hour factor of 0.85?

    (b) What would be the hourly capacity of an intersection with the same light cycle as in part (a) but located in the central business district of a city with a population of 1 million, with 30 local buses making near side stops each hour, the same turning and through truck–bus traffic as before, and designed for a service level D with a load factor of 0.7 and a peak hour factor of 0.90? The curb-to-dividing line distance is 24 ft.

15. A rapid transit line plans to operate cars that are 50 ft in length and have a total capacity, seated and standing, of 200 people. The trains will have an average acceleration–deceleration rate of 3 ft/sec/sec. Station stops will be of 30 sec duration. Compute the savings in number of trains required to move 60,000 persons per hour in 8-car trains vs. 10-car trains and the running speed required for each. Assuming these cars operate pendulum wise over a 10-mile route with stations spaced one-half mile apart, will one operation as opposed to another result in any savings in the number of cars required to move the given volume of traffic?

SUGGESTED READINGS

1. *Track Capacity and Train Performance*, a report of Subcommittee 1, E.E. Kimball, Chairman, Committee 16, American Railway Engineering Association, Bulletin 462, November 1946, pp. 125–144, A.R.E.A., Chicago 5, Illinois.
2. *Proceedings of the A.R.E.A.*, Vol. 22, 1921, pp. 744–759, Vol. 32, 1931, pp. 643–692, American Railway Engineering Association, Chicago 5, Illinois.
3. Unpublished Ph.D. theses, Civil Engineering Department, University of Illinois, Urbana, Illinois:

   Mostafa K. K. Mostafa, Actual Track Capacity of a Railroad Division, 1951.
   Meng-Te Chang, Effects of Traffic Capacity and Locomotive Performance, 1949.
   Wai-Yum Yee, Centralized Traffic Control as a Means of Accelerating Train Movements, 1943.
   Wai-Chiu Chang, A Study of the Traffic Capacity of Railways by the application of the Relation between Delays to Train Operation and Number of Trains Operated, 1941.
4. *Highway Capacity Manual*, Highway Research Board Special Report 87, National Academy of Sciences–National Research Council, publication 1328, Washington, D. C., 1965.
5. *The Transportation and Traffic Engineers Handbook*, Institute of Traffic Engineers, John Baerwald, Editor, Washington, D.C., 1976.
6. A.S. Lang and R.M. Soberman, *Urban Rail Transport*, Joint Center for Urban Studies, The Massachusetts Institute of Technology Press, Cambridge, Massachusetts, 1964.
7. Charles E. DeLeuw and William R. McConochie, *Exclusive Lanes for Express Bus Operation*, paper before the American Transit Association Western Regional Conference, San Francisco, 1963.
8. Robert Horonjeff, *Planning Design of Airports*, McGraw-Hill, New York, 1962.
9. *Airport Capacity Criteria Used in Long-Range Planning*, U.S. Department of Transportation, Federal Aviation Administration, Washington, D.C., December 24, 1969, AC 150/5060-3a.

10. *Airport Capacity, A Handbook for Analyzing Airport Designs to Determine Practical Movement Rates and Aircraft Operating Costs.* Airborne Instruments Laboratory, Deer Park, Long Island, New York, June 1969.
11. *Airport Site Selection,* Advisory Circular 150/5060-2, Federal Aviation Administration, U.S. Department of Transportation, Washington, D.C.
12. FAA Order 7480.1, *Guidelines for Airport Spacing and Traffic Pattern Airspace Areas,* Federal Aviation Administration, U.S. Department of Transportation, Washington, D.C.
13. *Bus Use of Highways—State of the Art,* Report No. 143 of the National Cooperative Highway Research Program, Transportation Research Board, National Research Council, Washington, D.C., 1975.
14. *Bus Use of Highways—Planning and Design Guidelines,* Report No. 155 of the National Cooperative Highway Research Program, Transportation Research Board, National Research Council, Washington, D.C. 1975.

Chapter 9

Performance Criteria – Quality of Service Factors

Level of service denotes the *quantity* of transportation required to meet a given demand. Quality of service reflects the *way* in which that quantity is made available, involving such matters as safety and dependability, flexibility, speed, and door-to-door travel time, comfort, energy economy, effects on the community and environment.

SAFETY AND DEPENDABILITY

Safety and dependability are so closely interrelated there is difficulty in discussing one apart from the other. There is an implied obligation that persons and goods entrusted to a carrier reach their destination in the whole, undamaged state with which the journey began. There is a similar obligation that the journey or trip be performed with reasonable dispatch and dependability.

Dependability Dependability refers to the safe, on-schedule movement and delivery of goods and passengers and freedom from delays and from loss and damage en route. It is one of the most important characteristics a carrier can offer. The nice interrelation of industrial-commercial elements requires that all coordinate perfectly. Parts and raw materials must arrive on schedule to insure continuity in manufacturing processes. The assembly line shuts down if short-term inventories are depleted by transport delays.

306

Speed of individual movement is usually not so essential as maintaining a steady flow of raw materials, parts, fuels, and lubricants. Transport delays are costly both to the public and to the carriers. Delayed perishables spoil. A long transit period means additional interest on inventory. Delayed newspapers, magazines, and newsreels lose their timeliness. Delays to persons mean lost opportunities, contracts unsigned, appointments missed, deathbeds reached too late. A carrier with a poor record for dependability stands in a difficult competitive position. Many shippers place dependability first in selecting a carrier.

Chapter 10 Terminals discusses the adverse effects terminal delays can have on dependability.

Vehicle Operator   Transport must begin with the vehicle operator—the pilot, engineman, or driver of private automobiles, trucks, or buses. The operator is a maker of decisions—speed, distance between vehicles, selection of route, and obedience to rules and laws are all under his control. Operators vary greatly in training, skill, alertness, and experience. Professional operators of trains, aircraft, buses, and ships receive extensive training in schools, on vehicle simulators, or through long apprentice-type service. Operators of private craft and vehicles must be licensed but receive considerably less training. Private aircraft pilots meet exacting licensing requirements; most states require auto drivers to pass a written and driving test; small pleasure craft on water require little if any such requirements.

Regardless of skill, experience, or training, operators are people and subject to human frailities and limitations. The best of training and experience can be negated when the operator is in poor health, under the influence of drugs or alcohol, is fatigued, worried or emotionally upset, or lacking a sense of responsibility. Even at his best, the operator has certain limitations. *Reaction time* is the time that elapses between perceiving the need to act (reduce speed, apply brakes, make a diverting move, etc.) and the moment that action is initiated. It includes the recognition of the need for action, deciding what is the proper action to take, establishing volition, and actually activating the nerves and muscles to take action. Reaction time may vary from 0.5 to 3+ seconds under normal, alert conditions; more if the operator is fatigued, drugged, intoxicated, or emotionally upset.

Experienced operators probably react faster because of their familiarity with what action needs be taken in a given situation. (They may also develop careless habits.) Here the factor of judgment enters in. An operator may react quickly but may make the wrong decision, take the wrong course of action—step on the gas pedal instead of the brake, for example.

The operator can be assisted by proper design of his or her operational environment. He should have to make only one decision at a time. In highway design this is achieved by use of clear and legible signs that impart only the information needed in plain and simple terms and sufficiently in advance of the time or place that action must occur. Advance warning signs should be used in advance of final signs. Channelization helps by directing the vechicle into the proper path for a turning movement or into the correct lane for future diversion. Enginemen on locomotives are reminded audibly of signal changes with cab signals or train control systems, and commercial aircraft have available warning devices to alert the pilot that he is on a collision course. Reducing the number of controls, dials, gauges, and indicators placed before the operator—the simplified control panel of an automobile, for example— aids in narrowing the range of decisions the operator must make and permits him to concentrate on more vital aspects of the task.

Operators must make other decisions. They determine their speed, whether to overtake and pass another vehicle, whether to change course, how close to come to another vehicle or to a fixed object. There is a normal tendency for drivers to veer away from a close object such as a row of parked vehicles, a fence, or a cut or tunnel wall. (Hence the need for wide shoulders on highways.) They will intuitively increase their distance from the vehicle ahead as speed increases—except when long habit has established a loss of fear and caution, with chain reaction collisions the not infrequent result.

A question before the engineer today is how much decision-making freedom should the operator be allowed. Certain programmed systems make the operator only a monitor; others include an override capability for emergencies. Still other systems act only as a check on the operator if he fails to respond to a situation or makes the wrong decision. With personal vehicles, there are few if any checks other than fear of the police or Coast Guard. As speeds get higher, less time is allowed for the operator to react. Simplification and automation would seem to be a desirable feature of very high speed design and operation.

Safety   Dependable, efficient operation and safe operation complement each other. Table 9.1 and supplementing data give a measure of the incidence of fatal accidents in transport.

A total of 128,700 pedestrians were killed and injured in 1974 with approximately 70 percent of the deaths occuring when pedestrians crossed or entered streets and 40 percent of them occurring between intersections. There were 8700 pedestrians killed, 64 percent in urban areas. Motorcycles were involved in 360,000 accidents leading to 3160 fatalities of motorcycle

TABLE 9.1A COMPARATIVE TRANSPORTATION SAFETY RECORDS[a] (PASSENGER FATALITY RATE PER 1000,000,000 PASSENGER MILES)

Carrier	1972		1973		1974	
	Fatalities	Rate	Fatalities	Rate	Fatalities	Rate
Cars and taxis	35,200	1.90	25,700	1.80	26,800	1.30
Buses	130	0.19	170	0.24	150	0.21
Passenger trains	48	0.53	6	0.07	7	0.07
Airplanes[b]						
Domestice scheduled	160	0.13	128	0.10	158	0.12
International scheduled	0	0.00	69	0.18	262	0.80
General aviation	1,426	21.00	1,411	19.00	1,290	17.00

[a] *Accident Facts*, National Safety Council, 1975 edition, Chicago, Illinois, p. 75.
[b] Ibid., p. 76.

TABLE 9.1B SOURCES OF HIGHWAY ACCIDENTS[a]

Vehicle Type	Fatal Accidents		Involved in Fatal Accidents		Accidents—All Types (millions)	
	Deaths	Percent	Number	Percent	Number	Percent
Passenger cars	—	—	41,700	70.4	20.6	82.0
Trucks, all types	—	—	11,500	19.9	3.4	13.5
Riders in autos on turnpikes	310	70.0	—	—	—	—
Buses, class I intercity	12	0.6	—	—	—	—
Truck tractors and semitrailers	—	—	3,300	5.7	0.44	1.7

[a] *Accidents Facts*, National Safety Council, 1975 edition, Chicago, Illinois, pp. 55, 56, 61.

riders or 14 deaths per 100,000,000 miles of motorcycle travel. Pedalcycle fatalities of 1000 occurred in 1974, 60 percent of them in urban areas. These data are also from *Accident Facts*, 1975 edition, pp. 55, 56, and 61.

Guidance Systems The ability to hold to the established route or roadway has obvious safety implications directly relative to the guidance system. The positive, inherent guidance of confining roadways, side rail or center rail guidance, or of flanged wheels on rails affords a greater measure of safety than dependence alone upon the skill and attention of driver-pilots to their task. See Chapter 4 where the several guidance systems and related modes are explained.

Susceptibility to Weather Ideal transportation should have all-weather dependability. This can be achieved only by placing the roadway underground or enclosing it completely. Pipelines, placed underground, are free (usually) from weather effects other than for a possible heating of the fluid in cold weather to facilitate flowability. This becomes a more severe problem if the line is placed above ground as it might be in permafrost areas. Belt conveyors are susceptible to weather if exposed but have no problems when enclosed or, as with many moving sidewalks, are located indoors. Rail sections of the route may be on or above the ground surface.

Airplanes are susceptible to weather conditions. Commercial airliners are able to rise above most bad weather to maintain schedules, but even these planes must delay takeoff when conditions are exceptionally severe. Landing rates are usually cut in half when instrument landing systems are in use; where not in use the airport may be overflown. Smaller craft cannot fly at all when high winds or fog prevail. Weather creates uncertainties in route and landings. Fogbound and heavily congested airports may force a plane to fly off route and perhaps land at some point far from its scheduled destination before fuel runs out. High winds alter scheduled times and may cause unscheduled stops for fuel at intermediate points.

The railroad technology is organized to experience a minimum effect from weather. Delays to trains en route may occur because of frozen train lines and difficulties in circulating air and steam through the train lines. A fleet of snowplows and switch-heating, snow-blowing, and snow-melting devices are kept ready to keep the track and switches open. Delays on the main line from snow-blocked track are rare. Cab signals and automatic train control or train stop reduce the hazards and delays of poor visibility; there is no guidance problem. Severe floods can cover or wash away the tracks and bridges, causing delays of a few hours to several weeks.

Although highways too rely on snow plows, salting or sanding, and other means, bad weather, especially ice, snow, or fog, reduces highway speeds or halts traffic entirely. Private automobiles and light trucks are especially dangerous to operate in rain, snow, and ice. Since they are light, they lose adhesion and face skidding hazards. Even the heavier semitrailers may skid and jackknife on slippery roads or steep hills. Like the railroads, highway departments organize workers and machines to keep the highways open, but, all too frequently, highways are still reported as closed or unsafe.

Weather can have a severely adverse effect on all shipping. Even with the aid of modern radar, collisions have occurred in fog, and ships have gone off course and become lost. Fog can also halt traffic in harbors, rivers, and locks. High winds and storms may cause coastal shipping to put into port, and heavy seas can spring plates and cause leaks or shift the cargo to a less stable position. Snow, ice, and flood conditions will close the inland rivers and canals for several days to several weeks each year and the Great Lakes are closed to navigation because of ice and storm conditions from early December through late March.

Aerial tramway cars are attached to the cable loop and cannot become lost or, as long as the cable is operating, even delayed. Sway from high winds will halt the operation for the storm's duration. Excessive icing can also halt operations and may even cause cable breakage when the increased weight and tension of the ice-coated cable are not fully included in the design.

Exclusive Right of Way   A carrier with a private, restricted right of way with all movements under its direct control has a high potential for safety. Conveyors, pipelines, and most rapid transit systems are in this category. Most of this advantage is lost for rapid transit, however, if its lines "surface" and its trains mingle with other traffic on public streets. The operation and use of railroad tracks and trains is generally exclusive and under the control of the operating agency. The high accident rate at highway grade crossings is evidence of a significant exception.

Aircraft, ships, and highway vehicles share their roadway media with other commercial operators and with private craft and vehicles. Human guidance is necessary with highway vehicles. Each year sees a high toll in death, injuries, and property damage. Private vehicles are the worst but by no means the only offenders. The intermingling, meeting, and passing of vehicles on the same roadway places full responsibility for safety on the driver. Lack of skill, inattention, or recklessness, even in trivial amounts, may culminate in a major highway disaster. All the skill and experience of a well-trained truck or bus operator may be set at naught by the carelessness of an untrained, irresponsible driver in a private car. The divided-lane design of the most recent highways isolates cars going in opposing directions. Passing still remains a significant hazard.

Ships and tows are dependent on human guidance. Collisions between vessels or with fixed obstructions, bridge piers for example, sometimes occur. The intermingling of private pleasure craft with commercial craft increases collision hazards.

Referring to congested ports and waterways in the United States, the National Transportation Safety Board has pointed out that the number of ships above 100 tons has increased significantly with sharp increases in size along with changes in speed and design characteristics, increased traffic density, and limited maneuverability and "...reduces effective safety margins for vessels confronting each other in restricted waters."[1]

[1]"Collisions Within the Navigable Waters of the United States—Consideration of Alternative Protective Measures," National Transportation Safety Board, 1972.

A combination of air congestion and high speeds has introduced collision hazards to an alarming extent. Crashes and near misses around airport areas are reported with increased frequency, especially between private and commercial craft. Efforts to bar private flying from commercial airports have not been fully successful.

Shock and Impact   Damage to lading and equipment can arise from shock and impact in transit and in terminals. Assembled unit systems, especially long railroad trains, are susceptible to stretching and buffing. A campaign to educate employees, improved methods of train makeup and operation, of stowage (including movable bulkheads for part loads), better suspension systems (trucks and springs), cushion underframe cars, and a more nearly precise control of speed in classification yards represent efforts to reduce shock and impact. Studies now underway to disclose the principles that govern track/train dynamics will aid in reducing this cause of damage.

Damage from shock and rough handling is not a factor in handling bulk cargos. Dangers of instability in both sea-and aircraft due to improper stowage and discharge of cargo have been noted in Chapter 4.

Highway vehicles are not as susceptible to shock and impact as railroads; this means less crating and packing. Consolidating small shipments in trailers or containers permits tighter stowage and reduces loss, breakage, and pilferage that accompanies frequent rehandling and transshipment of package and LTL freight.

Forward Motion   One principal characteristic of airplanes works against their safety and dependability; that is, the plane must remain in motion and its engines must function for it to sustain flight and avoid a disastrous crash. Planes do have a margin of safety in that a multi-engined craft can usually make a safe landing with one or more engines idle. It is a measure of the excellence of engine design and ground-crew maintenance that relatively few accidents occur from engine failure. Ships and tows experience difficulties without forward motion in swift currents or heavy seas. Adequate headway must be maintained.

Urban Transit   Urban rail and bus systems are probably the safest form of transport. Rapid transit trains are subject to many of the hazards common to intercity rail operation. Rail rapid transit enjoys the safety features of exclusive right of way and control over operations, positive rail-flange guidance, and all-weather dependability. Modern rail systems are generally protected by automatic wayside and cab signals, continuous train control, automatic train stop, dead-man control, and other fail-safe features. Not all systems are so equipped or, if equipped, may include an override feature that leaves final control of the train to the operator. Operators' errors have led to collisions and to excessive speed-related derailments. Nonenclosed intercar passageways have permitted riders to fall off trains. Doors opening and closing are another source of accidents. The former, occurring when the train is moving, may cause automatic brake application. At station platforms, trains cannot start until all doors are closed (although here, also an override feature may be present), and doors will not close if any part of a rider's body is within the doorway. Nevertheless, riders have occasionally been caught in closing doors and dragged along with the train. The space between train and platform ($\frac{1}{2}$ to $3\frac{1}{2}$ inches) (1.21 to 8.89 cm) has also led to accidents, catching peoples' feet or a cane or crutch.

The station platform becomes a hazardous area through the pushing and hurrying of crowds to get on and off trains and up and down stairs and ramps. Persons have fallen or been crowded or pushed off the platform and under trains. Suicides from jumping in front of trains are frequent enough to be a problem. Curtain or screened platforms with elevator-type doors are one proposed solution. Additional and expensive equipment is then required, especially for greater precision in stopping automatically controlled trains. Railings placed at the end of car locations can be of some help but, if set back 10 or 12 in., can increase the hazard if people crowd beyond the railings for faster boarding. A high incidence of assaults and "muggings" has required increased police protection and the installation of closed-circuit TV for surveillance of all parts of the station area. Station design seeks to eliminate dark corners where attackers can hide or operate unobserved. Nontrain hazards include loss of ventilation due to failure of ventilating equipment or to tunnel fires.

Accidents with mass transit by bus arise from loading and unloading, from being caught in bus doors, or from falling when the bus has a collision or makes a sudden start or stop. A high incidence of driver "mugging" has led to locked fare boxes, the withdrawing of all cash from the drivers, and a consequent requirement that riders must have exact change for their fare.

Several factors combine to give rail rapid transit, when properly maintained, a high degree of reliability—an exclusive right of way, a tunnel location free from weather constraints (even an elevated structure is less susceptible to weather than a surface location), an elaborate system of signaling for high density operation—in sum, a well-developed technology.

Bus transit is less dependable because buses in urban streets encounter all the delays common to street traffic—ice, snow, rain, traffic signals, and traffic congestion. Nevertheless, a surprisingly good record for dependability has been maintained. The use of exclusive bus lanes and bus-activated traffic signals may futher improve the record.

Pedestrians and Cyclists Accident records indicate that most pedestrian accidents occur when crossing streets, with a high percentage occurring between intersections. Corrective action includes the use of signing, such as school-crossing signs, to caution motorists, and crossing guards (again, usually for schoolchildren). A pedestrian "walk" phase in the traffice signal cycle offers some protection, especially when *all* vehicular traffic, including right turns, is stopped, but this delays vehicular traffic. Under-and overpasses for expecially busy streets and expressways are desirable. Subway stations are sometimes used for this purpose.)

A better alternative is the complete separation of vehicular and pedestrian traffic by the use of walkways extending through central business districts and between neighborhood centers. Streets can be closed to vehicles to form the walkway or specially constructed walks and routes can be utilized, preferably with occasional islands or rest areas with benches and even landscaping.

Pedalcycles cannot be intermingled safely with motorized traffic. Bikeways, similar to pedestrian walkways, can be constructed entirely apart from the street system. Another possibility is to close certain streets or routes to vehicular traffic, a proposal that usually draws protests from landholders along the streets so closed. A weak alternative is the posting of signs designating the least-traveled streets that form a route to a given destination as a bikeway.

THE FEDERAL ROLE

The Federal government is assuming an ever-growing role in transport safety, manifested in a variety of ways.

The national transportation Safety Board has recognized modal variations in safety vs. public policy by saying that when "...a choice of mode exists, government policy should encourage the movement of freight via the safest mode of transport".[2] The Board further referred to charts from a Department-of-Transportation executive briefing that said "...to shift a substantial amount of traffic from highway motor trucks to railroads...is a net saving for our society of approximately 550 deaths and 7300 injuries per year."

[2]Safety Recommendation 1-72-1 adopted by the National Transportation Safety Board, U. S. Deptartment of Transportation, Washington, d. c., May 1972, p. 2.

(a.) *Hours of service laws*. Recognizing an obvious relation between operator alertness and safety, laws have been enacted that limit the number of hours an operator may be on duty without rest. In water transport, licensed pilots of tow vessels may not be on duty "...in excess of a total of 12 hours in any consecutive 24-hour period except in cases of emergency."[3]

Workers engaged in train and engine service or in dispatching operations can have only 12 hours of continuous service without an intervening eight-hour rest period. Truck drivers in interstate operations are limited to 10 hours behind the wheel without eight hours of rest. Commercial, domestic airline pilots are permitted only eight hours of flying time per 24-hour period and at least 16 hours between flights. Before returning to duty pilots who have not flown for more than 90 days must pass a qualifying test in the aircraft type they are to fly.

(b) *Accident investigation*. The National Transportation Safety Board investigates aircraft crashes, major train wrecks, and disasters on the water to determine causes and make recommendations for greater future safety. (Investigation of air crashes is facilitated by a review of flight recorder data. Such recorders have been ordered installed on all commercial aircraft.) The Coast Guard, Federal Aviation Administration, and Bureau of Railroad Safety routinely investigate and report on accidents. There is no similar type of investigation for highway accidents. State or local police usually perform only a cursory investigation.

(c) *Regulation*. In part because of its accident investigations, the Department of Transportation advocates and enforces Congressional enactments and promulgates orders relating to the licensing, inspection, and physical condition of craft and vehicles. Aircraft inspection routines, the Federal Track Safety Standards (of 1972), Federal Motor Vehicle Safety Standards (involving such matters as required stopping distances and standards for wheel lockup for vehicles equipped with air brakes, etc. as well as speed restrictions) are typical of such regulation. Inspections are performed and unsafe aircraft, vessels, locomotive, or track can be ordered out of service until deficiencies are remedied. A problem of recent origin is the extensive damage and loss of life caused by accidents involving the transport of flammable, explosive, or lethal substances such as anhydrous ammonia, vinyl chloride, or wastes from nuclear fission. Procedures for identifying and dealing with these

[3]Public Law 92-339, H. R. 6479, July 1972.

substances when a wreck occurs and the design of safe containers are among the solutions being developed and tested.

(d) *Research and development* (R&D). The Department of Transportation has taken an active role in research and development. Testing and research have led to reports, rule changes, and orders for changes in equipment design for greater safety. The armed forces have long conducted R&D for aerial navigation and aircraft safety and dependability. Many devices arising from these efforts, for example radar and instrument landing systems, are contributing to the safety of commercial flight, a privilege not enjoyed by competing modes. One percent of the Highway Trust Fund is available for R&D work, some at least safety-related, and the 1965 and later funding of the High Speed Ground Transportation Act and of the Urban Mass Transit Administration's efforts have supported research at research centers, universities, and the DOT's High Speed Ground Transport Test Center at Pueblo, Colorado. See Figure 9.1.

OTHER SAFETY-DEPENDABILITY FACTORS

Grade Crossings Annual fatalities at rail-highway grade crossings ranged from 1548 in 1969 to 1233 in 1973. Of these, 82 were pedestrians in 1969 and 78 in 1973. There were 3259 nonfatal accidents; 37 involved pedestrians.[4] States have long been involved in ordering the installation of crossing protection devices at grade crossings and in sharing the capital costs (often on a railroad 10 to 20 percent—state 90 to 80 percent basis). The railroads usually bear the maintenance costs. On August 8, 1967, the U. S. Secretary of Transportation ordered an immediate Program to Reduce Rail Highway Grade Crossing accidents.[5] Guidelines were later developed and made available to the states; these included, among other things, procedures for establishing an inventory of rail-highway grade crossings and methodologies for computing a hazard index for all rail-highway grade crossings.

Hazard Index Crossing hazards are a combination of many factors—volume and speed of rail and highway traffic, pedestrian traffic, number of tracks and number of highway lanes, sight distances, condition of roadway, alignment of tracks and of highway, type of protection in use, customary

---

[4]*Accidents Facts*, National Safety Council, 1975, p. 77.
[5]U. S. Department of Transportation News Release, 8 August 1967.

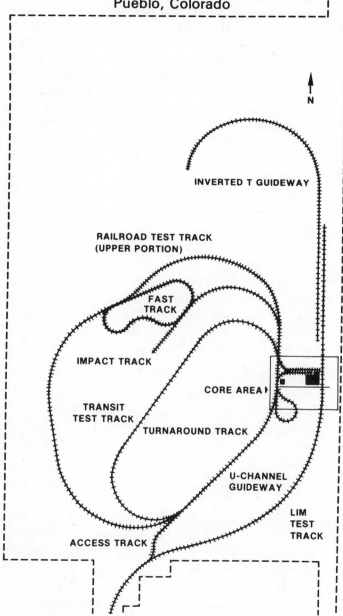

**TRANSPORTATION
TEST CENTER**
Pueblo, Colorado

N

INVERTED T GUIDEWAY

RAILROAD TEST TRACK
(UPPER PORTION)

FAST
TRACK

IMPACT TRACK

CORE AREA ▸

TRANSIT
TEST TRACK

TURNAROUND TRACK

U-CHANNEL
GUIDEWAY

LIM
TEST
TRACK

ACCESS TRACK

FIGURE 9.1 HIGH SPEED GROUND TRANSPORTATION
TEST CENTER, PUEBLO, COLORADO, (U.S. DEPARTMENT
OF TRANSPORTATION, WASHINGTON, D.C.)

weather conditions, gradient of highway, and other factors peculiar to local situations. The hazard index is dependent on the probability of rail-high-way conflict, the sight distances, and the relative protective value of the type of protection device in use. The Association of American Railroads has issued a *Bulletin No. 5, Recommended Standards and Practices for Railroad-Highway Grade Crossings Protection.* Studies of the AREA and others have shown that the familiar flasher light protection is effective at single-track grade crossings and the automatic short arm crossing gates combined with flashers are effective at multiple-track crossings. These are more effective than other crossing protective devices.

Typical of the hazard index formula is the one used by the state of Wisconsin.[6]

$$H.\,I. = T\left(\frac{V/20 + P'/50}{5}\right) + D + A_e$$

Where

$H.\,I.$ = hazard index

T = number of trains in 24 hours

V = traffic volume by highway in 24 hours; one train is considered to offer a hazard equivalent to 20 vehicles and to 50 pedestrians

P' = number of pedestrians in 24 hours

D = a sight distance rating

A_e = accident incidence = average number of accidents per year × 100

The foregoing is reduced to $H.\,I. = \dfrac{T \times V}{100} + D + A$ when pedestrian traffic is negligible.

The sight distance rating, D, for each quadrant is based on a seven-second vision of the train. By using the following tabulation, a factor is prepared for each quadrant, multiplied by the appropriate weighting factor; the sum of all four factors is used for D.

[6]Highway-Railroad Grade Crossing Data, Wisconsin Public Service Comminssion, Madison, Wisconsin, May 1965.

Urban Crossings			Rural Crossings		
Distance to Crossing (ft)	Factor	Safe [a] Speed (mph)	Distance to Crossing (ft)	Factor	Safe [a] Speed* (mph)
25	100	15	25	100	15
35	50	20	35	50	20
65	22	25	65	44	25
90	0	30	90	40	30
			150	30	35
			225	20	40
			300	10	45
			360	0	50

Weighting factor: 0.7—for worst quadrant
                 0.1—for each of other quadrants
*[a] Safe stopping distance speed for highway vehicle.

Problem Example

Find the Hazard Index (H.I.) for a crossing that has 20 trains, 500 highway vehicles, and 250 pedestrians in a 24-hour period. It also has an accident rate of 1 in 5 years. For a seven-second vision of the trains,

$$Q_1 = 150, \qquad Q_2 = 225, \qquad Q_3 = 300, \qquad Q_4 = 360$$

Solution   Distance factor 'D': $Q_1 = 0.7 \times 30 = 21$
$$Q_2 = 0.1 \times 20 = 2$$
$$Q_3 = 0.1 \times 10 = 1$$
$$Q_4 = 0.1 \times 0 = 0$$
$$\text{Factor } 'D' = \overline{24}$$

Accident factor 'A' = $\frac{1}{5} \times 100 = 20$

$$H.\ I. = 20 \times \frac{\left(\dfrac{500}{20} + \dfrac{250}{50} \right)}{5} + 24 + 20 = 164$$

Route Obstructions   Delays ensue when a route is obstructed whether purposely (as when maintenance is being performed), by an accident, or by forces of nature. It is traditional with railways, and on busy lines a practical necessity, to keep tracks open and trains running regardless of the cost or effort. A system of detour arrangements with other railroads keeps trains moving when tracks are completely blocked. The failure of a pipeline, belt, conveyor, or aerial tramway halts the entire operation unless (which is unlikely) parallel facilities have been constructed. The ubiquity of roads and highways makes detouring highway traffic a fairly simple matter.

Arrival Reliability On-time performance is a measure of dependability that can only be obtained by a combination of adequately designed facilities and the tight operation of well-maintained vehicles over equally well-maintained roads and tracks. Equipment and route capacities as well as signs and signals must be adequate to avoid congestion. An airport with an inadequate capacity causes delays as aircraft await their turn to take off or land. Speed restrictions due to poorly maintained track have been a source of delay to trains, and the rebuilding of sections of the interstates have slowed down highway vehicles. Equipment failures have also played a role in train delays, including those of Amtrak. Good maintenance and high morale among well-trained employees are prime ingredients in maintaining schedules.

Summary Table 9.2 contains the author's evaluation of transport dependability for the several modes. A quantitative index for each mode can be established by assigning point values to each qualitative term and giving a numerical value to the tabulation. The reader is invited to make his or her own evaluation.

TABLE 9.2 TRANSPORT DEPENDABILITY

Carrier	Movement	Freedom from Loss and Damage	Combined Effect
Railway	Good to excellent[a]	Fair	good
Highway—freight	Fair to good	Good	Good[b]
Highway—passenger	Poor to good[c]	Poor	Poor
Waterways	Fair	Good	Fair to good
Airways	Poor	Good	Fair
Pipelines	Excellent	Excellent	Excellent
Conveyors	Excellent	Excellent	Excellent
Aerial tramways	Good	Excellent	Good to excellent
Public transi—bus	Fair	Excellent	Good
Public Transit—rail	Excellent	Excellent	Excellent

[a]Primarily line movement; as pointed out earlier in the section, railroads are subject to terminal delay problems.
[b]For motor freight; poor to fair for private automobile traffic.
[c]Poor for city streets; good for interstate highways.

The accident rates of Table 9.1a and 9.1b indicate relative modal safety. Safety is considered of sufficient importance to promote federal action in passing hours of service regarding laws, investigating accidents, setting standards for the safe condition of rolling equipment and physical facilities, inspecting equipment and facilities, and instituting research and development, including studies and programs to reduce rail-highway grade crossing accidents. Other aspects of safety are found in Chapter 11, "Operational Control."

FLEXIBILITY

The ability to react or adapt to a variety of needs or changing conditions can have a deciding effect upon a carrier's utility. Flexibility appears in a variety of forms.

Volume   Certain modes handle large volumes of traffic efficiently. Personalized transport—autos, bicycles, and small aircraft—give effective individual movement but are not efficient in moving large numbers of people at the same time. Buses and rail systems are suitable for mass movement. Rail transit systems are effective in high density areas or where route flexibility of autos and buses is used to concentrate large numbers of riders from diverse locations at stations.

For freight movement the somewhat limited capacity of highway trucks and tractor-trailer combinations directs their area of capability to moving small shipments and packaged goods, especially where rapid movement over short distances is required. Road-haul units will have capacities of 10 to 20 tons for single trailers and twice that amount for so-called double-bottomed operations. Size and capacity are limited by state laws and by the bearing capacity of highway pavements.

In contrast, the pipeline, which is a continuous flow system, can operate efficiently only when moving very large volumes. The ICC has recognized this feature by permitting lines to require in their tariffs a minimum tender for shipment, sometimes 10,000 bbl but more often 100,000 bbl. Smaller shipments may be accepted and held for movement when accompanied by a large shipment of the same grade and specifications.

Bulk cargo ships and barges are generally limited to large volumes of bulk commodities or to manufactured goods in quantities, such as pipe. Barges assembled in tows can be varied in number to meet demand volumes. However, their slow speed limits the utility of barges and ships for moving merchandise cargo.

Railroads have wide volume flexibility. Their prime advantage lies in moving large volumes of anything—coal, grain, automobiles, lumber, or fluids. Being an assembled unit technology, a train can add or remove according to the volume of traffic. A single car may contain a full load or a partial load. The development of trailer-on-flatcar and container-on-flatcar services has opened the area of small lot and merchandise traffic to the efficiencies of large volume handling.

Commodity   Passenger traffic differs principally in requirements involving volume, speed, time, and comfort. Thus persons can and are moved by all modes except pipelines. Conveyor systems for people are presently limited to moving stairs and moving walkways, but other applications have been projected.

Freight commodities are more often mode-oriented by volume and speed requirements than by inherent types. Granular bulks move by those carriers with large volume capability; hence, by water, rail, or conveyor. Liquid bulks similarly move by barge, tankship, pipeline, or train. Pipelines are generally limited technologically to the shipment of liquids or of solids converted to a semiliquid state as in "solids" pipelines for coal and some ores. Perishables such as fresh fruits and vegetables, flowers, meats and frozen foods, and fruit concentrates move by rail or truck in protective equipment, mechanically refrigerated on fast train schedules, or with the rapid movement capability of trucks, and for long hauls the high speed of airplanes. Merchandise freight that also requires rapid movement is carried by highway, rail, or rail-highway, or by air.

Route Flexibility   The ability to travel directly from any given point of origin to a given destination is a highly useful characteristic. A highway vehicle can travel wherever a reasonably firm and smooth way is available. Not only can the three-million odd miles of highway be traversed to any city or village, but trucks can drive into fields for agricultural loading or to a job site to deliver construction materials. Trucks offer true door-to-door pickup and delivery service. They also enjoy road flexibility with two-way movement and the ability to overtake and pass. Route obstructions are seldom serious. Either the obstruction can be bypassed at a slow speed, or a detour on other network units can be made with minor delay.

The highway mode offers the same flexibility in moving people, a prime reason for its universal use. Route flexibility permits bus routes to be varied at little or no cost to meet the changing needs of traffic and commends their use to collector-distributor service for line-haul operations. Flexibility for autos is limited only by their ability to find parking spaces. Unlike the bus, autos can give door-to-door service.

Rail technology is at its best for movement through corridors. Individual service can only be extended to those patrons with a private side track. The rail network of the United States is presently composed of numerous independent operating companies. These companies function more or less as one system to the convenience of the shipping public. A standard gauge (4 ft $8\frac{1}{2}$ in.) (1.43 m) of track makes it physically possible for a car to move anywhere on the network. Cars are standardized as to couplings, brakes, other safety features, and car floor levels. A system of private ownership exists, but free interchange of cars between railroads is made possible by interchange rules regarding the mechanical condition of cars, Car Service Rules providing for the orderly return of cars to the owning road after a shipment has been unloaded, and car accounting and per-diem payment rules and procedures to compensate one carrier for the use of its cars by another carrier. These rules are promulgated and policed by the Mechanical and Car Service divisions of the Association of American Railroads and by the Car Service Bureau of the ICC. Passenger equipment, with minor exceptions that include suburban service, is owned by Amtrak and moves freely in Amtrak trains over the participating rail lines.

On single track, operating procedures, signals, and passing sidings permit two-directional train movements and overtaking and passing. Accidents and other track obstructions can halt or delay train movements, but detour arrangements (made in advance for a network) permit traffic to keep moving via other company routes or over other railroads.

For waterways route flexibility is limited to the presence of continuous navigable depths. Most of the large urban and shipping centers of the country have interconnected access by water, but all-water routes between any two remote points may be so circuitous as to render such movement impracticable. There are, generally, opportunities for two-way movement and for the overtaking and passing of tows, but there is little opportunity to detour if a route (or lock) is obstructed. If one route is closed by flooding, it is likely that all adjacent routes will also be flooded.

Conveyor systems and pipelines have extremely limited route flexibility. Two-way movement is impossible unless duplicate facilities are constructed and detouring is generally impossible.*[7] Thus a stoppage at one point in the systems halts the entire operation.

---

[7]*Consideration was given to carrying iron ore by conveyor over the route now occupied by the QNS&L Railway in Labrador and Quebec. One of the deciding factors was the inability to have two-way movement, that is, to bring in personnel, materials, and supplies to mines from which the ore was being mined and shipped.

TABLE 9.3 FLEXIBILITY CHARACTERISTICS

Characteristics	Railroads	Highways	Waterways	Airways	Pipelines	Conveyors	Aerial Tramways
Adaptability to terrain	Good	Good	Poor	Good	Good	Fair	Excellent
Extent of available routes	Good	Excellent	Fair	Good	Fair	Poor	Poor
Facility of movement	Good	Excellent	Good	Excellent	Poor	Poor	Poor
Adaptability to size and type of load	Excellent	Poor	Good	Poor_p	Poor	Poor	Poor
Ease of interchange	Excellent	Fair	Fair	Poor	Fair	Poor	Poor
Ease of entrance to and exit from service	Poor	Excellent	Good	Fair	Poor	Poor	Poor
Adaptability to varying sizes and volumes of traffic	Excellent	Poor	Good	Poor	Poor	Fair	Poor
Adaptability to varying types of traffic	Excellent	Good	Excellent	Poor	Poor	Fair	Fair
Continuity of flow in spite of local stoppages	Excellent	Excellent	Excellent	Excellent	Poor	Poor	Poor
General or overall flexibility	Good	Good	Fair	Poor	Poor	Poor	Poor

Aerial tramways also offer no route flexibility, but they do offer great adaptability to terrain. Mountainous areas, desert land, and broad river crossings that are not feasible to other land-based systems can be overcome with relative ease by cableways

Table 9.3 presents the author's evaluation of the foregoing and other flexibility factors based only on the technology of the several modes. Again, numerical ratings can be assigned to each of the terms to give flexibility indices. The term "overall flexibility" may not be significant. One particular factor can govern decisions. The adaptability of aerial tramways to rugged terrain or the need to move petroleum in mass quantities across an ocean or desert are such examples. The Alaska pipeline controversy indicates the variety of factors that can arise in selecting a mode for a particular purpose.

SPEED

Rate of travel is a factor both in level of service and quality of service. Speed as it affects the level of highway capacity and service was a subject in Chapter 7. Other aspects of speed are developed below.

Speed Types   Speed has many definitiona and connotations. *Maximum speed*, that obtained in brief spurts, may be of moment to record seekers but has only passing interest to transport. There is an occasional usefulness in knowing *spot speed*, attained at a given moment in time or place. Far more important is sustained *running or cruising speed*, a travel rate that can be maintained consistently for periods of long duration. Even more significant is *average speed* that takes account of accelerating and decelerating times and the effects of periods of reduced speeds. Most significant is the *overall speed* reflected in door-to-door travel time that includes the impact of station halt time, terminal, transfer and interchange time, and other usual sources of delays—traffic lights, stop signs, and congestion, for example. Table 9.4 shows typical speeds.

Speed Hierarchies   Such terms as slow, fast, or high speed have little significance when taken out of context. A series of running speed hierarchies can be identified in which a slow speed for one hierarchy may be quite rapid for another. One possible grouping with appropriate areas of application is given in Table 9.5. There will obviously be some overlapping between the extremes of each hierarchy. Table 9.5 indicates that very high speeds are really not that important. Speeds of 80 mph (128.7 kph) or less prevail in all urban situations and in many rural and intercity movements.

TABLE 9.4 TRANSPORT SPEED RANGES

Carrier	Normal Range	Practical Maximum	Speed-Trial Maximum
Railroads[a]			
Passenger, main line	75 to 100 mph[b]	150 mph	205 mph
Passenger, secondary	60 to 70 mph		
Passenger, branch line	30 to 50 mph		
Statistical average	mph	79 mph	
Freight, dispatch	50 to 90 mph	90 mph	
Freight, "regular"	40 to 50 mph		
Freight, slow or "drag"	25 to 40 mph		
Freight, steep grades	10 to 20 mph		
Statistical average	20.4 mph		
Passenger automobiles	50 to 75 mph	100 to 120 mph	631.6 mph
Trucks	40 to 70 mph	70 mph	
On grades	20 to 30 mph		
Steep grades and city traffic	7 to 20 mph		
Average	23 to 34 mph		
Ships, deep water			
Passenger liners	20 to 40 knots	35.59 knots	
Freighters	10 to 25 knots	30 knots	
Great Lakes bulk cargo	8 to 12 knots	20 knots	
Tows			
Lower Mississippi			
Upstream	6 to 8 mph		
Downstream	10 to 14 mph		
Upper Mississippi and Ohio			
Upstream	5 to 7 mph		
Downstream	8 to 10 mph		
Upstream, flood stage	2 to 4 mph		
Intercoastal waterways	6 to 8 mph		
Aircraft			
Single and twin-engined private planes	125 to 150 mph		
Twin-engined commercial	150 to 300 mph		
Four-engined transports	300 to 400 mph		
Three- or Four-engined jet transports	400 to 600 mph		
Military and experimental	1000 mph and up		2070 mph
Pipelines	1 to 5 mph		
Conveyors	300 to 800 fpm (3.4 to 9.1 mph)		
Aerial tramways	4 to 6 mph		
LIM vehicle			255 mph

[a]High speed requires excellence in track maintenance. The slower speeds of recent years reflect poor track conditions and the slower schedules engendered by deferred maintenance.

[b]Note: 1 mph = 1.609 kph; 1 knot = 1.852 kph.

TABLE 9.5 RUNNING SPEED HIERARCHIES

Hierarchy	Typical Speed Range, mph (kph)	Typical Modes	Appropriate Area of Application
I	3 to 20		
Urban	(4.8 to 32.2)	Walking, bicycle, automobile, bus, truck	Short distances: 1 to 5 miles (1.6 to 8.1 km) Central business districts campuses, shopping centers, intraairport, etc.
Rural and intercity		Pipelines, river tows, conveyors, bulk cargo ships, aerial tramways	Long distance hauls of 5 to 2000+ miles (8 to 3218 km) Intercity and interterminal moves except conveyors and aerial tramways which are usually no more than 20 miles (32.2 km) or less
II	20 to 45		
Urban	(32.2 to 72.4)	Automobile, bus, rail rapid transit, transit expressway, Personal Rapid Transit systems, trucks	Intermediate distances: 5 to 20 miles (8 to 32.2 km), principal main streets, urban expressways and urban rail routes in and away from the CBD; also urban expressways and suburban area routes
Rural and intercity		Tonnage trains, unit trains, and way freights	Long distance intercity: 20 to 2000+ miles (32.2 to 3218 km)
III	45 to 80		
Urban	(72.4 to 128.7)	Automobiles, buses, trucks, some freight trains, passenger trains	Metro and suburban hauls: 10 to 50+ miles (16.1 to 80.5 km), some urban expressways and rapid transit to suburbs
Rural and intercity		Same as for urban	Long haul intercity movements: 50 to 2000+ miles (80.5 to 3218 km), corridor service
IV Urban	80 to 125 (128.7 to 201.1)		Not applicable
Rural and intercity		Some automobiles, high speed trains (a few freight trains), small aircraft, helicopters	Corridor service and long hauls of 50 to 1000± miles (80.5 to 1609 km) with passengers, merchandise, express and mail
V Urban	125 to 300 (201.1 to 482.7)		Not applicable
Rural and intercity		Advance passenger trains, aircraft, linear-induction motor, and air cushion and magnetic suspension vehicles	High-speed ground transportation: corridor and feeder service of 50 to 500± miles (80.5 to 804.5 km)
VI Urban	300 to 1000 (482.7 to 1609)		Not applicable
Rural and intercity		Concepts for the future; Foa tube, jet aircraft, supersonic aircraft	Long corridor express service, both continental and intercontinental service

Far more important than fast running speed is dependable scheduling. Time and discomfort involved in accelerating and decelerating to and from high speeds work against their use for short hauls or where station stops are closely spaced. See the next section for the effects on speed of station spacing.

In Chapter 8, street and highway speeds were presented as level of service factors closely related to lane capacities. Of the six level of service categories, A through F, levels A, B, C, and D represent operating speeds, respectively, equal to or less than 60, 55, and 40 mph (96.5, 88.5, and 64.4 kph); level E has a 30- to 35-mph (48.3- to 56.3-kph) range; level F represents flow at less than 30 mph (48.3 kph). These categories can also be considered as representing various levels of quality of service as well as levels of capacity. In fact, one can start with the desired speed quality or level desired and then design the highway capacity to meet that speed requirement.

Surface vehicle speeds of 200± mph (322 kph) are technologically feasible and can probably be attained by progressively improving existing technologies. Concepts as yet unproven may lead eventually to systems capable of 200 to 500 mph (483 to 805 kph) or more. The costs of developing vehicles to operate beyond the 200-mph (322-kph) range are so great that a major increase in speed capability must be obtained to justify the expenditure. Whether such high speeds are worth the costs depends on the benfits they confer on society and raises questions with nice philosophical implications.

Station Stops and Spacing   The effect of station halt time on average running time and door-to-door time is obvious. The longer the stop the longer the time for a journey. The time to load or discharge passengers is a function both of vehicle and of station design. See Chapter 10 for a discussion of terminals.

Station spacing greatly affects running time, especially where stops are close together as with bus or rapid transit operation. In Figure 9.2, a train or bus is shown stopping at stations spaced one-half mile apart. Note that maximum running speed is achieved for only a short part of that distance because of the time and space required for acceleration and deceleration. The average speed is shown by the dashed line.

If the stops were spaced one mile apart, the average speed would be higher (dotted line) because more running can be done at maximum running speed. The wider station spacing could be achieved by placing the stations farther apart (or by simply closing every other one.) But this solution reduces service. Patrons at the intermediate points would now have farther to travel to reach a station, and some of them would

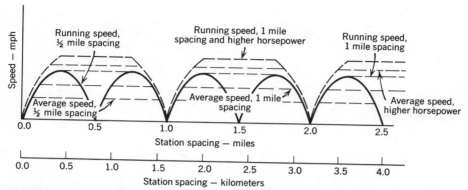

FIGURE 9.2 EFFECTS OF STATION SPACING AND HORSEPOWER ON SPEED.

undoubtedly turn to alternative modes of travel. Also the halt time at stations remaining open will be increased so as to allow the greater number of patrons to get on or off, possibly dissipating the time gained by the faster running speed.

The closer spacing (one-half mile in the sketch) can be retained by skip-stop operation with A and B trains whereby each train or bus stops at every other station. There is a full mile of running between stops by all trains or buses. Multiple tracks or the use of passing sidings permits express service that stops every third, fourth, or fifth station; this service is combined with the slower-moving local trains on other tracks. Another possibility has the train or bus at the end of the line stop at every station for the first several stations (enough to fill the vehicle), then run express to the final destination (the CBD). Another train or bus would start local pickup where the first had left off, then it too would express to its destination. For outbound moves the process would be reversed, expressing outbound and then making local station stops.

One other possibility for faster speed lies in the use (at additional cost) of motors with higher horsepower. These would permit more rapid acceleration and running as shown by the dotted lines in Figure 9.2.

The significance of station spacing on average running speed is shown in Figure 9.3. Regardless of the maximum possible running speed, low average speeds prevail until station stops are five miles or more apart. Thus, as noted earlier, there is little gain in designing vehicles with high-speed capabilities if the frequency of station stops prevents achieving that high speed.

Station stop time depends on the number of passengers to be loaded or discharged. This, too, is a function of station spacing and also of train

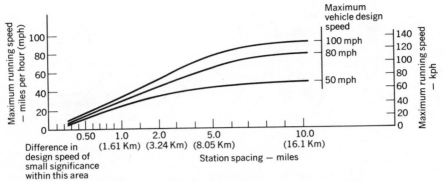

FIGURE 9.3 SIGNIFICANCE OF STATION SPACING VS. SPEED.

headway. More riders accumulate when stations are farther apart, fewer when trains are frequent and headways short. Stopping time can vary from 5 to 10 seconds for most bus stops (longer at transfer points) and from 20 to 40 seconds for rail transit.

Door-to-Door Time Riders are far more concerned with overall travel time (door-to-door) than with vehicular running speed. Running time often is but a small part of the total time spent in reaching a destination. That time will also include time spent in traveling to the station, transferring to other lines or modes, waiting to purchase tickets, checking in luggage and a security check, parking a car or retrieving it, and even elevator time at the destination building. Freight delays and movement time are discussed in Chapter 10.

A typical journey to work by rail transit might be represented by the following:

$$T = t_1 + t_2 + t_3 + t_4 + t_5 + t_6 + t_7$$

where

T = total door-to-door time in minutes

t_1 = travel time to rail station. It can vary from 2 or 3 minutes to 20 minutes or more depending on distance from home to station and mode of travel, whether by foot, bus, or car.

t_2 = time to park car and walk onto station platform, 1 to 5 minutes depending on size of the lot and difficulty in finding an open space. t_2 varies from 0 to a nominal 30 seconds if the trip is made on foot or the rider is delivered by a car that does not park.

$t_3 =$ wait for train. A five-minute "safety factor" can be observed here. (The train is assumed to be on time.)

$t_4 =$ Possible transfer to another line, 2 to 5+ minutes

$t_5 =$ Train movement time, as earlier set forth

$t_6 =$ Travel time to final destination, again a function of mode and distance. Mode will be by foot, bus, taxi, or rapid transit, 2 to 10 minutes.

$t_7 =$ Elevator time to office floor, 1 to 5 minutes

All this adds to 13 to 33 minutes exclusive of road time.

If auto travel is used for the entire trip, total elapsed time will include travel over local streets to an expressway, possible queuing delays at access ramps, congestion on the expressway (sometimes to a complete stop) as the CBD is approached, delays in getting off the exit ramps onto city streets, hunt for a parking place or waiting for service at a parking garage, and, finally, a walk or ride (by local transport) to the final destination and elevator. There are countless variations in this and the foregoing rail transit procedure, but the pattern is obvious. The reader is invited to chart his own travel pattern.

Airline travel has its own door-to-door patterns: an initial journey by car, bus, taxi, or rapid transit of 15 to 60 minutes, depending on the distance, mode, and congestion encountered. In most large cities, the wise traveler starts for the airport no less than an hour before plane departure time. Check-in should occur at least twenty minutes before flight time. It may take as long to travel the 5 to 20 miles, or more to the airport as would a flight of several hundred miles. At the arrival end, an added delay is caused by waiting for checked luggage and the arrival of a surface vehicle.

It follows that the location and degree of proximity of airports, rail stations, and bus depots to traffic origins and destinations are of great importance in the reduction of overall travel time. Rail and bus stations located in the CBD reduce significantly access time from that area, and one or more stops at suburban stations can do the same for riders in outlying areas. Because of access and exit time, high speed trains can be competitive with air travel over distances of 100 to 200 miles or even more.

Door-to-door time can also be reduced by locating important buildings and activities close to airports, rail, and bus stations. These matters are of significance in urban planning.

Acceleration and Deceleration   The rate at which a vehicle can increase or decrease speed bears heavily on its overall travel time. The higher average speed represented by the dotted line in Figure 9.2 is due at least in part to a more rapid rate of acceleration.

In earlier chapters it was shown that horsepower and propulsive force vary with the speed, in part because propulsive resistance also varies with the speed. For a given horsepower, propulsive or tractive force varies inversely with the speed. Thus for a fixed horsepower one can have speed or haul a heavy load, but both cannot be done at once. In transport design and operation hauling capacity is gained at the expense of speed and vice versa. As a train or truck climbs an incline, grade resistance lessens the speed. On descending a grade, the force of gravity tends to accelerate the speed. For a given set of constraints, maximum speed or balancing speed occurs when propulsive force and resistance are equal. In effect, acceleration is a form of resistance because additional horsepower (and fuel) and propulsive effort are needed to overcome the resistance of higher speeds.

A frequent acceleration problem is to determine the distance and time required to accelerate from one speed to another. Equations can be derived using the principle of kinetic energy, that is, work performed in moving a mass at a given speed.

$$F_a \times S = \tfrac{1}{2}mv^2$$

where

$S =$ distance in ft over which acceleration occurs.

$F_a =$ force available for acceleration. It is equal to the tractive effort minus the resistance at v fps

$m =$ vehicle mass equals weight in lb divided by the acceleration due to gravity of 32.2 ft/sec^2

$v =$ speed in fps

Solving for S and converting v to mph and w to W in tons

$$S = \frac{66.8 \; WV^2}{F_a}$$

For wheeled land vehicles, additional energy is required to impart rotative energy to the wheels. This is usually taken as an additional 5 percent, whereby

$$S = \frac{70 \; WV^2}{F_a}$$

The distance to accelerate from an initial speed V_1 to a final speed V_2 can be represented by

$$S = \frac{70 \, W \left(V_2^2 - V_1^2 \right)}{F_a}$$

To determine the time for acceleration, the value of S is taken in terms of speed and time from which

$$F_a \times vt = \tfrac{1}{2}mv^2$$

By following the same derivation procedures as for distance, the time in seconds to accelerate from one speed to another becomes

$$t = \frac{95.6\,W\,(V_1 - V_2)}{F_a}$$

In terms of unit acceleration utilizing a force per ton,

$$S = \frac{70(V_2^2 - V_1^2)}{F_a'} \qquad \text{and} \qquad t = \frac{95.6(V_1 - V_2)}{F_a'}$$

where

$$F_a' = \frac{F_a}{W}$$

But acceleration does not take place at a constant rate. Its graphical representation is curvilinear rather than a straight line. A practical solution to the equation lies in taking successively small increments of speed, 0 to 2, 2 to 4, 4 to 6 mph, etc., and computing the individual distances and times, using the average speed and accelerative force over each velocity range. Speed-distance, speed-time, and time-distance curves may thus be plotted from a solution of the two basic equations. By applying appropriate portions of these curves to a railraod, highway, or monorail profile, we can compute and graph the entire speed, gradient, time, and distance relations and the performance of the motive power in making a run. This is of considerable value in establishing schedules, predicting performance, and studying and designing profiles and motive-power units.

The accelerating capabilities of various recent automobile designs are shown in Figure 9.4.

Vehicle performance programs have been developed, using electronic computers to perform the many computations involved. For example, in 1964 Dr. R. W. Drucker developed a train performance calculator program at the University of Illinois that has been used by several railroads. See Figure 9.5 for an example.

These equations may be used in the same way to determine time and distance for deceleration. Deceleration is indicated by negative values for time and distance if the initial and final velocity subscripts are adhered to as given. F becomes the decelerating force and, for unbraked deceleration, is equal to the retarding effects of propulsive and grade resistances. The same factors are additive to braking force when speed is reduced by brake application. The equations remain the same when computing braking

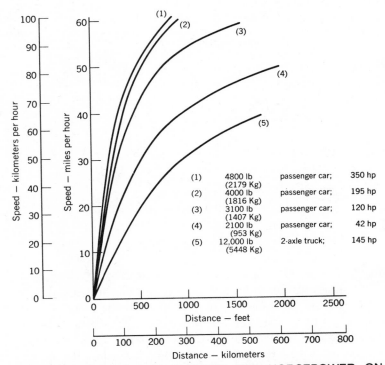

FIGURE 9.4 EFFECT OF WEIGHT AND HORSEPOWER ON AUTOMOBILE ACCELERATION. (BASED ON TABLE 2.5, P. 21 OF THE *TRANSPORTATION AND TRAFFIC ENGINEERING HANDBOOK*, JOHN BAERWALD, EDITIOR, INSTITUTE OF TRAFFIC ENGINEERS, PRENTICE-HALL, ENGLEWOOD CLIFFS, NEW JERSEY, 1976.)

distance except that F_a now becomes F_b, the braking force that includes the force of the brakes plus the decelerating effects of grades and propulsive resistance.

Dr. L. K. Sillcox points out that maximum braking force is obtained if the wheels continue to turn when the available coefficient of friction between wheel and rail or tire and pavement is 100 percent, that is, when the force resisting sliding is equal to the weight that the vehicle imposes on the roadway. Under most favorable conditions, seldom reached, a well-sanded rail provides an adhesion value of 40 percent of the car weight; 50 percent is a practical maximum under existing design. This limits deceleration to 6.4 ft (1.95 m) per sec per sec, or 4.4 mph (7.08 kph) per sec.

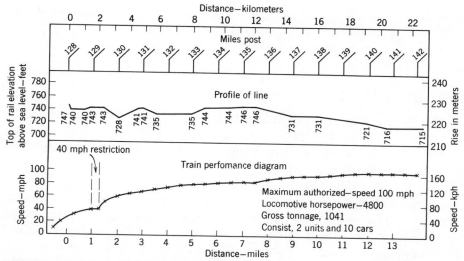

FIGURE 9.5 TRAIN PERFORMANCE DIAGRAM. (BASED ON TRAIN PERFOR-
MANCE CALCULATOR PROGRAM DEVELOPED BY DR. R.W. DRUCKER,
UNIVERSITY OF ILLINOIS, 1965.)

Automotive tires have gripped concrete highway surfaces with an adhesion
approaching 100 percent, permitting a deceleration of 32.2 ft per sec per
sec (equal to the acceleration of gravity).[8] Values of 40 to 80 percent are
more often in effect. On dry cement concrete, coefficients of friction for
rubber tired vehicles range from 0.78 at 20 mph (32.2 kph) to 0.76 at 40
mph (64.4 kph). Similar values for wet pavement are 0.40 to 0.30. At 60
mph (96.5 kph) and 80 mph (128.7 kph) 0.30 and 0.27 are suggested values
for all pavements and conditions.[9] However, for all types of wheeled
vehicles, the actual value is a function of the speed and materials of wheel
and surface and whether the surface is wet, dry, oily, or glazed. Total
braking distance is also a function of driver perception time and brake-application time. Perception time varies with the individual and circumstance
from 0.50 to 3.0 sec. Brake-application time for highway drivers and
vehicles varies from 0.20 to 1.7 sec. For railroad trains, brake-application
distance is usually taken as $1.47V_i nt_b$ where V_i is the initial speed in mph

[8]L. K. Sillcox, *Mastering Momentum*, 2nd edition, Simmons-Boardman, New York, 1955, p.
33.
[9]*Transportation and Traffic Engineering Handbook*, Institute of Traffic Engineers, John E.
Baerwald, Editor, Prentice-Hall, 1975, Table 2.8, p. 28.

and t_b is the brake-application time per car in seconds; n = number of cars in the train. For modern high-speed, quick-acting airbrakes, the application may proceed serially from one car to the next at a rate as low as 0.10 sec per car.

CARE OF TRAFFIC

Closely allied to safety but in a different category, nevertheless, is the quality of care given to traffic, a question of comfort for passengers and of loss and damage for freight. The design details for meeting these needs are beyond the scope of this book.

Comfort   Although ride comfort starts with a smooth roadway or track, springing and suspension systems of the vehicle must be designed to reduce shock, impact, sway, and vibration to a minimum. Shock absorbers and other damping devices combined with adequate springing are required. Nosing of air cushioned vehicles, stabilizing devices for ships and aircraft, and reduction in sidesway with monorail systems are further aspects of the problem.

Quiet adds to ride comfort. Rubber-tired subway trains, as on the Paris Metro and Montreal subways for example, aim toward quiet operation. Proper maintenance of wheels and track can do the same for steel-wheeled systems. Rubber tires on pavements give a noticeable hum that can become annoying, especially when pavements are grooved to reduce skidding tendencies. Careful attention to audio insulation and to placement of power plants can reduce engine noise as evidenced by the relatively quiet interiors of jet aircraft.

Temperature control also contributes to comfort. Heating in the winter and air conditioning in the summer have become accepted features. Railroads were early users of air conditioning. Air conditioning is especially helpful in subways where heat given off by motors combined with hot summer humidity can create uncomfortable ambient conditions. Temperature control has made bus travel far more enjoyable and is an absolute necessity for air travel at stratospheric altitudes. Air conditioning in private autos is currently a problem because it results in greater fuel consumption.

Temperature control must be combined with ventilation. Many persons confined inside a vehicle will quickly exhaust the air supply. Air must be renewed continually for health and comfort.

Cushioned seats aid in reducing the effects of vehicle vibration and impact. Cushions are desirable, at least for long distance travel, but uncushioned plastic is often used for the short hauls of rapid transit. Seat belts are required on aircraft and autos for safety and may become a feature of future high-speed ground transport. The width and position of seats as well as the very presence of seats is a consideration. Subways, where travel time is usually short, may provide only a token number of seats supplemented by handholds for the much greater number of standees. The number of seats—two vs. three abreast—juxtaposes comfort to the load factor.

Seating comfort merges into privacy as a comfort and service quality factor. First class service has always combined de luxe appointments and additional space with privacy, as in the first class section of an aircraft or the private bedroom in a railroad car. One of the attractive features of automobiles is the privacy given to the individual using it.

Amenities include the availability of food and drink, clean rest rooms, considerate, courteous attendents and crews, pleasing decor, telephones, TV, and radio. As a general rule the degree of comfort and the amenities available vary with the length of the journey. Short haul service offers a bare minimum; long haul services can be quite elaborate.

Freight Loss and Damage   The protection of freight against damage from shock and impact requires almost as much attention to springing and suspension as it would for passengers. Freight must be securely packaged and then braced or stowed in the vehicle or craft so as to prevent movement and shifting en route. Excessive vertical movement, for example, can vibrate even well-packed fruit to the point where the fruit interior becomes mushy as if it had been frozen. Single unit vehicles are less susceptible to these effects. Household goods usually move by truck with a minimum of crating. Trains have serial run-in and run-out of slack between individual units unless, as in modern rapid transit systems, each unit is motored so that all units start at the same time and are tightly coupled to the adjoining car.

Certain traffic must also be protected from spoilage en route, a question of temperature control. Meats, fresh fruits and vegetables, and frozen foods may require ventilation, cooling, or freezing. Ships, trailers, containers, or rail cars with mechanical refrigeration meet these needs. Heating may be needed in low temperatures. Livestock must be promptly brought to their destination or to feeding, watering, and resting places along the way.

Always the traffic must be protected against pilferage or theft. The opportunities for theft are a function of openness of shipments and frequency of rehandling. A trailer or container can be loaded and sealed for direct delivery to the final destination, but when the vehicle has to be opened and the contents transferred to another vehicle the chances for theft, breakage, or misplacement are vastly multiplied.[10]

ENVIRONMENTAL EFFECTS

The effect of transportation on the environment emerged in the 1960s as part of the larger national problems, long in the making, of environmental quality and ecological response. Transport operations can have a generally adverse impact upon environment through the production of air, water, noise, vibratory, land and visual pollution and through the destruction of community values. The several modes contribute to pollution in differing degrees. Federal concern with the problem culminated in creation of the Federal Environmental Protection Agency in December 1970 and in the promulgation of rules and standards for pollution control. Among other measures, The National Environmental Policy Act of 1 January 1970, requires in Section 102 that all federal offices file a detailed statement on the environmental impact of their actions, including the licensing and financing of federal projects.

Air Pollution   Exhaust emissions from the prime movers of vehicles are the principal sources of air pollution from transportation. Pollutants emerge in the form of hydrocarbons, CO and CO_2, nitrogen oxides, compounds of sulphur, aldehydes, and particulate matter. Carbon monoxide, CO, is toxic; continued exposure to it at concentrated levels found in the cities can temporarily impair mental and physical performance. Carbon dioxide, CO_2, may react with oxides of nitrogen in sunlight to form toxic products termed "oxidants" that can cause eye irritation and contribute to respiratory diseases, especially for children, and create allergical problems. The functioning of biological systems may also be altered with a destructive effect on vegetation, especially through the creation of photochemical smog.[11] Cellulose, marble, dyes, and polymers may also be affected. High concentrations of lead appear in the bloodstreams of people, such as taxi drivers, who are exposed to urban exhaust-gas concentrations. Symptoms of lead poisoning appear with 60 micrograms of lead per 100 grams of blood; as much as 10 to 30 grams per 100 grams are now found in the blood of persons in urban areas.[12]

---

[10]Containerization has reduced waterfront thefts, but some ports report thefts of entire containers.

[11]John T. Middle, Acting Commissioner of Air Pollution Control Office, "Planning Against Air Pollution," *American Scientist*, Vol. 59, No. 2 March-April 1971, pp. 188–194.

[12]Ibid.

Eye irritation from the "oxidants" can appear at concentrations of 200 to 600 micrograms per cubic meter (0.05 to 0.30 ppm). Odor and the irritating properties of diesel exhausts are related to the concentration of aldehydes. The threshold of odor occurs at 0.2 to 0.3 ppm by volume and of nasal and eye irritation at one part per million. The dirt that exhaust emissions spread over the surroundings is in addition to the harmful effects on health.

The automobile is considered a prime source of about half of the air pollution in urban areas. It contributes more than half the hydrocarbons, nearly half of the nitrogen oxides, and two-thirds of carbon monoxide released in air in the United States each year.[13] Table 9.6 gives amounts of pollutants emitted by various modes of transport. That same source attributes 16.7 percent of all air pollution in the United States to transportation. The concentration of automobiles in urban areas creates pollution where it harms the most people. Diesel trucks and buses make a noticeable contribution but are fewer in number than automobiles. Diesel locomotives and towboats also produce emissions but in lesser quantity than highway vehicles and usually in less urbanized localities. Railroad locomotives are said to pollute the atmosphere only $\frac{1}{3}$ to $\frac{1}{6}$ as much as highway transport per ton mile of freight.[14] This is because of lower fuel consumption, the result of lower propulsive resistance, and more nearly complete combustion. Worth considering is the possibility of reducing air pollution by a more extensive intermodal piggyback operation that combines long haul economy and the relatively lower pollution rate of rail haul with the flexibility of trucks in terminal pickup and delivery.

The Federal Clean Air Act of 1963 and amendments of 1967 and 1970 have led to rules that required the elimination of hydrocarbon emissions from crankcases in 1968 model automobiles and call for a reduction by 1975 of carbon monoxide and hydrocarbon emissions to 90 percent less than 1970 models and, as amended, by 1977, a 90 percent reduction of oxides of nitrogen emissions over those of 1971.[15] What effect the so-called energy crisis of 1973–1974 will have on enforcing these standards is yet to be determined.

Procedures and devices are available or undergoing development to reduce harmful automotive exhausts to harmless gases. Much can be accomplished simply by maintaining a well-adjusted and tuned motor, a necessary step with any system. One type of control, now being developed to meet the 1977 deadline of the Clean Air Act, is a catalytic-type device

[13]Ibid.

[14]*Illinois Central Gulf News*, November 1973, pp. 8–11.

[15]A one-year suspension of 1977 emissions control standards was granted in early 1975. Further extensions have been requested.

TABLE 9.6 AIR POLLUTING EMISSIONS FROM TRANSPORTATION SOURCES[a]

Mode	Carbon Monoxide (CO) Millions of Tons	Percent	Particulates Millions of Tons	Percent	Sulphur Oxides (SO_x) Millions of Tons	Percent	Hydro Carbons (HC) Millions of Tons	Percent	Nitrogen Oxides (NO_x) Millions of Tons	Percent
Motor vehicles Gasoline fueled	96.8	8.8	0.3	37.5	0.2	18.2	16.9	85.4	7.6	67.9
Diesel motor vehicles	1.0	0.9	0.1	12.5	0.1	9.1	0.2	1.0	1.1	9.8
Aircraft (total emissions)	2.9	2.6	0.1	12.5	0.1	9.1	0.4	2.0	0.4	3.4
Railroads	0.1	0.1	0.1	12.5	0.2	18.2	0.1	0.5	0.1	0.9
Ships	1.7	1.5	0.1	12.5	0.3	27.2	0.3	1.5	0.2	1.8
Nonhighway use of motor fuel	9.0	8.1	0.1	12.5	0.2	18.2	1.9	9.6	1.8	16.2
Total of all sources	111.5	100.0	0.8	100.0	1.1	100.0	19.8	100.0	11.2	100.0

[a]Based on 1972 *National Transportation Report*, U. S. Department of Transportation Office of the Assistant Secretary for Policy and International Affairs, Washington, D. C., July 1972, Table III-47, p. 67.

that uses a chemical catalyst to oxidize carbon monoxide. A reduction of 60 percent for CO and one of 80 percent for other hydrocarbons are sought.

In another system a direct flame afterburner reduces CO and CO_2 exhaust content by combining a heat source such as a spark plug or a heat exchanger (with or without additional fuel supply) to bring about combustion in the muffler. Either of the foregoing may be combined with an additional air supply causing the exhaust manifold to act as an afterburner.

Any antipollutant device must function and maintain combustion under all types of driving. The cost of these devices is a stumbling block to their use as is the need to keep them as well as the rest of the engine properly tuned and adjusted. A reduction in the lead content of gasoline aids in reducing lead as a pollutant.

The use of kerosene fuels in jet aircraft can lead to unburned portions of fuel emitted as black smoke. A four-engine jet will emit many pounds of pollutants during each takeoff. The odor of kerosene is prevalent in all large jetports; the vapor is offensive both to the eyes and nostrils. Good airport design properly calls for sealing the passenger areas on the apron side and placing air conditioning intakes away from the aprons. Visibility is also affected by intensified smog conditions. Los Angeles Airport leads the nation in instrument landings, followed by O'Hare and Boston's Logan Airport.

As evidence of the seriousness of such pollution, suits against 23 airlines were filed by the Attorney General in November 1969. Smoke emissions have been reduced by the installation of afterburners, first scheduled for 1980 but later advanced to 1972 at an estimated cost of $15 million. There is presently no practical solution to the emission of poisonous nitrogen oxides.

Other sources of air pollution from transport, which are relatively minor in nature, include steel dust from brake shoe and rail abrasion, sparks from locomotive and towboat exhausts, fumes from mechanical refrigerating units, fumes and gases from overturned trailers and derailed rail cars, and drift from chemical weed spraying. Dust is often disseminated by the loading, unloading, transfer, and the storage of granular bulks such as coal, ore, grain, fertilizers, and aggregates. Evaporation and spill from the fuel tank and the evaporation of gasoline through carburator vents after the engine is shut off but still warm also contribute. Pollution comes from power plants that provide electric energy for rapid transit and electrified railroads. These plants can, however, be located away from heavily populated areas.

Water Pollution   Land transport makes few direct contributions to water pollution. Wastes from automotive and locomotive shops and servicing facilities, such as discarded lube oils, fuel spills, and wasted radiator coolants, must be given treatment before discharge into sewers and waterways. So must wastes from wash racks for rail, bus, truck, and airline equipment. Retention tanks similar to those used on aircraft and buses must be used on trains and ships to prevent the discharge of human wastes onto tracks or into water. By 1980 all waste waters must be recycled.

Oil spills arise from cleanouts, leaks, or from collisions of oil-carrying vessels. The dumping of bottom sludge and wastes from tankers has been prohibited by the 1970 Act and shipowners face liability for cleanup of oil spills from these and other causes. The discharge of polluted water from ballast tanks is still a problem. The fear of settlement, breakage, and leaks in permafrost areas delayed construction of the Alaska Pipeline from Prudhoe Bay to Valdez. Spilled oil and chemicals from overturned trailers and rail cars have contaminated local water supplies and groundwater.

A program to clean United States waters was established by the Federal Water Pollution Control Act of 1970 and its 1972 amendments and by a Federal Court decision in 1966 extending the Refuse Act of 1899 to outlaw not only debris that might obstruct navigation but industrial discharge of pollutants into navigable waters.

Noise   Noise is a nuisance that regularly accompanies transport operations. It interferes with the reception of desired sounds. It varies in intensity and duration with different carrier modes. Extremely high noise levels can cause physical damage to the human body, that is, mechanical damage to hair cells and other parts of the ear; prolonged periods of exposure to lesser noise levels can result in an eventual metabolic change.

Noise is usually expressed in decibels. A decibel is the least detectable sound. It is defined as a sound intensity producing an impact of 0.0002 dyn/cm^2. Exposure to 85 dB or more for eight hours can be damaging. Levels of 70 dB are common. Ordinary office noise is in the 55 to 60 dB range and crowded city streets may reach 95 dB. Table 9.7 gives values generally accepted for various transport and modal situations. The values are mostly expressed in A-weighted noise levels corresponding to average values at a 50-ft distance under normal operating conditions at typical speeds. For airplanes, the distance is 1000 ft (305 m) from the glide path.

The high rating for jet aircraft, 160 dB-A ± is expected and obvious. It is one reason why airports make poor neighbors for residential zones, hospitals, or schools. Diesel locomotives and trucks are at approximately the same noise level, 80 dB-A. The noise of rubber tires on pavements is possibly more annoying than noise from engines. One should note that trains and locomotives pass only intermittently whereas a major truck route

TABLE 9.7 TRANSPORT SOUND RANGES[a]

Normal conversation	54 to 60 dB
Telephone conversation becomes impossible	80 dB
Busy city streets	90 to 95 dB
Passenger cars: 40-70 mph	65 to 70 dB-A
Diesel and other heavy duty trucks: 50-60 mph	75 to 85 dB-A
Railroad trains, intercity	70 to 80 dB-A
Rapid Transit trains	85 to 95 dB-A
Aircraft, general	75 to 95 dB-A
Aircraft, jet	100 to 160 dB-A
Motor buses	70 to 85 dB-A
Motorboats	75 to 80 dB-A
Ships	55 to 68 dB-A
Motorcycles and snowmobiles	85 to 95 dB-A

[a]Based in part on "Transportation Noise and Noise from Equipment Powered by Internal Combustion Engines," Table B-1, p. B-3, U. S. Environmental Protection Agency, Washington, D. C., 31 December 1971.

may experience an almost continuous passage of vehicles. Ships are the quietest of surface modes. Pipelines are noiseless, the only sounds being those of the occasional pumping station. Retarders for control of car speeds in rail yards emit a high-pitched screech as the brakes are pressed against the sides of car wheels. Sound-absorbing barriers such as mounds of earth or insulating fences of sound-absorbing modular construction around the retarder areas have been of some help. Similar barriers have been suggested along highway routes through populated locations. Extended use is made of such barriers in Tokyo.

Noise problems are of two types: internal noise that affects employees and passengers and external noise that affects the adjacent community and environment. Internal noise can be handled by proper sound insulation of buildings and vehicles. Airport employees working near planes must wear ear coverings to avoid damage to their hearing. Protection against external noise is more difficult to provide. It includes the muffling of vehicle engines, consideration as to the location and scheduling of operations, and the use of sound insulating barriers in certain areas.

The Noise Abatement and Control Act of 1970 was passed by Congress to evaluate the health hazards of noise, summarize the state of noise-suppression technology, and make recommendations to Congress for appropriate control legislation.

Environmental Statement Federal agencies proposing new legislation or other major action that may affect the environment must present an *environmental impact statement* to the President's Council on Environmental Quality and must also give the statement wide publicity to Congress and the public. A draft must be circulated to the Environmental Protection Agency and to other appropriate federal, state, and local environmental agencies for their comments. The Council on Environmental Quality advises the President on the best course of action on the basis of the evidence. The building or financing of highways, airports, waterways, or railroads by Federal agencies is within the scope of this requirement.

Energy Tardy public recognition that the sources of conventional energy are dwindling has added another dimension to the problems of transportation. Until science shows the way to economical uses of new energy sources, economy and careful planning are required to make the best use of what is still available. The relative energy demands of various transport modes in moving a given volume of traffic becomes pertinent.

Thermal efficiency has been discussed in Chapter 5. The horsepower-per-net-ton ratio and deadload-to-payload ratios are also factors that affect overall fuel economy. These are related, in turn, to the unit propulsive resistance for each type of transport. The question arises: how much fuel is consumed for each type of transport? It gives rise to another question: what modes of transport should be encouraged in the interest of conserving national resources?

The author has computed values in Table 9.8 that are based on a largely theoretical consumption of fuel, an inherent technological characteristic, in which work done by propulsive force over propulsive resistance is related to the foot-pounds of work produced by one BTU (778) and then related to the BTU content per gallon of fuel oil (approximately 130,000) and to the thermal and mechanical efficiencies of the prime movers (approximately 0.25 percent for internal combustion-type engines). Such an analysis takes no account of wastage, evaporation, poor engine maintenance or operation, of effects of grade, route circuity, curvature, or road surface, speed, acceleration, idling time, or variations in energy content of the fuel. This accounts for the differences between the author's figures and those of Table 9.9. Nevertheless, the author's data are useful measures of relative modal differences and potential. The data show the relative value of various modes in a descending order of economy: waterways, pipelines, railways, and highways, with airways a poor last.

TABLE 9.8 FUEL CONSUMPTION PER TON MILE (THEORETICAL)
(LEVEL GRADE; 130,000 BTU PER GALLON)

Mode	Speed (mph)	Unit Resistance (lb)	Gallons per Ton Mile	Gross Ton Miles per Gallon	Net Ton Miles per Gallon[d]
Railroad[a]	55	8	0.0020	500	375
Tractor-trailer[a]	55	56	0.0143	70	50
River tow[b]	8	3	0.0012	806	690
Great Lakes bulk cargo carrier[b]	10	2	0.0008	1250	833
Airplane[c]	300	240	0.0716	14	4.4

[a]Including a 82-percent mechanical and a 25-percent thermal efficiency.
[b]Including a 50-percent mechanical and propellor efficiency and a 25-percent thermal efficiency
[c]Including a 70-percent mechanical and propellor efficiency and a 25-percent thermal efficiency.
[d]Based on the average payload to empty weight ratios from Table 5.3.

TABLE 9.9 ENERGY EFFICIENCY OF TRANSPORT MODES[a]

Mode	Freight			Passenger		
	Ton Miles per Gallon	BTU per Ton-Mile	Joules per Kilogram/ Kilometer	Passenger Miles per Gallon	BTU per Ton-Mile	Joules per Kilogram/ Kilometer
Pipeline	300	450	330			
Waterway	250	540	390			
Bus				125	1000	710
Railroad	200	680	490	80	1700	1,100
Truck.auto	58	2300	1,700	32	4250	2,800
Airplane	3.7	37000	27,000	14	3700	6,400
Assumed: 136,000 BTU per gallon						

[a]Eric Hirst, "Energy-Intensiveness of Transportation," *Transportation Engineering Journal, Proceedings of the American Society of Civil Engineers*, February 1973, pp. 111–122 (especially p. 122).

Table 9.9 contains relative energy (fuel) consumptions of the common modes of transport. Such values vary considerably with the method of computation and the number of variation factors included. Frequently quoted data is based on national totals of fuel consumed by each mode and related to estimates of passenger miles or revenue ton miles and to the usually accepted load factors for each mode. Circuity of route is an element to be considered in making comparisons of modal fuel economy.

James M. Symes, then President of the Pennsylvania Railroad, is quoted in excerpts from his testimony before the Subcommittee on Surface Transportation as presenting the following tabulation of the amount of fuel necessary to move 100,000 tons (90,700 tonnes) of freight from New York to San Francisco.[16] The effects of circuity of routing are evident.

Fuel	Rail	Highway	Water	Air
Gallons	832,300	3,366,300	4,346,100	21,801,200
Ratio to rail	1.0	4.0	5.2	25.0

The question of how far a given tonnage can be moved on one dollar's worth of fuel might be asked.[17] The following answers have been given to the question of how far 40 tons can be moved on one dollar's worth of fuel.[18]

A 40-ton freight car	151 miles (243 km)
A 40-ton railroad coach	116 miles (187 km)
Two 20-ton trucks	11.9 miles (19 km)
Twenty 2-ton automobiles	2.9 miles (4.7 km)
A 40-ton, four-engined airplane	3.1 miles (5.0 km)

A more recent evaluation of the relative amounts of energy consumed by the various modes in conducting the nation's transportation business is found in Table 9.10.

Land Use   A major factor in both community values and its effects on the environment is the amount of land required to conduct transportation.

The scarcity of land, especially in and near urban areas, and the high value attached thereto raises the question of the relative land use of the

[16]James M. Symes, "The Next War: We Could Lose It on the Rails," *Railway Progress*, March 1958, p. 7, Federation for Railway Progress, Washington, D. C.
[17]At 1957 fuel price!
[18]*A Ten-Year Projection of Railroad Growth Potential*, prepared by Transportation Facts, Inc., for the Railway Progress Institute, Chicago, Illinois, 1957, Table 8.

TABLE 9.10 ENERGY CONSUMPTION BY UNITED STATES TRANSPORTATION[a]

Mode	Amount of Energy (percent)		
	1950	1960	1970
1. Automobiles	38.0	51.4	54.2
2. Trucks	16.6	19.8	21.1
3. Railroads	25.2	4.9	3.3
4. Airplanes	1.7	7.5	10.8
5. Buses	1.1	1.0	0.8
6. Nonbus urban mass transit	1.0	0.3	0.2
7. Waterways, freight	3.6	2.8	2.5
8. Pipelines	0.7	0.9	1.2
9. Other	12.1	11.4	5.9
	100.0	100.0	100.0
Total transportation energy consumption (10^{15} BTU)	8.7	10.9	16.5

[a]*Draft Environmental Impact Statement Pool 22, Upper Mississippi River 9-foot Navigation Channel*, U. S. Army Corps of Engineers, Rock Island District, Rock Island, Illinois.

several carriers. When the carrier's route is provided by the state, as is the case with highways and canalized waterways, the problem of withdrawal of large amounts of land from the tax rolls is a matter of serious concern for local communities and taxing authorities.

The basic land requirement of most transport systems is for right of way. Railraods, aerial tramways, highways, and others need a strip of land varying from 30 to 200+ ft (9.1 to 61.0 m) in width, extending from terminal to terminal. Where there are deep cuts or fills, an even wider strip must accommodate the necessary cut or fill width and attendant drainage. On level ground the minimum width is $w = b + 2sd$, where b is the width of subgrade, s = rate of side slope, and d is the depth of cut or fill. Sufficient additional width is desirable for side and intercepting ditches. For a canal, b is the base of the canal channel.

Table 9.11 presents the typical right-of-way requirements of straight route mileage for various modes. The productivity of that use can be obtained by dividing route capacity by land area to give the number of passengers per mile per hour per acre that can be carried.

In addition to route right of way, land is also required, often in large quantities, for terminal and interchange facilities. Large railroad yards may require areas 2 to 5 miles long and $\frac{1}{2}$ to 1 mile wide. When laid in subways and tunnels, railroads do not interrupt land surface but can create congestion for subsurface structures and utilities.

Modern superhighways and expressways require additional land at interchange points and service areas; as much as 20+ acres are not uncommon for an interchange. Motor freight and bus stations and parking lots require a relatively small amount of land, but often that amount is needed in congested city centers where land is scarce and expensive.

Airways make only nominal demands on land use for route markers and radio range stations, but may require from 10 to 20 acres for small fields up to thousands of acres for large fields. O'Hare Airport occupies 6554 acres, the Dallas–Ft. Worth Airport is designed to eventually be 17,000 acres.

Community Values   The many effects of a transportation system on a community are far-reaching. The most obvious, the benefits of movement and accessibility, have already been discussed. The concern here is with the impact made on a community by reason of its physical presence and operation.

The extent of land use has already been noted. Entry into an urban area by a new route, be it rail or expressway, can be disturbing. An existing route may be a barrier to local development. For a new route, land must be acquired and cleared for a right of way. In the process dwelling units,

TABLE 9.11 TYPICAL LAND REQUIREMENTS FOR RIGHT OF WAY (ACRES PER MILE DEVOTED EXCLUSIVELY TO TRANSPORTATION)

Carrier	Width of Right of Way (in Feet)	Acres per Mile
Railroad	30 to 100	4 to 12
Highway	30 to 300	4 to 36
Waterway	100 to 300	12 to 36
Airway	—	Nominal
Pipeline	1 to 6	Nominal
Conveyor	25 to 100	3 to 12
Aerial tramway	25 to 100	Nominal

Note: aerial tramways and pipelines, like power transmission lines, may maintain a cleared right of way of 50 to 100 ft in width for ease in maintenance and access and to prevent trees and high shrubs from fouling the cars and cables.

many that have been "home" to their occupants for many years are razed. Small shops, stores, and industries are destroyed or move far away, often with loss of work opportunities. These losses can be given a dollar value. Playgrounds are preempted. Trees and open spaces are sacrificed and buildings of historical or artistic value disappear. In all fairness it should be noted that what is lost may equally represent disreputable, run-down, rat-infested tenements and litter piles.

If taken to be part of a publicly owned system, the land is also removed from the tax rolls and may become an item of expense to the community rather than a source of revenue. Against this disadvantage must be weighed the increased value of adjacent land and the taxes and new business thereby created.

A finished roadway structure can shut off access from one part of a neighborhood to another. It can block out pleasant vistas and become a source of visual pollution. It has, however, the potential for a narrow but long parkway, as evidenced by the landscaping along and under the BART System's elevated structures.

Transportation is vital to a community, but that is no reason to ignore these and other impacts a specific proposal may entail. A little forethought in planning can permit a location that avoids some of the adverse effects or may lead to a design that is an attractive addition to the scene rather than a nuisance and eyesore. A location should be developed that is in harmony with its physical and social environment. The work by Ian McHarg in this area should be familiar to any planner of a transport route location.[19] It is just possible that a community might sometimes decide that the benefits a proposed system would confer aren't worth the costs in dollars or in evnironmental and human values.

QUESTIONS FOR STUDY

1. What would have been the effect on transportation in general and rail transportation in particular if more than one gage of track were used in North America?

2. Highway and automotive engineers foresee the day when automobiles and trucks will be automatically guided on principal highways by inductive coupling to an electrified cable laid in the middle of each traffic lane. Discuss the full effects this could have on automotive flexibility and safety.

3. To what extent is crating and packaging an important technoeconomic factor in each mode of transport? Use specific commodity examples in your discussion.

[19]Ian McHarg, *Design with Nature*, American Museum of Natural History, Doubleday Natural History Press, Garden City, New York, 1969.

4. What problems incident to flexibility had to be met in designing double-decked rail rapid-transit equipment?
5. What problems would be involved in setting up a system of trailer interchange among trucking lines similar to that for car interchange between railroads?
6. What force is required to accelerate a train from 60 to 70 mph in 100 sec and how far will it run meanwhile? (Omit incremental steps but indicate how such steps could be used for greater accuracy.)
7. How far might an automobile travel before actual deceleration begins when the driver makes an emergency application of his brakes at a speed of 60 mph?
8. Explain why the maximum practical speed is not a good criterion for determining speed of traffic flow for a carrier.
9. What is the role of guidability in establishing the safety status of a carrier?
10. Assuming farm land is being taxed at $30 an acre, what are the losses in tax receipts for a publicly built right of way 50 ft wide and 300 miles long?
11. Explain the problems in the speed requirements of a rapid transit service within the core city area, such as the CTA in Chicago, and that serving the suburbs such as the BART system or the Lindenwold Line.
12. Compute the transportation productivity of land-passengers per area per mile per hour—for rapid transit, commuter railroad, motor bus, 747 aircraft, and the private automobile.
13. Develop a rating system by which the transport mode that is most favorable to the environment on an all-around basis might be established. What problems arise in attempting such an evaluation?

SUGGESTED READINGS

1. L. K. Sillcox, *Mastering Momentum*, Simmons-Boardman, New York, 1955, Chapter 1, pp. 1–94.
2. W. W. Hay, *Railroad Engineering*, Vol. I, Wiley, New York, 1953, Chapter 10, pp. 134–150.
3. Frank M. Cushman, Transportation for Management, *Service Characteristics of Carriers*, Prentice-Hall, New York, Chapter 3, pp. 41–90.
4. Hermann S. D. Botzow, Jr., *monorails*, Simmons-Boardman, New York, 1960.
5. Hoy A. Richards and G. Sadler Bridges, *Traffic Control and Roadway Elements—Their Relationship to Highway Safety/Revised*, Chapter 1, Railroad Grade Crossings, published by the Automotive Safety Foundation in cooperation with the U. S. Bureau of Public Roads, 1968.
6. W. J. Hedley, "The Achievement of Railroad Grade Crossing Protection," *American Railway Engineering Association, Proceedings*, Volume 50, 1949, pp. 849–864.
7. A. Sheffer Lang and Richard M. Soberman, *Urban Rail Transit*, M. I. T. Press, Cambridge, Massachusetts, 1964.
8. Eric Hirst, "Energy Intensiveness of Transportation," *Transportation Engineering Journal*, Proceedings of the American Society of Civil Engineers, February 1973, pp. 111–122.

9. Anthony V. Sebald, "Energy Intensity of Barge and Rail Freight Hauling," Center for Advanced Computation Technical Memorandum No. 20, May 1974.

10. Edward T. Meyers, "Energy—Are the Railroads on the Right Track?" *Modern Railroads*, August 1973, pp. 41–48.

11. Martin Wohl, "A Methodology for Evaluating Traffic Safety Improvements," papers and discussion of the 1968 Transportation Engineering Conference: *Defining Transportation Requirements*, The American Society of Mechanical Engineers, New York, 1960, pp. 175.182.

12. *"Air Pollution Control for Urban Transportation,"* eight reports, ISBN 0-309-02199-5, Transportation Research Board, National Academy of Sciences, Washington, D. C., 1973.

13. *"Traffic Accident Analysis,"* four reports, ISBN 0-309-02272X, Transportation Research Board, National Academy of Sciences, Washington, D. C., 1974.

14. *Proceedings* 1974 *National Conference on Railroad-Highway Crossing Safety*, sponsored by U. S. Department of Transportation, 19-22 August, 1974, held at U. S. Air Force Academy.

15. L. A. Hoel, R. L. Lepper, R. B. Anderson, G. R. Thiers, F. DiCeasare, T. E. Parkinson, and Jon Strauss, *Urban Rapid Transit Concepts and Evaluation*, Research Report, Transportation Research Institute, Carnegie-Mellon University, Pittsburgh, Pennsylvania, 1968.

16. *Energy Primer: Transportation Topics*, Office of Research and Development Policy, Transportation Systems Center, U. S. Department of Transportation, Washington, D. C., 1975.

17. "Coal Transportation: Unit Trains—Slurry and Pneumatic Pipelines" from CAC Document 163 Final Report, *The Coal Future*, by Michael Rieber, S. L. Soo, and James Stukel, Center for Advanced Computation, University of Illinois, Urbana, Illinois, May 1975.

Chapter 10

Terminals

TERMINAL FUNCTIONS

A study of techno-cost relations might logically follow the materials of the preceding chapters. However, a glance (and only that) at the more important aspects of terminals and operational control will give fuller understanding and depth to the cost relations.

Definition and Function Terminals have been variously defined. In a limited sense, a terminal is simply the beginning or the end of the line for a transportation operation. The term also applies to specific structures used for transportation purposes. The author here considers terminals as the sum total of facilities and their locale where road-haul traffic is originated, terminated, and/or interchanged before, during, or after the road-haul movement, including the servicing of facilities for the vehicles and equipment in which the traffic is moved. Such a grouping of facilities does, true enough, usually occur at the end of a route, but it also occurs frequently at one or more intermediate points along the route.

Other terminal functions with appropriate facilities include holding and reconsignment, storage and warehousing, classification, concentration, and loading and unloading.

Terminals are as important in the transportation picture as line haul. In fact, terminal problems often surpass those of line haul in extent and complexity. Line haul, furthermore, has significance only if there is traffic to move. A terminal is the operational origin or destination of that traffic,

352

or the point to which it is usually brought from outlying areas for consolidation prior to road movement or for distribution to those outlying points following a road haul.

From the standpoint of time alone, terminals possess more significance than line haul. The average daily movement of a railroad freight car is about 57.4 miles. This is a distance that can normally be run in 1 hour. In other words, a freight car spends only 1 hour out of 24 in road movement. Most of the remaining 23 hours are spent in or at some terminal activity or facility—in yards, at the shipper's or consignee's door, in transfer, on repair tracks, etc. Bulk-cargo carriers will spend about 15 percent of their time in port on the Great Lakes but defer maintenance and repairs to the three winter months, when they don't sail at all. General-cargo ships may spend as much as 50 to 65 percent of their time in port. Airplanes spend 30 to 60 minutes at turnaround and principal stops per two hours of flying time; longer stops are required for classified repairs and maintenance. Much motor truck operation is confined solely to terminals, and a trailer will often require as much time for stripping and stowing as to make an overnight run.

The variety of terminal facilities needed or available and the investment in them is extensive. Railroad yards, freight stations, wharves, transit sheds, grain elevators, produce terminals, icing docks, car dumpers, tank farms, coal and ore docks, and shops and servicing equipment are among the more obvious of such facilities. Parking lots provide automobile storage; servicing occurs at the corner gas station.

The railroads' investment in terminal facilities is over $1 billion. Yard tracks constitute about 17 percent of railroad track miles. A partial listing of the terminal facilities owned and constructed by the Port of New York Authority, including airports, motor freight terminals, bus terminals, wharves, office buildings, and PATH, the commuter rail line running under the Hudson River to New Jersey, amounts to well over $100 million. Adding the highway bridges and tunnels giving access to and through the terminal area increases the total to over 350 million. The airports of the United States have been valued at over $4 billion and an average value of $3000 per car space has been placed on the parking lots of the land. Corresponding investment in terminals could be cited for other cities and other transport modes.

Types of Traffic   The types of traffic passing through a terminal have important effects on the operation and the facilities required (this is also true for road haul). One may distinguish between commodity and traffic types and the particular needs of each. *Perishable* fresh fruits and vegetables must be moved rapidly, kept cool by ventilating or icing in summer and warm by charcoal heaters in winter. Precooling plants for cars and trucks, icing docks, and produce terminals are typical of the facilities required. Frozen foods, fruit concentrates, and meats must be kept frozen by mechanical refrigeration, both in cars and trucks and in terminal buildings. Bananas and other perishables require holding and reconsignment yards and extensive communications so that cars may be diverted, en route, to the best markets. *Granular bulks* require grain elevators, coal docks, car dumpers, and conveyor loaders for storage and transfer. Hulett unloaders and bridge-mounted bucket cranes unload granulars from ships. *Liquid bulks* such as petroleum are stored in tank farms. *Manufactured goods* have great variety and high value, may require lifting equipment for transfer, and frequently require specially designed vehicles for their transport. Rapid movement is usually desirable. *Livestock* must have wayside loading pens and intermediate or terminal unloading chutes, pens, and water and feed facilities.

Coal, grains, and other dusty products pose problems of neighborhood nuisance and the hazards of explosion and fire. The latter hazard applies equally to petroleum and chemicals. Livestock movements are accompanied by offensive odors. The list of commodities and the special problems presented by each in terminal handling could be expanded almost indefinitely.

The classification of freight into carload (CL)—or truckload—and less-than-carload lots (LCL), sometimes classified as merchandise, has already received comment in Chapter 3. In addition to commodity types, one may further distinguish between general and bulk cargos, especially in water transport. General cargo, as the name implies, includes all types of cargo—processed, semiprocessed, and manufactured goods, and small shipments of what is usually termed bulk freight. Bulk freight applies principally to raw materials being moved in quantities and requiring special facilities for handling, transfer, and storage. Coal, ore, grain, petroleum, sulfur, molasses, etc. are a major part of this category, usually handled at private docks.

A new traffic category has arisen with the introduction of containerization and trailer-on-flatcar (sometimes called *flatback* traffic). Special designs of terminals are required for loading and unloading trailers and containers moving by rail. Elaborate container-ports serve waterborne phases of this movement.

Loading and Unloading   In addition to the obvious originating and terminating functions, terminals perform a variety of other services. A principal terminal function is that of loading and unloading the transport unit. Terminals at the Upper Lake ports load ships with ore, grains, and aggregates. Those at the Lower Lake ports provide unloading facilities plus car dumpers to load coal into some of the upbound vessels. Freight and transit sheds perform loading and unloading services for cars, trucks, and ships. Grain elevators, petroleum docks, etc. carry out these functions for specific commodities.

Traffic Concentration   Concentrating traffic permits efficient and economic handling. Only trucks and airplanes, because of their small individual capacity, are suitable for the individual movement of single small shipments. Even with these it is helpful if freight can be concentrated in one freight house, or passengers at one airport ramp. In this respect, every freight facility, whether in a metropolitan center or at country cross roads performs a terminal function. The country grain elevator concentrates grain from many farms so that freight cars can be loaded and sent on their way without the delays attendant on piecemeal loading. Similarly, grain is concentrated at terminals in train loads, tow loads, or ship loads in the secondary grain markets and elevators at Minneapolis–St. Paul, Duluth–Superior, Chicago, Winnipeg, and Thunder Bay, and at the export elevators at Seattle, Montreal, New York, Galveston, and Houston. In addition to the grading, drying, blending, and storage functions performed, the grain elevator brings about rapid loading and unloading of ships and cars, keeping turnaround and terminal time to a minimum. Most recently, grain concentrating elevators of high capacity have been established in rural areas, close to the source of production, from which unit trains move grain to milling and export centers in 70- to 100-car lots. Elevator storage is a prime function in the marketing of grain.

Ore and coal docks concentrate quantities of those commodities, thereby permitting rapid loading of ships and cars for quick turnaround and efficient use of equipment. Where such docks are not available, stock or surge piles perform the same function albeit with somewhat less efficiency.

LCL freight is similarly accumulated and concentrated into large tonnages in freight houses of piggyback companies, motor freight lines, airlines, and freight-forwarding agencies. General cargo is concentrated for water movement in transit sheds and other waterfront storage areas or in loaded cars in nearby supporting and holding yards.

Another aspect of traffic concentration is the act of pickup or collection. Individual freight shipments must be brought to the freight house or transit shed by rail or highway, or barge. Local LCL freight may be brought to the freight house in the shipper's own trucks, in the vehicle of an independent for-hire trucker, or in the vehicle of a pickup-and-delivery service provided by the line-haul carrier in its own or contracted vehicles.

Car lots moving by rail are concentrated in railroad classification yards for makeup into trains. Individual cars for loading are ordered by the shipper through the freight agent, usually from that same yard, and brought to the shipper's door (or to a public delivery yard) by local switching service called switch runs. Switch runs also take the loaded cars back to the classification yard. In river traffic, individual barges are loaded at industrial docks or public-transit sheds and brought together to make a tow, either by small pusher tugs, or by the principal towboat. Often one industry has enough traffic to load a complete tow.

Concentrating Less Truck Load (LTL) freight in freight houses permits rapid loading of trailers. Also, trailers may be left for loading while the tractors perform service elsewhere.

The concentration of passengers at airports, railway stations, and passenger piers by private automobile, taxi, bus, and local rapid-transit trains needs more than this brief mention.

In making detailed deliveries, the reverse of the foregoing operations prevails. Inbound freight is unloaded and concentrated at transit sheds, freight houses, railroad yards, and in stock piles and tank farms preparatory to detailed distribution to the consignees by switch runs, delivery services, or the consignees' own trucks or agents. Warehouses, elevators, tank farms, and stock piles hold quantities of goods in central locations for detailed distribution over longer periods of time as required.

Interchange   Much freight reaching a terminal is destined for another point and requires transfer to a similar or different mode of carriage to complete the journey. This is a function of classification yards from which carload transfers to other railroads are made, of the transfer platforms of freight houses, and of transfer piers and lighters. Railroads coordinate the interchange of bulk granulars with ships at ore and coal docks, grain elevators, etc. Pipelines interchange with ships and barges and with railroads through tank farms and flexible nozzles at loading docks (Figure 10.1). Airports aid in the transfer between airlines and between air and ground transport. Rapid-transit stations permit transfer from one line to another.

Classification   One last highly important terminal function is classification. This reaches its highest state of development in railroad classification yards, where cars are sorted into groups of like destination (or grade, commodity, or similar grouping). From there the cars are placed in trains of appropriate destination. A freight house or transit shed performs a similar sorting or classifying function for the shipments there tendered so

FIGURE 10.1a PETROLEUM CONCENTRATION AND INTERCHANGE–
PETROLEUM CONCENTRATION IN TANK FARM AT BEAUMONT, TEXAS.
(COURTESY OF *THE OIL AND GAS JOURNAL*, TULSA, OKLAHOMA, AND TEXAS
EASTERN TRANSMISSION CORPORATION.)

FIGURE 10.1b PETROLEUM CONCENTRATION AND INTERCHANGE–TANK-CAR
LOADING RACK. (COURTESY OF *THE OIL AND GAS JOURNAL*, TULSA,
OKLAHOMA, AND THE ARKANSAS FUEL OIL COMPANY, SHREVEPORT, LOUI-
SIANA.)

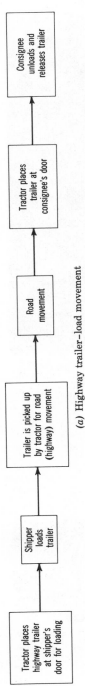

(a) Highway trailer–load movement

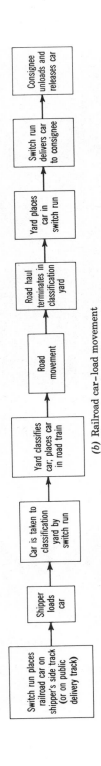

(b) Railroad car–load movement

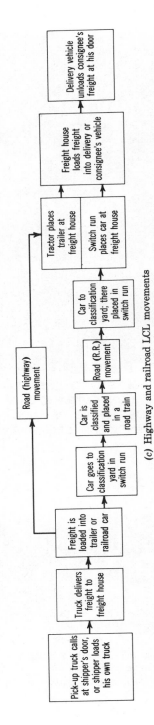

(c) Highway and railroad LCL movements

FIGURE 10.2 TERMINAL OPERATIONS AND COORDINATION. *NOTE:* ALL OPERATIONS EXCEPT ROAD MOVEMENT ARE TERMINAL FUNCTIONS OR OPERATIONS. ROAD MOVEMENT MAY BE INTERRUPTED BY ONE OR MORE HANDLINGS THROUGH CLASSIFICATION YARDS OR ACROSS FREIGHT (TRANSFER) HOUSE PLATFORMS. (*a*) HIGHWAY TRAILER-LOAD MOVEMENT. (*b*) RAILROAD CAR-LOAD MOVEMENT. (*c*) HIGHWAY AND RAILROAD LCL MOVEMENTS.

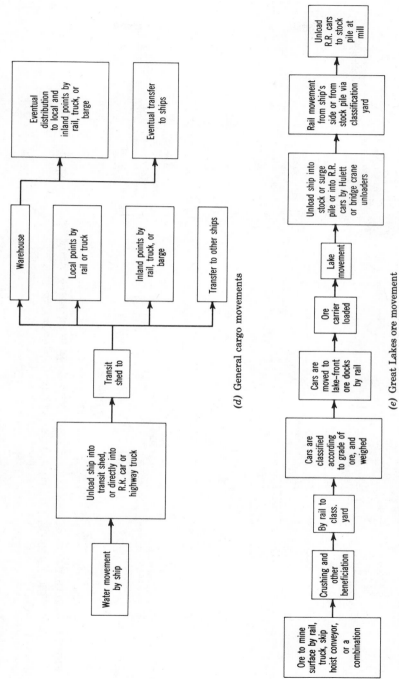

(d) General cargo movements

(e) Great Lakes ore movement

(d) GENERAL CARGO MOVEMENTS. (e) GREAT LAKES ORE MOVEMENT.

359

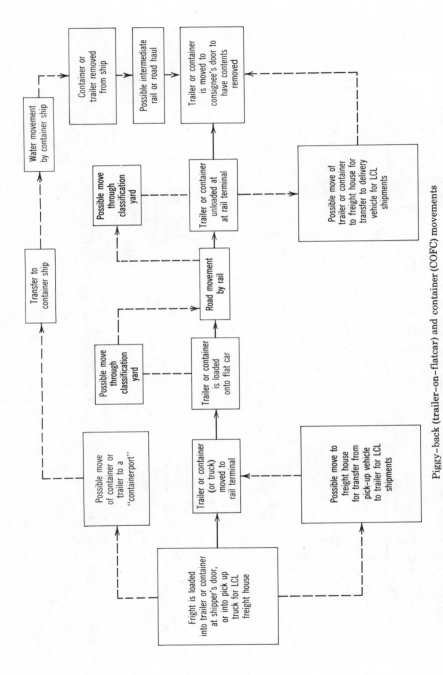

Piggy–back (trailer–on–flatcar) and container (COFC) movements

(f) PIGGY-BACK (TRAILER-ON-FLATCAR) MOVEMENTS.

that LCL freight of like destination is placed in a car or truck going to that destination.

Figure 10.2 shows the relation between road and terminal movement.

Storage and Warehousing   These too are terminal functions. Grain is stored in elevators. Warehouses hold imports awaiting customs inspection, quarantine periods, or being held for detailed distribution over an extended time period. Strategically located warehouses serve as break-bulk points permitting the lower rates of car-load and truck-load movement into a distribution area from which local distribution can be made in small lots. Warehouses also permit ready access to needed goods from nearby areas.

Reconsignment   The final destination of goods is not always known because of a critical market or other conditions. Goods are billed to an intermediate destination and held there by the carrier until final destination instructions have been received. The shipment is then forwarded. Reconsignment, most frequently practiced by railroads, calls for holding tracks or yards, switch engines to service those tracks, and necessary communications and supervision. Reconsignment in transit may be combined with storage or milling, for example, in transit.

Servicing and Maintenance   Vehicles and craft must be fueled, cleaned, inspected, and given running repairs. Water tanks and sand boxes must be filled, oil changed in the engines, and food and other supplies replenished. Terminal facilities for these operations include airplane hangers, mobile fueling and servicing units, locomotive terminals (engine houses), dry docks, garages—and the ubiquitous service station on the corner.

Interface   One of the most important functions a terminal performs is making the transportation system and its services available to the shipping and traveling public. It is the interface between the user and the carrier. It is also the interface between carriers of like mode and between carriers of different modes. The interface problem of coordination is discussed later in the chapter.

PROBLEMS AND CHARACTERISTICS

Comprehensive Planning   Most terminal problems in the past have been solved piecemeal. Railroads have built and expanded yards as their local traffic plan dictated, but only recently has this been done with regard to other yards in the system. Parking facilities often are placed where land is available with little regard to accessibility, interference with traffic flow, or projected land uses. Terminal planning should be an integral part of comprehensive planning. It should relate the terminal facilities and functions to the proposed land uses as well as to the transportation system. Comprehensive terminal planning is made difficult by a lack of tradition and experience and by a mixture of private, municipal, county, state, and federal ownership and responsibility. With comprehensive planning, conflicts can be resolved before they become set in construction and land use patterns established to guide the planning of all elements.

Facility Design   An obvious problem is the operational and engineering design of the various terminal facilities. Such design must satisfy a variety of requirements as to (a) mode, (b) traffic types, (c) required capacity (a reasonable current peak load with provision for expansion), (d) community and regional planning, (e) relation to other parts of the transportation system, (f) operational speed and efficiency, (g) effects on the environment, and (h) service to shippers—last named, but hardly the least important.

Operations   Plant design must be directed toward facilitating terminal operations and the relation of those operations to line-haul movements. Door-to-door travel time, both for freight and passengers, merits special consideration. Terminal delays, unfortunately, have been a major factor in poor door-to-door performance as well as in poor equipment utilization. Vehicles and traffic must move rapidly through terminals. Coordinated layout with a minimum of back haul, cross haul, and duplication of facilities and routes must be planned and operated to maximize vehicle utilization.

Turn-around Time   Terminal efficiency and equipment are reflected in the overall time for a vehicle to be placed empty for loading, moved to its destination, and unloaded ready for another load. Line-haul time is usually a small proportion of the total time thus consumed. For tramp steamers and gypsy truckers, efficiency is reflected in the time to arrive in a terminal at the place of unloading, unload, take on another load, and be under way again. These times will vary greatly in individual cases, but average times are known for some carriers. A railroad car, for example, requires 10 to 14 days. Ships will be in port varying lengths of time depending on tonnage—4 to 12 hours for bulk cargo carriers on the Lakes, 6 to 10 days for an 8000-ton (7256-tonnes) general cargo ship. Truck trailers will arrive loaded at a freight station in the morning and go out loaded again that night when traffic is moving freely. The differences in time reflect in part the efficiency of the terminal (for example, ships may have to await their turn at a wharf) or the technological characteristics of the carrier of which size is of considerable importance. A railroad car that has to pass through one or more intermediate terminal yards before reaching the yard that will classify it for road haul is not going to show a favorable terminal-time figure. A poor harbor channel may require a ship to lie at anchor outside the entrance waiting for favorable tides and currents. Flexibility in detailed terminal movements is an inherent techno-operational advantage for trucks and automobiles; lack of that flexibility is a disadvantage for railroads. Long hours of ground crew inspection and maintenance are behind the excellent safety record and low point-to-point flight time of commercial air lines.

Turnaround time has an obvious relation to the amount of equipment required. Shortening the average turnaround time of a railroad freight car would be equivalent to adding thousands of additional cars to the existing car fleet. Cutting turnaround time in half would permit a truck, ship, or airplane line to handle the same volume of traffic with approximately half the number of vehicles.

Availability of equipment in terms of ownership, physical condition of that equipment, and suitability for a particular type of traffic are contributing factors in equipment supply. Of considerable import is the effort made by local supervision to keep equipment moving forward. Making a supervisor cost-responsible for the time that a car, trailer, barge, or aircraft is under his jurisdiction could have a salutory effect.

One element in turnaround time beyond the direct control of those carriers affected is the abuse of equipment—railroad cars, barges, motor trailers—by shippers and consignees. Where a cargo unit—car, trailer, container, or barge—is placed for unloading on Friday at a plant working a five-day week, the unit will remain idle and under load until an unloading crew reports for duty Monday morning. Once unloaded, the units may be held several days to receive an outbound load, the shipper fearing a unit will not be available when he orders it. Other abuses include holding the cargo units for storage, ordering more units than are immediately needed, leaving a unit dirty and filled with dunnage after unloading. All these and other delays increase turnaround time and the number of units required to give service. Single unit carriers are less subject to this delay than assembled unit carriers. Pipelines, conveyors, and aerial tramways do not have the problem at all.

Terminal Costs   The total costs of transportation may be divided into (a) road-haul costs and (b) terminal costs. Within certain limits, road-haul costs will vary with distance hauled, the unit cost usually decreasing as the distance increases. Terminal costs bear no relation to the distance to be hauled. The costs of terminal service will be the same whether the pay load is being moved 10 or 1000 miles in line haul. Thus a carrier with a high percentage of terminal costs in proportion to line haul can encounter financial difficulties. The ICC has calculated the 1957 terminal costs for Eastern railroads as $64.38 for an average gondola car with a 47.5-ton load (43.1-tonnes).[1] The line-haul cost was $0.33 per mile. Terminal costs were thus equivalent to a 195-mile (314-km) line haul. However the terminal costs would have remained the same had the car made only a 50-mile (80.5-km) line-haul move. The carrier would then have had to earn $1.62 per car mile to recover its costs. The dollar amount may vary but the principle is the same for all modes of transport including the often-excepted pipelines, conveyors, and aerial tramways.

Terminals versus Land Use   Terminal location in relation to land use plagues both transportation and urban planners. Ideally located terminal facilities are close to traffic sources. Railroads have a competitive advantage with in-town locations giving proximity to sources of traffic. With lack of adequate zoning in the past, industry and commerce have located in widely scattered parts of many urban areas. The result has been a complex crisscrossing of rail, truck, and canal routes.

A large city with an established land use pattern can do little to improve the situation except at great expense. Rail lines usually were there first and require that they be "kept whole" if asked to move their tracks or facilities; that is, the rail line should bear no cost for the move and there should be no loss of traffic, of traffic potential, or of services presently being rendered. If such loss occurs, there must be full compensation.

Street level access routes pose obvious problems of grade crossing hazards and delays and of traffic congestion. Placing rail lines and expressways on elevated structures or in open cuts gives some relief but tends to divide the community and poses access problems across the route. Drainage and refuse accumulations are problems in open cuts. Tunnel access gives surface relief but is costly and brings problems with underground water, gas, and sewage mains and other utilities; there can be little or no access to industry along the route. Transportation corridors placing

[1]*Railway Age*, June 1, 1950, p. 38, Simmons-Boardman Publishing Company, New York.

more than one mode on the same right of way, as with rapid transit in the median of an expressway, have been used successfully. Air rights may be utilized in the form of buildings over tracks or expressways. Parking garages may be placed underground. Consolidation and abandonment of duplicate facilities have proven advantageous. Union truck and bus terminals reduce cross-haul and duplicate routes. Terminal railroads provide for the unbiased access of several rail lines to shippers without duplication of trackage. Conflicting with accessibility to traffic sources is the need for accessibility to intercity routes—to main rail lines, intercity highways, and to harbor areas.

The development of industrial parks where access to transportation is combined with other needed services and utilities is a forward step in land use planning.

Effects on Environment   Terminal facilities and operation make serious contributions to all forms of pollution—air, water, noise, and visual. Auto exhausts from parking areas; dust from coal, grain, and ore dumping; noise from the impact, concussion, and retarder shriek of rail yards; discharges from ships in harbors; the "Chinese Wall" effect of access routes; and the rumble of trucks going to and from freight and trailer-on-freight-car terminals exemplify possible pollution sources. Terminals can, however, contribute to environmental improvement. Rail yards can serve as transition areas between noncompatible land uses. Union freight and passenger stations reduce the number of structures and the amount of street congestion from excess auto and truck movements. Underground parking garages may help sustain areas of gardened beauty in urban centers, and street level parking lots can be landscaped to provide an attractive open area amid city buildings. The need is for foresight and a resolve to protect and improve the adjacent land.

Coordination   Coordination is sufficiently important to devote the following section to that subject alone.

COORDINATION PRINCIPLES

Definition and Significance   The ideal transportation is usually considered to be that of loading a shipment into a container (freight car, ship, truck, etc.) or into a transport system (pipeline, conveyor, etc.) at the shipper's door and performing no other operation than a single road haul before the freight is turned over directly to the consignee at his door. Such an ideal is not always realized, possible, or even desirable in practice.

Each type of carrier possesses certain inherent technical and economic advantages and disadvantages. These advantages often can best be realized, or the disadvantages overcome, by combining two or more carriers to perform a joint or coordinated transportation service. Coordination can bring about faster or more dependable service for the shipper and economies for the transport agencies, some of which may be passed on to the public through lowered rates. Concern must be directed to the complete door-to-door movement of persons and goods. The mechanics of operation that bring about coordination are usually focused on terminal functions and facilities.

Union rail and bus stations, the collection and delivery by trucks of line-haul freight, and the travel to and from an airport by automobile, bus, taxi, or helicopter are examples of coordinated transport services. Many other services are not so well known to the general public. Each arrangement is a problem to be settled on its merits. The reader can gain some concept of the possible savings through coordination in only one branch of the industry by reading the many schemes proposed by Federal Coordinator of Transportation Joseph B. Eastman in his reports submitted to the Congress in 1936.[2] It is probable that the administrative savings occasioned by coordination have been overemphasized. At the same time, the possibilities for more economical, rapid, convenient, and dependable service through coordination are far from realization.

Coordinative Factors   What are the factors or combination of conditions that bring about coordination?

Extension of Service   This has been one of the prime factors. Airplanes cannot take off and land in the center of a large city or its central business district. They must extend their service into that potential source of traffic by the use of ground transportation. The flexibility of trucks is used by railroads to extend their service to the shipper's door. Feeder service to line-haul routes is a form of coordination. The intervening presence of a large body of water may require joint rail or highway and water service. Rail movements between Wisconsin and Michigan use a ferry service across Lake Michigan to avoid the roundabout land route and congestion in the Chicago area. A railroad mine spur may penetrate rugged terrain as far as possible and then use a cableway to reach areas across terrain impassable by rail.

---

[2]Fourth Report of the Federal Coordinator of Transportation on Transportation Legislation, 74th Congress, 2nd Session, House Document No. 394, 1936.

Speed The use of combined modes of transport to produce a speedy passage is exemplified by the surburban dweller who drives his car to a parking lot on a commuter or rapid transit line and rides the train into town.

Convenience The desire to offer a more convenient and competitive service has led to such types of coordination as truck pickup and delivery for rail and highway motor freight and the substitution of truck delivery for trap- and way-car delivery from break-bulk points. Coordination facilitates concentration by permitting a large- or bulk-shipment carrier to accumulate large quantities of freight in stock piles, tank farms, or warehouses and by using other means of transport to make small-lot distribution from there.

Economy The simple urge for economy may, in some instances, bring about coordination. Two types of economy—land use and financial—are likely to be involved. The joint use of the same right of way by different types of transport illustrate land-use economy. Rail transit lines have been placed in the median strip between highway lanes. Some pipelines have been laid on railroad rights of way. In Wuppertal, Germany, a monorail is suspended over a waterway.

Coordination facilitates concentration by permitting a large- or bulk-shipment carrier to accumulate large quantities of freight in stock piles, tank farms, or warehouses and by using other means of transport to make small-lot distribution from there.

Financial economy may be the reason for a union rail or bus station or for a terminal service for all railroads that is provided by a single terminal system within an urban area. The capital, operating, and administrative costs of duplicate facilities are thus avoided.

Several or all of these factors are usually present at the same time to bring about a particular type of coordination. Economy in land use, for example, usually (but not always) leads to financial economy as well.

Limitations on Coordination Coordination should be used only when it makes a real contribution to overall economy and efficiency of movement. Combination frequently involves interchange or rehandling of equipment and lading. Each handling or interchange increases the possibility of loss or damage. Transshipment of coal, for example, by means of a car dumper to a ship and then via bucket crane back to a railroad car can break the initial lumps into smaller particles; this means the loss of tonnage through dust and leakage of fine particles. Interchange design should be based on a minimum of rehandling and reverse movements. By interchanging barges, cars, or trailers, large shipments can be moved from one carrier to another without disturbing the contents. Trailer-on-flatcar and containerization similarly reduce the need for rehandling smaller shipments and the attendant potential for loss, damage, and pilferage. Time and cost factors will vary with the situation, facility, commodity, and system of operation.

TYPES OF COORDINATION

Coordination is carried out between different companies of the same mode (intracarrier) and also between carriers of different modes (intercarrier). It usually takes one or more of the following forms.

Joint Use of Terminals   The union depot for rail, bus, or airline, the union freight station for rail or motor freight, and the transit shed serving both river and lake traffic are examples of joint terminal use. Advantages include convenience of transfer, provision of better facilities than an individual carrier might be able to afford, avoidance of cross hauls and duplication of facilities, and land use economy.

The economies of a union passenger station arise from the tax, interest, maintenance, and operating expenses avoided at individual stations. Not all of this can be counted as net saving, however, as each tenant must contribute to the costs of the joint facility, but not in the same proportion.

A union freight station saves mileage and the number of vehicles required. It reduces the number of calls made at the door of an individual shipper. The Port of New York Authority estimated a saving from use of its lower Manhattan motor-freight house of 1,830,000 truck miles per year, 15,600,000 tire miles, and 336,000 gal of gasoline. At the same time, the over-the-road truck capacity was increased about 20 percent because of faster terminal turnaround. A shipper served by the Newark terminal had had up to 18 trucks a day with an average load of 1900 lb (863 kg) each call at his shipping-room door. With union-depot operation, four trucks of 8500-lb (3859-kg) load each were all that were needed.

A single freight house might not have enough capacity to justify mechanization, whereas the combined traffic of several carriers could serve as an adequate warrant. For example, the use of fork-lift trucks with palletized loads has been estimated as effecting a savings of 30 to 65 dollars per carload.[3] The same principle holds for other types of union-station operation.

[3]"Better Materials Handling, etc." *Railway Age*, August 10, 1959, Simmons-Boardman Publishing Company, New York.

Disadvantages include the scarcity of large land areas often required in central locations for the enlarged capacity of a joint facility. The coach yards of rail terminals, for example, may therefore have to be located some distance from the station with expensive deadhead running to and from the yards. The attempt to handle commuting and line-haul traffic in the same station may cause excessive congestion at the platforms during commuting rush hours, when line-haul operations will also be in progress. Additional platforms may have to be built to stand idle and unproductive most of the day. (The recent reduction in number of passenger trains under Amtrak operations facilitates the concentration of passenger services for a city in one union station with one set of coach yards and maintenance and servicing facilities.)

Other disadvantages are some loss of individual carrier identity, loss of economic advantage from an individual location, difficulties in apportioning user costs and charges, and excessive charges required for small operators who may not require the elaborate facilities provided.

Coordination of Schedules   Passenger and freight schedules of one carrier may be arranged to connect with those of other carriers. This more often occurs where the carriers share a union terminal or where one carrier acts as a feeder or bridge line for another. Railroads coordinate their schedules as a routine practice where they interchange large volumes of freight traffic. Entire trains may be moved from one railroad to another on predetermined schedules.

Interchange of Equipment   Truck and barge lines interchange trailers and barges to a limited degree. This practice and its advantages are most fully realized in railroad operation, where cars and contents move freely from one carrier to another. Such interchange saves rehandling costs and loss of time in transferring freight. The necessity of standardized features to permit interchange is again noted. Disadvantages include the difficulties of getting equipment returned, unfavorable equipment account balances with carriers having a deficiency in ownership of cars, trailers, or barges, maintenance of foreign or off-line equipment, determining and collecting user costs for using and maintaining foreign equipment, and empty return moves.

Trackage Rights   Railroads offer additional opportunities for intracarrier coordination. When two rail lines are approximately parallel, trackage rights granted by one carrier permit the other to use its tracks. The excess capacity usually available in most track layouts permits additional trains to be added with little or no extra capital outlay. If capacity is a problem, two parallel single tracks may be paired to give the advantages of double-track operation, or the institution of Centralized Traffic Control may permit one track to carry the added load and the other to be abandoned. Savings due to abandonment lie in the recovery of scrap value of rail and fastenings and a saving in taxes ($500 to $5000/mile) and maintenance ($3000 to $4000/mile). There may also be a saving on operation if more favorable grades are encountered on the host's tracks and if train order and block stations can be closed. Payment may be on a per-train, a gross-ton-mile, a per-car, or a straight-percentage-of-operating-cost basis. Trackage rights may also be ordered by the ICC to permit one carrier to gain access to a traffic territory over another's tracks.

Joint Use of Right of Way   The possibilities in the joint use of rights of way are only recently receiving adequate attention. From time to time, proposals have been made that a highway be built over the tracks of a railroad right of way on an upperdeck or viaduct. Similarly, railroads over highways have been proposed. The loads for the latter are heavier to support, up to 80,000-lb (36320-kg) axle loadings for railroads as contrasted with 18,000- to 20,000-lb (8172- to 9080-kg) axle loadings for trucks. As an offsetting factor, a railroad structure need only provide 14 to 16 ft (4.3 to 4.9 m) of clearance over the highway whereas a highway structure would have to allow 22 to 24 ft (6.7 to 7.3 m) of clearance for the railroad. The railroad grades are usually more suitable than highway grades. In fact a railroad couldn't operate over some of the grades that would thus be established even by high-class highways. Either arrangement poses problems of access.

The center mall or median strip of expressways and superhighways can be used as railroad rights of way, primarily for rapid-transit systems. A two-track, standard-gage system would require 29 to 32 ft (8.8 to 9.8 m) of median strip width, 13 to 16 ft (4.0 to 4.9 m) more than required by minimum AASHTO standards for interstate highways in urban areas but well within the 40 ft (12.2 m) of width recommended for expressways and the highest-class construction. Additional land can be obtained far more easily and cheaply when the highway land is being purchased than it could if obtained separately. Several of the Chicago expressways have center malls with rail rapid transit. An unfortunate aspect of this coordination is that urban, and other, expressways may parallel an existing railroad already prepared to give service.

Another aspect of rail-highway locations is the desirability of keeping noncoordinated rail and highway routes far enough apart through industrial zoning to permit adequate development of industry on both sides of each. A highway and railroad closely paralleling each other prevent or seriously hinder any industrial access or development between them Various separation distances have been proposed ranging from 300 ft to a half mile.

Pipelines are adaptable to location on rail and highway rights of way, especially rail. Pipe of the Southern Pacific Pipe Line Company has already been placed alongside the tracks of the parent company. The railroad or highway provides easy access to the line during construction and for maintenance when it is operating. Where severe development curvature is present, the pipeline may cut across country, bypassing one or more of the curve loops. The lines should always be buried deep enough to avoid breakage and a disastrous fire in the event of train derailment.

Great Lakes Rail-Water Coordination Outstanding examples of end-to-end movement are the handling of iron ore and coal via the Great Lakes. Ore is brought from the pits to the surface by skips, belt conveyors, trucks, or rail. At the surface, it is crushed and loaded into railroad cars, classified as to grade of ore by cars, moved to the lake front, and dumped into the pockets of the ore docks. Each ore-dock pocket is 24 ft (7.3 m) long and each ore car is 24 ft (7.3 m) long, just covering the pocket. These dimensions also correspond to the 24-ft (7.3-m) hatchway spacing on most ore boats. Because some boat hatches are spaced on 12-ft (3.7-m) centers and in order to provide loading flexibility, chutes from the ore-dock pockets are spaced on 12-ft (3.7-m) centers. The movement of ore trains, classification into various grades of ore by cars in the classification yard, and dumping and blending of ore at the ore docks are all coordinated with the arrival and departure of the ore boats. Ore-boat movement is, in turn, closely watched and scheduled by ship-to-shore radio telephone.

At the other end of the run, ore-unloading facilities are located at Gary, Indiana Harbor, Detroit, Toledo, Sandusky, Lorain, Cleveland, Ashtabula, Conneaut, Erie, And Buffalo. Gigantic bridge cranes and Hulett unloaders, taking 10 to 20 yd at a bite, unload the vessels into stock piles or into railroad cars for rail movement to the steel mills of Gary, River Rouge, Cleveland, Canton, Youngstown, Pittsburgh, and adjacent areas. See Figure 10.3.

From these same Lake Erie and Lake Michigan ports, coal brought by railroad to the lakes moves north and west, often in the same boats that bring down the ore. Car dumpers pick up the cars of coal and pour their contents into the hold of a vessel, like sand is poured from a toy sand bucket. At Detroit, Duluth, Superior, and intermediate ports, bridge cranes

FIGURE 10.3a ORE-TRANSFER EQUIPMENT—ORE RUNNING FROM ORE-DOCK POCKET TO HOLD OF GREAT LAKES BULK-CARGO CARRIER. (COURTESY OF PITTSBURGH STEAMSHIP DIVISION, (NOW GREAT LAKES FLEET) U.S. STEEL CORPORATION.)

and similar specialized devices unload the coal into stock piles or into railroad cars for further land haul.

Service Railroads   Service railroads, acting as terminal, belt-line, interchange, switching, and bridge companies, are organized and owned jointly by several railroads, by a municipality, port authority, or industrial group

(b)

FIGURE 10.3b ORE-TRANSFER EQUIPMENT—HULETT UNLOADER AT DOCKS OF PITTSBURGH AND CONNEAUT DOCK COMPANY, CONNEAUT, OHIO. THIS MACHINE HAS JUST DUMPED A BUCKET OF ORE INTO A CHUTE THAT WEIGHS THE RAW MATERIAL BEFORE DEPOSITING IT IN THE RAILROAD CAR. (COURTESY OF PITTSBURGH STEAMSHIP DIVISION, (NOW GREAT LAKES FLEET) U.S. STEEL CORPORATION.)

to reduce congestion and duplication in crowded areas or to give unbiased service to all shippers in an urban, industrial, or port area. Examples of these are the Kansas City Terminal Railway, the Terminal Railway Association of St. Louis and the Belt Railway of Chicago that operates the Clearing Yard for interchange of traffic between the several carriers entering Chicago as well as performing a belt-line function.

A single railroad may perform all switching services for several railroads in a port or industrial area, charging so much per car switched. Other roads may perform a similar service in other parts of a city on a reciprocal basis. Unfortunately the result is sometimes poor service for the shipper who is not giving the switching line the road-haul business. Such poor service is often hard to pinpoint but may appear as delays in settling claims, delays in placing and moving cars, and difficulty in getting an adequate supply of cars, especially in times of car shortage. A railroad is sometimes given the road haul in order to overcome these deficiencies. Belt railways or switching lines, whether owned jointly by the railroads or by a municipality, port, or industry are planned to serve connecting carriers and shippers alike without bias.

Tugs, carfloats, towboats, and local trucking lines perform functions that correspond in many ways to those of terminal railroads.

Ferry, Piggyback, and Containerization   The term "piggyback" is a crude but popular term used to describe a type of coordination that has come into common use, especially on railroads. It is simply a form of ferry service in which the transport unit of one carrier is moved by another. The system of rail and automobile ferries across Lake Michigan and the movement of railroad and highway vehicles by car floats and ferries in New York and San Francisco harbors are examples. Railroad cars are also ferried between Atlantic and Gulf Coast points and between West Coast ports and Alaska. Entire trains between London and Paris move by ship across the English Channel. The trailer-on-flatcar (TOFC) service of the railroads finds a counterpart in water carriers that haul highway trailers. More frequently, the water carriers provide a container service whereby the container, equivalent to a highway trailer without the wheels, is transferred to and from the ship to the highway or rail for land movement and vice versa.

With TOFC or piggyback, highway motor freight trailers are loaded at shipper's door or at a freight house, brought by tractor to a railroad loading ramp or yard, placed on flatcars, and hauled in trains to a destination terminal. The trailers are then unloaded and terminal delivery is made to the consignee's door by tractor over city streets.

With the line-haul time of merchandise trains approaching or even exceeding that of passenger trains, TOFC line-haul service is faster than highway haul, while the terminal times are practically equivalent. Much of the expense and nuisance of highway haul—traffic congestion, personnel problems, traffic violations, accident hazards, delays in small communities, and restrictive limitations on weight and size—are avoided or markedly reduced. Truck operators can also reduce their pool of tractors and drivers and their tractor maintenance. The requisite fast overall train schedules can best be obtained with unit trains that avoid intermediate yard delays.

Five systems of TOFC (with variations) are generally recognized.

Plan I. Vehicles of common highway carriers are hauled by rail. The shipper deals with the highway carrier, which, in turn, deals with the railroad.

Plan II. Only the railroad's highway trailers are carried. The public deals directly with the railroad or its highway subsidiary.

Plan III. Anyone's trailer, including those of the individual or private trucker, is carried by rail.

Plan IV. An intermediate or forwarding agency or broker secures the freight, loads it into its trailers and onto its own flatcars, and turns it over to the railroad to haul. The "third party" may be a company such as "Trailer-Train" owned by one or more railroads, but the shipper deals with the third party.

Plan V. Plan I plus joint rail-highway rates with the highway carrier. The shipper deals with either trucker or railroad.

The advantages of TOFC cited earlier accrue mainly to the trucker. Railroads, in hauling common-carrier and privately owned trailers, gain in securing a portion of the revenue that would otherwise have gone entirely to a competitor. When railroads haul their own trucks, additional advantages accrue. If merchandise traffic is picked up in railroad trailers and taken directly to the loading ramp, terminal makeup, transfer, and break-up, which now give truckers a big delivery-time advantage, will be avoided and the speed advantage of railroad line haul will be maintained. Big metropolitan freight houses with their high operating costs are being retired by many railroads or turned over to TOFC companies for their operation. One railroad estimated a saving of $7 per ton by keeping merchandise freight entirely out of freight houses.[4]

Dr. L. K. Sillcox, pointing up these savings (1959 figures), has quoted testimony of the Pennsylvania Railroad (ICC Docket 32533) as to costs of LCL loading across an inland platform at their Thirty-Seventh Street Station, the

[4]Nancy Ford, "Piggy-Back Spreads Out," *Modern Railroads*, December 1955, p. 69, Modern Railroads Publishing Co., Chicago.

originating terminal:[5]

Cost to load car of 7.9 tons at $4.72 per ton	$37.29
Less tariff charge of $4.09 per ton	32.31
Net cost to load car of 7.9 tons	$ 4.98
Float bridge—37th St.	2.63
Floating—37th St. to Harsimus Cove	33.82
Float bridge—Harsimus Cove	5.13
Switching cost—Harsimus Cove	13.44
Total per car (excluding switching at 37th St.)	$30.00

Dr. Sillcox then made a comparison with the same operation under Plan III (piggyback) at the Kearny, New Jersey, piggyback terminal:

Loading two trailers	$9.46
Clerical and supervision expense—	
Kearny Truck-Train Terminal	5.68
Switching loaded car to road train	
(includes cost of placement)	4.23
Total	$19.37

Since this loading would average 25.9 tons, the foregoing is equivalent to loading three box cars at Thirty-Seventh Street. Dr. Sillcox thus finds a cost of $180 for conventional practice as compared to the foregoing $19.37; a savings ratio of 9.3 to 1. This is a per-ton saving of $6.24 as compared with the $7.00 per-ton saving quoted earlier. Not all merchandise operations may be as complex as the Thirty-Seventh Street operation, but the pattern is discernible.

Costs of operating a tractor-trailer combination have been estimated at $0.30 to $0.40 a mile for a 15-ton (13.6-tonne) load, with a median of $0.35. Not all of this can be saved by piggyback because expenses of trailer ownership, insurance, terminal-area operations (after the rail haul), and other overhead expenses will amount to about 20 percent ($0.07 cents of the $0.35 median cost). Also deductible is the cost of the rail haul, approximately $0.16 per trailer mile with trailers loaded two to a flatcar. This leaves a possible saving by piggyback of $0.12 per mile for the 15-ton (13.6-tonne) load.

If the railroads haul only their own or trailers of common carriers, Plans II and I, there will be an assured pool of tractors and drivers to move the trailers at the end of each rail run. This may not always be true for private or Plan III truckers for whom a protected parking area would have to be provided. Problems of nonstandard equipment may arise with private operators.

[5]"Trucks by Train," printed account of address presented at the Canadian Railway Club, Montreal, Quebec, April 13, 1959, by Dr. L. K. Sillcox, pp. 14 and 15.

Railway-owned trailers (Plan II) replacing the waycar for local delivery of LCL freight, can be hauled piggyback to the break-bulk point under load. The trailer then proceeds by highway to make detailed deliveries. The way freight is used only for handling carload business.

Not only does TOFC offer relief to freight operation, it also suggests the possibility of eliminating classification-yard service for Plans I, II, and III loads. Cars brought from line haul to the unloading ramps are reused in place for outbound trailers. The trailers can be loaded in whatever station order or blocking is desired without reference to individual cars. The loaded cars are then made into trains from the loading tracks or, if the total number is small, into a complete block (or complete blocks) to set independently on a train after its other consist has been assembled. The economies gained from these procedures are self-evident if one assumes yard costs of $10 to $20 per car.

A variety of methods is available for loading and unloading. For small traffic volumes, end loading can be used; this enables trailers to be pushed or pulled on or off the rail car with tractors via fixed or portable ramps from pavement to car floor level. Portable ramps bridge intercar space. The system is slow and lacks flexibility. Eight trailer car lengths is about the limit for one track. Trailers can be unloaded only in the same (reversed) sequence as loaded. Trailers may also be handled by lift-on/lift-off devices such as giant fork-lift trucks or by vertical-lift trucks with straddle yokes that fit over the tops of trailers and grasps underneath. Where large numbers of trailers and cars are involved, traveling gantry or straddle cranes are used that move up and down a line of cars on steerable rubber-tired wheels, straddling two tracks with an intermediate driveway. Arms or grapples, positioned by a gantry crane over the trailer have the capability of rotating the trailer into proper position on the car or on the pavement.

Originally, ordinary 35-ft (10.7-m) trailers were used on standard 40-ft (12.2-m) flatcars equipped with hold-down devices. True economic viability came with development of a two-trailer car. Today's 85 to 89-ft (25.9- to 27.4-m) car can carry two 40-ft trailers (or a 35-ft and a 45-ft trailer). Cross-sectional dimensions are generally 8×8–$8\frac{1}{2}$ ft. Longitudinal slots in some car decks provide both stability for the trailer and additional clearance for its standard 12-ft, 6-in. height (3.8 m). The car must have stable, quick-in-application hold-down devices, shock-absorbing elements, and conformity to A.A.R. interchange rules. See Figure 10.4.

TOFC has several technoeconomic problems. (a) The 35-40 ft trailer capacity may be too great for some shippers so that their small shipments must still be consolidated across a freight-house platform—also the trailers are too large to place in an airplane or in other than specially designed

FIGURE 10.4 PIGGY-BACK TRAILERS IN PLACE. (COURTESY OF THE ERIE (NOW CONRAIL) RAILROAD.)

ships. (*b*) The length of the car creates derailing tendencies, especially when coupled to a short car and under heavy draft on sharp curves. (*c*) The height of trailer above top of rail, up to 17 $\frac{1}{2}$ ft, creates clearance problems that have led some railroads to make major changes in tunnels and bridge portal clearances, even to transforming tunnels into open cuts. (*d*) The high van with its undercarriage creates additional train resistance. (*e*) TOFC requires the hauling of one vehicle upon another, thereby decreasing the payload to an empty-weight ratio of 3 to 1 or worse from a possible 4 to 1.

Trailers are moved to and from ships either with lift-on, lift-off or roll-on, roll-off loading and unloading.

Containerization A solution to the last problem exists in use of containers, boxes similar to trailers but without the undercarriage. These come in many sizes from 350 ft³ to 2560 ft³ (9.8 to 72 m³) and from 20 ft (6.1 m) to 40 ft (12.2 m) in length. The 35- and 40-ft (10.65- and 12.2-m) containers are usually 8 ft (2.44 m) square. Containers may be simple boxes that are set on truck bodies, on flat cars, or in gondola cars, or in ships' holds or for placement on demountable truck bodies or chassis for land movement. Only the largest aircraft can carry a standard container, but smaller units that fit the fuselage contour are in use. Individual freight items loaded in the container need not be rehandled from the time that the container is closed and sealed by the shipper until the seal is broken by the consignee.

The requirement that may be imposed by government regulation to break this sequence for custom inspection is one of the administrative obstacles to development of this service in foreign trade. There is need for more streamlining of the paper work. Such a door-to-door procedure represents the final step in eliminating large-scale freight house operations. The ultimate step in this direction is an all-purpose or universal container that can be used equally well for shipment by rail, highway, ship, or airplane. Containers must possess structural strength to withstand the wracking of lift-ons/lift-offs.

In foreign trade shipment by containers is becoming a well-established practice. Special designs of fast-sailing container ships, valued at $30 million or more, costing $20,000 to $23,000 a day to operate and carrying 500 to 1200 containers with container-handling cranes and holds and decks designed for this purpose, ply across both the Atlantic and Pacific and through the Carribean. On shore a number of container ports with wharves, handling facilities, tracks, pavements, and holding and classification areas have been developed, especially at New Jersey points and Baltimore.

Land Bridge   A recent innovation in containerization is the land bridge concept whereby traffic between the Far East and Europe or the East Coast of the United States is transferred from ship to rail to move COFC over a relatively short, rapid land route to Gulf or Atlantic Coast points rather than through the Panama Canal. Transit time for certain ships between Japan and the Atlantic Seaboard via the Canal is 32 days. By transshipping containers to rail cars at Western ports the transit time to the Eastern Seaboard can be reduced to 16 days.[6]

There are no technological obstacles to this procedure. Unbalanced traffic does, however, pose the problem of returning empty containers and of who will pay that cost and deal with administrative and customs problems and regulations. Eastern Seaboard ports and longshoremen object to the loss of loading and handling operations where shipments from the East Coast to Japan are loaded into rail cars for transshipment at Pacific ports, thereby eliminating the ship loading and sea passage via the Panama Canal. The empty container problem has caused United States railroads to show only a small interest in the domestic use of containers.

How Coordination Occurs   Although generally desirable, many apparently sound schemes for coordination have never been put in effect, for the

[6]"PC Shapes Flatback's Third Generation," *Modern Railroads*, April 1973, p. 62.

following reasons.

1. The carriers feel they would be losing competitive advantages and their identity by giving up established routes, locations, and services when entering into combinations with other carriers.
2. Laws prohibiting the immediate reduction in personnel as a result of consolidations make estimated savings unattractive.
3. The lack of immediate and extensive saving, plus a certain apathy in some quarters, make the carriers loath to abandon present methods in favor of vaguely defined savings and a period of trial and development.

Most coordination has been the result of self-interest. The use of motor trucks for pickup and delivery service by the railroads arose from the competition of the truckers. Trackage-rights agreements derive from hoped-for economies, access to new traffic, or other means of self-interest.

Vertical integration is the combining of several unlike agencies into one. The Canadian Pacific and Canadian National railways directly or through subsidiaries control and operate railways, airways, trucks, buslines, and steamship lines.

In private industry, the oil and steel companies, for example, such operations are sometimes carried on directly, but more often by means of holding or subsidiary companies. Under existing statutes and regulatory procedures, vertical integration offers only limited possibilities to common carriers. The Interstate Commerce Act as interpreted by the ICC and the courts prohibits ownership of one carrier type by another except as supplementary and incidental to the carrier's normal operation.[7] There are a number of instances where rail carriers do operate over-the-road truck and bus service. These operations were begun before the passage of the Motor Carrier Act of 1935, and, by the "Grandfather Clause" of the act, are permitted to continue—but no new operations can be instituted.

TERMINAL FACILITIES

Design and operation of specific facilities, as these reflect technological characteristics, are of some interest at this point. Typical of terminal facilities are ports and harbors, railroad freight yards, and the ubiquitous freight house and transit shed for rail, truck, water, and air transport. Freight houses are used to consolidate small lot freight to common destinations for loading into (or unloading from) vehicles going to (or from) that destination.

[7]ICC vs. Parker, 326 U.S. 60 (1945). Rock Island Purchase of White Line Motor Freight Co., 40 I.C.C. 456 (1946).

Freight Houses   Freight-house planning is based on a design tonnage from which the requisite floor or platform space, berths, car spots, tracks, and truck tailboard room are developed. The design tonnage may be the anticipated yearly average, reduced to a daily or shift basis, plus a factor—usually 15 to 20 percent—to cover peak periods and traffic increases. Design tonnage may also be based on the expected arrivals of an estimated or scheduled number of ships, trucks, trains, or other units of a given capacity. Location should be near intercity routes, away from congestion, close to traffic sources, and with room for expansion.

For an outbound-only rail or truck house, the floor need be only wide enough to receive freight and move it across to the waiting line-haul vehicle, 30 to 50 ft (9.1 to 15.2 m). Inbound freight is customarily allowed 24 to 48 hours of free storage, so more space is required. The area may be calculated as so many square feet per ton of freight—130 ft^2 per ton has sometimes been used—but depends on the weight-volume relation of the type of freight handled. Live loads of 300 to 500 psf are usually considered. About 80—90 percent additional space is needed for aisleways, the higher value when forklift trucks are used.

Platforms are generally placed about 3 ft 9 in. (1.14 m) above the top of rail and 4 ft 0 in. (1.2 m) above the tops of pavements, 4 ft 6 in. (1.37 m) being allowed for local delivery trucks.

For rail freight, the platform length is based on the number of car spots required on one side of the platform and tailboard room for trucks and trailers along the other side. For motor freight alone, tailboard room is provided on both sides. See Figure 10.5. The tonnage for an LCL car varies from 6 to 10 tons and about twice that for trailers. This tonnage, divided into the daily or shift tonnage, indicates the car spots required per day or shift. With tailboard room satisfied, an economic balance is obtained between the annual costs of construction and operation for varying lengths of platform and corresponding number of spots per track and length of track. The optimum design will give the length for minimum combined costs.

Tailboard may be determined in several ways. The AREA has recommended 1.12 tons (1.02 tonnes) per foot of platform length in their manual. Thus a daily average of 100 tons (101.6 tonnes) would require approximately 90 ft of platform length or ten 9-ft (2.74 m) spaces. By state law, trucks cannot exceed 8 ft in width. The 9-ft allowance provides only a 1-ft clearance between any two maximum-width trucks. For ease and safety in maneuvering, this width should be increased to 10 or 12 ft, (3 or 3.7 m) and the platform correspondingly lengthened. If service and maintenance are to be performed at the platform (thereby reducing terminal time), then as much as 14 ft (4.3 m) of space per truck is desirable.

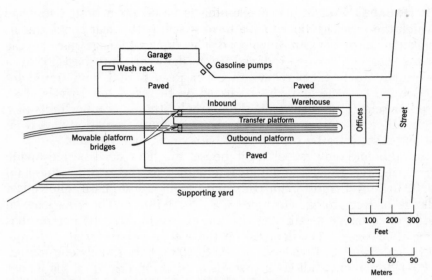

FIGURE 10.5 TYPICAL FREIGHT-HOUSE DESIGN.

Trailer spots should be the length of the overall (tractor-trailer) vehicle, 35 (10.7 m) to 55 ft (16.8 m) with an apron width for maneuvering of approximately the same length as the anticipated vehicle length. Parking space should be available for trailers along with servicing area for tractors and trailers. Lighting and fencing may be required for security. Entrance and exit gates should be placed to avoid conflict with heavy traffic routes and to avoid left-hand turns against traffic. Scales to weigh the vehicles may be included.

The roof support system is designed to maximize the amount of unimpeded floor space. Doors will be of the roll-up or sliding types, and, if the truck does not enter the house, a covered apron platform 8 to 10 ft in width will give flexibility to freight movement and truck spotting. The cover should extend about 3 ft over the trailer to protect stripping and stowage operations from the weather. Platform movements are made with hand trucks, push carts, cart trains towed by electric tractors, forklift trucks, or drag lines (moving cables to which carts can be attached or detached). A cooperage shop to repair damaged goods and containers is advisable.

Pickup and delivery is by one of two methods. The *swing* procedure has the vehicle follow a prescribed route each day. With the *call-in* system the driver makes a certain number of calls upon the advice of the dispatcher then calls in for more instruction. Freight must be received before an established cut-off hour to be loaded and dispatched that day. Trucks (and rail cars) are usually scheduled out in late afternoon or early evening for delivery the next (or second) morning.

Transit Sheds   A transit shed is a freight house for waterborne general cargo. It performs the same function as a freight house with concentration assuming more importance. It operates on a longer turnover cycle and often with heavier weights of traffic. Floor capacities are based on the number of ships to be served during a given period. The floor space must not only accommodate, with appropriate stacking, what is required to load the expected ships upon arrival but must also hold the cargo that is off-loaded from the ships before they can take on new cargo. Covered-space requirements can be reduced by the anticipated tonnage that can be stored on open wharf areas or that will be loaded directly into railroad cars, lighters, barges, or trucks.

A transit shed designed to handle two 8000-ton ships (common cargo capacities are 6000 to 10,000 tons) would have to provide space for accumulating up to 16,000 tons of outbound freight. In addition, space would be necessary for another 16,000 tons to be off-loaded before the outbound tonnage could be put aboard (minus tonnages that are given open storage or handled directly from another carrier to or from a ship's hold). A total of 32,000 tons would thus have to be accommodated at a rate of 300 to 500 lb of cargo per square foot plus about 80 to 90 percent of the loaded space for aisle-ways and space between stacks. A volumetric relation of 40 to 60 ft^3 per long ton gives 56 to 37 lb per ft^3.

It is usual for United States sheds to permit 5 days free storage for accumulating outbound cargo and 10 days free distribution for inbound tonnages. About 80,000 to 120,000 ft^2 (7432 to 11,040 m^2) of covered transit-shed space are required per ship berth; note again that not all cargo has to be under cover.[8]

If, in the foregoing example, another two ships were expected 5 days after the first two were unloaded, the shed would also have to accommodate what still remained undistributed from the first two ships. Cargo to be held beyond the 10-day free storage period is usually taken to a warehouse, which should be reasonably close to the transit shed (and is sometimes included in the same structure). About 2.5 to 5.0 percent of all general cargo handled annually in the United States passes to a warehouse from or through the transit shed, and 10 to 20 percent of that will be warehoused on the waterfront for about 3 months.[9]

[8]Maurice Grusky, "Harbor Engineering," in R. W. Abbett, *American Civil Engineering Practice*, Volume II, Wiley, New York, 1956, Chapter 21, pp. 78–81.
[9]Ibid.

Ports and Harbors Harbors provide a safe anchorage, protecting ships from the open seas. Typical harbor locations are mouths of rivers, natural bays and inlets, the interiors of coral reefs, and manmade breakwaters and tidal basins. Tide variations, currents, silting, and wave action are problems that must be considered in the design and operation of harbors.

A 26-ft (7.92-m) depth of harbor is said to be adequate for about 80 percent of the world's shipping. Hence the selection of a 27-ft (8.23-m) depth for the St. Lawrence Seaway and connecting Great Lakes channels. The newest Soo Lock has a 30-ft (9.14-m) depth (requiring 35-ft (10.67-m) channels). Depths of 36 ft (10.97 m) are necessary to dock the largest passenger liners. Maximum tonnage oil tankers and deep-water ore carriers require 40 to 65 ft (12.19 to 19.8 m) with 105 ft (32 m) needed for projected future designs.

Deep-draft vessels may dock at high tide in shallow harbors, then wait for another high tide before casting off. Another solution to shallow harbors is to anchor the ship away from shore in deep water and to shuttle cargo between ship and shore in shallow-draft, light-tonnage lighters or barges. For very deep-draft tankers, an alternative is to attach a high capacity container to the sea floor along with a mooring buoy anchored in deep water. The tanker can tieup to the buoy and discharge its cargo into the submerged tank. The tank is connected to shore facilities by pipeline for "landing" the cargo. An alternative scheme places a group of mooring buoys around a pumping station for direct pumping to a tank farm on shore. The Deep Water Port Act of 1973 authorizes the DOT to license deep-water port construction and operation beyond the three-mile territorial limit.

Where exceptionally high tides prevail, those exceeding 6 to 12 ft (1.83 to 3.66 m) variation, tidal basins may be built. These are landlocked basins bordered by wharves. There is a gate at the entrance that is closed during low tide to maintain a constant high-tide level in the basin. Ships enter and leave only when the gates are open at high tide. The deck of a wharf is usually kept about 7 ft (2.13 m) above high spring tide.

A port combines harbor protection with facilities for concentrating cargos to be loaded, for loading and discharging cargo, and interchanging with other carriers; it also provides places where the ship may take on fuel and other supplies and be repaired.

The principal feature of a port is the wharf across which goods to and from the ships are moved. The wharf may be of the pier type built over the water, or a quay built on shore, or a filled-in extension of the land into the water. A wharf may also be partly on land, partly over the water, depending on topography of the shore line. A transit shed is a usual feature of most wharves. Other elements include an apron between the ship and shed with one or more railroad tracks and paved for land vehicles, tracks and pavements in or at the rear of the shed and, sometimes, a rail-mounted gantry or portal crane of the rotating or turret type on the dockside apron. Apron widths vary from 18 to 40 ft (5.49 to 12.19 m). Figure 10.6 shows the cross section of a typical wharf layout.

Piers are used where shore space is limited or where there is ample room in the harbor channel. Quay-type wharves find use where channels are narrow or where land and shore line are plentiful. Berthing is much simpler with quay systems parallel to the stream. Slips are the open spaces between adjoining piers or docks.

The capacity of a terminal must be based on the pattern of ship arrivals. Pier and wharf lengths are functions of the number of ships to be berthed there at one time. The usual length of vessel to be accommodated must be determined in each situation, but for a normal 10,000-ton (9070 tonnes) vessel of 520-ft (158.5-m) length, 600 ft (182.9 m) of berth space would be required. (Containerships vary from 500 to 1000 ft in length.) United States design is based on vessels with 5 cargo hatches, loading or unloading 300 long tons per hour for an 8-hour working day, 200 days per year.[10] The longer the pier, the wider it must be to accommodate the increased tonnage moving shoreward, or vice versa, from the far end. Recommended widths for a pier with one ship on each side are 350 ft (106.7 m) minimum (for one-story sheds) and 450 to 500 ft (137 to 152 m) desirable. With two 10,000-ton (9070 tonnes) ships on each side, a minimum width of 450 ft (137 m) is required, but a width of 500 to 600 ft (152 to 183 m) is desirable. Included in these widths are 30- to 40-ft (9.1 to 10.2 m) aprons. Marginal wharves should have widths similar to those used for a two-ship pier.[11]

Also essential to a port is ready access to land transportation for the assembly and delivery of cargo, including supporting yards for rail and marshalling areas for highway vehicles; it should be near principal rail and highway routes. There must also be servicing facilities for fuel, food, and repairs.

Mechanization   The movement of freight from ship to shore, on aprons, in transit sheds, and in motor freight and railroad freight house was once performed by hand. Most of that drudgery has been removed, accidents and damage reduced, and economic speed and efficiency brought forth by mechanization. For very small houses, the old hand truck still prevails, but

[10]Ibid.
[11]Ibid.

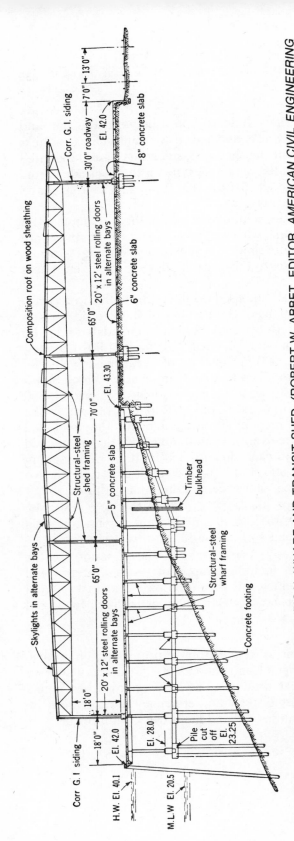

FIGURE 10.6 CROSS SECTION OF TYPICAL WHARF AND TRANSIT SHED. (ROBERT W. ABBET, EDITOR, *AMERICAN CIVIL ENGINEERING PRACTICE*, VOL. II, WILEY, NEW YORK, 1956. PP. 21–82 AND 21–83, FIGURE 67.)

in many freight and transit sheds, small tractor-pulled trucks are used for detailed movements. The introduction of pallets to which shipments may be attached for handling as a unit permit the use of fork trucks for moving and stacking the pallets with a more economical use of floor space. Where large volumes of freight prevail, mechanical conveyors move individual trucks around a closed path. A continuous moving chain, placed either above or below the floor, has hooked projections to which the handle of the truck may be attached. The truck is moved along to its proper location and unhooked from the chain.

Where heavy or oddly shaped cargo is to be handled, special rail-mounted gantry, turret, or other types of cranes of 5 to 20 tons (4.5 to 18.1 tonnes) capacity may supplement the foregoing. For unusually heavy lifts, tractor cranes up to 50 tons (45.35 tonnes) may be kept available. Rail-mounted locomotive cranes will lift up to 250 tons (227 tonnes). A few of the largest ports have floating cranes of similar capacity.

TOFC Terminals   A TOFC terminal, whether by itself or as a part of a containerport complex, is designed for the type of loading or unloading to be used. For end loading, tracks should not exceed 8 two-trailer rail cars. Longer tracks make too long a tractor run. Depressing the tracks below ground level assists by reducing the approach gradient, but drainage problems may be created. Tracks should slope toward the ramp end and cars held under air to compress the couplers and prevent movement. Regardless of the loading system, tracks should be equipped with electric or air outlets for use of power wrenches in tying down the trailers. Walkways between tracks at or near car floor level will aid in the tie-down and release operations.

With moving straddle-type gantry operation, the crane should span at least two tracks and a 40-ft (12.19-m) driveway. Paths for the steerable rubber-tired wheels are usually painted on the pavement. See Figure 10.7.

Trailer parking is provided with spaces preferably slanted at a 45- to 60-degree angle for ease in parking. Driveways, craneways, and parking areas are paved for all-weather operation. Trucks scales should be long enough for a 45-ft (13.72-m) trailer. Fencing to prevent pilferage and lights for night operation are other requisites. There should be easy access to main highways and rail yards and away from rush hour traffic that might affect truck schedules. Trailer parking space for both on- and off-loading should be provided.

Containerports   Facilities to transfer trailers and containers between ship and land modes are being concentrated in containerports. These are

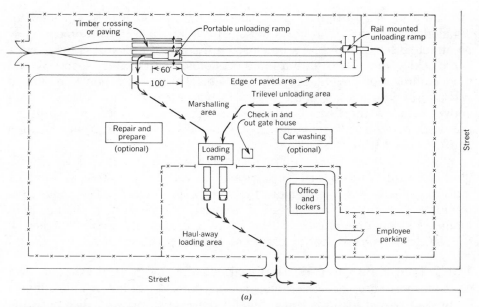

FIGURE 10.7A CROSS SECTION OF A TOFC-COFC VTERMINAL. (*MANUAL OF THE AMERICAN RAILWAY ENGINEERING ASSOCIATION*, CHAPTER 14, "TERMINALS," 1972, P. 14-3-21 AND BULLETIN NOVEMBER-DECEMBER, 1975, P. 124.)

equipped with a wharf and dockage, lifting and transfer equipment, TOFC loading/unloading ramps, and container trailer storage and classification areas. An example is the Sea Land Company's containerport with a parking capacity of 10,000 containers at Elizabeth, New Jersey and the Canadian National's port at Sidney, Nova Scotia.

Containerships may be equipped with lifting devices, but the port will usually have bridge-type cranes capable of movement along the wharf for on/off loading. Ships especially designed for trailers or containers on chassis can be handled by roll-on, roll-off methods across hinged ramps or portable gangways. Tracks embedded in the wharf apron pavement permit direct ship transfer to rail cars and trucks on the apron area. Containerships with speeds up to 33 knots vary in capacity from 200 to 1200 containers and from 450 to 950 ft (137 to 290 m) in length.

Port capacity is a function of vessel scheduling, capacity, and rate of cargo handling. An average of 20 to 30 containers per hour (2 to 3 minutes per unit) can be used where adequate cargo is already at hand. The lower

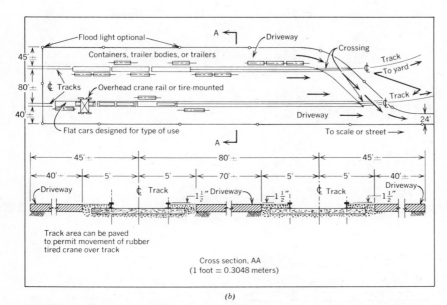

(b)

FIGURE 10.7B SUGGESTED RAMP LAYOUT OF A TOFC-COFC TERMI-
NAL. OVERHEAD LOADING (*MANUAL OF THE AMERICAN RAILWAY
ENGINEERING ASSOCIATION*, CHAPTER 14 "TERMINALS," 1972, P.
14-3-23.)

figure is more realistic because of delays. The number of berths depends on
the size of vessels handled and the arrival-departure schedules.

Storage and classification or marshalling areas can be based on the area
occupied by one container 8 ft^2 (2.43 m^2) and 35 to 40 ft (10.67 to 12.20 m)
long, whether stacked one, two, or three deep, with an additional 80 to 90
percent of space added for driveways and space between containers. The
Port of New York and New Jersey Authority has designed a containerport
on the basis of 35 acres per berth for containerships of 1000 capacity.
Movement within the area is made by containers on tractor-pulled chassis,
by forklift trucks, or by straddle buggies. Container inventories may be
recorded by mobile Automatic Car Identification type scanners moving up
and down the parking area when containers are equipped with ACI coded
strips. See Figure 10.8.

Additional facilities will include a TOFC loading/unloading yard, fencing, and lighting to discourage pilferage and for night operation, truck
scales, and a port office.

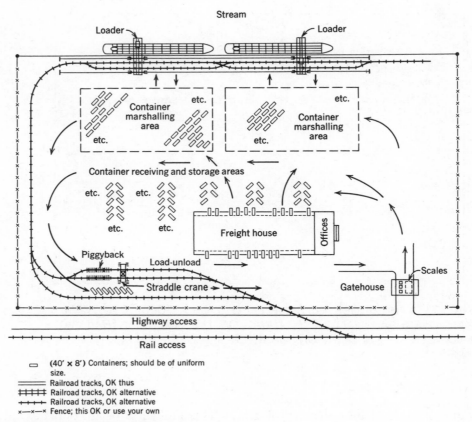

FIGURE 10.8 CONTAINERPORT.

A suitable number of chassis must be available on which containers are placed for movement about the area by tractor. Trailers are moved on their own wheels. Alternatively, forklift trucks or straddle buggies may be used for container movement and for stacking containers in two or three levels.

As containerport traffic increases more and more land area is needed for storage and marshalling. More tractors, chassis, and forklift trucks are also needed and the distances to move the traffic grows longer. The associated costs may require a solution with lower land and distance requirements. An automated container storage has been suggested that would consist of a modular structure with stalls into which containers would be stored and retrieved by an automatic and computer-guided elevator tower system. The structure would stack as many as 10 levels above ground to give high density land use. One design, with a projected storage capacity of 1000/forty-ft (12.19 m) units and 2000 twenty-ft (6.10-m) units (3000 total) is said to require no more than 3.5 acres of land.[12]

Yards Railroad yards serve the varied purposes of storage, holding, reconsignment, public delivery, and supporting industrial, water-front, and switching activity. The principal type of yard, and its function, is the classification yard, which also performs a concentration function by accumulating enough cars to fill out a train. Classification includes the receiving and breakup of trains and the sorting and classifying of the cars into new trains for road haul, transfer to other yards or railroads, or local delivery. These last two functions are distribution factors. A large classification yard usually contains three yard units: the receiving yard, into which trains are moved from the main line preparatory to sorting; the classification yard proper, where the sorting or classifying into blocks of common destinations takes place; and the departure yard, in which the sorted groups or blocks are made into trains and held pending main-line movement. Small yards consist of only one general yard, certain tracks within that one yard being assigned to receiving and departure purposes. See Figure 10.9.

The track lengths of the receiving and departure yards are based on the number of cars in the average- and maximum-length trains (at 50 ft per car) plus length of locomotive (50 to 60 ft) per unit, and caboose (40 ft); 200 to 300 ft are added as a factor of safety in stopping. The number of tracks is based in part on arrival rate of inbound trains but is governed fully as much by the rate of classifying, the rate at which cars can be taken from the receiving yard and sorted into groups or blocks, and assembled into trains in the departure yard. The final assembly of trains involves combining the blocks from various tracks in station orders, front-to-rear or rear-to-front depending on the set-out operation employed.

In the classifying unit, each track is assigned a given "destination," and cars going to that destination, and no others, are sorted onto it. Classifying rates vary with the method of operation. Small yards are flat switched, that is, the engine pushes and pulls a cut of cars in and out of the various tracks, sorting the cars onto the proper tracks at a rate of 30 to 60 cars per hour. Gravity yards make use of an artificial hump placed between the receiving and classification yard over which cars are pushed at a speed of about 3

[12]"Vertical Storage and Retrieval of Containers," by Alfred Hedifine, President, Parsons, Brinckerhoff, Quade, and Douglas, *American Import and Export Bulletin*, No. 1, Vol. 73, July 1970.

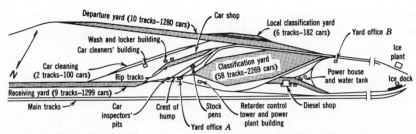

FIGURE 10.9 PLAN OF A CLASSIFICATION YARD. (COURTESY OF *RAILWAY SIGNALLING AND COMMUNICATION*, SIMMONS-BOARD-MAN PUBLISHING CO., NEW YORK, MARCH 1955.)

miles per hour, uncoupled, and allowed to roll into the assigned track. Gravity yards are seldom used where less than 1500 cars a day are to be classified.

In manually operated gravity yards, now largely obsolete, a rider mounts each car and controls its speed into the body track with a brake club in the handbrake wheel. The classifying or humping rate for such a yard has been 60 to 120 cars per hour.

Semiautomatic yards have retarder units placed on leads to groups of tracks. The retarders control car speed by pressing brake shoes, electrically or electropneumatically, against each side of the wheel as the car moves through the retarder. Tower operators judge the cars' speeds and, by push-button arrangement, control the amount of braking pressure that will be applied. In this type of yard, classification rates vary from 100 to 180 cars per hour.

Completely automatic yards use electronic and radar devices to measure the weight and speed of each individual car and automatically apply the proper retardation. The electronic computer in which the control is set thus takes into account temperature, wind effects, rolling resistance, and, by additional measurement, the distance the car must roll to couple to the cars already standing on a particular track. The safe coupling speed that will not cause damage to car or contents is about 4 mph. The operator needs only to push a button indicating the proper track for each car, and the switches are aligned and control exercised from there on by the electronic control. The most recent designs can feed the consist of a newly arrived train into a computer, possibly located hundred of miles away; the computer automatically prepares and distributes a switch list complete with proper cuts and track designations. Classifying rates of 200 to 300 per hour can be obtained when there are no delays due to external causes. It should be noted that technical know-how is already available to operate the locomotive pushing cars over the hump by remote control and uncouple them. Thus the human element can be removed almost entirely from this phase of the operation.

An essential part of every yard is a comprehensive system of communication. Use is currently made of telephone, loudspeaker, and talkback systems, radio, inductive telephone, teletype, television, and pneumatic tubes. Consists of trains are sent by teletype from the point of origin to the destination terminal so that switch lists and consists for the new trains to be built can be prepared in advance of the train's arrival.

The problem of yard location in an urban area includes proximity of the yard to industrial, waterfront, and other traffic sources and to other rail routes, land values and availability, taxation, and opportunities to expand. These are obvious and highly important. More significant to rail operation, however, is the location with respect to other yards on the system. Involved here are problems of traffic patterns, scheduling, and how far in advance of final destination, and in what detail sorting should occur. The trend today is for a few large strategically placed yards equipped with all modern car-handling devices.

Rail Passenger Stations   A rail station complex will include the platforms and platform tracks, concourse and other access ways, ticket sales, baggage and checkroom facilities, waiting rooms, rest rooms, and other amenities such as restaurants and sales booths, parking areas, and access through covered walks or tunnels to streets and to local modes of transport.

Stations for rail transport are of two general types—*stub* and *through*. The through station is, in effect, a way station with arriving trains continuing through to the stations beyond. The stub station is found primarily where the trains terminate their runs. Some stations have both stub and through tracks.

At stations with light traffic, the platforms are usually adjacent to the main tracks. Where traffic is heavy, especially where many commuter trains operate, the main line tracks will be augmented by platform tracks out of or diverging from main tracks or terminal lead tracks. The tracks that connect the platform tracks to the principal leads or to the mains are termed "throat tracks." A general rule is to have a 2.5 to 3: 1 ratio between a throat track and the number of platform tracks it serves. See Figure 10.10a.

There must be enough platform tracks to serve all trains scheduled to arrive or depart at a particular time plus a few additional to take care of off-schedule and extra trains. A track may also be reserved to park business cars or other special equipment.

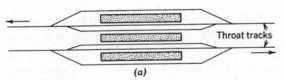

FIGURE 10.10A PLATFORM TRACKS:
"THROUGH" PLATFORMS AND TRACKS.

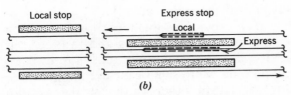

FIGURE 10.10B PLATFORM TRACKS: RAPID
TRANSIT LOCAL-EXPRESS SERVICE.

At wayside stations and rapid transit stations, the platform may be placed between two tracks, one track for movement in one direction; alternatively, two platforms are used, one platform for each track, the tracks being between the platforms. Figure 10.11 shows a cross-sectional view of a station on the Market Street Subway in San Francisco. Note the two levels of rail.

High density rapid transit lines, operating both local and express trains on separate tracks, have platforms between each pair of local and express tracks. Passenger can then cross the platform from a local to an express train (and vice versa) at express stops.

Minimum length of platform is based on the longest train anticipated (car length × number of cars plus locomotive) plus two or three additional car lengths for emergency situations and to provide a factor of safety in stopping the train. Additional length may be added to anticipate longer trains of the future.

Platform widths vary from 20 ft (6.1 m) if baggage trucks also use the platform to 13 ft (3.96 m) for passengers only. Wider platforms than these minimum are found in practice, especially on rapid transit lines.

The problem of locomotive release and servicing arises with stub end stations. This can be met by having additional track length beyond the stopping point and a crossover, so that the locomotive can be cut off and run through an adjoining track to the enginehouse. Many European and British stations use a transfer table to move the locomotive to the adjacent track.

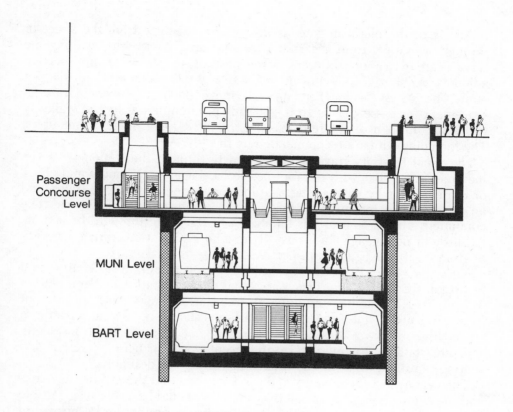

Scheme of typical station on Market Street Subway.

FIGURE 10.11 SCHEME OF TYPICAL STATION ON MARKET STREET SUBWAY, SAN FRANCISCO MUNICIPAL RAILWAY AND BAY AREA RAPID TRANSIT DISTRICT. (COURTESY OF *NOTES*, SPRING 1975, P. 1, PARSONS, BRINCKERHOFF, QUADE, AND DOUGLAS.)

Platforms can be either at rail level or car floor level. The latter are preferred when large numbers of passengers are to be on- or off-loaded. Using the rail level platforms, 1.8 seconds per passenger is required with 24-in. (61.44-cm) doors in commuter and rapid transit cars, 1.4 seconds per passenger with 3.4-ft (1.04-m) doors. Car floor level platforms reduce these times to 1.1 seconds per passenger with the 24-in. door and 0.8 seconds per passenger with the 4-ft (1.22-m) door. Entry to the platforms can be by turnstile or gate: 50 passengers per minute through a 5.3-ft (1.62-m) radius turnstile, 46 per minute through a gate.

Access to the platform gates is through a concourse from the street or through walkways from the station service area. When the waiting and service areas are on different levels, ramps or stairways are needed adjuncts. A 32-in. (81.28-cm) moving stairway will move 5000 people per hour, one 48 in. wide (1.22 m) will move 8000. An average rate of movement for passengers is 15 ft (4.57 m) per minute per foot of walkway width for through or intercity passengers, 30 ft (9.14 m) for commuters. The foregoing data are recommended by the AREA.[13]

The usual intercity traveler moves slowly through the station area. He may not be familiar with the routing, he has baggage to handle and check or retrieve, he may have a long wait for connections or delayed trains, and he may require information, food, and a comfortable place to sit. The commuter, on the contrary, is familiar with the route through the station, has little or no luggage, and is usually in a hurry. He wants direct access to or from local streets and transport.

These two types of traffic should be kept separate to avoid conflict and confusion. In large stations such as Grand Central Station or Penn Station in New York, commuter and intercity trains arrive and depart on different levels. In smaller stations, separate platforms should be used and traffic so routed that the two lines of movement do not cross. In some instances, separate stations are in use. In any event, clear, concise direction and routing signs and other means of channelization are desirable.

The coach yard for cleaning and servicing cars and the making up trains and the engine house for locomotive servicing should be as close as possible to the station at major and destination terminals. Trains can be turned on balloon or loop tracks or on wye tracks. Tracks may extend beyond stub stations on rapid transit lines for a trainlength or more to permit trains crossing over to the adjacent track for a return move.

Bus Stations   Stations for buses vary from tiny shelter sheds (if any) at outlying bus stops to elaborate multilevel structures in large urban areas. Much that was said about train stations—the need to keep commuter and intercity traffic separate, rates of movement, service facilities, amenities, and signing—applies equally to bus terminals.

Station platforms are usually of two types: those designed for berthing parallel to the platform and those in which the bus occupies a sawtooth berth or standing. Parallel berths are usually associated with commuter

[13]*Manual for Railway Engineering (Fixed Properties)*, Chapter 14, Table 1, p. 14-2-1, and *A.R.E.A. Proceedings*, Vol. 37, 1937, pp. 317.318, both published by the American Railway Engineering Association, Chicago, Illinois.

service where speed of movement is required for the riders and to obtain a high rate of equipment utilization.

Sawtooth berths are again of two types. One type forms a flat angle with the platform and has a pullout pocket in front of the bus as in Figure 10.12. This, too, can be used for commuting service because it permits a quick departure of the bus from the platform. The second sawtooth arrangement makes a greater angle with the platform and forms a pocket or standing for the bus outside the flow of traffic. The bus must back to enter the traffic flow. This arrangement is useful in providing many berths with a minimum of platform space. It is used primarily for intercity service where baggage must be on- and off-loaded.

The number of berths for commuting service is a function of total passengers per hour to be handled, loading and off-loading rates, and frequency and capacity of buses in use. The number of berths, N, required for a given number of passengers per hour, J, is

$$N = \frac{J(Bb + C)}{3600B}$$

where

$$J = \text{passengers per hour boarding at station}$$

$$B = \text{boarding passengers per bus in peak 10 to 15 minutes}$$

$$b = \text{boarding service time in seconds per passenger}$$

$$C = \text{clearance time between buses (door closing to door opening time)[14]}$$

For intercity buses, there must be enough berths to hold all scheduled arrivals and departures at any given time. The length of time a bus remains at its berth will vary with its schedule, amount of baggage, number of riders, and policies regarding crew changes and layovers. The intercity bus station should be close to traffic sources, but also it should have direct access to intercity routes.

[14]*Bus Use of Highways: Planning and Design Guidelines*, National Cooperative Highway Research Program Report 155, Transportation Research Board, National Research Council, Washington, D.C., 1975, p. 41, Table 19.

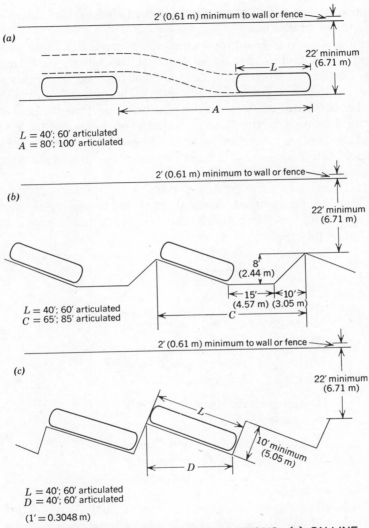

$L = 40'$; 60' articulated
$A = 80'$; 100' articulated

$L = 40'$; 60' articulated
$C = 65'$; 85' articulated

$L = 40'$; 60' articulated
$D = 40'$; 60' articulated

$(1' = 0.3048 \, m)$

FIGURE 10.12 BUS BERTH CONFIGURATIONS. (*a*) ON-LINE PLATFORM. (*b*) SAWTOOTH: PULL OUT. (*c*) SAWTOOTH: BACKOUT. (*a* AND *b* ARE BASED ON TRANSPORTATION RESEARCH BOARD, *BUS USE OF HIGHWAYS*, 1975, FIGURE 21, P. 40.)

Robert T. Aangeenbrug

FIGURE 10.13 BART REGIONAL RAPID TRANSIT STATION AT DALY CITY. (COURTESY OF PARSONS, BRINCKERHOFF, QUADE, AND DOUGLAS OF SAN FRANCISCO AND NEW YORK, PARSONS BRINCKERHOFF-TUDOR-BECHTEL, GENERAL ENGINEERING CONSULTANT TO BART WITH PBQ & D, INC. IN CHARGE OF THE DALY CITY STATION.)

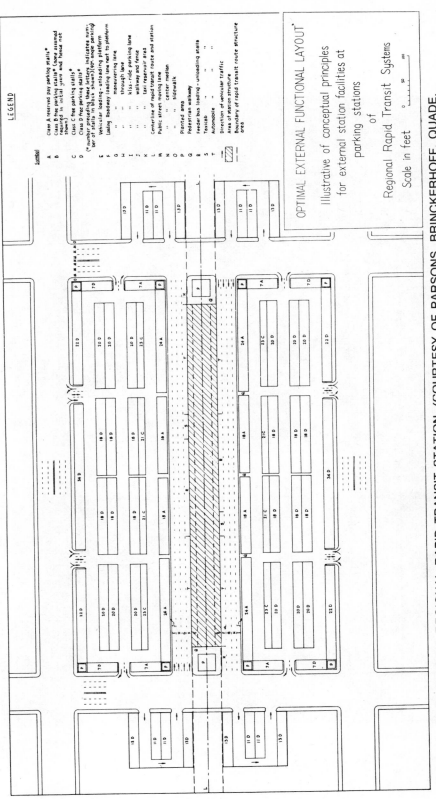

FIGURE 10.14 PLAN OF A REGIONAL RAPID TRANSIT STATION. (COURTESY OF PARSONS, BRINCKERHOFF, QUADE, AND DOUGLAS, FROM A PAPER BY HENRY D. QUINBY, HIGHWAY RESEARCH BOARD RECORD 114, 1966.)

Transit Centers   A new type of intermodal terminal facility is developing for urban transit, an outgrowth of the commuter railroad park-and-ride station. The transit centers are located at or near the ends of rapid transit lines and suburban railroads. Transfers between train or rapid transit, local buses, intercity buses, express buses, and automobiles are facilitated. A major element is the all-day parking area with close access to the line-haul mode. Additional facilities include covered train or bus platforms, covered walkways, moving stairs and sidewalks, bus and taxi load/unload ramps, and a kiss-and-ride circle for meeting and discharging passengers. Ticket sales, newsstands, rest rooms, and a limited amount of waiting area are often included. Adequate signs are important for directing newcomers to their destinations, trains, and buses. The stations are located at important junctions and should be far enough away from the CBD to make the interchange and train or express-bus trip worthwhile. At the same time they must be close enough to the CBD to serve a large demand area. That demand area can be greatly increased by feeder buses centering on the station. See Figures 10.13. and 10.14.

Examples are found in the 95th Street station of the CTA in Chicago where rapid transit, automobiles, city, and intercity buses meet and interchange with rapid transit trains. The Jefferson Park station on the Kennedy Expressway adds to the foregoing a direct bus connection serving O'Hare Airport. In Jersey City, the Journal Square Transportation Center provides for inclosed and weather-protected transfer between the PATH system trains and buses from 29 routes coming into the Square; a 600-car parking capacity is also provided.

Parking   Parking lots and garages are terminal facilities that perform a short-term storage function. They vary in form and complexity from street-level lots to complex multilevel structures. Most lots and structures are designed for driver parking, many have attendent parking, and a few have a mechanized parking system.

Ideal parking places the user in close proximity to his or her ultimate destination. Parking in a central area (CBD) gives direct access to that destination. Many buildings have allocated one or more floors or levels for parking by occupants and patrons. Parking in the central area can be difficult to reach over congested CBD streets and may represent an uneconomical use of land area except as a temporary measure or where an open spot (preferably landscaped) is desired in a building-congested area. An alternate location is at the edge of the CBD but a second mode of transport, a shuttle bus service for example, may be needed for access to CBD midpoints. Periphery parking is designed to keep cars away from the

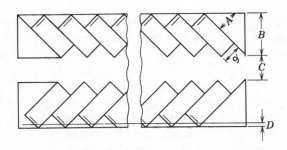

Parking Angle (A)	Depth of Stall (B)	Minimum Aisle Width (C)	Wheel Stop Clearance (D)
30°	17.6'	12'	2.7'
60°	21.2'	17'	3.0'
90°	19.0'	25'	3.3'

(1' = 0.3048 meter)

FIGURE 10.15 TYPICAL DIMENSIONS FOR PARKING LOTS.

CBD by providing parking facilities at outlying transfer points with mass and rapid transit.

The facility itself should be placed for easy access from expressways, collector-distributor, or principal main streets, but the entrance and exit should be away from congested streets and from the need to cross opposing lanes of traffic.

Lot capacity will be based on surveys that determine total demand, peaking, and rate of turnover. The space-hour (one space used one hour) may be used as a measure of demand and use. Demand should be determined throughout the day in 15 to 30 minute intervals to establish the percent of parkers that stay 15 minutes, 1 hour, 2 hours, a half day, a full day, etc. A 15-percent excess of available space over demand is desirable except where a known number of assigned or rented spaces is involved. The size of the lot or structure is a function of number of spaces required, the size of each space, angle of parking, and driveway-maneuvering areas. More cars can be parked per lineal foot of driveway with 90-degree parking with the number decreasing as the parking angle decreases. Room for wheel stops and for wall fenders must be provided. A common size of space is 8 × 18 ft (2.44 ×5.49 m), but this can be varied according to the size of cars to be accommodated. For shopping centers 9ft (2.74 m) widths are recommended. Driveway space should equal length of space plus 20 ±

TABLE 10.1 CRITICAL DIMENSIONS OF 1973 MODEL AUTOMOBILES[a]

Dimension	Range
Wheelbase	94.2 to 129.0 inches (239.2 to 327.6 cm)
Overall length	164.1 to 235.3 inches (416.8 to 597.7 cm)
Overall width	69.0 to 79.9 inches (175.3 to 202.9 cm)
Overall height	47.7 to 62.8 inches (121.2 to 159.5 cm)
Minimum running ground clearance	4.3 to 7.8 inches (10.9 to 19.8 cm)
Tread width (center to center of tires	51.5 to 64.5 inches (130.8 to 163.8 cm)
Bottom of front bumper to ground	4.4 to 18.4 inches (11.2 to 46.7 cm)
Turning diameter—wall to wall (outside front)	34.7 to 49.9 feet (88.1 to 126.8 cm)
Turning diameter—curb to curb (outside front)	31.5 to 46.0 feet (80.0 to 116.8 cm)

[a]"Parking Dimensions 1973 Model Cars," *Engineering Notes*, N. 731, Motor Vehicles Manufacturers Association, Detroit, Michigan.

percent. Figure 10.15 shows typical dimensions for common parking angles. Critical dimensions are car length, width, height, wheel base, and turning radius. Table 10.1 shows the range of these dimensions for 1973 model automobiles.

In multilevel garages, parking may be either along the access ramps on a flat floor with floor-to-floor ramps at the end(s) or at midpoint. Ramp parking permits easy grades and circulation but may involve long driving distances that reduce flow capacity. Ceiling heights (floor to floor) have varied from 10 to 12 ft (3.05 to 3.66 m) and ramp grades from 8 to 11 percent; lesser grades are preferred. Lighting and ventilation must be provided for enclosed areas and drainage to remove rain and snow drip. Underground garages may require the relocation of subsurface utilities. There should be a surge area off the street to hold waiting vehicles and, for attendent parking, a waiting room for patrons. If transient vehicles are a small portion of the total, the design should keep those separate from all-day parkers. Illuminated signs may be used to indicate to drivers those areas with open slots.

Airports The principal feature of an airport is its runways, a prime necessity in landing and takeoff for all present-day airplanes. A large, well-equipped airport will also have hangars for storage, inspection, and

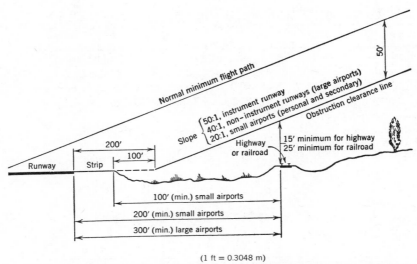

FIGURE 10.16 VERTICAL CLEARANCE FOR APPROACH PATHS. (ROBERT W. ABBETT, EDITOR, *AMERICAN CIVIL ENGINEERING PRACTICE*, VOL. I, WILEY, NEW YORK, 1956, P. 5-06, FIGURE 2.)

maintenance, fuel and oil facilities, fire-fighting equipment, hard standings for airplane parking, and taxiways leading from hangars, terminal building, or standing area to the runways. An operations center, control tower, offices, freight, and express platforms, ticket sales, baggage and waiting rooms, loading ramps and lounges, amenities for the passengers, and adequate automobile parking are also included. These facilities, in addition to their other functions, serve to concentrate passengers who are interchanging or transferring from one flight to another in plane-load quantities.

Airports require extensive land areas and clear air space with good visibility approaching those areas. Proximity to traffic sources and to main roads are further considerations. The large runways and approach areas and the noise nuisance of large jet operations are likely to place airports designed for their use in the regional category, with conventional-type feeder service to adjacent communities. Clearance and glide paths for various types of operation and classes of runways are shown in Figure 10.16.

Runway Length   Airports and the required length and allowable weights were classified earlier according to type of service: personal, secondary, feeder, trunk line, express, continental, intercontinental, and intercontinental-express. Today's classification is on a threefold basis: local interest

TABLE 10.2 AIRPORT CLASSIFICATION CRITERIA[a]

Airport Category	NASP Codes	Public Service Level (Annual Enplaned Passengers)	Aeronautical Operational Density (Annual Aircraft Operation)
Primary system		>1,000,000	
High density	(P1)		>350,000
Medium density	(P2)		250,000 to 350,000
Low density	(P3)		<250,000
Secondary system		50,000 to 1,000,000	
High density	(S1)		<250,000
Medium density	(S2)		100,000 to 250,000
Low density	(S3)		<100,000
Feeder system		Less than 50,000	
High density	(F1)		<100,000
Medium density	(F2)		20,000 to 100,000
Low density	(F3)		<20,000

[a]*National Airport System Plan*, Federal Aviation Administration Advisory Circular No. 150/5090-2, FAA, U.S. Department of Transportation, 25 June 1971.

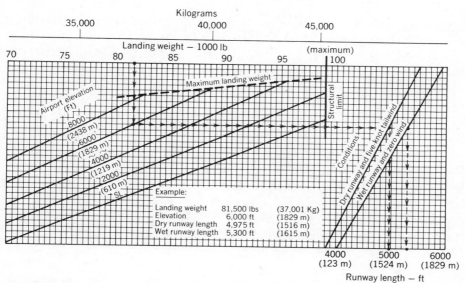

FIGURE 10.17a AIRCRAFT PERFORMANCE CURVE, LANDING (DOUGLAS-DC-9-30 SERIES). (*RUNWAY LENGTH REQUIREMENTS FOR AIRPORT*, AC 150/5325-4 CHG 8, FEDERAL AVIATION ADMINISTRATION, WASHINGTON, D.C., 8 NOVEMBER 1967, P. 20.)

airports, national system of airports, and military airports. The National
Airport System Plan designation is that shown in Table 10.2; it is based on
traffic volume.

Runway length is a function of many factors: whether the runway will
be used for landing, takeoff, or both, the type and weight of aircraft,
airport elevation, and anticipated wind and weather conditions. These
relations are presented graphically in the typical aircraft performance
curves of Figures 10.17*a* and 10.17*b*. The distance factor shown in Figure
10.17*b* reflects the cruising range established by the division of takeoff
weight among the empty weight of the aircraft, the payload, and the fuel
load.

Utility airports for small, personal aircraft—up to twin-engine piston

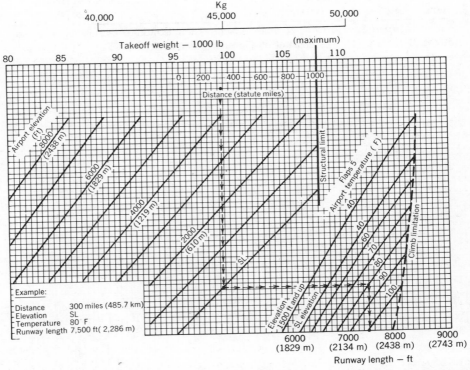

FIGURE 10.17b AIRCRAFT PERFORMANCE CURVE, TAKEOFF (DOUGLAS
DC-9-30 SERIES). ('*Runway Length Requirements for Airports*', AC 150/5325-4 CHG 8,
Federal Aviation Administration, Washington, D. C., 8 November 1967, p. 21.)

craft—are grouped into three categories according to weight and engine characteristics. Lengths vary from 2400 to 3500 ft (731.5 to 1066.8 m) at sea level and a temperature of 80 degrees F. Lengths must be increased at higher temperatures and altitudes. Minimum strip widths range from 100 to 150 ft (30.48 to 45.72 m) and runway widths from 50 to 75 ft (15.24 to 22.86 m). Strip width beyond the runway is a minimum of 200 ft (60.96 m). For more detailed information see *Utility Airports*, Federal Aviation Administration, AC 150/533-4A, November 1968. Other factors include airport altitude (roughly a 7-percent increase for each 1000 ft (304.8 m) above sea level), temperature (based on normal maximum for the hottest month in degrees F), and the weight-distance relation, that is, the distance (based on performance curves) is that distance the aircraft can fly from the takeoff airport to the next with a maximum payload and a minimum load of fuel.

Runway Capacity   Runway capacity is a function of the type of aircraft, prevailing winds and other climatic conditions, and is expressed separately in terms of Visual Flight Rules (VFR) and Instrument Flight Rules (IFR) operation. Practical capacity (PC) is reached when delays to departing commercial-type aircraft average 4 minutes during two normal adjacent peak hours of the week; 2 minutes of delay represent PC for small aircraft. Use is also made of the practical annual capacity (PANCAP). Air terminal capacity is further dependent on the number and configuration of runways and the number and configuration of taxiways to and from those runways. Practical capacity is further related to weather by assuming a 90-percent VFR and a 10-percent IFR operation.

Four traffic mixes are assumed. Mix No. 1 contains 90-percent Type D + E (light 1- and 2-engine piston planes) and 10 percent Type C (executive jet and 2-engine piston planes). Mix No. 2 has 30 percent Type B traffic (2- and 3-engine jets, 4-engine piston craft, and turboprops), 30 percent Class C, and 40 percent D + E. Mix 3 has 20 percent Class A (4-engine jets and larger), 40 percent Class B, 20 percent each for classes C and D + E. Mix 4 is 60 percent Class A and 20 percent each of Classes B and C.

For a single runway, 90 percent VFR, and twin-engine piston planes or lighter (Class 1 mix), the hourly capacity is 99 takeoffs and landings per hour using VFR, 53 using IFR, with a PANCAP of 215,000. Corresponding capacities for Mix 2 are 76, 52, 195,000; for Mix 3: 54, 44, and 180,000, and for Mix 4: 45, 42, and 170,000. For parallel runways 5000 ft (1524 m) or more apart, the combined capacities are double those for a single runway. Various typical runway configurations are shown in Figure 10.18. The reader is referred to reference 25 in the Suggested Readings of this

Runway Configuration		Runway Configuration	
Layout	Description	Layout	Description
 Single runway (arrivals — departures)		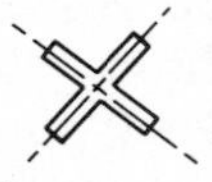 Direction of ops	Two intersecting in middle
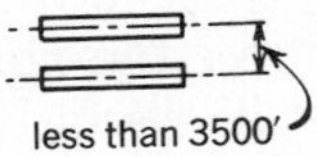 less than 3500'	Close parallels (IFR dependent)	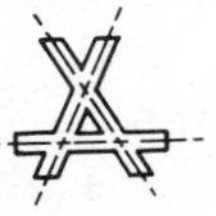	Three intersecting
 3500' to 4999'	Independent IFR approach/departure parallels	 5000' or more	Parallel and intersecting intersection points at ends
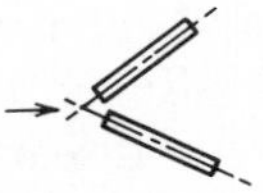	Open V, dependent, operations away from intersection	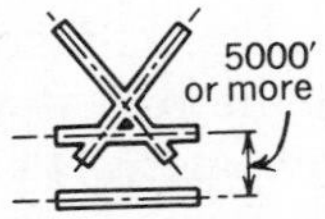 5000' or more	Parallel and intersecting
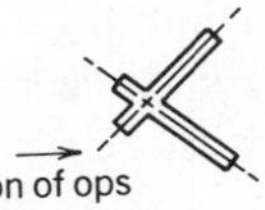 Direction of ops	Two intersecting at near threshold	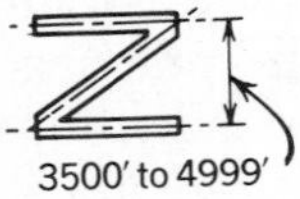 3500' to 4999'	"Z" configuration and parallel with both intersecting

FIGURE 10.18 TYPICAL RUNWAY CONFIGURATIONS. (*FROM AIRPORT CAPACITY CRITERIA USED IN LONG-RANGE PLANNING*, FEDERAL AVIATION ADMINISTRATION, WASHINGTON, D.C., 8 NOVEMBER 1969, PP. 6-10.)

chapter for the capacities of more complex layouts and the traffic mixes from which the foregoing data were obtained.

Runway orientation must be such as to permit takeoffs and landings with or into the wind and with 95 percent of such operation free of cross-wind components exceeding 15 mph (24 kph) (20 to 30 mph (32 to 48 kph) if only heavy transports are involved).[15] This requires a study of prevailing winds, usually plotted on the radial lines of a so-called wind rose, which shows the percentage of time winds of different velocities blow from the radial directions. To achieve freedom from cross winds, more than one

[15]Phillips Moore, "Airport Engineering," in R. W. Abbett, *American Civil Engineering Practice*, Volume I, Wiley, New York, 1956, pp. 5.11.

runway may be required for alternate use as the wind changes direction. Runways should also be placed in such a way as to avoid conflict between two being used simultaneously, especially if one or more are designed for instrument landings—as at least one should be at the larger ports.

Access to the planes from the terminal building affects the overall speed of transport and the passengers' convenience and comfort. Characteristically the planes are reached via a concourse and then out into the weather. Moving sidewalks have shortened concourse transit time and effort in some ports. At large airports planes are brought close to overhanging galleries that have telescoping walkways that can be extended to the plane door. Airports with more than one terminal building use various types of buses and people movers to facilitate passage from one building to another. Passengers' luggage is carried by conveyor from check-in points to baggage rooms for loading into capsules or containers that are later moved by tractor train or truck to planeside and into the plane by conveyor or lift-truck platform.

An additional requirement for today's society is a provision for conducting security checks on persons and their carry-on luggage to prevent bringing weapons or other devices on board that could be used in a hijacking attempt.

The Dallas–Fort Worth air terminal, the world's largest, opened in 1973, has a series of multilevel terminal buildings adjacent to a 10-lane central highway. Parking areas near each building reduce walking distance to a minimum. Beyond the buildings are major runways parallel to the central highway with parallel taxiways. Transportation between buildings is provided by an automated, electric, rubber-tired, Airtrans system using a troughlike guideway. The cars seat 17 persons (with room for 24 standees) and average 17 mph (27 kph).

Service and Repair   Service and repair facilities are usually provided by the carrier except in water transport, where heavy repairs are made by contract. Drydock companies give the ship's hull, rudder, and propellers (as well as its internal fittings) a thorough overhaul. Similarly, lighter vessels and barges are hauled from the water on marine railways. Running repairs on rail equipment is performed in engine houses and car-repair shops with major repairs being made in centrally located heavy-repair or back shops. Airlines maintain elaborate inspection and repair facilities in company hangars at airports. Bus and truck maintenance is performed at company garages at principal terminals: some of it is contracted. A few large truck lines have erected intermediate maintenance garages along their routes.

Locomotive fuel, sand, water, and lubricants are placed at the engine-house site and at other convenient terminal and on-line points. Fuel for water transport is usually obtained at privately owned shore points. At airports, fueling is performed by an oil distributor having a concession contract with the airport. Trucks and buses use their terminal facilities for fuel and oil but also make use of regular gasoline stations along the road that perform a similar function for privately owned passenger cars.

QUESTIONS FOR STUDY

1. A railroad inbound freight house is planned to handle 100 tons per day. What area of floor space should be provided? How many car spaces?
2. Assuming platform length has been designed on the basis of adequate tail-board space for the delivery trucks to distribute the inbound freight of Question 1, how many tracks will be needed and how many car spots on each track if the tracks are pulled only once a day and average car contents weigh 10 tons?
3. What is the floor space required of a transit shed to handle two 10,000-cargo-ton ships at a time, assuming all the cargo will first be delivered to the shed and stacked in two layers?
4. How many receiving yard tracks would be needed to hold 80-car trains arriving at the rate of two per hour during one 8-hour shift (with light, scattered arrivals throughout the rest of the day), using a classifying rate of 120 cars per hour?
5. If the 120 cars per hour of Question 5 are uniformly distributed in a ratio of 10, 10, 20, 20, and 40 percent over each of five classification groups or blocks, how many tracks and of what standing-room length would be needed in the classification yard to handle the traffic of the 8-hour shift?
6. How many tracks would be required in the departure yard, assuming 100-car trains outbound and the trains moved into the departure yard as soon as sufficient blocking becomes available? Assume each train spends 1 hour in the departure yard having the air tested, the caboose and locomotive attached, and waiting for orders to move onto the main track.
7. Explain the presence of grain elevators, ore docks, and coal docks in the light of the principle of concentration. What do these facilities have to do with questions of car, truck, and barge supply?
8. Given a containerport that receives two 800-unit containerships each week that discharge and load their full capacity. Forty percent of the traffic arrives and departs by highway, 60 percent by rail. Both inbound and outbound traffic average two days in the marshalling area. Develop a set of specifications for the number of berths, the length of wharf, the size of the marshalling (in acres) of the marshalling area, and a suggested type and capacity for the TOFC portion of the port. Mention any other features that should be considered.

9. A union truck terminal receives 180 tons of freight daily inbound from 90 semitrailers that carry an average of 20 tons per vehicle. How many delivery vehicles will be needed to make local distribution of that 1800 tons if each truck averages 5 tons per load and requires 4 hours for each round trip? What reduction in number of vehicles would be made if a delivery vehicle would carry two additional tons?

10. What are the advantages of having industrial activities concentrated in a limited number of zones in an urban area?

11. Diagram or show schematically all the coordination involved in bringing iron ore from a mine in Labrador to a steel-mill furnace in Pittsburgh, Pennsylvania.

12. Diagram or show schematically possible systems of coordination in moving petroleum from Texas oil fields to New Jersey refining points.

13. What are trackage rights and what conditions are conducive to this type of coordination? Why?

14. What advantages do belt lines or switching railroads offer as compared with reciprocal switching?

15. Using the approximate cost data contained in this chapter, tell how much circuitry of line would have to be saved to justify moving railroad cars in a 30-car-capacity ferry across a body of water.

16. A flatcar weighing 18 tons with a load capacity of 60 revenue tons holds two highway trailers, each containing 20 revenue tons and weighting 8 tons apiece when empty. What changes in deadload-to-payload ratios are made in the flatcar capacity by this use for piggybacking? If the result is disadvantageous, how can it be improved?

17. Explain how the development of Plans II TOFC could lead to the eventual elimination of large-capacity freight houses. What would be the possible effect of Plan IV?

18. A single runway is being designed for a traffic consisting of 20 three-engined jet planes, 144 twin-engined transport planes, and 30 light twin-engine piston planes. What practical peak hourly and annual capacity would this runway provide if the airport is open 18 hours each day under (*a*) VFR operation and (*b*) IFR operation?

19. Determine by use of performance charts the minimum length of dry runway required for a DC-9-30 series aircraft weighting 80,000 lb at an elevation of 2000 ft, an average hot month temperature of 80 degrees F, and a weight-flight length distance of 400 miles; also determine the minimum length of landing runway for the same aircraft with a 5-knot tail wind on (*a*) a dry runway and (*b*) a wet runway with zero wind.

SUGGESTED READINGS

1. "Yards and Terminals," Chapter 14, *Manual for Railway Engineering (Fixed Properties) of the American Railway Engineering Association*, American Railway Engineering Association, Chicago, Illinois.

2. E. W. Coughlin, *Freight Car Distribution and Car Handling in the United States*, Association of American Railroads, Washington, D.C.
3. *Principles of Freight Terminal Operation*, American Truckers Association, Washington, D.C., 1950.
4. Wilbur G. Hudson, *Conveyors and Related Equipment*, 3rd edition, Wiley, New York, 1954, Chapters 1, 12, 13, 14, and 22.
5. "Design of Ore Docks," *Proceedings of the A.R.E.A.*, Vol. 36, 1935, p. 255 ff., American Railway Engineering Association, Chicago, Illinois.
6. Harold M. Mayer, *The Port of Chicago and the St. Lawrence Seaway*, University of Chicago Press, Chicago, Illinois, 1957.
7. C. L. Sauerbier, *Marine Cargo Operations*, Wiley, New York, 1956.
8. R. W. Abbett and E. E. Halmos, "Harbor Engineering," in R. W. Abbett, *American Civil Engineering Practice*, Volume II, Wiley, New York, 1956, Chapter 21, especially the section entitled "Marine Terminals" by Maurice Grusky.
9. *Manual for Railway Engineering (Fixed Properties) of the American Railway Engineering Association*, American Railway Engineering Association, Chicago, Illinois, Chapter 14, "Yards and Terminals."
10. "Hump Yard Systems," *American Railway Signaling Principles and Practices*, Signal Section, Association of American Railroads, Chicago, Illinois, Chapter 21.
11. Alonzo DeF. Quinn, *Design and Construction of Ports and Marine* Structures, McGraw-Hill, New York, 1961.
12. Phillips Moore, "Airport Engineering," in R. W. Abbett, *American Civil Engineering Practice*, Volume I, Wiley, New York, 1956, Chapter 5.
13. R. Horonjeff, "Planning and Design of Airports," McGraw-Hill, New York, 1962
14. D. C. Wolfe, "Huge Oil Pier Built in Open Water," *Engineering News-Record*, May 25, 1950, pp.34-39, McGraw-Hill, New York.
15. "The New Orleans Union Passenger Terminal," various articles, *Railway Age*, Simmons-Boardman, New York, April 26, 1954, pp. 22-31.
16. "P. R. R. Unveils New Ore-Unloading Terminal," *Railway Age*, March 15, 1954, pp. 45-47.
17. "Modern Lake Port Transfer Facility Open for Business at Toledo, Ohio," *Railway Age*, May 1, 1948, pp. 32-37.
18. "Phosphate—from Train to Ship," *Railway Age*, August 14, 1948, pp. 60-63. "Canton Railroad Expands Ore Docks," *Railway Age*, December 15, 1952, pp. 52-53.
19. Henry D. Quimby, "Coordinated Highway-Transit Interchange Stations," *Highway Research Record 114*, Highway Research Board, Washington, D. C.
20. Alfred Hedifine, Consultant-Associate, Parsons, Brinckerhoff, Quade, and Douglas, "Storage and Retrieval of Containers," *American Import and Export Bulletin*, Vol. 73, No. 1, July 1970.
21. Gene Dallaire, "Dallas-Fort Worth: World's Largest, Best-Planned Airport," *Civil Engineering*, American Society of Civil Engineers, July 1973, pp. 53-61.

22. *"Transportation and Parking for Tomorrow's Cities,"* Wilbur Smith and Associates under commission from the Automobile Manufacturers Association, July 1966.

23. "Opportunity Park Garage," *PBD&Q "Notes"*, Parsons, Brianckerhoff, Quade, and Douglas, *Engineers*, Summer 1972, pp. 3-11.

24. Walter C. Boyer, "Containerization—A System Still Evolving," *Journal of the Waterways, Harbors, and Coastal Engineering Division, Proceedings of the American Society of Civil Engineers*, November 1972, pp. 461-473.

25. *"Airport Master Plans,"* Federal Aviation Administration, U. S. Department of Transportation, Washington, D. C., February 1971.

26 A. W. Thompson, "Evolution and Future of Airport Passenger Terrminals," *Journal of the Aero Space Transport Division, Proceedings of the American Society of Civil Engineers*, Vol. 90, No. AT2, Proceedings Papers 4070, October 1973.

27. "Intermodal Transfer Facilities," *Transportation Research Board Record 505*, Transportation Research Board, National Research Council, TRB, 1974, ISBN 0-309-02298-3.

28. *Use of Containerization in Freight Tnsport. Highway Research Record No. 28*, Highway Research Board, National Academy of Science, Washington, D. C., 1969.

29. Eric Rath, *Container Systems*, Wiley, New York, 1973.

Chapter 11

Operational Control

FUNCTIONS OF CONTROL

Definition and Application Operational control is the regulation exercised over vehicles and traffic to attain the maximum in safe and efficient utilization of plant and equipment. The control may be simple or highly complex.

Control is exercised to achieve safety and dependability of movement. Contact (collision) between vehicles must be avoided. Coupled with this is the need to move the vehicles as quickly as possible with the least delay for early and prompt arrival. These goals are sometimes conflicting. For example, vehicles cannot travel too closely together with safety at high speed. Responsible operators make safety paramount although the stress of competition and heedlessness of individuals sometimes work to the contrary.

In addition to safety, dependability, and speed, there is the goal of maximum realization of traffic capacity. Involved here are maximum-tonnage ratings (loadings)—based on principles earlier described—maximum permissible speeds, scheduling, and the effective use of route capacity. Principles governing speed and route capacity have also been considered earlier.

Control includes keeping movement records of all the vehicles—trains, airplanes, ships, trucks, buses—directing their movement with regard to meeting and passing (where appropriate), calls at sea and at airports, and

414

pickups and setouts of cargo units and traffic. Control also includes forwarding information for operational and planning purposes regarding movements underway and contemplated.

Means of operational control include rules and regulations, standard operating procedures, use of signs and pennants, signals, communications, records, and reports. Closed-circuit television and electronic computers and data systems perform surveillance, vehicle identification, and fully automated control of all or a portion of a system and provide information that is needed for day-to-day decision making.

Supervising Agencies Operational control is exercised by different groups under different situations. Railroads, conveyors, cableways, and pipelines maintain their own control. Private automobiles are subject to the safety rules of the road, signs and lights placed by communities and highway departments, and supervision by local and state police. Trucking and bus lines, in addition, control their operations through their own dispatchers and supervisors. Shipping control emanates from admiralty law, lights and markers set out by U. S. Army Corps of Engineers, the supervision of the U. S. Coast Guard, and from the ship line's own dispatchers and supervisors. Airlines are controlled by company dispatchers, by terminal control-tower personnel, by runway lighting, various radio-type navigation aids, and by Air Traffic Control Centers.

A major problem has been how much control should be made automatic and how much left to the individual operator. In some systems the operator may have only a monitoring function, sometimes combined with the capability to override the automated controls in the event of an emergency or change of plan. The trend thus far has been to give the operator or dispatcher as much electronic and automatic help as possible but to let him make the final decisions. The problems involved are the increasing complexity of control apparatus, sonic and supersonic speeds of some carriers, and increasing traffic densities in media of limited extent versus the reaction time and ability of the human mind and body to comprehend and make decisions. The solution to these problems lies within the realm of human engineering, where the psychologist and engineer meet on a common ground.

COMMUNICATIONS

Use Primitive systems of operations rely on customs, codes of rules, signs, pennants, and orders. Modern control of operations cannot exist without an adequate system of communications. Telegraph, and later telephone, sufficed for early transport, but today's operations require extensive systems of telephone, radio, microwave, teletype, and even television. Ship-to-shore and ship-to-ship telephones provide contacts on the Great Lakes, inland rivers, and, to some extent, the high seas. Railroads and pipelines use leased commercial lines with carrier circuits superimposed to give more paths or channels or they install their own lines and carrier channels. Microwave is used by these modes to provide reliable communications over long distances regardless of weather.

Pole lines are susceptible to damage by avalanches and snow slides, by floods, by wind, and by ice storms. Railroad operations often have been brought to a complete standstill for several hours to several days because of such disasters. Reliance in the past has been placed on amateur and police radio for emergency communication but microwave installations, being relatively free from such interruptions, are replacing wire lines, beginning in those areas most frequently experiencing severe storms.

Dispatching   A primary use of communication is for dispatching—directing and keeping track of vehicular movements. Taxicabs, contractors' trucks, service vehicles, industrial fleets, and terminal and over-the-road motor freight vehicles are under the guidance of central dispatchers who maintain radio contact with each vehicle, record movements on a log sheet, and direct the vehicles to successive tasks and new traffic contacts as required. Without radio, the drivers call from checkpoints during road haul or as the assigned task is completed.

Shipowners and agents at ports of call are given advance letter, cable, and radio information concerning the expected arrival and cargo of deep-water ships. This is later verified and revised by radio reports from the ship itself as it nears the port. Where difficult harbor approaches (or strong unions) prevail, the ship takes on a local pilot at the harbor (or canal) entrance to guide the vessel in and out. Orders and requests for tug service are made via ship's radio or on-shore agents.

Within the harbor, the harbor master assigns berthing locations and priorities at public wharves and anchorage locations in the roadstead. When fog or storm conditions make harbor operations unsafe, he may require all ship movements to halt. He keeps informed of ship movements and locations by radar as well as by port documents and radio contact with individual ships.

Bulk-cargo vessels on the Great Lakes maintain contact with each other, with their terminal and intermediate offices (which direct the ship to the desired ports), and with lock and Coast Guard personnel. Knowing the arrival time, speed, type, and capacities of ships makes it possible for ore and coal docks, grain elevators, etc. to be prepared with sufficient cargo and supplies on hand for prompt loading and refueling and for minimum time in port. At the destination point, sufficient railroad cars, barges, or storage space will be held available.

Operational Control—Airways Flights along and across the Federal System of Airways are under the direct supervision of 20 air traffic control centers strategically located across the country.[1] Before leaving the ground, the pilot files an approved flight plan and adheres to that plan while in the air. He makes regular check-ins to the Control Centers and to his own company dispatchers, who log the progress of the flight, stating his location and conditions of flight. Commercial airlines usually have their own dispatchers in direct touch with each flight and also coordinate activities with the Control Centers. Weather information, air-lane conditions, and the ceiling and traffic at the next point of landing are given to the pilot. Permission to make changes in the flight plan as to destination, altitude, or direction must be secured from the appropriate Control Center.

As the pilot nears a point of landing, the flight comes under the control of an airport control tower, he then receives landing (or stacking) instructions on weather, ceiling, priority of landing, and runway assignment.

Airport Traffic Control Towers, using light signals, radar, and ground level radar, control aircraft in and on the immediate vicinity of airports. Flights are picked up and monitored when they are within about 35 miles (56.3 km) of an airport control tower. Full control is assumed when the aircraft is about 7 miles (11.3 km) out. Local control is exercised over VFR traffic in and near the airport pattern. Ground control directs movements of aircraft while they are on the ground. Arrival and departure clearances and instruction to maintain separation of arriving and departing IFR flights are given.

The "speed-of-sound," high-flying jet transports use principally three transcontinental routes, 40 miles in width, between elevations of 24,000 and 35,000 ft to separate them from military and defense aircraft that also fly at those altitudes. The jets are separated by radar observation from installations of the Air Force Defense Command with special Air Route Traffic Control personnel assigned specifically to the control and safeguarding of jet commercial transport. No aircraft can enter or cross these lanes without an Air Route Traffic Control clearance unless the plane has a transponder, a radar beacon that gives on a scope in the aircraft a radar picture of the surrounding air space, more positively and readily

[1]Seven more serve Hawaii, Alaska, and United States territorial routes.

identifiable than simple radar scanning. For a military plane not so equipped to enter or cross the commercial jet lanes is a court-martial offense. Flights in and out of major airports are directed by 386 Airport Traffic Control Towers.[2]

Air Route Traffic Control Centers are primarily responsible for directing all aircraft operating under Instrument Flight Rules in controlled air space. They so direct traffic as to keep a 1000-ft (304.8-m) vertical separation between and a 10-minute separation from planes at the same elevation. They advise aircraft of potential hazards to flight, transmit weather advisories, and advise of anticipated delays, and give navigation assistance —permission for change of course or altitude—to avoid thunderstorms. Individual centers control IFR flights within their jurisdiction and coordinate with adjacent centers to maintain a safe and orderly flow of traffic.

Train Dispatching   Railroads operate under one of the most elaborate of dispatching systems. A train dispatcher will have control of all train movements in an assigned territory of 100 to 500 miles (161 to 805 km). Operators at train-order and block-station offices along the line report to him by wire the time each train passes his office. The dispatcher records the times for each train on a train sheet or log. Trains run according to timetables, rules, and signals. Situations arise regarding the movements of extra trains, handling of traffic, and emergencies not covered by the rules. The dispatcher then issues train orders to the trains involved, handed on to the train and engine crews by the wayside operators. The dispatcher can supersede any previous orders or rules and can annul or change schedules with a train order. Dispatchers and wayside operators can also have direct radio communication with train and engine crews for the exchange of information on the progress of the run.

Centralized Traffic Control   Centralized Traffic Control (CTC) provides for the direct operation of all switches and signals over a given territory varying from a few to hundreds of miles in extent. The dispatcher sits before an illuminated track diagram and panel control board, Figure 11.1. The diagram shows by lights the location and movement of trains and the position and aspects of switches and signals. By controlling these elements through buttons and levers on the panel board, the dispatcher is able to effect the most advantageous meets between trains at the various sidings, often without delay to either train. The system is applicable to

[2]"Air Transport 1974," Air Transport Association of America, Washington, D. C., 1974, p. 32.

FIGURE 11.1 "TRAFFIC MASTER" CENTRALIZED TRAFFIC CONTROL SYS-
TEM, BALTIMORE AND OHIO, AKRON, OHIO. (COURTESY OF GENERAL
RAILWAY SIGNAL COMPANY, ROCHESTER, NEW YORK.)

multiple-track operation but has found its principal use in increasing the
traffic capacity of single-track lines, often deferring indefinitely the need
for double tracking or permitting the reduction of multiple-track to single-
line operation.

Early CTC systems used a pair of wires from each switch and signal
mechanism to the control board. With any number of units and distances,
the cable and equipment became too unwieldy. Distance of no more than
20 to 30 miles (32 to 48 km) could be controlled. The introduction of coded
impulses sent over one pair of wires with code-following relays at each
switch and signal responding only to its own code, greatly increased
capacity, and lengthened distances over which control could be extended.
Even greater distances became available by superimposing additional
carrier paths on the line wires. The same susceptibility to wind and ice
storms, experienced by communication lines, holds true for signal-control

lines. One solution has been to bury the wires underground. However line wires can be eliminated entirely by use of radio and microwave. Radio signals in the very-high-frequency range (VHF) are modulated to carry codes as did the line wires. Each field station must have code following, receiving, and sending equipment and a source of power supply. It should be noted that not only is an activating code sent to the field equipment but a response must be transmitted back by each switch and signal location to indicate on the control board that it has responded properly to the coded instruction.

Recent developments permit a route to be set by manipulating only a few panel levers at a master console. The console activates all switches and signals involved in the proper sequence instead of moving individual controls for each switch and signal as in earlier models.

Yard Operation   In yard operations, train consists are sent by teletype to the next yard before the train's arrival. Switch lists indicating the proposed assigned track for each car (by car numbers) are prepared and sent by teletype to the hump-master's office. Or electronic tapes containing the same information serve as inputs to a computerized control that automatically sets up the tracks and routes leading to them for each car. Inbound trains are advised by radio of the receiving yard tracks to enter. Television scanning permits recording the car numbers and initials as inbound and outbound trains enter and leave the yards, serving as a prime record or as a check on electronic transmission of train consists. Yard locomotives are in contact with the yardmaster by radio, inductive telephone, or talkback loudspeaker systems and can report progress of their work, explain delays, and receive instructions on the spot. Car inspectors with walkie-talkies maintain similar contact. Transfer and switch runs use radio to keep the yardmaster advised of their progress, expected time of arrival, number of cars being brought to the yard, and cause and extent of any delay. Pneumatic tubes carry waybills from the inbound yard to the yard office and from the yard office to the departure yard.

Other Railroad Controls   Railroad mainline switches or turnouts as well as movable point crossings, drawbridges, and signals are usually controlled, and routes set up, at switch or interlocking plants. The operations of switch, lock, and signal controls are so interlocked that the movement of each can only be made in a safe and predetermined order. Interlocking is designed to require all parts of a route and the signals governing it to be properly aligned and set with no conflicting routes and no conflicting signals. Routes must be set and all obstructions such as derails, open bridges, and open switches removed before a signal can be cleared to permit movement of trains. Early interlockings had the operating levers only mechanically interlocked. Modern plants depend on all-relay control and have, in addition, interlocked protection of the switches and signals outside the plant building by means of interlocking and control circuits superimposed upon the track circuits. Lower costs, more pushbutton operation, faster operation, and less maintenance are obtained with solid-state (transistorized) systems to transmit controls and indications.

Where a switch plant becomes complicated with many crossovers and diverging routes, the manual setting up of a route is time-consuming and cumbersome. Entrance-exit systems of interlocking permit an entire route with all switches and signals to be aligned automatically merely by the pushing of a button at each end of the desired route.

An innovation in switch control is having a route automatically set up by a train as it approaches a diversion point. The switch-operating mechanism responds inductively to a particular frequency emitted by a small transmitter. Trains taking one route emit a frequency for that route to which the switch responds. Trains requiring the switch in the opposite position emit a different frequency and get a corresponding change in switch position.

The radio control of trains has been successfully carried out in field operation. The stage is thus set for a completely automatic operation of trains, especially those in rapid-transit service, with perhaps an attendant on the train but with no train or engine crews. A variety of inductive systems are also available.

Automatic Car Identification   A recent development, Automatic Car Identification or ACI, uses an electronic trackside scanner to read car numbers and initials that are represented by different arrangements of red, white, and blue strips on the sides of cars. The scanner is lighted by the approach of the train and sends a beam to the side of the passing car. Light from the strips or labels is reflected back to the scanner head, which analyzes and checks the color code. These data are then transmitted to a computer storage for the programmed preparation of switch lists and passing reports, and to provide information on the location and movement of any particular car.

This system of identification is also being applied to container terminals. Scanners mounted on small trucks move up and down a container/trailer holding or marshalling area; all units on hand are scanned and recorded on tape or transmitted via radio.

Centralized Transport Control   The dispatching of oil shipments by pipeline is similar in many ways to train dispatching. A dispatcher is in direct contact with and has control over the intermediate stations along the line. He or she sends orders, patterned after train orders, to each station regarding time, volume, pressure, temperature, viscosity, etc. that will govern a particular movement of traffic. The station operators keep him and adjacent stations informed of the movement and operational details at their stations. An upstream station, for example, cannot begin pumping until the station next in line, downstream, has advised of its readiness to receive and handle the flow. In the event of an emergency, the upstream station may be ordered to stop the flow and accumulate what it is receiving in storage tanks. Systems of remote control, corresponding to CTC for railroads have been developed for one-man control of an entire pumping operation.

Where centralized control is used for pipeline operation, the panel board before the dispatcher contains data regarding the operation of pumps at the booster stations, position of valves, bearing temperatures, pressures at selected points, etc. Flow information is shown by illuminated flow arrows. Telemetering and switching equipment indicate instantaneous changes in the type of material being pumped and permit diversions to required destinations. This type of Centralized Transport Control provides the means for individual control and operation of valves and pumps through computerized and automated control and routing of products.[3] The application of these same methods and principles to one-man or automated control of conveyors and aerial tramways is equally possible.

Systems of centralized control may be used to operate belt conveyors such as the complex systems that convey grain from point to point in a grain storage terminal. Control is exercised from one central board or console. The control elements are so interlocked that the first conveyor flight will not operate until all flights in the planned route are operating and ready to receive cargo.

Surveillance and Monitoring   Traffic on urban expressways has been expedited in several cities by means of various systems of surveillance; these include the use of helicopters, watchers stationed at critical locations, closed-circuit TV, and automatic sensor systems. The Chicago Expressway Surveillance Project combines the sensors of an automatic detection system with ramp monitoring and control that are exercised through a digital computer at 48 ramps along 90 miles (144.8 km) of expressway. Ramp

[3]"Centralized Transport Control for Pipe Lines," from *Photographs of Progress*, prepared for the University of Illinois by the Union Switch and Signal Division of Westinghouse Airbrake Company, Swissvale, Pennsylvania.

signals, responsive to various metering rates, restrain entry when express-way lanes are congested and permit entry when traffic density reaches a suitably low level. With multiple lanes, traffic counts for any lane can be obtained by means of inductive loop detectors in each lane. These may be supplemented by overhead ultrasonic detectors.

Traffic flow data evaluated through surveillance procedures may be used to flash messages on illuminated signboards to advise drivers on speed, lane, or route changes or of emergencies ahead.

Emergency communication between drivers and traffic control agencies can be set up via wayside telephones, either conventional or pushbutton speakers and, more recently, by use of Citizens Band radios. Wayside phones too often suffer from vandalism.

Bus operations are usually controlled by inspectors riding the buses or following their routes in automobiles. Radio, installed in some vehicles, enables a dispatcher to maintain communication with individual units. Radio contact is an essential feature of the Dial-a-Bus systems whereby buses can be directed to specific patron requests. The Chicago Transit Authority has been experimenting with an Automatic Vehicle Monitoring System (AVM) that uses radio equipment and a central on-line computer to monitor the operation of portions of its bus system. Electronic sensor "signposts" placed at intervals along the routes report the time of passing of a vehicle and its location. Two-way radio communication and a secret driver emergency alarm are provided. Better adherence to schedule is given by the system, which also increases safety, deters crime, and alerts supervision to operating difficulties.

Both intercity rail and urban rapid transit have been operated without human control. This is accomplished through track circuits or through computer control combined with track circuit or trackside sensors. Computer/sensor control of more exotic systems such as the Transit Express-way and a variety of personalized transit systems are similarly automated. All of the foregoing systems of signals and controls have been functionally and economically improved by the introduction of transistorized solid-state devices.

Nondispatching Communications   There is, of course, a wide area of nondispatching communication involved in transportation. Traffic solicitation, weather reports, reservations for passenger space, ordering cars, tracing, passing reports, maintenance operations, and administrative matters require a constant use of all available facilities. Shippers and carrier personnel issue holding and reconsignment orders on traffic already en route or in temporary storage. Consists of ships, trucks, and trains are sent ahead by radio, teletype, or computer from the point of origin to the point of destination. Stores and purchases make demands for prompt, personal contacts.

Electronic information systems collect data from all over a system for transmission to a central computer where data are stored and processed into meaningful reports for the guidance of day-to-day operations and managerial decision making. Such systems are especially useful in supplying information on the location of vehicles, cars, ships, and planes. An electronic data collection and processing system called TRAIN has been developed by the AAR to help its Car Service Division in the distribution of cars throughout the United States. The TRAIN system when combined with ACI provides an exceedingly useful tool for the rapid and efficient distribution of cars.

INTERVAL CONTROL

Interval Systems   A primary function of operational control is to prevent contact in the form of collisions between vehicles, especially where the operator of one vehicle may be inherently unaware of the presence of the other—as with trains, high-speed aircraft, or ships in a fog. Three general systems are followed to provide a safe interval between vehicles: (a) the time-interval system, (b) the space-interval system, and (c) the see-and-be-seen system. New systems still under development include the so-called "rolling block," by which a following train or vehicle "senses" by radar or through the guideway circuitry the presence of a train or vehicle ahead and speed controls are initiated to maintain a safe distance. Somewhat similar is a system in use by the BART rapid transit system whereby the location of a train is made known to a central controlling computer by its presence on the rails in relation to a series of inductive cable loops laid parallel to the track.

The see-and-be-seen system which is found mainly in rural highway, waterway, and slow-moving-aircraft operation, leaves the avoidance of contact up to the individual pilots and drivers. This places more demands on the operator's skill than can be expected from human effort at very high speeds, where disaster may occur before the driver or pilot is able to act. It also is of little help when fog, storm, or darkness obscures visibility. At such times, ships can rely on radar to warn them of other vessels. This is not always enough, as evidenced by occasional collisions between radar-equipped ships.

The time-interval system keeps vehicles separated by a predetermined number of minutes. Distancewise, the separation is variable, depending on the speed of the vehicles. Timetable scheduling is a familiar example of control by time interval. In air transport, aircraft of the same elevation and direction are supposedly kept 10 minutes apart. In a preceding chapter, the corresponding distances were noted as varying from 30 miles (48.3 km) at 180 mph (289.6 kph) to 100 miles (160.9 km) at 600 mph (965.4 kph). Railroad rules have specified that trains must stay 5 to 10 minutes apart (varying with the railroad) and also set time clearance by which slow or opposing trains must clear the main track to let a superior train overtake or meet and pass. Rapid-transit trains and buses may be spaced on 2-to10-minute headways.

The time-interval system has certain obvious defects. Timetable scheduling is too inflexible and does not allow for extra, nonscheduled movements. (European railways get around this by setting up in advance the maximum number of schedules or paths the route will accommodate. Extra movements are then assigned one of the paths or schedules.) Furthermore, the time interval is difficult to maintain. A faster-than-prescribed speed closes the gap between one vehicle and the vehicle ahead. A slower speed allows the following vehicle to catch up. An unscheduled stop, deviation from the route, or speed variation for any reason immediately destroys the time protection.

For pipelines and conveyors, the time interval between shipments is completely at the dispatcher's discretion. The extreme flexibility of highways makes the time-interval factor largely dependent on the volume of traffic and the psychological reactions of the drivers. Time intervals are also established indirectly by the coordinated timing of successive traffic lights and the light cycle. Ships at bottleneck points may be spaced by specific instruction or by the time it takes to pass through a lock or narrow channel. The time interval seems to have the most significance in air and rail transport.

Space Systems   A space-interval or block system has been devised to overcome defects of the time-interval system. The route is divided into sections or blocks and only one vehicle at a time is permitted in a block except under special precautionary measures. Entrance into and through a railroad block is governed by block signal indications. The signals may be operated manually by block station-train order operators or automatically by the presence of the train. Staff systems and the station-to-station working of European and other railroads are adaptations of the block concept.

The block system has application to modes of transport other than railroads. The signal lights at street and highway intersections are a modified space or block-interval system. A further highway adaptation is the signaling of traffic lanes for one-way movements at certain hours of the day or for no movement at all in the event of emergencies on bridges or in tunnels.

Car spacing on aerial tramways is definitely a space-interval system although it may be predicated on a time-interval calculation. The geometry of cableways is one of distance.

It has been suggested that the block system has air-transport application. Vertical space intervals or layers have long been established. The air lanes would be divided into space intervals, longitudinally, and only one plane at a time would be permitted in that interval. An overtaking plane would have to reduce speed and/or stack or circle until the block ahead had cleared before it could proceed. Radio warnings beamed vertically from the ground would mark the block limits, in conjunction with Air Traffic Control Center supervision. This proposal is simply a means of maintaining more effectively the time intervals now prescribed. There are obviously a number of problems to overcome in its application.

The relations between vehicular intervals and route and traffic capacity have been indicated in an earlier chapter.

SIGNALS

A signal is, essentially, just another means of communication. It is a method of giving prompt, on-the-spot, concise information to the drivers of vehicles, pilots of ships and aircraft, and enginemen of trains. Where traffic is light, the way is wide, and speeds are slow, "rules of the road" supplemented by oral or written directions (orders) may suffice. For dense, high-speed traffic, signals are an invaluable aid in maintaining safe movement at maximum capacity. Signals have no direct application in the operation of pipelines, conveyors, or aerial tramways.

Waterways   In water navigation, signals have a long-established place. Channels are indicated by buoys with lights, bells, or whistles attached for night indication. Danger points—shoals, submerged rocks, etc.—are also marked by similar buoys or by lighthouses. Lighthouses also mark headlands and harbor or channel entrances as navigational "fixes" or bearings. The development of powerful lenses and dependable lights for these devices has called for the best of engineering in the fields of applied light and optics.

At locks on inland waterways, the lockmaster is charged with the immediate control and management of the locks. Precedence for locking is established by rule subject to change by the lockmaster for better utilization. The vessel arriving first is the first to lock through, but vessels belonging to the United States and commercial vessels may receive precedence in that order. Passenger boats have precedence over tows. Pleasure boats, when practicable, shall be locked through with other craft. Arrival posts are placed on shore or below the locks to establish a point of arrival.

Whistle signals are used both by vessels and by the lockmaster. For high density traffic, whistle signals are supplemented by flashing lights. To control the use of single locks or the landward lock of double locks, flashing lights (1 sec on, 2 sec off) are displayed on or at each end of the guard wall. To control the use of riverward locks, interrupted lights (1 sec on, 1 sec off) are similarly displayed at each end of the intermediate wall.

Again a red, yellow (amber), green sequence is used.

Red indicates that the lock is not immediately available; vessel must stand by.

Amber signifies that the lock is being made ready; vessels may approach under full control.

Green signifies that the lock is ready for entrance; proceed. All locks and dams must also display navigation lights.

Three green lights displayed vertically mark the *upstream* end of the intermediate guard wall when that extends farther upstream than the others.

Two green lights displayed vertically indicate the *downstream* end of the guard wall.

One *red* light is displayed at each end of the land (guide) wall, both upstream and downstream.

Range lights, set in line with prescribed courses, are a help in aligning a ship or tow on its proper course. River tows may depend largely on these in some river-crossing situations. See Figure 11.2. The pilot aligns the tow with a range marker (perhaps setting his searchlight beam on it at night) and steers toward the marker, then picks up another light where the next crossing is required.

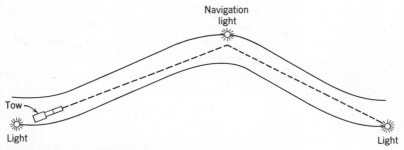

FIGURE 11.2 NAVIGATION LIGHTS ON A RIVER.

Railroad Signals   Railroads maintain the most elaborate and complex system of signaling of all the carriers. Running on fixed paths, often with opposing trains in front, the tremendous momentum of a heavy fast freight train requires adequate clearance and warning to permit safe control. In light-traffic areas, operating rules, timetable authority, and supplementing train orders are sufficient. Where traffic of high volume and density is to be moved at high speed, a system of blocks and automatic block signals must be used.

Block-signal indications are given by the position of semaphore blades, colors of lights, patterns of lights, or a combination of light and pattern. The simple signal sequence behind an occupied block (and in front of it too, in single-track territory) is *Stop* (or more likely *Stop and proceed* at 15 mph (24 kph) looking out for train, broken rail, or other obstruction), *Approach* [which calls for a speed reduction to 30 mph (48 kph) or to one half the maximum authorized speed, whichever is less], and *Proceed*, at maximum authorized speed, Figure 8.3*c*. These indications are given by a single light, semaphore blade, or light pattern. For more elaborate systems giving information about the second or third block ahead or establishing safe speeds for turnout or crossover moves, two or three blades, lights, or light patterns are used in combination to give a greater variety of *aspects* (the appearance of the signal) and *indications* (the message conveyed by the signal).

Track Circuits   A basic element in the control of any system of automatic signals for railroads or for switch and interlocking control or highway-crossing protective device is the closed-track circuit. See Figure 11.3. Closed circuits are used so that any failure in the circuit will cause it to

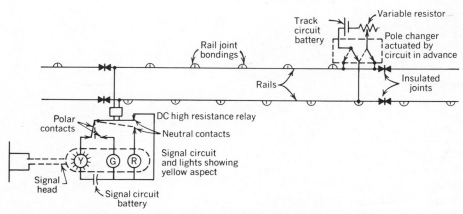

FIGURE 11.3 THREE-ASPECT CLOSED TRACK CIRCUIT.

open and give a restrictive indication. This ability to "fail safe" or "fail red" is a characteristic of all United States systems of railroad signals and interlockings. It is, actually, a fundamental concept in most designs of automatic equipment, including pipeline controls.

Simple closed-track circuits are insulated from those of the adjoining blocks, and a current is fed through the rails from a d-c battery or a-c transformer to a high-resistance relay. The up or down position of the relay opens or closes circuits that operate the semaphore blades, lights, or light patterns. When a train is in a block, the relay is shunted, because the wheels and axles of the train form a path of lower resistance than that through the relay. The armature holding the signal-control contact falls away to give the aspect for a "stop" indication. When the block is cleared, the relay is reenergized and the armature is lifted to close a circuit through either a green or a yellow light.

Various systems are used to give the intermediate yellow or "Approach" indication. One uses a polarized relay that makes contact "right" or "left" with an auxiliary armature depending on the polarity of the circuit. Polarity is established by a pole-changing device that responds to a change in the polarity of the circuit ahead. When the block ahead is occupied, the relay is deenergized and a red light shows (Figure 11.3). When the second block ahead is energized, the corresponding polarity causes the polar armature to make contact through the yellow light circuit. When the block

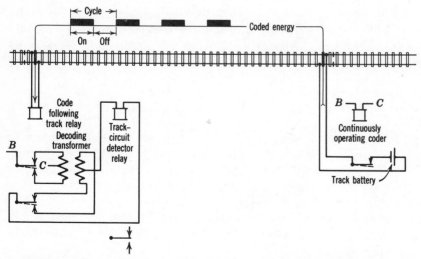

FIGURE 11.4 CODED TRACK CIRCUIT. (*"UNION" CODED TRACK CIRCUIT CONTROL*, BULLETIN 157, JULY 1943, UNION SWITCH AND SIGNAL DIVISION, WESTINGHOUSE AIRBRAKE COMPANY, SWISSVALE, PENNSYLVANIA, P. 10, FIG. 4.)

ahead has cleared, the polarity is again changed and the polar contact is made through the green light.

A variation of the closed-track circuit is the coded-track circuit, which makes use of codes of different frequencies to operate code-following relays and to control, in turn, the signal circuits. See Figure 11.4. The coding device that sends the various frequencies is, in turn, activated by a feedback from the circuit ahead, that is, the one that is occupied or one that has already been activated by the occupied circuit. In addition to increased sensitivity, lowered cost, less interference from stray currents, and greater flexibility, coded circuits make possible the usual types of cab signal indications displayed within the cab of the locomotive. Train stop devices, which make brake applications if the enginemen fail to respond to restrictive indications are also worked by coded circuits. The code is picked up inductively by sensor coils mounted on a bar just above the track and laterally under the locomotive pilot location. Both cab signals and automatic train control are thus activated.

TRAFFIC CONTROL DEVICES

Traffic control devices are placed on or adjacent to streets and highways by the authority of properly empowered officials or a public body in order to promote the safety and utilization of road and street capacity by giving directions, warnings, and commands to vehicle operators and pedestrians. Such devices include signs, signals, lane delineation, and channelization. Recommendations for design, installation, and use of these devices are found in *The Manual of Uniform Traffic Control Devices* developed by ASSHTO and the National Committee on Uniform Traffic Control Devices from which much of the following is derived. The *Manual* has been adopted by the Federal Highway Administration "...as the National Standard for application on all classes of Highways open to public travel...."[4]

State and local officials must conform to a *State Manual*, which is in substantial conformance with the *FHA Manual*. Through the authority granted by a Congressional Act of 1966, the Secretary of Commerce has indicated that all control devices on streets and highways should conform with the standards issued or endorsed by the FHA.

Effective control devices must (1) fulfill a need, (2) command attention, (3) convey a clear, single meaning, (4) command respect, (5) give adequate time for a proper response.[5]

[4]*The Manual of Uniform Traffic Control Devices*, U. S. Department of Transportation, Federal Highway Administration, Washington, D. C., 1971, p. ii.
[5]Ibid., p. 3.

Signs Regulatory information, warning of one-way movement, of hazardous conditions and directional guidance are given by signs. More specifically, regulatory signs include STOP and YIELD, signs that set maximum speeds, and a movement series that relate to turning, alignment, exclusion, and one-way traffic. Design features include shape, color, symbol, and legend. The first three are most significant in permitting a ready, immediate comprehension of the message being conveyed.

Shape Special significance is attached to the geometrical shape of signs.
  Octagonal—STOP (exclusively)
  Equilateral triangle, point downward—YIELD
  Round—advance warning for railroad crossing
  Pennant-shaped, longitudinal, horizontal—no passing zone
  Diamond—existing or possible hazard warning
  Rectangle, long dimension vertical—regulatory
  Trapezoidal—recreational area guidance
  Pentagon—school advance and crossing
Color Code: background for symbols and legends
  Red—STOP or prohibition
  Green—permitted movement and directional guidance
  Yellow—general warning
  Orange—construction and maintenance
  Black or/and white—regulation
  Brown—recreational and scenic guidance
  Purple, strong yellow-green, light blue—presently unassigned

Symbols Markings indicate the message being conveyed and are a relatively new addition. A diagonally barred arrow, for example, indicates a prohibited turn. The *Manual* suggests, however, that legends be continued for a significant period of time to allow the public to become familiar with the symbols.

Legend The legend is a word message on a sign such as STOP or ONE-WAY, or a speed designation. The *Manual* recommends clear, block lettering for legibility. Messages must be concise and not susceptible to misinterpretation. Signs should be reflectorized for visibility at night or illuminated by special lights if ambient lighting is inadequate. Examples of shape, symbols, and legends are given in Figure 11.5.

Placement Regulatory signs are usually placed where there is to be a required response. Signs to warn of existing or potentially hazardous

Regulatory signs

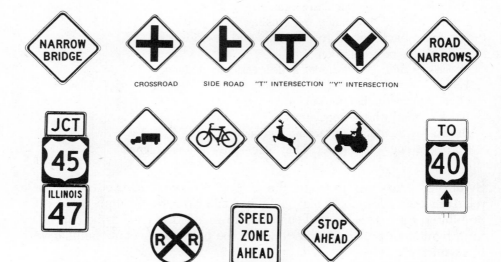

Warning signs

FIGURE 11.5 TYPICAL TRAFFIC CONTROL SIGNS. (FROM *RULES OF THE ROAD*, STATE OF ILLINOIS, SPRINGFIELD, ILLINOIS, 1976.)

conditions are of especial importance because of their effect on safety. These signs are placed far enough in advance of the hazard point to permit drivers and pedestrians to take appropriate action at prevailing speeds under prevailing conditions: 250 ft (76.2 m) for low-speed urban conditions, 750 ft (228.6 m) for rural locations, and 1500 ft (457.2 m) for high-speed roads and freeways.

Signs are positioned laterally and vertically as shown in Table 11.1 at approximately right angles to and facing the direction of traffic being controlled. Grades and curves or mirror reflection occurrence may require tilting or turning the sign slightly to improve readibility.

Reflectorized delineators may be placed on the outside lanes of one-way and two-way roads the full length of the road or at sections when there is danger of confusion.

Markings   Delineation of lanes and movements by pavement markings, both longitudinal and transverse, assists traffic flow without diverting the driver's eyes from the pavement. The painted lines are an inexpensive and useful form of channelization even though they lack durability, can be obliterated under snow and ice, and lose visibility when wet. Markings are used to delineate traffic lanes, turning lanes or pockets, center lines, no pass zones, pedestrian crossings, approaches to railroad crossings, and parking or no parking areas. Markings are sometimes supplemented by legends or symbols. See Figure 11.6.

A color code recommended by the Traffic Control Manual contains the following:

Yellow.   This color separates flow in different directions; it marks the left boundary in hazardous locations.

White.   This separates lanes for traffic in the same direction; it is also used to indicate right-hand edge of outside lane.

Red.   Red indicates roadways that are not to be used or entered by the viewer.

Black.   This color supplements the other colors when contrast is needed.

Line widths indicate degrees of emphasis: 4 to 6 in. for normal use, 8 to 12 in. for restrictive intersections. Double lines (two normal lines) are used for maximum restriction. Solid lines are restrictive; broken lines are permissive. Broken lines usually have a 3 to 5 ratio between segment lengths and gaps between. For night visibility lines can be augmented by reflectorized buttons where ambient lighting is inadequate. For other types of channelization, see the chapter on geometrics.

TABLE 11.1 SIGN POSITIONS[a]

Roadway Type	Position	Height	Lateral
All roadways in general	Right-hand side of road		Sign post or overhead support: $>6\,ft$[b] from edge of shoulder; of no shoulder, 12 ft from edge of traveled way
Rural		At least 5 ft bottom of sign to near edge of pavement	
Urban (business, commercial, residential)		7 ft to bottom of sign (1 ft less for secondary sign below primary)	1 ft from curb face with limiting side walk width or where poles are close to curb
Expressways		7 ft—minimum 8 ft—major sign 5 ft—secondary sign below major sign 6 ft—route markers, warning and regulatory 5 ft—signs placed 30 ft from edge of nearest traffic lane	>2 ft from usable roadway shoulder or unmountable curb >10 ft from edge of nearest highway lane Large guide signs >30 ft from nearest traffic lane
Expressways (Overhead signs)		17 ft over entire width but 1 ft over minimum clearance established by a structure, etc.	>6 ft on connecting roads or ramps at interchanges

[a] *Manual on Uniform Traffic Control Devices for Streets and Highways*, U.S. Department of Transportation, Federal Highway Administration, Washington, D. C. 1971, pp. 21–27.
[b] One foot $=0.3048$ meter.

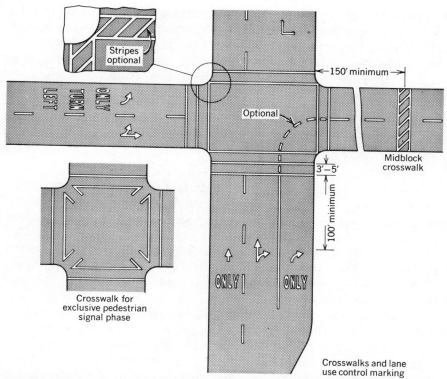

Stripes optional

150' minimum

Optional

Midblock crosswalk

3'–5'

100' minimum

Crosswalk for exclusive pedestrian signal phase

Crosswalks and lane use control marking

FIGURE 11.6 TYPICAL LANE AND CROSSWALK MARKINGS. (FROM *MANUAL ON UNIFORM TRAFFIC CONTROL DEVICES FOR STREETS AND HIGHWAYS*, STATE OF ILLINOIS, SPRINGFIELD, ILLINOIS 1976.)

Traffic Signals   Street and highway traffic signals are the most commonly known and used system of signals. Safe passage for cars and pedestrians is secured at intersections, but they are also used for speed control, for turning movements, and to open or close ramps leading to expressways. Signals find use in converting expressway lanes of the morning into outbound lanes in the afternoon. The basic effect on capacity of signalized intersections has been discussed in Chapter 8.

The *aspect* of a signal is the appearance—a red, yellow, or green light, arrow, or similar signal—to the beholder. The *indication* is what that aspect means: "proceed" for a green aspect, "caution-prepare to stop" for a yellow aspect, and "stop" for a red aspect. These aspects are displayed on a signal head to which may be added green arrows to designate turns or "proceed" for specific lanes or a "walk" aspect at busy intersections to assist in the safe crossing of pedestrians. Variations of the foregoing may be developed at complicated five- or six-way intersections. The signal head is normally mounted on a post to the right of the lane governed, but it may also be cantilevered outward over the lane or suspended above the lane from cables or from a signal bridge. The prime requisite is clarity—certainty of observation and without confusion as to the lane(s) and movement being controlled.

The entire sequence of green, yellow, and red (and variations) is termed the cycle length. The green or proceed phase has a duration that ranges from 15 to 30 seconds or more depending on the volume and speed of traffic flow and on the importance of each street. The next interval is usually a yellow aspect of 4 to 8 seconds that warns the driver of an approaching change to a red or stop aspect. The opposing signal face(s) present a red aspect during the foregoing sequence. The cycle is often further lengthened by introducing a left (or right) turn phase and a "walk" phase during which all vehicular traffic should stop (but often is alerted only to caution). The "proceed" phases need not be equal, a long "proceed" often being given to the more important street, usually the one having a preponderance of traffic.

At an isolated intersection, a light can function as a *fixed-time* signal that maintains the same cycle and phases throughout the day unless changed. The signal controller may have several dials to give different cycle and phase time lengths for the morning and evening peak periods and for off-peak traffic. Dial selections are made manually or by a remote master control.

Signal Cycles   To move traffic safely and with a minimum of delay to all vehicles, the green phase must be of a length that the capacity to volume ratio on each approach will, during peak periods, accommodate all the traffic that has accumulated during the preceding red interval and all that has arrived during the green phase. Durations should be proportional to the lane volume of demand on each approach. The actual phase length must include the time it takes for waiting vehicles to initiate movement—a starting delay time of 1.5 to 3.8 sec., often taken as 2.5 sec—plus the time it takes for the remaining accumulation and later arrivals to enter the intersection. The latter is a function of the intersection capacity and the flow rate when cars enter the intersection at a rate of 2 to 2.5 sec per vehicle per lane, usually taken as 2.1 sec. Short cycles reduce the delays to waiting vehicles but, because of starting time delays, short cycles are likely to incur more lost time and accommodate fewer vehicles per hour.

Other factors affecting the length of the green interval are left turns, the addition of a turning interval, and a pedestrian "walk" interval. The turning interval length can be based on the number of turns as a percent of total traffic in conjunction with the start and entering rates just mentioned. The pedestrian interval can be the factor that determines the green interval duration. The minimum duration for a pedestrian interval should include a 5-sec start time plus the time it takes to cross the street. If one assumes an average walking speed of slightly less than 3 mph, the speed becomes a commonly accepted rate of 4 fps. The minimum pedestrian portion of the green interval becomes 5 sec plus $W/4$, where W is the street width in feet. This value can be further reduced by the length of the yellow interval. (W could also represent the distance from the curb to a safety island or median, thereby requiring the pedestrian to utilize two cycles to complete the crossing.)

The yellow interval allows vehicles committed to the intersection to clear it before opposing traffic is released and also warns approaching vehicles that the signal is about to change and that a stop must be made. The stopping distance is the driver's perception time (0.5 to 1.5 sec) plus time to decelerate $= v/2a$ where a is the rate of deceleration taken as actual or approximately 15 fps^2. In practice, the yellow interval is usually taken as 3 to 5 sec (unless special conditions, such as a three-approach intersection, prevail or the street is unusually wide).

The reader is referred to Chapter 8, the section on intersection capacity, for procedures to determine approach capacities. A fairly high load factor, 0.8 to 0.9, representing peak hour demand may be used. Traffic surveys are used to establish volumes of traffic flow on intersection approaches. Usually peak flow for 15-min intervals is recorded. The use of the foregoing is illustrated in the following somewhat simplified problem example.

Problem Example

Consider a right angle intersection, both streets 2-way, 40 ft in width with parking prohibited on both approaches. The opposing approaches are A and B. One can assume a 4-sec clearance (yellow) interval for each. (If greater precision is required, the deceleration time and the time it takes to clear the intersection can be computed as described in the preceding paragraphs.)

In this example, the time required for pedestrian movement will be computed as follows:

Minimum start time	5 sec
Walk time at 4 ft/sec; $40 \div 4$	10 sec
Total pedestrian time	15 sec
Utilization of 4-sec yellow	4 sec
Minimum green for pedestrians	11 sec

Using approximate values from Figure 8.8 with a load factor of approximately 0.3, the hourly approach values per hour of green time are 2650 vph for both A

and B approaches. From field survey data, the average peak hourly volumes on a typical weekday are $A = 600$ vph and $B = 400$ vph. The volume to capacity ratios for each approach are:

$$V/C \text{ ratio } A = 600/2650 = 0.22$$

$$V/C \text{ ratio } B = 400/2650 = 0.15$$

Phase B, the smaller of the two ratios would normally be assigned the minimum green time based on pedestrian crossings of 11 sec, but the use of the accepted 15-sec minimum indicates 15 sec as the green interval duration for B. In order to keep the capacity to demand relations in proportion, the larger ratio is divided by the smaller, that is

$$(V/C \text{ ratio} A) \div (V/C \text{ ratio } B) = 0.22/0.15 = 1.5$$

The green interval for route A is thus made 1.5 times as long as for route B: $1.5 \times 15 = 22$ sec. The complete cycle thus becomes:

Cycle for Approach A	Cycle for Approach B
Green— 22 sec	Green— 15 sec
Yellow— 4 sec	Yellow— 4 sec
Red— 19 sec	Red— 26 sec
Cycle— 45 sec	Cycle— 45 sec

Because the cycle length is usually set in 5-sec intervals, the foregoing would, when necessary, be rounded off to 45 sec with 19 sec of red for approach A and 26 sec of red for approach B.

Refinements not including the above would account for the ability of the established cycle to handle the volume of demand. The probability of vehicles arriving bunched is sometimes assumed to be a function of volume of demand. To determine the volume capability and the optimum phase length, the reader is referred to works on traffic engineering, especially to the *Transportation and Traffic Engineering Handbook* of the Institute of Traffic Engineers.

A quick check on volume adequacy based on an average or uniform rate of arrival at the intersection is made as follows. In one hour of 3600 sec there will be 80 cycles with 22 sec of green per cycle. With a starting delay of 2.5 sec and an average time of 2.1 sec for succeeding vehicles to enter the intersection, the maximum number of vehicles entering on one green interval side of the intersection will be

$$2.5 + (n-1)\, 2.1 = 22$$

$$n = 10.3 \text{ (take as 10)}$$

where $n =$ number of vehicles entering the intersection in one green interval per lane per direction or 40 vehicles for 4 lanes.

$\langle 40 \text{ vehicles} \times 80 \text{ intervals} = 3200 \text{ vehicles} \rangle$

This cycle can handle the total number of vehicles per hour on Approach *A*. Approach *B* can be similarly checked.

Traffic Activation   *Semitraffic actuated signals* are sometimes used when side street traffic is markedly less than for the street intersected. A green phase is initiated for the main street. It continues green until a vehicle approaching from the side street is detected by a sensing device. The side street signal then turns green after a preset time lag after which the main street "proceed" is again restored. If other side street vehicles appear after the first one, additional time is given for each vehicle up to a prede- termined number before the "proceed" on the side street is withdrawn. The sensing device may be a mechanical treadle, a photoelectric "eye" (usually placed overhead), or other electronic device. The mechanical treadle is not favored because of possible impaction by dust or snow and ice and by corrosion from salt water off of icy streets.

A *fully actuated signal* has sensing devices on each street and, when traffic is heavy, functions much like a fixed time cycle system.

The fully traffic actuated signal may have a controller-sensor system responsive to the instantaneous *volume-density* characteristics of the traffic flow. Arrival times, headways, and waiting times are "sensed" and stored in the controller's "memory" from which continual adjustments are di- rected to the controller, varying both phase and cycle times to give a maximum response to short-term changes in traffic flow.

Several patterns of operation are available for combining a series of intersection lights into a coordinated system along a particular street. One of the oldest, long active on 5th Avenue in New York, is the *simultaneous system* in which all signals along an extended portion of a street change and show the same aspect at the same time. It finds its greatest use where blocks are short and signals are spaced closely together.

With an *alternate system*, every other signal (or small sequence of signals) will show a green aspect while those in between will show red. This works well where blocks are of approximately the same length. The cycles can be adjusted to permit a more or less continuous movement of a vehicle traveling at a posted speed.

With the so-called *progressive system* a common cycle length in a series of intersections is used, but the stop and proceed phases at individual intersections are adjusted to the needs of cross traffic. The stagger on the cycles again permits continuous movement at posted speeds.

Signal cycles at successive intersections can be coordinated to permit a nonstop flow of traffic at posted speeds. The cycle is offset by the time necessary to permit movement from the preceding intersection to give an

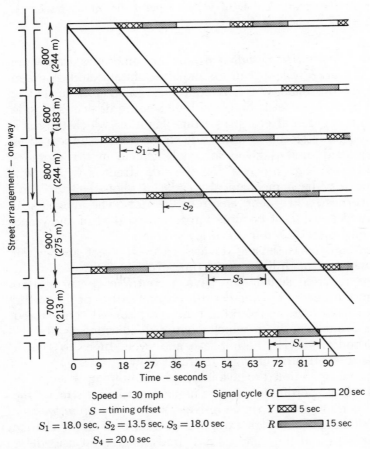

FIGURE 11.7 COORDINATED SIGNAL TIMING.

uninterrupted sequence of green signals. Figure 11.7 illustrates such an offsetting.

Other Traffic Controls A variety of signs was described earlier. The STOP and YIELD signs control entry from side streets onto more important streets and at intersections where traffic is heavy enough on both streets to warrant protection. Other controls include one-way streets, parking controls that prohibit, permit, or set a time limit, speed controls, pedestrian controls (WALK or DON'T WALK), bus lane controls, and controls for turn and lane use. Criteria for the use of these controls are given in the 1976 edition of the *Traffic and Transportation Engineering Handbook* of the Institute of Traffic Engineers.

NAVIGATIONAL AIDS

The Need for Navigation Aids From the earliest days of sailing, mariners have faced the problem of determining where they are. Classical solutions to this problem are, even in modern times, based on relating the ship's position to that of the sun or stars. These methods are of little help when the heavens are obscured by fog or storm. The problem becomes acute as the ship nears land with treacherous shoals, reefs, and obscured harbor entrances. The ship must then be very sure of its position. Radio direction finders and radar have added to the safety of sea transport. Radar permits a shadow on the radarscope of objects in the immediate vicinity, including the shore line and other ships. A radio-range compass permits a ship to determine its position, its latitude and longitude, by taking a bearing on two or more radio-range stations of known location (identified by the code transmitted), using directional antenna to determine the bearings from the ship to the stations. These have been developed to a state of almost complete automation. Sonar and other sound and radio-impulse echo devices permit taking accurate soundings as well as determining the distance from shore and other ships.

Waterways Navigation charts showing course, channel, depths, obstructions, lights, and similar data are prepared for inland waterways by the U. S. Army Corps of Engineers. Navigational aids are the responsibility of the U. S. Coast Guard, which places and maintains buoys or markers alongside channels to mark obstructions, turns, and the point at which channels divide. Most tows are now equipped with radar, thereby maintaining continuous movement when visibility is poor as with rain or foggy weather. Bridge-to-bridge and bridge-to-shore radio are now required. Waiting time is reduced by establishing radio communication with lock crews and with other tows when a meeting impends in narrow channels.

Airways Early airways were marked by towers on the ground equipped with revolving lights. By following the intersection of the paths of two adjoining lights, a pilot could find his way by night. This advantage was lost when fog or storm rendered the lights invisible. The towers have largely been superseded by radio beams except at airports.

Aerial navigation shares these problems with marine navigation plus the problems of altitude and high speed and the necessity of keeping the aircraft in flight. Thus the blinding effects of fog and storm are more serious for aerial navigation. At approximate sonic and supersonic speeds, the pilot is practically flying blind since his high speed does not permit a sequence of decision making based on visual observation. The air congestion, especially around airports, is a complicating factor.

The country is crisscrossed with a network of routes marked by radio-range stations on the ground that send identifying signals and by directional radio beams with which a pilot can align himself and follow to his destination. Navigation is thus possible without ground visibility. See-and-be-seen flight is kept at low altitudes, while the high-speed transports fly above all other commercial routes under Instrument Flight Rules (IFR).

There are many electronic devices available that add to the safety and reliability of flying, but not all of these are afforded on even the commercial transports. Radar enables a pilot to obtain first-hand information about weather conditions ahead, and, therefore, once he has secured permission to vary his flight plan, allows him to change his course and avoid much rough weather and rough air. A transponder type of radar gives the pilot an unusually clear picture of other aircraft in the vicinity. Another device progressively traces the route a pilot is flying on a map before him.

The primary facility for civil aviation in the National Air Space System is the Very High Frequency Omni Range (VOR). The use of very high frequencies reduces atmospheric interference. Directional information (azimuth location) is generated and transmitted to aircraft from VOR stations along the selected flight route. A rotating signal is sent out of phase with a fixed reference signal one degree for each degree change in the azimuth of the rotation. The aircraft's omni receiver measures the phase difference and indicates the azimuth bearing. A course deviation indicator enables the pilot to stay on the electronically selected course to the next station. An automatic direction finder gives an accurate reference to ground installations, and distance-measuring devices give the distance from the aircraft to any radio station along the route. Ground-sited location markers light bulbs on the instrument panel as the plane flies over. Appropriate position information is also relayed to the Air Traffic Control Center.

Instrument Flight Rules (IFR) are used at all times by commercial flights and by other planes properly equipped when visibility is less than three miles. Aircraft not equipped for IFR fly according to Visual Flight Rules (VFR), which is permissible only when visibility is three miles or more.

From about 35 miles out, the aircraft is under the jurisdiction and monitoring of the Airport Traffic Control Tower and under direct control from 7 miles. At the Tower the position of each flight is followed on radarscopes and given a runway assignment and clearance for landing or directed into a "holding" pattern to await its turn to land. See Figures 11.8 and 11.9. Runway markings identify the edges of runways and of the threshold and touchdown zones. White center lines give directional guidance.

When traffic is heavy, the Airport Traffic Control assigns aircraft to holding patterns that are flown around two or more radio compass marker beacons with a 1000-ft (304.8-m) altitude separation between aircraft. As one plane lands, each craft in the pattern descends to the next lower 1000-ft level and continues the procedure until its turn to land.

At night, the approaches to the runways and the edges of the runways will be marked by lights, blue lights, and blue-white flashing lights. Runways equipped for instrument landing may have rows of lights arranged at decreasing altitudes as the runway is neared, and transverse to it, in the approach zone; this aids the pilot in establishing his or her altitude visually and in relation to the runway. The pilot is told by the color of light pairs whether he or she is on, above, or below the proper descent path.

In making an instrument landing, the pilot follows a radio beam with a sloping approach path much as he followed a similar radio beam across the country. He may receive a sound or a visual indication of adherence to the glide path depending on the system in the cockpit. Coming from a bottom position in a holding pattern, the plane sweeps into a final approach at approximately 1200 ft (365.8 m) of elevation and picks up a precise ground approach path generated by electronic signals so the pilot can line up a runway and descend to a point where approach lights can be seen. Cockpit instruments tell the pilot of the position of the craft relative to the required flight path. The plane is flown in such a way as to center an instrument bar over a marker on the panel board. When centered, the plane is on the correct flight path. Electronic aids include an outer electronic marker beam, a VHF glide slope beam transmitter, and a VHF localizer beam. See Figures 10.5 and 10.6.

The control tower, observing and plotting the plane's approach on radar will also give helpful advice to the pilot to aid in orientation. Landings have been made by instrument alone, with the controls electronically locked onto the glide path beam, and no human control exercised at all. Cruising can also be guided automatically by an automatic pilot.

There is need for futher development to prevent collisions in the crowded airspaces around airports. Collision Avoidance Systems (CAS) of various kinds are being investigated and used. One system sends precisely timed signals at 7-sec intervals from an on-board computer that surrounds

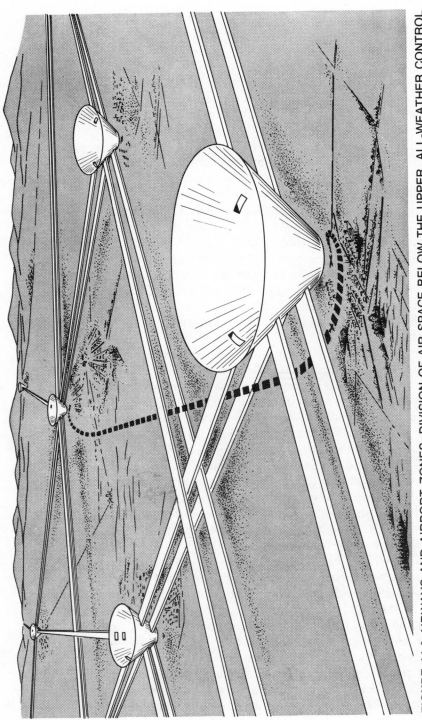

FIGURE 11.8 AIRWAYS AND AIRPORT ZONES. DIVISION OF AIR SPACE BELOW THE UPPER, ALL-WEATHER CONTROL ZONE. CONES OF CONTROLLED AIR SPACE FUNNEL TRAFFIC BETWEEN THE UPPER CONTROLLED ZONE AND AIRPORTS ON THE GROUND. SLANT AIRWAYS (NOT SHOWN) LEAD FROM THE BOTTOM OF THESE CONES TO THE AIRPORT SURFACE. RIBBONS OF CONTROLLED AIR SPACE LINK THE CONES WITH EACH OTHER. DASHED LINE SHOWS SEPARATE, NONCONFLICTING PATH OF SEE-AND-BE-SEEN AIRCRAFT. (FINAL REPORT BY PRESIDENT'S SPECIAL ASSISTANT, EDWARD P. CURTISS, WASHINGTON, D.C., MAY 10, 1957, P. 14.)

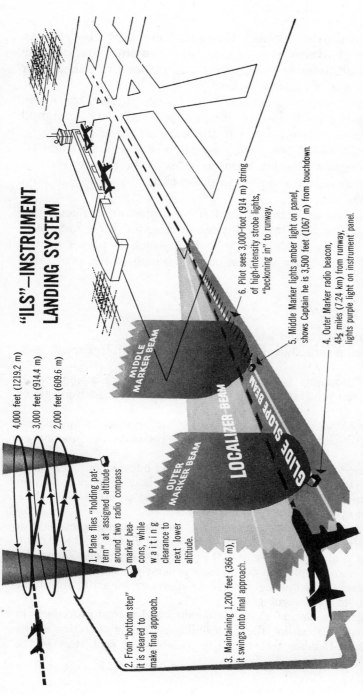

"ILS"—INSTRUMENT LANDING SYSTEM

4,000 feet (1219.2 m)

3,000 feet (914.4 m)

2,000 feet (609.6 m)

1. Plane flies "holding pattern" at assigned altitude around two radio compass marker beacons, while waiting clearance to next lower altitude.

2. From "bottom step" it is cleared to make final approach.

3. Maintaining 1,200 feet (366 m), it swings onto final approach.

MIDDLE MARKER BEAM

OUTER MARKER BEAM

LOCALIZER BEAM

GLIDE SLOPE BEAM

6. Pilot sees 3,000-foot (914 m) string of high-intensity strobe lights, "beckoning in" to runway.

5. Middle Marker lights amber light on panel, shows Captain he is 3,500 feet (1067 m) from touchdown.

4. Outer Marker radio beacon, 4½ miles (7.24 km) from runway, lights purple light on instrument panel.

FIGURE 11.9 SCHEMATIC VIEW OF LANDING OPERATIONS. (COURTESY OF *WELCOME ABOARD*, AMERICAN AIRLINES, APRIL 1957, PP. 46 AND 47.)

445

each plane with a protective "bubble." When one bubble touches another an audiovisual alarm gives a 60-sec warning. The computer analyzes the courses, speeds, and altitudes of the approaching planes and indicates or initiates the best collision avoidance maneuver.

Traffic control systems are available that coordinate air traffic by recording air space assignments and rejecting automatically any attempted simultaneous assignment of two aircraft to the same air space.

A further aid to instrument landings is a system of so-called "Microvision" whereby radio signals emanate from the sides of the runway to form an image of the runway before the pilot in the cockpit. Landings can then be made with zero visibility.

Collisions have occurred on the ground from confusion as planes taxi between runways and loading/unloading ramps. Ground level radar and automated systems have been developed to keep track of planes on the ground and to protect ground route assignments similar to the protection given in the air.

QUESTIONS FOR STUDY

1. What is operational control and what are its goals?
2. What role does communication play in operational control? Explain with specific examples.
3. Differentiate, with examples, between the three methods of obtaining operating intervals between vehicles. What are the advantages and disadvantages of each?
4. What are the problems of "human engineering" in relation to operational control and what developments have brought these problems into prominence in transportation?
5. Prepare a time-space diagram for a route where street intersections are spaced respectively 600, 800, 700, and 900 ft apart to show how signal coordination can be obtained to give a continuous green indication at a posted speed of 25 mph with a signal cycle of G-15, Y-5, and R-20 sec.
6. Design a signal cycle for the intersection of two one-way streets, each 40 ft in width, parking permitted on both sides, with a load factor of 0.8. Traffic surveys indicate a peak hourly flow of 800 vph on one approach, 500 on the other. Pedestrian movements are 60 and 30 persons on the A and B approaches, respectively. What changes would an allowance for a 10-percent left turn interval make in the foregoing?
7. Using a flow chart presentation, show the nominal and normal procedures in conducting a commercial airline flight, including a holding operation at the destination terminal. Indicate the control responsibility at each stage of the procedure.

8. Explain, with a diagram, the uses of lights in river navigation and the movement through locks.
9. What factors have led to the adoption of microwave for railroad and pipeline communication?
10. Explain how train orders are used in railroad operation.
11. Sketch a track circuit, explain its operation, and define its significance in railroad signaling and operation.
12. How does Centralized Traffic Control (CTC) differ from ordinary dispatching for (*a*) railroads, (*b*) pipelines?
13. If you were setting up a CTC system for a belt-conveyor operation, what kinds of information and controls would you place on the control board before the dispatcher?

SUGGESTED READINGS

1. *The Standard Code* (of Operating Rules, Block Signal Rules, Interlocking Rules), Association of American Railroads, Chicago, Illinois.
2. *American Railway Signaling Principles and Practices* (separately bound chapters, especially Chapter III, "Principles and Economics of Signaling"), Signal Section, Association of American Railroads, Chicago, Illinois.
3. Harry W. Forman, revised by Peter Josserand, *Rights of Trains*, Simmons-Boardman Publishing Company, New York, 1974 edition.
4. Edmund J. Phillips, Jr., *Railway Operation and Railroad Signaling*, Simmons-Boardman Publishing Company, New York, revised edition.
5. *Elements of Railway Signaling*, Handbook 50, June 1954, General Railway Signal Company, Rochester, New York.
6. L. R. Allison, "A Modern Cab Signaling and Train Control System for Railroads," *Transactions of the American Institute of Electrical Engineers*, Paper No. 59-252, 20 February 1959.
7. Martin Wohl and Brian U. Martin, *Traffic Systems Analysis*, McGraw-Hill, New York, 1967.
8. Paul K. Eckhardt, "A New Centralized Control System to Handle Complex (Pipe Line) Dispatching," *Petroleum Engineer*, January 1955.
9. Irving Conklin, *Guideposts of the Seas*, Macmillan, New York, 1939.
10. *Manual on Uniform Traffic Control Devices*, U. S. Department of Transportation, Federal Highway Administration, Washington, D. C., 1971.
11. *Transportation and Traffic Engineering Handbook*, John E. Baerwald, Editor, Institute of Traffic Engineers, Prentice-Hall, Englewood Cliffs, N. J., 1975.
12. *Traffic Signals*, Transportation Research Board, REC 445, Transportation Research Board–National Research Council, Washington, D. C., 1973.
13. *Freeway Operations and Control*, Transportation Research Board, REC 388, Transportation Research Board–National Research Council, Washington, D. C., 1972.

14. *Traffic Control Guidance*, Transportation Research Board, REC 503, Transportation Research Board–National Research Council, Washington, D. C., 1974.
15. Donald R. Drew, *Traffic Flow Theory and Control*, McGraw-Hill, New York, 1968.
16. *Instrument Flying, Handbook*, Federal Aviation Administration, U. S. Department of Transportation, Washington, D. C., 1971.
17. Report of the Task Force on Air Traffic Control, report of *Project Beacon*, Richard R. Hough, Chairman, Federal Aviation Agency, Washington, D. C., October 1961.

Chapter 12
Cost of Service

Cost as a Determining Factor The technological features of an undertaking are seldom seen in true perspective unless set against an economic background. The cost of providing (or obtaining) transportation service usually governs the final selection of carrier type except where restrictive technological conditions prevail—as where a cableway is selected because rugged terrain rules out other types of carriers. Even then, cost is probably still the controlling factor. A rail line might have been selected if the enormous costs of heavy tunneling and development had been acceptable. It is usually the overall greater economic cost of any other alternative that underlies the technological advantages of the one selected.

Private carriers are interested directly in the cost of providing service. Investment in route and equipment and the costs of operating and maintaining the service are paramount. A common or contract carrier, in addition, is interested in costs as a factor in determining the rates to be charged the shippers. The persons who are buying transportation from a carrier look to the rate alone as their cost without any direct interest in the cost to the carrier.

Preceding chapters have discussed the technological features, the advantages and disadvantages of the several carrier types. These characteristics aid or hinder each in performing its transportation function. Differential abilities to provide flexibility, speed, safety, dependability, and economy vary accordingly. The most efficient, technologically speaking, should be able in most instances to reflect its technological advantages in lower costs of overall operation and therefore in lower rates.

449

Traffic will and should flow to the carrier that extends the lowest rate cost to the shipper, but this is subject to qualification. Low rates, for instance, may not compensate for the attendant costs of slow delivery. The high cost of speedy airline service may be economically less expensive when a much-needed spare part is being transported to prevent delays in an important enterprise. There is a class of passenger traffic that is willing to pay a premium for luxury or speed in transit. On the contrary, however, many shippers will be attracted to a low freight or passenger rate quotation is spite of a high overall expense because they are not aware of the full economic cost. Similarly, carriers through incomplete cost evaluation have hauled traffic at rates that did not cover even the out-of-pocket costs of the service.

Rates and attendant problems are therefore of great importance. It matters little how technically sufficient a carrier is if its services are not utilized by the public because of its real or apparent high rates.

Rate differences, insofar as they relate to costs of service, may reflect accurately the technological characteristics. However, nontechnological factors of competition, regulation, subsidation, and even political policies enter into the formation of transportation charges and may be of more significance than technological factors in determining the rate a carrier will, can, or is permitted to impose.

Cost as a Basis for Rates   Under the classical theory of transportation charges, the rate is based on cost of service. Cost of service is understood as the actual expenses, both direct and indirect, including a reasonable margin of profit. Rates will thus vary between carriers because their cost of performing the transportation services will differ. Rates will also vary with commodities and the differing types of service required. High-speed rail service for perishables, for example, incurs more service cost and commands a higher rate than a slow-speed movement of bulk coal. The engineer has the continuing problem of assuring technological practices that will produce the minimum in costs.

A different situation is present, however, when there is an extensive lack of correspondence between the rates and the costs of service. In these instances, rates may have been adjusted to demand or to lack of demand. In the latter instance, rates may have been adjusted downward until just sufficient to cover the marginal costs of service. Examples of pricing to meet a lack of demand are frequently found outside of transportation. Power companies usually grant lower offpeak rates to consumers. Transportaion companies may offer lower than customary rates to utilize surplus capacity—as in offering excursion and weekend rates. In some instances, with railroads for example, rates are based on the value of service to the shipper, charging "what the traffic will bear"; this means that not all traffic bears its full share of the cost.

The importance of cost as a factor for the private carrier where profit on transportation is not a motive is self-evident. The nature of private and common carrier costs alike will be developed in the following analyses. It is the determining and assigning of costs to different types and classes of service that gives rise to many problems relative to rates for public carriers.

Various criteria have been used over the years as a basis for establishing transportation rates. For a time, the courts and later the Interstate Commerce Act made "a fair return on a fair value" (of the investment) the rate base. That is, the regulatory bodies permitted rate structures that would enable the carriers to earn a reasonable return on the evaluation of a prudent investment in plant and equipment. Much argument developed as to whether that valuation should be on the basis of initial cost, reproduction cost, or some compromise between the two. Another criterion has been the credit basis, whereby a carrier would be allowed a sufficient return so that it could pay dividends, thus maintaining its credit and its ability to attract capital in the money markets. Still another consideration has been the effect of rates on the movement of traffic. Outright aid has been extended, usually to products of agriculture, under "emergency" conditions, in the form of reduced rates. On other occasions, regulatory bodies have refused to permit rate increases because they thought the proposed rate would overprice and drive away the traffic on which it was imposed. Carriers, too, must consider this possibility. Attention is also given by regulatory bodies to the effect of rates on other carriers. Fear has been expressed by such bodies that lowering certain rates could undercut the prices and traffic of a competitor.

It is not the function of this book to inquire into the merits of these several criteria, some of which are highly debatable. The point to be stressed is that regardless of the importance attached to these criteria, rates must reflect costs and cannot long be kept below the level of costs without incurring disaster.

Cost of service is therefore a basic element in and the foundation of any rate structure. A carrier's rates would tend toward a common value for all commodities if costs of service for each commodity were equal. Cost differences permit and justify differential pricing, as exemplified by the higher rate charged for less-than-carload-lot freight as contrasted with carload-lot freight, or for plate glass as opposed to sand. This is not always the case, however, The lower unit costs of a long haul over a short haul are generally not permitted by regulatory bodies to be reflected in common carrier rates.

Cost is a term that needs more definition, especially when it is considered as a factor in the rate structure. One can speak of average costs, overhead costs, avoidable costs, out-of-pocket costs, engineering costs, fixed costs, variable costs, capital costs, and operating costs. These will be defined and the importance of each noted in subsequent paragraphs.

Capital Costs and Operating Costs   *Capital costs* are the costs of providing the initial plant and equipment and of additions to or betterment of those facilities. Such costs can be divided conveniently into two principal groups: investment in route and structures and investment in equipment. Some carriers have an investment primarily in equipment, for example, airlines. Other agencies have investments only in way and structures, for example, turnpike authorities. Railroads and pipelines have large investments in both. The split in ownership of plant and facilities in certain modes makes cost determination difficult and leads to anomolies, misunderstandings, and misstatements regarding the true costs of such carriers. Capital costs include interest charges on the money invested. Usually the money is borrowed and interest is considered as a charge against income, not as an element in operating costs.

Operating expenses, the costs of conducting the transportation business, include the following:

Route maintenance, the costs of maintaining roadway and track, pavement and subgrade, rivers and harbors, channels and dams, aerial-tramway cables and towers, pipelines, etc., and the structures appurtenant to them. Railroads refer to these costs as maintenance of way and structures. Traditionally, railroads, pipelines, conveyors, and aerial tramways build and maintain their own routes. Highways, river channels, canals, harbors, airway guides, and airports are built and maintained by governmental agencies.

Equipment maintenance includes all costs of maintaining motive power and rolling stock—cars, locomotives, trucks, tractors, trailers, automobiles, buses, airplanes, ships, barges, pumps and compressors for pipelines, conveyor power equipment and belts, and cars and machinery for aerial tramways.

Transportation costs are all the costs of conducting transportation. Principal items are fuel and power, wages of vehicle crew, terminal costs, and the wages of those directing vehicle movements. In the case of aircraft, this last cost may be largely borne by governmental and airport traffic-control agencies. Highway traffic control is partially directed by state and local government traffic divisions and police, while control of water navigation is shared by company dispatchers with the Coast Guard and local harbor masters.

Traffic costs are the costs of traffic solicitation, advertising, publishing rates and tariffs, and administration.

General costs and **miscellaneous** take in all general office expenses, legal advice, accounting, and the salaries of general officers and their staffs.

Table 12.1 gives some *average costs* and *revenues* for various carrier types. Because these are average costs, the overall expenses of operation are represented. The cost for any one class of traffic may differ considerably from the average. As will be seen later, actual costs may not vary directly with traffic volume.

TABLE 12.1 TYPICAL OPERATING COSTS AND REVENUES

Carrier	Cost per Net Ton (or Passenger) Mile (Cents)	Revenue per Net Ton (or Passenger) Mile (Cents)
Railroads:		
Freight	1.10[c] to 1.76[a]	2.410[a]
Passenger—intercity (Amtrak)	14.882[a]	5.609[a]
Motor freight	2.0[c] to 8.0[f]	
Inland waterways	0.45[c]	0.175 to 0.300[b]
Airways—freight	22.7[d]	19.8[d]
passenger	0.086[d]	0.073[d]
Bus—intercity		3.6[e]
mass transit		8.3[e]
Pipelines	0.22[c] to 0.27[f]	
Conveyors	1.80 to 2.50[g]	
Aerial tramways	5.0 to 12.0[g]	

Based on data from various sources:

[a]*Yearbook of Railroad Facts*, Economics and Finance Department, Association of American Railroads, 1976 edition, Washington, D.C., pp. 9,34,62.

[b]*Weekly Letter* published by The American Waterways Operators, Inc., Washington, D.C., Vol. XXIV, No. 45, November 11, 1967, p. 1.

[c]Peter Penner, *Summary of Transport Characteristics for Vehicular Freight Transportation, 1971*, Center for Advanced Computation, University of Illinois, Urbana, Illinois, 1974, pp. 1.8.

[d]*Air Transport 1975*, Air Transport Association of America, Washington, D.C., 1975, pp. 12, 16, 17, 20.

[e]"More Miles to the Gallon," published by the Association of American Railroads, Washington, D.C., 1971.

[f]"A Railway Trust Fund," Bruce Hannon (Director, Energy Research Group, Center for Advanced Computation, University of Illinois, Urbana, Illinois), April 1974, p. 12.

[g]Author's estimate.

Fixed and Variable Costs   The most important cost distinctions in this study are those between *fixed* and *variable* costs. Fixed costs are incurred with little or no relation to the volume of traffic moving and may even continue to accrue when no traffic at all is moving. General office expenses fall into this category. Salaries of the president, his vice-presidents, their staffs, the accounting department, etc. continue whether 10 vehicles or 20 are operated; whether there is a constant flow of traffic throughout the year or seasonal peaks and depressions. Bridges must be repainted and inspection carried out as long as any traffic moves. Tunnels must be maintained for one vehicle as well as for 10 vehicles. Locks and dams must be maintained and manned without regard to volume of river traffic. Fixed costs are also referred to as overhead costs.

Variable costs, by contrast, fluctuate in harmony with traffic fluctuations. A truck driver's wages are paid only if there is a truck to drive—and the truck is ordered for the road (or purchased or rented in the first place) only if there is traffic to move in it. An airplane or freight-train crew are not called unless traffic is sufficient to fly the plane or operate the train economically. Fuel is consumed only as the vehicle operates and in proportion to the load carried by the vehicle.

Fixed and variable costs are time-oriented. Short run refers to a time period short enough for capacity *to remain fixed*. Long term or long run refers to a time period long enough to *produce changes in capacity*. In the short run there is no change in the number of tracks, pavement lanes, locomotives, trucks, buses, or barges. Terminals and control systems experience no capacity changes; variations in traffic are met by a more or less intensive use of existing facilities.

In the long run changes in capacity occur. Increased traffic requires additional lanes of highway, more parking spaces, more tracks and signals, and more or larger aircraft, trucks, buses, cars, and locomotives. Fixed costs are increased by the capital and maintenance costs of the new facilities and total direct costs may also increase.

Cost changes due to variation in output (capacity utilized) are identified as *incremental costs*, a measure of the addition to total costs as a consequence of additional output. Conversely, *decremental* or *avoidable* costs measure the cost of reducing output. *Marginal cost* measures the costs of producing one more unit of output (or the saving from producing one less unit). Other costs will be defined as encountered.

Automobile Costs   Long and short term periods have cost significance for automobile owners. Most car owners are only concerned with short term costs—fuel, oil, tires, and repairs. These per mile costs average about

TABLE 12.2 AUTOMOBILE OPERATING AND OWNERSHIP COSTS[a]

Vehicle	First Cost Depre- ciated	Maintenance, Accessories, Parts and Tires	Gas and Oil (Taxes Excluded)	Garage, Parking, and Tolls	Insur- ance	State and Federal taxes	Total Cost
Standard	4.2	3.4	3.2	2.0	1.6	1.5	15.9
Compact	2.9	2.7	2.6	2.0	1.5	1.2	12.9
Subcompact	2.3	2.5	2.0	2.0	1.5	0.9	11.2

[a] L. L. Liston and R. W. Sherrer, *Cost of Operating an Automobile*, Federal Highway Administration, U.S. Department of Transportation, Washington, D.C., April 1974, p. 1.

eight cents. Studies have been conducted by the Federal Highway Administration that include purchase price (a capital cost) and depreciation for a suburban-based car extended over 10 years of use and 100,000 miles (160,900 Km). These data, presented in Table 12.2 show that per mile costs of operation are considerably more than commonly accepted. Details of the computations can be found in the table reference. Car ownership costs can only go higher as rising fuel costs and inflation take their toll.

Joint Costs   Joint costs are noted here because of their close relation to fixed costs. Joint costs are those costs of production that are shared by two or more products produced simultaneously, one usually being a by-product of the other. Hides, as a by-product of meat packing, are an example; in transportation, passenger and freight traffic, as on a passenger-carrying cargo ship or aircraft. This is open to some criticism as it is possible to allocate many costs separately and distinctly to each class of service. Return loads obtained on a normally one-way haul are a better example. A trailer hauling a one-way load must make a return journey to its point of origin whether it returns empty or loaded. If a return load is secured, the transportation so produced may be thought of as a by-product of the outbound load. The costs of the two loads—out and return—are joint.

Railroad freight and passenger traffic are often said to produce joint costs. These are however, more properly termed *common costs*, whereby the same plant and equipment produce two or more products, that is freight service and passenger service. Dams probably fall into the same common costs category. A dam may produce power, flood control, irrigation, and slack-water navigation. When the dam is erected for navigation and one or more other purposes, the costs of the dam are jointly shared by the several services provided. The question arises as to how those costs should be apportioned. No entirely satisfactory formula has yet been devised. One method takes account of the additional height of dam required to provide additional services. The defects of this criterion are obvious. A similar problem arises in apportioning the costs of subgrade and rights of way when a rapid-transit line is laid between the two lanes of a modern expressway. The use of the same highway by pleasure as well as by commercial vehicles is still another example of common costs.

Direct and Indirect Costs   Direct costs, sometimes termed "out-of-pocket" costs, are incurred directly by and are specifically attributable to an individual operation. The direct costs of operating an airplane are the wages of the crew, the fuel and oil consumed, landing fees, and the running repairs made between flights. Indirect costs are those incurred by the entire airline operation and are attributable to an individual flight only by some more or less arbitrary method of accounting distribution. Costs of hangar and repair facilities, accounting, sales effort, and general offices fall into this category. There is therefore a close relation and similarity between direct and variable costs and between indirect and fixed costs. The terms often are almost synonymous in transport operation.

Variable-Cost and Fixed-Cost Carriers   A resume of operating characteristics of all carrier types indicates that each type has a certain percentage of fixed costs and a complementary percentage of variable costs. These percentages vary so widely, however, with the carrier types that one can, by a crude generalization, divide the types into fixed- and variable-cost categories. It is important to understand the reasons and justification for such a generalization.

The principal factor determining the percentages of fixed versus variable costs is the amount of *fixed plant* devoted to transportation service. Before a pipeline, for example, can pump even one barrel of oil, a line of pipe must be laid from the initial to the terminating end of the route. It must be of a diameter sufficient to meet the anticipated usual peak capacity and must be wrapped and protected against corrosion, weighted in place at river crossings, and placed in a ditch on a previously secured right of way. Pumping and booster stations must be installed at intervals and gathering lines and intermediate and terminal tank-storage capacity provided. A communication system and access roads are necessary. The system must then be officered and manned and a supporting organization of sales, accounting, maintenance, legal, and stores services provided. These costs—*threshold costs*—capital investment and the interest thereon—are incurreed before the first oil is shipped. The costs of maintaining this plant and organization will continue to accrue whether the line operates at 50 or

100 percent capacity. Much of the expense will continue even if the line is temporarily shut down. Interest charges, general offices, stores, communication, and some operating and maintenance expenses will continue. Clearly this is an example of a carrier with a high percentage of fixed costs. The variable costs would be those of fuel and power, more being required for capacity pumping than for less-than-capacity pumping (but not in proportion), and the increase in maintenance due to increased wear and tear on equipment. Since this is a 24-hour-a-day operation, all other costs would be continuously incurred regardless of quantity of traffic.

Railroads are cited as another example. Before traffic can move between the initial and final terminals, a stable, well-drained subgrade and a cleared, fenced right of way are required. Bridges and tunnels, deep cuts and fills, highway crossings, and grade separations are also required. Rails, ties, and ballast must be laid to form track. Side tracks, sidings, and yards are necessary adjuncts. Stations, engine houses, shops, and fuel, water, sand, and similar facilities are built. A system of train direction with communications and signals is installed to permit safe and efficient movement of trains. Cars and locomotives, equal to the anticipated normal peak traffic must be made available. Again, a supporting organization—operating and maintenance personnel, general offices, stores, traffic (sales), legal, and other services—is necessary whether the line carries 10 trains a day or 30. A major percentage of costs are fixed and continuing without regard to the volume of traffic moving.

In contrast to the fixed-cost carrier is the one with a major percentage of variable costs and only a few fixed costs. Motor-truck operation may be cited as an example. First and foremost, the truck operator has no investment in or maintenance expense for the right of way or pavement. These are provided by the states. He does, of course, contribute to highway costs but principally through gasoline and tire taxes. These costs are incurred only as a vehicle is operated. The trucker needs no wayside fueling or maintenance stations. The local gas stations and garages meet these needs. His unit of operation, the truck, is small, and the number can be readily adjusted to the volume of traffic. Some truckers even rent their vehicles, thereby obtaining the maximum in adaptability to traffic volumes. No signals and no elaborate or expensive systems of communication are required. When communications are used, leased lines or radio are used. Station facilities usually are not elaborate or expensive. The orgainiztion supporting such a system is not elaborate either. There are some fixed costs for management, accounting, and solicitation and advertising and some investment in and maintenance of freight stations, garages, and vehicles. With a few companies, these facilities have become rather extensive. Such

TABLE 12.3 PERCENTAGE OF OPERATING COSTS THAT VARY WITH TRAFFIC

Carrier	Percentage Variable	Percentage Fixed
Railroads	25–50	75–50
Motor freight carriers (trucks)	80–90	20–10
Buses	80–90	20–10
Airlines	10–50	90–50
Pipelines	30–40	70–60
River and canal barges	50–70	50–30
Great Lakes freighters	50–70	50–30
Conveyors	20–30	80–70
Aerial tramways	20–30	80–70

companies are usually subject therefore to a higher percentage of fixed costs. It has been estimated however that no more than 10 percent of an average trucking company's costs are fixed; the remaining 90 percent are variable.[1]

Conveyors and aerial tramways fall into the fixed-cost category. A complete plant must be built and maintained regardless of volume of traffic. Barge, small airline, and bus operators are variable-cost carriers, along with truck operators. Large ship and airline operators form an intermediate group. These last are variable-cost carriers insofar as they have no direct responsibilities for building or maintaining a routeway or extensive terminal facilities. However, the great cost of even one ship or airplane establishes a basic capital and operating-cost burden that is incurred regardless of whether the ship or plane is fully loaded or almost empty.

Table 12.3 is an attempt to summarize this techno-economic characteristic for the several carriers, based on the author's studies and estimates.

Single-unit-type carriers usually are in the variable-cost category. Assembled and continuous-flow types almost invariably have a major percentage of fixed costs. Carriers utilizing small-capacity units are more likely to experience variable costs than those with large capacity units, the small capacities being more easily adapted to fluctuations in traffic. Nevertheless, the obligation, or lack of it, to provide the way and structures is usually the deciding factor. The difference between a fixed-cost and a variable-cost carrier may be an arbitrary function of ownership. If the

[1]C. A. Taff, *Commercial Motor Transportation*, Richard D. Irwin, Homewood, Illinois, 1955.

highway pavements and the trucks operating on them were combined in one ownership, the trucking operation would revert to a total fixed-cost category. If a proposal to have the Federal Government assume ownership of all railroad tracks and structures were implemented, the operating railroad companies would become more or less variable cost carriers.

Unit vs. Average Costs   The distinction between total costs and unit costs should be obvious. Unit costs, however, are frequently taken as the average, that is, the total costs divided by units of output. This may serve as crude approximations, but such costs can give misleading results. Referring to Figure 12.1a, the cost of a volume of traffic demand D_1 on an average cost basis would be the total cost divided by the units of output represented by D_1, OA/OD_1. For a volume of demand output D_2, the average cost would be OB/OD_2. The total costs will not follow a straight line increase because of the economies of large scale operation but will be parabolic giving a cost OC for demand D_2. An average cost based on cost OB would overstate the cost beyond the evaluation based on OC.

Average costs have the disadvantage of obscuring wide variations in individual costs. This could be critical. The high costs of expedited traffic averaged with the much lower costs of moving grain or aggregates is not a true indication of costs for either.

Excess Capacity and Effect on Rates   A concomitant feature of fixed costs is that of surplus or unused capacity. When a carrier is newly formed and for some time thereafter, a fixed-cost carrier is likely to have more capacity available than is actually used. A ship's hold may not always be filled, requiring it to make part of its run in ballast. The pipeline does not always pump to full capacity. The tracks of a railroad will hold more trains, the locomotives can haul more cars in a train, and more tonnage can be placed in an individual car. Also more cars can be moved through the yards.

If total costs are fixed, it is evident that the more traffic that is handled, the lower will be the unit cost of transport. The lowest unit cost would be reached, at least in regard to the fixed portion of costs, when the maximum capacity of the system had just been attained. It is also evident that fixed costs and excess capacity are found at several levels and stages in a carrier's development. Thus a railroad may eventually have its initial one track filled to capacity. Any increases in volume of traffic would lead not to lower but to higher costs owing to delays and interference of the increased number of trains. A pipeline may reach the stage of having its lines and pumps working at their maximum throughput. When that stage is

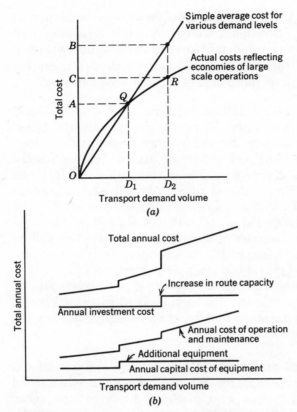

FIGURE 12.1 TOTAL COST RELATIONS. (*a*) UNIT VERSUS AVERAGE COSTS. (*b*) SHORT RUN VS. LONG RUN EFFECTS.

reached, the railroad may add a system of CTC to its single track or build an additional track, thereby raising its maximum traffic capacity to a new level. Similarly, the pipeline may install heavier pumps or add another line of pipe. Again a new stage or level of capacity, costs, and development is reached. See Figure 12.1b. It is this sequence of events that has caused some writers, J. M. Clark for example, to say that 90 percent more or less of any carrier's costs are actually variable.[2] Over a long period of time and from this point of view, Clark is correct. However, for the short and

[2]J. M. Clark, *Studies in the Economics of Overhead Costs*, University of Chicago Press, Chicago, Illinois, 1923, p. 268.

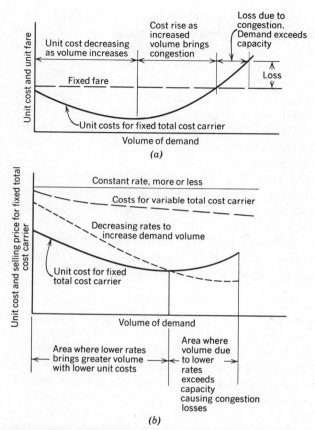

FIGURE 12.2 INCREMENTAL EFFECT OF VOLUME OF DEMAND ON UNIT COSTS OF FIXED AND VARIABLE COST CARRIERS. (a) RISING DEMAND VERSUS FIXED FARE. (b) EFFECT OF LOWERING RATES TO INCREASE DEMAND.

intermediate term and for any one level of capacity, that is, where no additional capital costs are involved, the fixed-cost concept holds true.

With a total fixed-cost mode where unit revenues are constant as with a fixed-rate (fare) public transit system, unit costs will decrease as volume of traffic increases until a point is reached at which capacity is attained or exceeded. Delays and congestion will then add to unit costs and these will begin to rise until the unit revenue or fixed fare is exceeded and losses ensue. A rise in costs due to inflation can have the same effect. See Figure 12.2a.

With no variable costs or expenses (all costs fixed), unit costs would vary directly with the volume of traffic, increasing as traffic declined and decreasing as traffic increased. In this situation, prices on some units might safely be set at less than overall average costs of operation (but greater than out-of-pocket costs) because the revenue thus obtained would help to cover a portion of fixed costs. There are no carriers that fall completely within the 100 percent fixed-cost category although a pipeline or a conveyor system is close to it. See Figure 12.2b. As before, the traffic generated by the lower rates may exceed the capacity of the systems while the costs will be greater than the revenues.

Where there are no fixed costs (all costs variable), expenses would increase directly with increases in traffic and decrease directly with decreases in traffic. Unit costs would remain approximately constant. Under these circumstances, the rate, that is, the selling price, can never be set lower than the unit cost, at least for an appreciable time, if the carrier is to remain solvent and in business. Truck, bus, and airline operations, the low capacity, unit-type carriers, come closest to this category. See Figure 12.2b.

A third possible group is that in which a large portion of the costs are variable but a significant percentage are fixed. In this situation, some rates will be set well above the cost of service for traffic that can afford to pay a high rate, but traffic that cannot be had at the higher rates will be priced at something less. This practice is similar to dumping surplus goods on a foreign market. Lower rates for off-peak loads in power production and for off-season periods at resorts and lower rates given by contract and private truckers for return loads are based on the same principle. All transport agencies are subject more or less to this situation, but railroads, pipelines, conveyors, and large ships are typical of the category.

Another characteristic of transportation, as opposed to a manufacturing industry, is that the product produced—transportation—cannot be stored. The service performed by a vehicle moving over its route is either utilized then in moving payload or is lost forever. Hence it behooves transport operators to keep their operations as near to capacity as possible.

From a slightly different point of view, one might assume a situation where there is enough traffic to cover the fixed costs of the plant and the direct or out-of-pocket costs of the actual carriage of existing traffic. The carrier would, however, make more money if its full capacity were being utilized. It therefore tries to get enough traffic to fill that capacity. Because the present traffic is already covering the actual costs, the carrier will not feel obliged to charge the same high rates as obtained from the traffic already being carried. Evidently the existing rates have brought in all the traffic that can be attracted by those rates. A lower rate, nevertheless, might attract a different category of traffic. The carrier would be willing to charge any rate that would cover the out-of-pocket or direct costs and give some small profit. In fact, even if the present traffic and rates are not covering all the fixed costs, a lower rate on *additional* traffic would be acceptable if it brought in traffic that would pay its direct costs and *make some contribution to fixed costs*. Such manipulation can be made safely only after taking account of the effects on traffic elsewhere in the system.

Effects of Cost on Competition   If the ability to establish rates were fully utilized, an excess-capacity carrier would have a system of rates somewhat as follows:

1. High rates on commodities that could afford to pay the high rates and would ship by that type of carrier in any event.
2. A lower band of rates for traffic that could not pay the high rates (or could find a competitive service with lower rates) but would be willing to utilize the carrier in question if lower rates were offered.
3. Successive bands of rates (sometimes becoming numerous to the point of a different rate band for each commodity) attracting traffic at different levels of ability or willingness to pay. The only limitation would be that the rate would have to cover out-of-pocket costs and make some contribution, however small, to fixed costs and/or profits.

A variable-total-cost carrier is thus placed at a competitive disadvantage. The proportion of fixed costs is so small that there is little leeway in setting prices. As explained earlier, the unit cost (and therefore the unit selling price or rate) remains approximately constant. Unit costs remain about the same regardless of how much traffic is secured because total costs are incurred in direct proportion to the volume of traffic. A fixed-total-cost carrier with a high excess capacity might possibly set rates so low (still covering costs) as to drive a competitor out of a particular traffic field or even out of business entirely. A strong carrier might deliberately set some rates below cost for a time for the same purpose, a form of "cut-throat competition."

In setting up a private transport system, these same principles should be kept in mind. A fixed-cost carrier might be selected where there is the likelihood of a large volume of traffic to bring about low unit costs. Where the volume of traffic is small, a variable-cost carrier would be more appropriate so that a high overhead would not continue to be incurred when the system was not being fully utilized.

Costs versus Regulation   Guided largely by fear of possible resort to the foregoing cut-throat competition, the regulatory bodies, especially the ICC, have tended to restrain the freedom of fixed-cost carriers to keep them from exercising this inherent ability to cut rates. The ICC recognizes that cost of service is a governing factor. However, they define cost of service as the average unit cost for all types and classes of traffic, "on the basis of normal operation considering *fully allocated costs*."[3] Not only must each ton-mile of traffic bring in a revenue sufficient to cover direct costs, it must also bear a proportionate share of indirect, overhead, or fixed costs. This limits the fixed-cost, excess-capacity carrier's use of a powerful competitive weapon. As might be expected, the policy has been the subject of much heated debate, expecially by the railroads, who feel themselves to be the chief sufferers. It is a problem that has not yet been resolved completely by legislation. Economists generally are likely to favor the marginal-cost concept of pricing.

Differential Pricing   Under conditions of true competition, rates tend to approach the actual costs of service. Note in this connection that the foregoing paragraphs have shown how the average cost of service decreases for the fixed-cost carrier as the volume of traffic increases and vice versa. On the one hand, the carrier cannot haul freight at a loss, that is, below the cost of service; on the other hand, the pressure of competition prevents rates that are significantly higher than the costs of service.

Where conditions of true competition do not exist, differential pricing is likely to arise. Interference with competition arises in a number of situations.

Monopoly   Monopoly pricing can arise where there is inelastic demand, that is, the service must be used regardless of price. With elastic demand, a carrier must compete with rates and services to hold or attract patronage that would otherwise use a more attractively priced service or would not use transport at all. Where there is a complete or partial monopoly, a tendency exists to take advantage of it by charging rates well above the costs of service—although there is a limit to which this is profitable. Even where competition exists, monopolistic conditions arise. A shipper with a side track on only one railroad is practically subject to that road even though he does have the right to route his freight over a connecting carrier.

---

[3]"Surface Transportation," Hearings before a Subcommittee of the Committee on Interstate and Foreign Commerce, House of Representatives, 85th Congress, testimony of Hon. Howard Freas, Commissioner, Interstate Commerce Commissioner, April 1957, p. 65.

A railroad can show favoritism by faster movement and prompter furnishing, switching, and spotting of cars. A shipper in a town served only by a truck line is entirely dependent upon that carrier. A truck line with its freight house centrally located in a traffic area enjoys a monopolistic advantage over one with a more remote location. Few if any carriers can be called truly monopolistic today, but most of them have certain localities in which their dominance is almost completely monopolistic.

Concerted action   Traffic pools, price-fixing agreements, and other restraining influences may ensue at times. Generally these practices are restrained under the various antitrust laws and as a matter of public policy, except when carried out as a permitted feature of government regulation.

Subsidies   Subsidies, usually from public sources, interfere with true competition by neutralizing the cost-of-service advantages or disadvantages. Providing roadways, airways, and shipways, granting land for rights of way, and making deficit payments to airlines and outright grants to ship lines are ways in which subsidies have been granted and have thereby influenced the cost to a carrier of the service it provides. Subsidized roadways, waterways, airways, and airports go a long way toward making the carriers that use these facilities variable- rather than fixed-cost carriers.

Subsidies are sometimes unintentional and hidden. If the rates on one commodity are less than the cost of carrying it, that commodity is being subsidized by rates on other commodities. Losses on suburban rail service are thus subsidized by higher rates imposed on freight service to cover the passenger losses.

Joint and Common Costs   The joint costs, already defined, may also affect costs of service and lead to differential pricing. There is always a tendency to have the stronger of the two or more services carry the weaker, sometimes obscuring the actual costs incurred.

Private Operation   From the standpoint of overall costs, private ownership of transport offers many advantages to an industry equipped to haul its own products. There need be no element of profit included in the costs, the operation is not subject to government regulation, and the size of the operation can be tailored to the needs of the industry served. When a private carrier is organized as a separate company and gives service to more than one unit in a large organization, when it makes a nominal purchase of goods carried, with resale at the time of delivery, or when it hauls its owner's freight from other industries, a question arises as to whether or not it is not actually performing contract or even common-carrier service and therefore subject to regulation. This is an unresolved problem.

Peak Demand Effects   A carrier with extreme peaking constraints, urban rail or bus transit for example, may encounter critical costs problems; especially rail transit with a large total fixed cost. In the short run, urban transit may experience lower unit costs as traffic volume increases. The demand volume might, however, grow so large eventually as to cause the cost curve to turn upward as in the dotted portion of Figure 12.2. An increase in capacity—for example, more buses, more rapid transit cars, additional tracks and signals—is required to meet the increased demand.

The need for such increased capacity is usually imposed by peak period commuter traffic, moving, more or less, between 7 to 9 a.m. and 4 to 6 p.m. A high capacity capability is then required, but because of low off-peak demand much of that capacity stands idle, earning no revenue, but incurring maintenance and interest charges 20 out of every 24 hours.

If a conventional "average" fare is imposed, the peak period riders pay too little, the off-peak riders pay too much; all costs are probably not recovered. If the single fare is set high enough to cover all peak period costs, the off-peak riders are still paying for capacity excess to their needs. They are, in effect, subsidizing the peak period riders in both instances.

Several solutions have been tried or proposed.

1. A one-level "average" fare with deficits made up by public subsidy, a frequently used solution.
2. A one-level fare sufficiently high to cover all costs. As noted, the off-peak rider pays too much. Also the high fare prices many riders out of the market.
3. A two-level fare for peak and off-peak riders. This is more equitable than the foregoing but incurs difficulties in fare collection and administration.
4. A no-fare service. Costs are borne by the community tax base.

There are several arguments in support of partial or complete (no-fare) subsidy.

(a) Public support of mass transit systems reduces the need for even greater expenditures for expressways and parking.
(b) Public support is a development cost that brings labor and patronage into a central business district with an overall gain to the community.

(c) Some means of transport must be available to those incapable of using automobiles—the aged, those under age, the infirm, and the poor, for example.
(d) Public transportation should be provided as a means of conserving energy and reducing pollution.
(e) Transportation is a municipal service that should be supported by and available to all in the same way that street lighting and police and fire protection are provided.

The one major argument against public support is the classical concept that the immediate user and beneficiary of transportation should pay the costs.

An extension of the foregoing problem statement and the pros and cons of suggested solutions to intercity freight and passenger carriage should present no difficulties.

Governmental Regulation   Government regulation is directed primarily toward rate making in an effort to prevent cut-throat competition and discrimination between persons, places, or things.

State and Federal commissions have set rates on other than a cost-of-service basis for various reasons: to aid drought-stricken areas, to assist new or critical industries, to foster new types of transportation, or to equalize the competitive position of carriers, commodities, or communities. On the other hand, regulatory bodies generally have not recognized certain cost-of-service features as being permissible. For example, the unit rate for a long haul is required to be the same as for a short haul included in the long haul although the unit cost of making a long haul is less than for the short. There has been considerable ICC reluctance to grant train-load rates as contrasted with car-load rates although the train-load costs are less.

The railroads also claim that the ICC's *fully distributed cost* concept prevents taking advantage of the reduction in unit cost that comes with increased volume of traffic and that the ICC holds an "umbrella" over competing modes by preventing rate reductions by the railroads below the level of their competition. This argument represents an unresolved "gray area" of conflict.

Composition of Rate   From the cost-of-service standpoint, rates usually contain two principal elements. The first and obvious element is that of line-haul service. This is the cost of providing and maintaining the route and/or equipment and of hauling the goods from one point to another. The second element is the cost of terminal service. This includes costs of freight-house, tank-farm, airport, or transit-shed operation, of classifying and sorting cars in yards, of delivering cars and LCL shipments to final terminal destinations, and of providing passenger-station and ticket-selling facilities. Transfer and placement of cars by other carriers, tug service in berthing ships, and movement of passengers from town to and from airports are supplemental services for which an additional charge is made. Certain other services may be included in the line-haul rate, but reflected in its magnitude, such as icing of refrigerator cars and in-transit privileges. Terminal costs, it should   be noted, are fixed for any one shipment regardless of length of line haul. Minimizing terminal costs is essential to transportation economy.

The entire subject of rates and pricing is highly complex and controversial. Only the more obvious aspects of the problem can be included in the scope of this book.

QUESTIONS FOR STUDY

1. What is the significance of cost of service as a basis for transportation charges as contrasted with other criteria?
2. Distinguish between capital costs and operating expenses; between fixed and variable costs, between short-run and long-run costs.
3. What factors tend to make a transport agency a "fixed"- or a "variable"-total cost carrier?
4. Explain fully the concept of "excess capacity" and relate this to fixed and variable costs.
5. How does the fixed-variable-cost concept introduce hazards in the use of average costs?
6. A railroad can move 5000 net tons a day over its 700 miles of line at a total cost of one cent per net ton mile. Assuming 50 percent of the costs to be variable, plot the unit costs for moving 10,000, 15,000, and 20,000 over the same line.
7. A motor freight carrier moves 400 net tons of freight per day over a run at a cost of 4 cents per net ton mile. What would be the cost per net ton mile of moving 600 tons over the same territory? 200 net tons?
8. In what situations may a rate be influenced by other than cost-of-service factors?
9. How does the fixed- versus variable-cost characteristic affect the competitive position of each type of carrier?
10. To what extent is a private carrier concerned with the characteristic of fixed-variable costs?

SUGGESTED READINGS

1. J. M. Clark, *Studies in the Economics of Overhead Costs*, University of Chicago Press, Chicago, Illinois, 1923.
2. Kent T. Healy, *The Economics of Transportation in America*, The Ronald Press, New York, 1935, Chapters 10 to 13, on the cost and pricing of transportation.
3. G. M. Wellington, *The Economic Theory of Railway Location*, 6th edition, Wiley, New York, 1914, Chapter V ff., on operating expenses, pp. 106.185.
4. W. W. Hay, *Railroad Engineering*, Wiley, New York, 1953, Chapters 3, 4, and 12.
5. C. A. Taff, *Commercial Motor Transportation*, 1953 edition, Richard D. Irwin, Homewood, Illinois.
6. *Surface Transportation*, Hearings before a Subcommittee on Interstate and Foreign Commerce, House of Representatives, 85th Congress, April 2, 3, 4, 5, and 11, 1957, Government Printing Office, Washington, D.C.
7. The Interstate Commerce Act, Government Printing Office, Washington, D.C.
8. The Civil Aeronautics Act and the Federal Aviation Act, Government Printing Office, Washington, D.C.
9. Meyer, Peck, Stenanson and Zwick, *"The Economics of Competition in the Transportation Industry,"* Harvard University Press, Cambridge, Massachusetts, 1964.
10. Earnest C. Poole, *Costs—A Tool for Railroad Management*, Simmons-Boardman Publishing Corporation, New York, 1962.
11. W. J. Baumal et al., "The Role of Cost in the Minimum Pricing of Railroad Service," *The Journal of Business of the University of Chicago*, Vol. XXXV, No. 4, October 1962.
12. *A Guide to Railroad Cost Analysis*, Bureau of Railway Economics, Association of American Railroads, Washington, D.C., 1964.
13. *The Eonomic Analysis of Railroad Roadway*, An Interim Report on Development of Methodologies and Procedures, prepared for the Federal Railroad Administritration Contract DOT-FR-30028 by the Bureau of Transportation Research, Southern Pacific Transportation Company, J. H. Williams, Project Head, March 1974.
14. Ann F. Friedlander, "The Dilemma of Freight Transportation Regulation," Brookings Institution, Washington, D.C. 1969.
15. Alan J. Montgomery, Staff Analyst, Association of American Railroads, "Motor Carrier versus Intermodal Costs—The Highs and the Lows," a speech presented at the National Rail Piggyback Association, January 1975.

Part 4

Planning
for
Use
and
Development

Chapter 13

Transportation Planning: Goals and Processes

PLANNING REQUISITES

Transportation planning is comprised of a variety of problems and procedures. These will vary with the level at which the planning is conducted and the type of need to be met. It may be the detailed location of a specific route, the selection of for-hire services by an industrial traffic department, or the planning of an entire system of private transportation by an industry. The planning may also be for the establishment of a pattern of Federal interstate highways, waterways, or airways, for the integration of various modes of transport in an urban area, or—on rare occasions—for the overall needs of the nation among all modes of transport.

Who Does the Planning Planning is conducted at many levels by a variety of agencies. Private companies, including those that provide and those that use transportation, make decisions for their corporate transportation needs in board rooms using data and alternative proposals developed by consultants or more often by special committees of their personnel or by their traffic, industrial engineering, or systems planning departments. The planning is usually compact and private. Notable exceptions are where private companies become involved in publicly financed projects such as the East St. Louis Terminal Studies, the Penn Central reorganization, and the introduction of high-speed Metroliners for the Northeast Corridor, or a grade crossing separation.

For private companies, the profit motive predominates and responsiveness to public need is usually only in relation to its effect on profits or through conformance to requirements of laws and regulatory agencies. Alternative plans will be subject to even closer financial scrutiny than is often accorded public projects. The public, nevertheless, may be served through the efforts of private concerns to meet competition.

Public Planning   Public transportation has been variously performed by federal, state, regional, county, and municipal bodies. Consultants have frequently performed the actual work, but many such bodies have established their own planning departments or commissions for that purpose.

The Federal Aid Highway Act of 1962 required any urban area of 50,000 population or more (after 1 July 1965) to carry on a continuing comprehensive transportation planning process to receive approval for any Federal Aid Highway project. Thus much of modern planning has been highway-oriented. State highway departments often have the responsibility of determining the need for highway transportation. The department or its consultant might take account of the existing system of federal, state, and county roads and future federal proposals, prepare projections of population, agricultural, industrial, and commercial growth, and project from them a plan (or plans) for the improvement of existing roads and for new road construction that will form a comprehensive system to meet projected future needs of the state. State departments of transportation may also be responsible for the rail, air, waterway, *and* highway needs of their states. When urban areas are involved, the state may make inventories of facilities, travel flow and traffic patterns while the local community develops inventories and forecasts of land use, population, and the local economy; or, the local community may perform all of the data-collecting and analyzing tasks or have a consultant do the work.

Federal involvement in planning is seen in the planning processes conducted by the Bureau of Public Roads for the Interstate Highway System and by planning for the formation of Conrail. The several National Transportation Reports are further examples of planning and data collection at the Federal level to determine the future transportation needs of the nation.

Planning on a regional basis is exemplified by the Chicago Area Transportation Study (CATS). The study was first established in 1955 by an agreement between Chicago, Cook County, and the State of Illinois. The State would provide $1\frac{1}{2}$ percent of Federal planning funds to the undertaking. CATS was later expanded to encompass the six counties comprising the Chicago Standard Metropolitan Area. Its duties, among others, were to determine policies, gather data and forecasts of data, recommend programs for improvement, and render technical assistance to local units of government. The original plan proposed by CATS receives a continuing revising and updating.

On a local level, the cities of Champaign and Urbana, Illinois organized the Champaign-Urbana Urbanized Area Transportation Study (CUUATS) in 1965 to provide transportation planning for the area under the Federal Aid Highway Act of 1962 requirements. Its responsibilities have been expanded to include air travel and mass transportation and its jurisdictional boundaries were expanded to coincide with those of Champaign County. Staff functions, formerly rendered by the Illinois Department of Transportation, have been transferred to the Champaign Regional Planning Commission. CUUATS performs surveys, develops plans and cost studies, and reports these to the Regional planning Commission and to decision-making bodies in appropriate jurisdictions. CUUATS has two committees, a policy committee composed of a group of officials who determine the direction and priorities of study and a technical committee comprising informed professionals who provide technical information and recommendations to the Policy Committee. Citizen input, among other ways, is obtained through a citizens advisory transportation committee.[1]

Steps in Planning   Planning and the implementation of the plans usually follow a general pattern according to the steps briefly listed below. Obviously the detail or need for each step will vary with the project.

1. Recognition of Need

Need can be current and acute—a bad congestion situation, lack of access to a new subdivision or shopping center, an intersection with a high accident record. In contrast, the need may not be apparent until a survey has been made to establish present and future need. The Federal Aid Highway Act of 1962 quickly pointed out a need for planning to those communities wishing to share in Federal funds. A breakdown in rail service in the Northeast led to the planning that led to Conrail.

2. Planning Goals

Planning must have direction and specific purpose. Goals that represent community values and objectives needed to attain

<hr>

[1]*Annual Report*, Champaign County Regional Planning Commission, Champaign, Illinois, 1975, pp. 26–27.

those goals must be established. Planning goals represent the direction in which a society—regional, state, or national—wishes to move. A community principally concerned with economic progress will have the attraction of commerce and industry as its goal and will design a transportation system to accommodate those activities. A town concerned with functional efficiency may demand straight, direct thoroughfares, whereas one concerned with aesthetic qualities will tolerate a fair amount of congestion or circuity if the beauty of ancient trees and buildings can thereby be preserved. Goals represent general desires of a community.

3. Objectives   Objectives are the means by which goals are realized—by building an expressway (or by not building it), by introducing scheduled bus service or turning to a demand-responsive system, or by reorganizing bankrupt railroads into a Conrail System. *Criteria* are used to quantify objectives. Establishing a 15-percent excess of parking capacity over demand represents a criterion for parking capacity development. The parking development represents, in turn, the objective used to fulfill the goal of no congestion in the downtown area.

4. Demand Surveys   The demand survey establishes a data base from which planning can proceed. Traffic loads to be carried by the system are largely a function of land use and population. Surveys develop the growth history and present status of population, land use, industry, commerce, existing transportation systems and uses made of those systems. More detail on specific techniques are found in the next chapter.

5. Demand Analysis; Projections   Once the demand has been established, the traffic is apportioned over existing routes and modes using traffic assignment procedures detailed later. Present capacity is compared with present demand and excess and deficit capacities are noted. Modal choice may be taken into account at this point. Traffic demand is projected for the future, assignments to routes are made, and capacity excesses and deficits again noted.

6. Design of Solutions   All possible solutions should be considered and the most promising two or three developed in detail.[2] Selection of mode, design of location and network, level of service contemplated, and economic cost for each alternative are prepared. Social and environmental consequences are noted.

[2]In planning for a revision of the rail network on the Niagara Peninsula, over 60 different networks were given an initial evaluation; 12 of these were studied in detail.

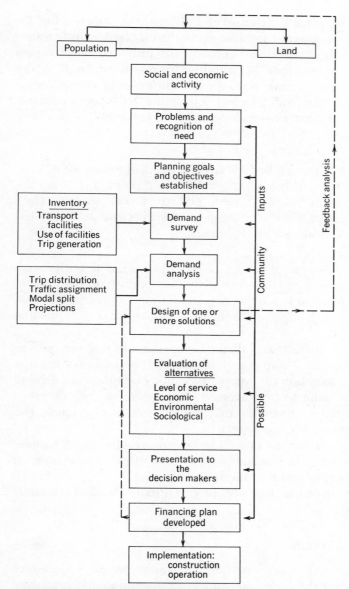

FIGURE 13.1 PLANNING PROCEDURES.

7. Evaluation of Alternatives   The several alternatives selected for detailed analysis must be evaluated for and during presentation to decision-making bodies. The evaluations should consider the utility or effectiveness of the alternative solutions, that is, whether they will accomplish the intended objectives. The economic cost of each must also be computed as well as the social and environmental costs. All significant consequences of each alternative solution must be developed. Public acceptability is a very important criterion.

8. Presentation   The planning body is seldom the decision-making body. Recommended plans and viable alternatives are presented to the appropriate planning board, county board, city or metro council, state legislature, or U. S. Congress for acceptance and authorization. Suggested methods of financing should also be included.

9. Implementation   Once a plan has been approved and authorized, procedures must be established for financing that permit preparation of final plans and designs, land acquisition, bid-letting, and construction, followed by the final step, operation. Figure 13.1 illustrates a possible planning sequence.

The breakdown into separate and distinct steps has been made only as an aid to study and analysis. There is a close interrelation and coordination between all steps. The selection of a modal type, for example, may depend almost entirely on the route to be traversed—or vice versa. The development of an adequate plan of financing may govern the entire project. The engineer should be involved in each step.

Initial planning can indicate the project to be unwarranted or unfeasible. In these events, planning has accomplished a worthwhile purpose in preventing a wasteful expenditure of funds and effort.

Several of the foregoing nine steps will be given more detailed consideration in following pages.

GOALS AND OBJECTIVES

As noted earlier, planning goals are general summaries of community values such as improving the economic status of the community, providing more job opportunities, or enhancing the quality of living. Objectives are the specific ways in which (in the area of transportation) those goals will be realized. Problems of defining demand characteristics, modal selection, and effects on values and environment are involved.

Planning Goals The very act of recognizing a need for improved transport may set planning goals. A primary goal would be the satisfying of that recognized need. The extent and kind of demand must be established and met. There are, in addition, general goals applicable in whole or in part to all transportation planning.

It can be assumed that a specific level of service must be established to meet estimated demand. *Adequate capacity, speed, frequency of service, and accessibility* are inherent planning goals. An appropriate *quality of service* must also be selected. Quality of service is often related to demand. A poor quality service can bring public rejection.

The compass of planning can be summarized by stating that the goals of transportation should be to provide a safe, door-to-door service, readily available at all times, offering all-weather dependability. The service should have a minimum disruptive and adverse effect on the environment and on the community and its values. It should offer a reasonable degree of comfort—all of these at reasonable cost.

Table 13.1 contains a listing of goals that might pertain to any planning effort. There are many others, often entirely local or peculiar to a particular situation.

Since one function of transportation is to connect population with land use, an additional goal is to assure that all intended land uses and populations have interaccessibility. Where it is desired not to develop a piece of land, the withholding of transportation can assist materially.

Where a privately owned and financed transport service is undertaken, a proper goal is making a profit. Profits will not be forthcoming, however, if the foregoing factors are not included in the planning. The service must be useful and attractive. More specifically, the transport needs of the individual shipper or traveler must be determined and understood, especially if the proposed service will be competitive with another. For commercial transport this effort is often referred to as merchandising rather than planning. The end result should be the same, a reasonable accommodation to the patron's need.

Establishing Demand The recognition of need may carry with it a fixed level of demand. The amount of production anticipated from an oil field determines the capacity of the pipeline that carries its output. The level of service on an ore-hauling railroad is based on the mining engineer's estimate of annual output (and sales) from the mine. An industry's estimate of its production and sales determines the need for the use of its own or of common carrier transport.

TABLE 13.1 TYPICAL GOALS FOR PUBLIC TRANS-
PORTATION PLANNING

1. Promote flow of traffic (relieve congestion)
2. Reduce travel time
3. Improve safety
4. Reduce costs of service
5. Give access to all land uses
6. Give access to a specific land use
7. Increase frequency of service
8. Make service more accessible
9. Serve the aged, infirm, minors, and other nonauto users
10. Provide all-weather service
11. Maintain existing land use patterns
12. Change existing land use patterns
13. Maintain or enhance tax base
14. Reduce pollution—air, water, land, visual, and vibratory
15. Preserve historic buildings, parks, and views
16. Preserve aesthetic features, trees, for example
17. Preserve or improve ecological balance
18. Recognize needs of all groups and interests
19. Preserve community values and neighborhood integrity
20. Maintain or create work opportunities
21. Maintain or increase dwelling unit availability
22. Meet industrial and commercial demands
23. Conserve energy
24. Make raw materials available; also finished products
25. Widen market opportunities

Planning a new route or service may require a more complex procedure. New highways, for example, require a study of the economic, social, and environmental status of the region to be served and the impact the proposed facility will have on these aspects. The new facility may attract traffic from existing routes and modes. It can be a stimulant to industrial, commercial, or agricultural growth; it can retard growth elsewhere.

Traffic sources or generators must be identified and their trip or freight generating potential quantified. The distribution of traffic, tons of freight or person or vehicle trips, to destination points and the modes and routes used for that movement must be determined. Peaking characteristics and their effect on required capacity are needed information. A discussion of inventory surveys, origin-destination studies, trip determinants, traffic assignment procedures, and trip distribution models common to urban and regional transportation planning is reserved for Chapter 14.

Present demand is analyzed and projected to some future date—20 years hence or, as of today, the year 2000. A traffic pattern, actual or estimated, is developed to include a quantitative evaluation of what is to be moved, where it will originate, where it will terminate, and the volume of movement by traffic type, route, mode, and terminal. Modal choice must be evaluated. There are problems in obtaining accurate current demand. The problems of projecting these elements to some future date are even more difficult and the results less certain.

Selection of Mode   A selection of mode must be made when an urban, regional, or other transport system is being expanded. Shall the city install a rapid transit system (as has Washington, D. C.) or extend its system of expressways? Shall automobiles be barred from certain streets so that a bikeway network can be established? How should state funds be divided between the needs of rail, water, and highway transportation? Technological and cost factors are involved.

The selection of mode is an obviously important aspect of the transportation planning process. Too often the subject is approached in the public mind (and not infrequently in the professional mind as well) with a mixture of ignorance, emotion, bias, and vested self-interest. It bears repetition that the proper goal in transportation is service to society, directly or indirectly. Mode in whatever form is merely a means to that end. The mode to be selected is the one that offers a maximum degree of utility, that is, is best suited to the task at hand, the one that can deliver the required level and quality of service at an acceptable level of economic, social, and environmental cost.

Once a demand level has been determined, various modal alternatives can be considered to establish their capability of providing the required capacity with a frequency of schedules and movement, over a route or network that will make the service available where and when desired.

The quality of service must also be considered. Demand is in part quality-oriented. A mode or system that meets all requirements for capacity, frequency, accessibility, speed, and cost may still be unacceptable because of its lack of dependability, excessive use of energy, or its adverse effects on the environment.

Utility   Planning goals aim for safe, convenient, rapid, door-to-door service, with all-weather, specified volume-specified commodity dependability, at a reasonable cost. The ability of any mode to perform such service is measured by its degree of utility or usefulness. Utility, in turn, is a function of a mode's technoeconomic characteristics. Often the modal utility of one carrier can best be realized in combination with those of another carrier—containers moving by land and water, trailer-on-flatcar, or pipeline-tank truck or barge combination, for example. See the section on cost-effectiveness.

Balanced Transportation Utility leads to the concept of balanced transport, the using of each mode in its area of optimum utility. In any other use the utility might be marginal, its use supported by other than engineering cost and capability. Balanced transportation does not mean equality in use or funding but rather the providing of services to meet various demands according to the technoeconomic ability of a mode to deliver that service. Priorities in land use and allocation of funds, the analysis of modes on the basis of the ability to provide desired levels and qualities of service, and the assignment of modes to appropriate areas of usefulness are all involved. It may also mean that special encouragement is warranted for one mode or another when its development and use has, for a time, been subordinated to the development of other modes.

Utility versus Technology The technoeconomic characteristics that contribute to utility have been stated earlier and include such cost-related items as propulsive resistance, propulsive force, payload-to-empty weight ratio, thermal efficiency, response to gradients, flexibility, safety, speed, capacity, effects on the environment, dependability, incidence of pollution, cost, and many other factors. One could evaluate each mode under consideration on the basis of each of these factors. Because many are reflected in construction and operating costs, the evaluation can be made primarily on a cost basis and then given further evaluation on the basis of safety, dependability, flexibility, effects on environment, pollution, and energy requirements. One of several possible procedures for evaluation is in the form of Table 13.2. An analysis of modal and other alternatives is presented in Chapter 15.

Modal Utility The way in which technological characteristics are reflected in a modal system utility is of significant interest. Aerial tramways have high initial construction cost, due primarily to their long, heavy cables. They have low productivity and poor flexibility in cargo type and traffic volume but are adaptable to terrain where no other type of transport can penetrate with reasonable ease and economy.

Conveyors, with expensive belting, also have high initial costs that require an offsetting high volume of traffic. These combine with low route and cargo (principally granular bulks) flexibility and the need for duplicate flights for two-way movement.

Pipelines within the field of fluid transport, are analogous to conveyors. Water, sewage, gas, and petroleum and its products have been the only fluid commodities moved in sufficient quantities to warrant extensive pipeline construction. Pipeline movement of suspended solids has broadened the area of pipeline utility, but the additional energy needed to crush ore or coal into small particles and the large volume of water required for suspension (later to be removed through the use of more energy) pose problems in the use of limited energy and water resources.

The net ton miles per airplane hour have not been high despite a high speed factor. Limited flexibility in takeoff and landing for commercial transports and uncertainties due to weather are additional handicaps. Nevertheless, the technological advantage of speed gives overwhelming favor to air transport for a large segment of the traveling public and to a somewhat restricted group of shippers.

Waterways offer favorable horsepower, productivity, and cargo flexibility. But slow speed, lack of route flexibility, and weather susceptibility have limited inland water transport to commodities for which speed is less important than quantity movement. Ocean movements of all commodities continues, but air transport has acquired most of the passenger business and some general cargo.

Despite low productivity, a high accident rate, and weather susceptibility, the private automobile has become a principal carrier of passengers with a hold on the American fancy a little out of proportion to its technological advantages. Motor trucks carry a high percentage of the nation's freight. The outstanding feature of highway transport is route flexibility that permits its use for a journey of a few blocks or coast-to-coast as desired. The truck's usefulness in terminal movements has earlier been noted. The ability to move a trailer load directly from the door of the shipper to that of the consignee is a significant advantage. The use of trucks for hauling bulk commodities is generally advantageous only when the distance is short, as in feeder operation. Three hundred 20-ton trucks would be required to haul the contents of one 6000 net ton freight train; 1000 to move the contents of one Great Lakes bulk cargo carrier.

Railroads offer low propulsive resistance combined with the dependability of movement arising from wheel-flange guidance, and flexibility in route and movement, cargo type, and cargo volume. Technologically speaking, there is no field of transportation, with the exception of detailed terminal movements, where railroads cannot perform a serviceable function, but their maximum efficiency is displayed in the mass movement of bulk freight in unit trains.

TABLE 13.2 TECHNOLOGICAL UTILITY

Carrier	Principal Technological Advantages	Field of Usefulness
Railroads	Minimum propulsive resistance and general flexibility, dependability, and safety	Bulk-commodity and general-cargo transport, intercity; of minimum value for short-haul, intraterminal freight traffic; commuting traffic in densely populated areas and intercity passenger movement for 50 to 300 miles (80.45 to 482.70 km); also long-haul passenger transport where speed is not paramount.
Highways	Flexibility, especially of routes; speed and ease of movement in intraterminal and local service	Individual transport; also transport of merchandise and general cargo of medium size and quantity; pickup and delivery service; intraterminal and short-to-medium intercity transport; feeder service
Waterways	High net-ton-mile-per-ship (or towboat)-hour productivity at low horsepower per ton.	Slow-speed movement of bulk, and low-grade freight where waterways are available; general-cargo transport where speed is not a factor or where other means are not available; not generally useful for domestic passenger service.

Fuel Economy   The cost of fuel is one of the big items in transportation operating expense. Economy dictates the need to get as much transportation, net ton-miles per ton or gallon of fuel, as possible. The energy shortage has made fuel economy a national problem. An initial factor is thermal efficiency as discussed in Chapter 5. High efficiencies are obtained with a central steam or diesel generating plant feeding electrical energy to electric locomotives, trolley buses, and rapid-transit trains. Turboelectric drive on ships approaches the same efficiency. Direct utilization of diesel-engine drive also makes for high thermal efficiency. Direct steam drive in ships and locomotives gives the lowest. Losses in power transmission also enter the picture as in the 50-percent losses in ships' propellers and

TABLE 13.2 (Continued)

Carrier	Principal Techno- logical Advantages	Field of Usefulness
Airways	High speed	Movement of any traffic where time is a factor—over medium and long distances; traffic with high value in relation to its weight and bulk; helicopter feeder and taxi service
Pipelines	Continuous flow; maximum dependability and safety	Transport of liquids where total and daily volume are maximum and continuity of delivery is required; have potential for use in movement of suspended solids when adequate fluid supply is available
Belt conveyors	Same as for pipelines	Transport of granular bulks where total and daily volumes are a maximum and continuity of delivery is required, 0.10 to 100 ± miles (0.16 to 160.0 km); have potential future development as carriers of urban passengers
Aerial tramways	Not greatly affected by terrain; low horsepower per net ton-mile	Useful only where terrain makes any other type of transport uneconomical or impossible; perform a feeder service
Personalized rapid transit	Lightweight vehicles and guideways, low-speed, low-capacity vehicles giving privacy	Airport, shopping centers, CBDs, networks for communities of 10,000 to 500,000 ±

shafting and the 20- to 40-percent losses in airplane propellers.

The horsepower-per-net-ton ratio and deadload-to-payload ratios also enter into overall fuel economy. These are related, in turn, to the unit propulsive resistance for each type of transport through fuel consumed simply to overcome friction and move dead weight. Other factors contribute to energy consumption: the amount of idling time for motive power, extent of empty mileage, typical load factors, and circuity of routes. The load factor must be considered both in terms of present practice and what could be achieved by direction and incentives. The actual consumption of fuel per passenger mile and per ton mile are the critical values, posing the question of what modes of transport should be encouraged in the interest of energy conservation. Quantitative evaluations of fuel consumption by various modes are given in Chapter 8.

Mr. A. K. Branham, Research Associate, Joint Highway Research Project, Purdue University, is reported as saying:

> The average automobile consumes approximately one gallon of fuel for each 15 miles. The average passenger load of the automobile is in the range of 1.5 to 1.7. If we assume 1.6 to be the average, then 1 gal of fuel will generate about 24 passenger miles. The average Class I intercity bus consumes about 1 gal of fuel for each 5.3 miles. The most popular size of intercity bus, when fully loaded, carries 37 passengers. Thus in assuming a full load, the bus may use 1 gal of fuel in generating 196 passenger miles. With a 50-percent load, this bus generates 92 passenger miles, or about four times that of the private automobile. For economy of our national resources can we justify intercity travel by private automobile?[3]

Users of energy can be listed in order of theoretical fuel efficiency as: pipelines, waterways, railraods, highways, and airways. Some studies ascribe to railways a greater fuel efficiency than waterways from the practical standpoint of route circuity. Obviously the relationship will vary for routes of different geographical location and distance. Chapter 8 contains a more detailed discussion on energy utilization.

Effects on Environment   The impact of transport modes on the environment is having increasing significance. These effects have already been given in Chapter 8, but principal impacts are here briefly summarized.

Highway vehicles account for a major portion of air pollution. Technological solutions are under development. A more ready solution is a reduction in the use of motor vehicles. Highway vehicles also contribute to noise, vibration, and sometimes to visual pollution.

Aircraft are a source of noise and air pollution, principally at takeoff and landing. The adverse effects of high flying jets and supersonic aircraft on the protective ozone layer surrounding the earth are not fully established, but the potential for hazard has created lively discussion. Noise, a major polluting effect of jet aircraft, is strongest within airport confines, but also affects areas several miles distant from the runways.

Oil barges and oceangoing tankers pose no pollution problems as such but can cause water and shore pollution, often of disaster magnitude,

[3] A. K. Branham, "National Transportation and Research Applied to Some Highway Transportation Problems," *Highway Research Abstracts*, April 1950, p. 19, Highway Research Board.

through careless dumping of waste water and sewage, bilge, and cleanout wastes, or through leaks and collisions.

Properly maintained and operated railroads produce some vibratory, noise, and water pollution (from engine house wastes), the last two usually originating in yard and terminal areas. The dangers from derailments involving cars with flammable, explosive, or corrosive contents, especially

TABLE 13.3 TYPICAL CONSTRUCTION COSTS PER ROUTE MILE
(EQUIPMENT INCLUDED EXCEPT WHERE SPECIFICALLY EXCLUDED)

Carrier	Cost per Mile of Route (dollars) (1 mile = 1.609 km)
Railroads:	
Track only[a]	100,000 to 200,000
Track, structures, and equipment	200,000 to 500,000
Rapid-transit—surface[a]	2 to 8 million
Rapid-transit—subway[a]	6 to 15 million
Highways:[a]	
24-ft rural highway (concrete, 2-lane)	100,000 to 400,000
Freeway (urban area)	6 to 12 million
Superhighway (expressway, freeway):	(Cost depends on number of structures.)
Maine Turnpike[b]	450,000
Pennsylvania Turnpike[b]	461,000 and 861,000
New Jersey Turnpike[b]	1,900,000
Ohio Turnpike[b]	1,250,000
Pipelines:	
Petroleum and gas	150,000 to 200,000
Coal (suspension process)	150,000 to 200,000
Conveyors:	
Belt-type	200,000 to 1 million
Two-way, Great Lakes to Ohio River proposal: 104[c] miles, 72-in. belt; 104 miles, 24-in belt; 27 miles, 42-in. belt	2.1 million
Moving sidewalk	1 to 1.5 million
Aerial tramways	70,000 to 120,000
Monorail:[a]	300,000 to 1 million
Alweg System, São Paulo proposal	2,500,000
Houston—experimental[c]	200,000 to 500,000
Los Angeles—proposed[c]	945,000

[a]Equipment excluded.
[b]Original construction cost.
[c]Proposed.

prevalent where low standards of track maintenance prevail, are a current problem receiving serious study for correction.

Oil pipelines, generally placed below ground, offer no environmental threat except when a break occurs in the line. Opposition to the Alaska pipeline has centered largely around the unique positioning of parts of the line above ground to avoid settlement in permafrost areas, thereby offering a possible obstacle to wild life migration.

Other environmental effects, involving all modes, include to varying degrees noise, vibration, motive power exhaust, extent of land use, visual pollution (lack of aesthetic quality), and intrusion of the roadway and its associated activities into residential neighborhoods, parks, and scenic areas, and adverse impacts upon the delicate ecology of wilderness and remote areas.

Technological Summary   Areas of technological utility are summarized in Table 13.2. These generally represent areas of economic utility as well. Nevertheless, evaluation procedures discussed in later pages should be used to establish the system cost effectiveness and the utility of any given set of goals and environment.

Construction Costs   Because initial construction or capital cost is often of prime importance in modal choice, some typical examples are given in Table 13.3. For urban expressways and rapid transit such costs are almost without meaning because of the mile-to-mile variations due to local conditions and land values. The examples given will suffice to indicate that such costs can be exceedingly high.

METHODS OF FINANCING

Importance   The ability to raise investment capital is a basic requirement for any project. The method of financing may have considerable bearing on the design of plant and equipment, location of routes, and rate of progress of the work. Land grants by Federal and state governments and gifts from local communities influenced the location of many miles of railroad. The lack of large amounts of capital may require a deferred or stage construction program. The use of Federal funds for highway construction and improvement is limited to routes of Federal choosing. In any case, a project should not be undertaken if funds are not available for its completion or if it is not going to be sound enough to attract investment capital.

Two primary types of financing are found in transportation enterprises: private financing and financing with public funds and credit.

Savings and Reinvested Income   The most conservative method of financing is to utilize savings. Whether the project is private or public, accumulated surpluses and reserves can be invested in the company's or government's own activities. If the return to be earned in the company's own business isn't attractive enough for reinvested savings or earnings, the wisdom of its continuance is in doubt. Furthermore, outsiders could hardly be expected to hazard their capital in such a venture. However, savings and reserves are seldom adequate for large projects, so additional resources must be had by other means.

Stocks and Bonds   Capital investment in the form of stocks and bonds has been a traditional method of financing for railroads, pipelines, and other commercial operations. No return need be paid on stock unless it is earned (and not always then) so that stock sale represents the lesser obligation for the carrier. Nevertheless, stock sale disperses ownership and will not bring the desired capital unless the project has a good credit position.

Bonds are a fixed obligation of the company. The effort to pay fixed charges and avoid bankruptcy often leads to deferred maintenance and accelerated deterioration of the plant and equipment. Service also suffers. There is no income or new capital to replace an obsolete or worn out plant or to take advantage of new developments. The burden of paying fixed charges is excessively great for an enterprise attempting to become established.

If a transportation company goes into receivership, the trustees usually concentrate first of all on saving the bond holder's investment by making up deferred maintenance and restoring the physical integrity of the property that has been pledged as security for the bonds. Maintenance engineers may find it more satisfying to work for a carrier with ample funds to carry on their work than for a solvent, economy-minded carrier!

Public Financing   The basis of most government financing is the general credit and taxing power of the government, whether it be Federal, state, or local. Traditionally, the Federal Government has provided waterways—constructed, lighted, policed, and maintained harbors, canals, locks, rivers, and dams—without cost to the user. Airways and navigational aids are also provided by the Federal Government. Construction and maintenance costs are financed through the general tax fund by Congressional appropriation, supplemented by an 8-percent "seat" tax.

Other government aid has included land grants to early railroads as an encouragement to build into undeveloped territory, direct subsidy payments to airlines (during their development stages, sometimes rather prolonged) and to the merchant marine, and loans and guarantees for loans, especially to the railroad industry in more recent times. Airports have been built for airlines, but for these, landing fees and hangar, etc. rentals are charged.

Except on Federal lands and reservations today and two noteworthy attempts in its early history, the Federal Government has not taken direct part in road building and maintenance. Since 1916, however, an increasingly large share of the construction and maintenance expense of the state roads has been met by Federal contributions. As noted in an earlier chapter, the current program of Federal interstate highway construction, although built under state supervision, is financed by Federal funds obtained by Federal taxes collected on the sales of fuel, oil, and tires. The program is one of user charges on a pay-as-you-go basis.

Taxes thus imposed are collected in a highway trust fund for further Federal highway construction. The Federal Highway Act of 1973 permits funding up to $800 million in the Act's third year for any type of transit by replacing, at a state's option, Interstate highway projects with transit projects.

In addition to Federal grants for national roads, states obtain highway funds from property taxes and general appropriations, but principally from state fuel taxes, motor-vehicle-registration fees, license fees, and from the revenues of toll roads, bridges, and tunnels. Towns and cities receive state aid in maintaining those streets that form part of a state or Federal route but may also impose local fuel and registration taxes, draw on the general fund, make special assessments, and utilize returns from parking meters and traffic fines.

Increasing reliance is being placed on Federal and other public funding by private as well as public transport agencies. The Federal Housing Act of 1961 made $25 million in capital grants available through the Department of Housing and Urban Development for mass transit demonstration projects. The 1964 Urban Mass Transportation Act authorized $375 million to assist state and local governments in financing (up to two-thirds of a project's cost) the capital costs of transit systems. The High Speed Transportation Act of 1965 and related enactments have made funds available for high-speed rail studies and services on the Eastern Seaboard. The Urban Mass Transportation Act of 1970 provides for capital expenditures of $10 billion over a 12-year period to state and local governments—again on a two-thirds Federal, one-third local basis. The Department of Transportation is permitted to guarantee funds over a five-year period instead of the previous one-year allocation, thereby permitting greater stability and continuity in planning and financing. The funds may be used for rapid transit and other transportation projects. Most recently, Federal aid has been given in the form of outright grants, loans, and loan guarantees to financially pressed railroads and airlines to continue operation. Funding for a complete reorganization of the Northeastern rail network is a part of this effort.

Also noteworthy are the referendum in the San Francisco Bay Area by which the taxpayers obligated themselves for $700 million to construct the Bay Area Rapid Transit System, the use of state funds to acquire and operate the Long Island Railroad in New York State, and funds allocated in New York, Illinois, and other states to provide additional service within the Amtrak system of passenger service.

These several examples are cited to indicate the growing extent and diversity of public support as a source of funding for transportation projects.

Government Bonds   The tax returns for any one period are seldom sufficient to permit the heavy construction that may be desired in that period (although some states have tried to operate their highway program on a pay-as-you-go basis). Money must be borrowed, usually in the form of state, county, or municipal bonds. Bonds are also issued by authorities or agencies that are created by and are instruments of the state. These bonds have as security both the general credit of the state and the taxes and other fees collected from highway users. The rates of interest are usually lower than commercial bonds, and their life may be as much as 75 years, but usually 30 to 40 years. Some issues fall due in a complete block at maturity. Others are designed for retirement in smaller blocks and over the life of the issue. Difficulties arise and programs are delayed when the bond issues required may exceed the statutory debt limit of the state or encounter other legal difficulties. Some states have made use, then, of the bond-issuing powers and credit of the counties by having the counties issue bonds and build roads as county roads. These may later be absorbed into the state system.

Several kinds of bonds may be utilized, the choice depending on the interest rates that must be paid, the sources and extent of amortization funds, and the condition of the market.

Bonds with General-Tax Support   The credit of the state is behind these bonds but no specific taxes or revenue are earmarked for their retirement.

Bonds with Vehicle-Tax Support   All or a certain percentage of taxes from license fees, fuel taxes, and tire taxes, from the state as a whole or from a specific section or portion of highway being financed, are set aside to support these bonds.

Revenue Bonds   Revenue bonds have gained much favor as a means of financing toll roads, bridges, and tunnels. The tolls received for use of the facility are pledged to the payment of the principal and interest until the bonds are retired. After that the facility may be thrown open to the public at no charge or at reduced rates sufficient only to cover annual maintenance charges. Such a charge should also include contribution to a sinking fund so that the highway can be rebuilt when worn out. Toll roads provide an ideal example of user-benefit payments.

Toll roads of superhighway type are generally feasible only in congested, highly populated areas, where sufficient traffic will be found to provide adequate revenues. If a new highway cannot attract enough toll-paying traffic to pay its way, the question might well be raised as to whether the road should be built at all, or at least as to whether a lower standard of design and construction should not be used.

Alternative Programs   A necessary phase of the financing problem, whether for private or government enterprises, is deciding whether to proceed on a pay-as-you-go basis or by borrowing money, that is, issuing bonds. The pay-as-you-go plan is based on deferring portions of the program for new construction or rehabilitation and doing each year only that amount of work for which funds are available from income or taxes. This means that the full advantages of a completed program must also be deferred for many years.

The alternative, borrowing money by issuing bonds, makes enough cash available and credit available in the present to complete the entire project. The advantages and possible savings are realized at once. The disadvantages lie in the heavy interest burden that the project must bear, especially in the initial years when, as a developing enterprise, it may be least able to do so.

There are offsetting features to the interest charges of borrowing. If, for example, a railroad, highway, or canal is to be rebuilt on a pay-as-you-go basis, then the old and unimproved portions must continue to be maintained and operated to bear the present and future traffic until those portions have also been improved. The fact that the route is being rebuilt indicates it to be inadequate, worn out, or obsolete. In any case, its upkeep probably involves excessive maintenance charges and equally excessive transportation costs through traffic delays, accidents, and administration. These excess costs must be balanced against the interest charges on bonds and mortgages to determine which of them will furnish the capital at the lowest overall economic cost. This is simply to repeat that the status quo is always an alternative that must be considered. The possibility of increased revenues arising from the improvements should also be included as an offsetting factor to the interest charges.

The decision will be influenced by the proposed duration of the bonds versus the duration of the pay-as-you-go plan. Average excess operating costs will be greater over a 20-year pay-as-you-go plan than over a 10-year plan. Total interest charges on 20-year bonds will be less than on 40-year bonds. Obviously, comparisons must be made both ways and in the light of anticipated annual funds from income or taxes.

An important problem in this analysis is the engineering-economic determination of what the excess construction and operating expenses will be with the pay-as-you-go plans of different durations. Piecemeal construction will usually be more expensive than would building a completed project at one time. Nice engineering judgment supported by as much historical cost data as possible must be used.

Another pay-as-you-go problem confronting the engineer is that of fitting the operation of the new portion of the route and operation to that of the old. A section of high-capacity, high-speed expressway pouring its traffic into narrow congested obsolete streets or roadways may create impossible traffic situations. Instrument-landing installations at airports will not bring about the desired improvement in safety and capacity until all aircraft are equipped with corresponding devices. Cab signals on a railroad cannot replace wayside signals until all locomotives and signal locations on a territory are completely equipped. This is not to imply, however, that stage construction (next section) is impossible or undesirable. The difference lies in the deliberate designing and planning to build completely, for stage construction, only that which is needed immediately, or a completed section of the whole. Pay-as-you-go involves the hazards of being forced to halt a project at an inconvenient spot and condition if funds for the year prove insufficient or the estimates of revenues or taxes have been overly optimistic. Pay-as-you-go, combined with careful stage-construction planning, has much to recommend it.

Stage Construction   Stage construction is similar in many respects to a pay-as-you-go plan. The method is applicable both to new construction and to rehabilitation. The initial program calls for doing only what is presently needed and to the minimum standards for the present needs. Improvements and additions will be made as required. A two-lane highway may be adequate at the moment. Five years hence a third lane will be added, and 10 years from now the fourth lane. Railroads and highways are sometimes laid out with sharp curves and heavy grades for initial construction economy. Grades and curvatures are reduced at a later date. It is possible to fall into serious error at this point. A railroad laid out with many heavy grades may have to be entirely relocated in the future to achieve operating economy. If, however, good construction and location principles are followed, the grade will be held as low as possible to some one spot and then the elevation will be overcome, all at one time. Later the grade can be reduced by relocation of the ruling grade only or by drilling a tunnel. At a later stage, a much longer and lower tunnel will be drilled to lower the ruling grade still more. The rest of the line will not need to be touched, having already been given proper location.

The overall economic advantages of stage construction are determined by comparing its costs with those of immediate construction to ultimate standards and requirements, a method very similar to the tests described for pay-as-you-go plans. Another factor to consider is the possible changes in traffic patterns and technology that may come later on and that should not be held back by a large investment in a fixed plant from an earlier day. These are decisions that can hardly be made without close study.

ALLOCATION OF COST

Principle of User Payment   Common carriers, contract carriers, and toll roads, bridges, and canals collect user payments for their service and from those payments meet the costs of operation and financing. Where toll roads, bridges, tunnels, and canals are built by public agencies, there is sometimes a question as to whether the tolls are adequate for full-cost recovery, but the question is largely one of degree, not of principle.

The engineer is more likely to become involved in cost allocations and recovery in connection with publicly owned highways, waterways, and airports. If there is no charge for use of the facility, as is the case with United States inland rivers and canals, the general taxpayer must bear the cost. Costs cannot be avoided. Although support of waterways by the general taxpayer has long been traditional, many engineers and economists do not accept that as a sound method of allocation. The St. Lawrence Seaway and the Panama Canal exact tolls (fees for the former are claimed by some to be inadequate), establishing a pattern for other waterways. Justification for free waterways lies in the concepts of public benefits conferred—the public good—and national defense. These factors are also present for railroads, highways, and airports—and have been urged as a basis for public support of the last two.

Conservative opinion holds, however, that only user benefits should be considered in making economic justifications and that costs, therefore, should be allocated only among the users. User-cost allocations are in keeping with a national system of competitive transport and offer a rough test of demand for the service. On this basis, user costs are collected from highway users in the form of fuel, vehicle, and accessories taxes, and from airlines and private air operators in the form of landing fees and rental for hangar, ticket sales, office, and other space occupied. Whether or not all costs are thus recovered or whether some remainder must be borne by the general taxpayer is a matter of debate. The majority opinion seems to hold that most but not all of the costs are user borne.

Differential Users A more difficult problem, involving engineering solutions, arises in allocating costs among the several classes of users of a facility. In an earlier chapter, it was pointed out that charges exacted by common and contract carriers are based primarily on costs of service. Those costs were known to vary for different classes of services, operations, and vehicles. Similarly, costs and the tolls or taxes in payment of costs for highways, waterways, and airports should be based on costs incurred in providing the various types of service for which there is a demand. Service, in these instances, usually refers to the load and size capacities of the facilities required by the users.

The engineering solution to the problem is not easy, especially in the case of highways. A series of full-scale tests conducted by the Highway Research Board for the Bureau of Public Roads and interested state groups have been conducted in Maryland, Idaho, and Illinois to determine the effects of different axle combinations and loads on various types of test pavements.[4] The Maryland tests and those conducted by the Western Association of State Highway Officials came to the same general conclusions.[5] Failure to the pavement was, in general, the result of pavement cracking, first being undermined by pumping and compression of the subgrade under repetitive load applications. A further general conclusion was that even the permitted minimum single-axle load of 18,000 lb will cause damage and that a pavement designed for 18,000 lb will suffer even greater damage under heavier loads than those for which it was designed. Tandem loads should not exceed single-axle loads by more than one and one half times to avoid excess damage. Nevertheless in December 1974 (S. 3934) Congress increased the allowable axle load on Federal aid roads from 18,000 lb (8172 kg) to 20,000 lb (9080 kg), increased tandem-axle loads from 18,000 lb (8172 kg) to 34,000 lb (15,435 kg) up from 32,000 lb (14,528

[4]*Road Test One—Md.*, Highway Research Board, Special Report 4, 1952.
[5]*The WASHO Road Test*, Part 2, Highway Research Board, Special Report 22, 1955.

kg), and increased the maximum gross load to 80,000 lb (36,320 kg) up from 73,200 lb (33,233 kg).

It would seem then that some method should be devised, not only to exact payments, but to allocate the costs equitably so that those vehicles causing the most user wear and tear, or for which maximum design is required, should bear a greater share of costs proportionately than the lighter vehicles, which do little damage or could be sustained on lighter construction.

While the determination of needs and costs of each class of vehicle, therefore, is primarily an engineering problem, it is made complicated by political factors and the belief in some quarters that providing a road for private passenger vehicles is a government function. Maryland, WASHO, and AASHTO road tests seem to indicate that passenger vehicles are probably subsidizing commercial vehicles to a significant extent.

QUESTIONS FOR STUDY

1. Explain how planning by a private concern might differ from that for public transportation.
2. Distinguish between community values, goals, objectives, and criteria in transportation planning. Give examples of each at the national, state, regional, and local municipality levels.
3. Define modal utility and identify the more important factors that should be taken into account in determining utility.
4. Compare "pay-as-you-go" plans and stage construction as to the meaning of each, bais of jusitifcation, and engineering problems involved with each. What would be the effect on financial arrangements and overall costs?
5. Define the purpose of each step in the planning process and cite problems common to each step.
6. A community of 80,000 wishes to purchase 10 new buses to enlarge its public transit fleet. What possibilities are open for financing the $40,000 cost of these buses? Explain the advantages and disadvantages of each.

SUGGESTED READINGS

1. E. L. Grant and W. G. Ireson, *Principles of Engineering Economy*, 4th edition, Ronald Press, New York, 1960 or later.
2. Richard M. Zettel, "Highway Benefits and Cost Allocation Problem," a paper presented to the Forty-Third Annual Meeting of the American Association of State Highway Officials.
3. *Social, Economic, Behaviorial, and Urban Growth Considerations in Planning*, Transportation Research Record 509, Transportation Research Board, National Research Council, Washington, D. C., 1974.

4. *Defining Transportation Requirements*, papers and discussions of the 1968 Transportation Engineering Conference sponsored by the American Society of Mechanical Engineers and the New York Academy of Sciences.
5. *Cost-Benefit and Other Economic Analyses of Transportation*, Transportation Research Record 490, Transportation Research Board, National Research Council, Washington, D. C., 1974.
6. A. M. Wellington, *The Economic Theory of the Location of Railways*, 6th edition, Wiley, New York, 1914, Preface and Chapter 1.

Chapter 14

Urban Data Collection
and Analysis

In the preceding chapter the process of transportation planning was dealt with, emphasizing the methods for evaluating alternative plans and systems. In this chapter, emphasis is placed on identifying the data required to develop alternatives and the techniques for securing and analyzing that data.

THE DEMAND FUNCTION

In any planning for transportation there are two basic steps: (1) establishing the demand for a given level and quality of service and (2) developing a plan of action that will satisfy that demand. The first is a study of needs; the second is a study of means. The techniques here presented are used extensively in urban transportation planning but have application to the regional and statewide levels as well.

Planning processes vary in detail with the purpose of the study and the type and size of community or study area. Figure 14.1 shows in graphical sequence the steps followed in conducting the Chicago Area Transportation Study. A general procedural pattern similar to that outlined in Chapter 13 can be identified.

498

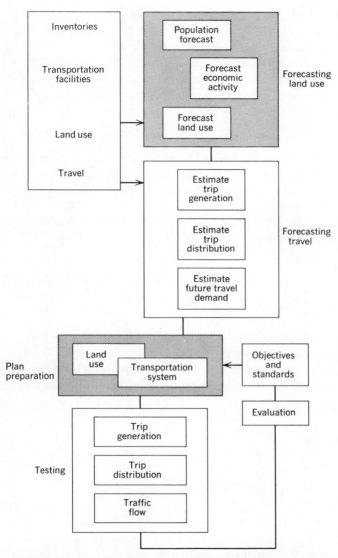

FIGURE 14.1 THE PLANNING PROCESS. (CHICAGO AREA TRANSPORTATION STUDY, VOLUME I, FIGURE 1, 1958.)

Sources of Demand   Data are collected to (1) establish existing transportation demand-use and (2) provide a basis for projecting future demand.

Demand is determined by the characteristics of population, land uses, and the amount of activity these two elements engender. Demand may also be a function of travel time, thereby relating demand, at least in part, to the technology of the transportation system in use or contemplated. A data base must be established from which to develop present and future demand.

Population   Total population is determined both for the entire study area and for each previously established zone or neighborhood in that area. The density of population, how concentrated, and the density gradient away from the urban core are required. Age, race, and economic status of the population, although not always utilized, contribute useful information. Data sources include census figures, city directories, and the home interview, which will be discussed later.

Land Use   For any study area, be it urban zone or neighborhood, an entire corporate city, or a region or a state, the intensity and type of land use arises from a combination of interrelated factors (Figure 14.2).

1.   The extent of growth and activity of the larger land area or region of which the study area is a part. A county or region within a state reflects to some degree the economic, agricultural, and social development of the state in which it is situated. These are outside or exogenous factors.

2.   Intensity of development of the study area itself. The population and type and extent of activity are in turn functions of:
(a) Past development policies and tax rates.
(b) Inherent characteristics of the land—terrain, productivity, natural resources, climate.
(c) Land use services—water supply, power supply, waste disposal facilities, zoning and environmental constraints.
(d) The transportation network.

Land uses represent activity centers and serve as traffic generators in amounts that depend on type and intensity of use. Residential areas may generate as much as 40 to 50 percent of all trips, but industrial and commercial work areas, schools, churches, hospitals, and places of entertainment and recreation also generate activity and therefore traffic. Land areas associated with each use must be identified as to location, area, and intensity of use. Intensity of use can be related to number of people per acre or per square mile, to the square feet of floor area per unit of land area devoted to a particular purpose, by the amount of business conducted

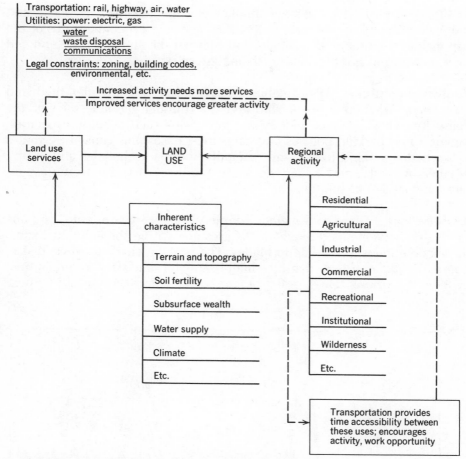

FIGURE 14.2 LAND USE DEVELOPMENT FACTORS.

(retail sales or type of manufacture, for example), or by the productive output (as from a manufacturing process) per unit of land. Land devoted to transportation purposes is included as well as public lands such as parks and public buildings.

Future demand must be determined in the light of developing and future land uses. This requires either simplifying assumptions (that future land uses will continue in the same proportion, for example), or land use models that indicate future land uses and the impact of transportation on those uses. The first procedure is empirical, possibly unwarranted and, therefore, lacks accuracy. The second is complicated, expensive, and requires inputs for which accurate data may be difficult to obtain.

Traffic Generation   The total number of trips generated by each land use is determined by purpose, destination, age, and economic status of the trip maker, the time of day, mode, and route of the journey as determined by some form of questionnaire or home interview.

Facility Inventory   All available transport facilities are identified and quantified. Automobile ownership and use (through gasoline sales or the home interview), streets and expressways with their widths, capacities, signalization, parking policies and capacities, buses, bus capacities, routes and schedules, rapid transit and commuter railroad schedules and routes, number of tracks, and rolling stock are necessary data for traffic assignment and design of solutions.

Traffic Flow   These data include simple volume counts of auto, bus, and rail traffic movements throughout the area and within the individual zones or neighborhoods, according to modes. The peaking characteristics, that is, hours of maximum flow density, should be determined and the quantities of flow for those hours noted. Figure 14.3 shows graphically typical

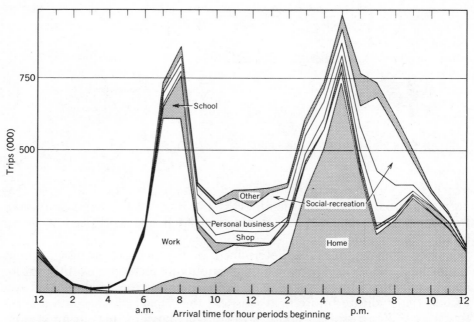

FIGURE 14.3 HOURLY DISTRIBUTION OF INTERNAL TRIPS BY TRIP PURPOSE. (CHICAGO AREA TRANSPORTATION STUDY, VOLUME I, FIGURE 15, 1958.)

peaking characteristics for Chicago traffic. Origins and destinations of traffic flow and the quantities moving from one zone to another by mode and route form the basis for planning. Origin-destination (O-D) data may already be available from a state agency or other studies but should be considered suspect if more than five years old. Again, the home interview, supplemented in special locations by cordon counts and screenline counts can be used. Travel times over various routes will be needed in the traffic assignment phase of the study.

History   The rates of population growth, land use development, development of public transit, growth of vehicle ownership and use, and any special factors contributing to the rates of growth or changes should be developed. These historical data are useful in giving perspective to the present demand situation, in showing trends, and in projecting demand levels into the future.

Presentation   The foregoing data as collected in survey notes and questionnaires form an unwieldy mass. It must be consolidated for ease in assimilation and analysis by the user. For small communities it may be consolidated by hand and presented in tabulations or on maps or charts of the study area, a form of graphical bookkeeping. Streets and rail lines are presented with widths of lines (or colors) showing capacities. A concentration of dots (or various colors) can show population location and densities. Land uses are usually denoted by indicative colors or cross-hatching and shading. Traffic flow and distribution can be shown between origin and destination by lines of a thickness corresponding to volume of demand or by showing volumes on the various routes—streets, expressways, tracks—by corresponding widths of line.

Usually, however, the mass of data is so voluminous that only by entering it into a computer bank can it be conveniently compiled, manipulated, and analyzed. For very large cities there is no other way to handle the data, although graphical bookkeeping will probably also be used as with small areas. Recent data processing equipment is now capable of three-dimensional presentations of certain data.

DATA COLLECTION: O-D SURVEYS

Traffic Counts   An obvious method of obtaining traffic-flow data is to count the actual number of persons and vehicles, trains, aircraft, or buses arriving at, departing from, or passing a given point. Such counts can be made manually, with hand-operated counters or tally sheets, or by use of automatic counters.

Cordon Counts and Screenlines The total volume of traffic entering or leaving an area, city, or region may be obtained by surrounding the area with interviewers and checkers, recording through questions and observation the number, kind, purpose, origin, destination, mode of travel, etc. of all traffic entering or leaving the area. Practically, such cordon points are usually limited to a few key points—junctions, yards, and interchange points for a railroad or rapid transit situation; airports and railroad and bus stations for passengers in and out of a city; principal streets or highways at the edge of the area for vehicular traffic, and the entrance or exit of bridges and tunnels bearing vehicular traffic. Every person on foot and every driver of a vehicle (or every fifth or tenth person or vehicle depending on size of sample required) may be interviewed. The records of weighing stations set up by highway patrols to check the weights of commercial vehicles have also served as check points.

Similarly, screen lines or stations are thrown across any series of routes to count the volume and character of traffic moving betweeen two points across the line. Thus flow between two city areas or between two adjacent route sections, communities, states, or regions can be obtained. Adjustments have to be made for traffic originating and terminating between two screen-line stations.

It is often easier (and more detailed information is obtained) if a large area is broken into smaller areas and the traffic flow between each of the subareas obtained. A city can be divided into zones by grid squares, neighborhoods, or into quadrants or sectors. Ideally, neighborhood lines should predominate, often based on local schools or other center of the land use activity, or determined by geographical boundaries. In very large cities, the divisions may be arbitrary, using a grid system in which each zone is numbered for ease in computer manipulation. One-quarter mile grid squares (grid lines at one-half mile intervals) have been found useful.

Home Interviews Cordon and screenline methods are a kind of questionnaire activity, but the utility of the questionnaire is most fully realized in door-to-door home interview surveys. Information is recorded on questionnaire forms concerning the travel habits and requirements for each individual in a household, usually for the current 24-hour period. See Figure 14.4. Each trip made (or to be made) during that 24-hour period is entered on a separate line along with origin, destination, route, purpose of the trip, by whom made, and mode of travel used; whether by private automobile, as a car passenger, or by bus, taxi, commuter train, or rapid transit. Interviewers call at every house in a block (or every tenth or thirtieth house depending on density size and on size of sample desired) and fill out the questionnaire form by personal questioning. Sample size has varied from 1 to 25 percent depending on size of city and distance from the CBD.[1] The CATS survey made in 1956 interviewed every thirtieth dwelling unit. Questionnaire forms may also be sent by mail or handed to persons at stations or check points. The response to mailings is seldom satisfactory or complete and is useless among those who have reading and writing difficulties.

Questionnaire data are transferred to punched cards or tapes for sorting and analysis by digital computers. Hand sorting is too slow and cumbersome except where only a few questionnaires are involved. Desirable analyses tabulations indicate the total number of trips of each type and mode generated by each zone; also the distribution of those trips to other zones in the study area. The latter is useful in applying growth factors and in the use of trip distribution models for projecting demand. The purpose and time of day for each trip-mode grouping will also be needed.

The use of questionnaires has been explained in terms of travel by persons. Similar questionnaire studies and office or shop interviews can be conducted to obtain the origin, destination, volume, and mode of conveyance for freight sent and received by industries and commercial houses. These establishments, however, often hesitate to make such information available. The aid of chambers of commerce can sometimes be enlisted to achieve the cooperation of reluctant firms and in providing some of the data. Legislation may be needed in some cases to achieve needed cooperation.

The home interview survey gives only a cross section in time. There is no time continuum. Hence care must be exercised in selecting the time for making the survey. Trip generation can vary with the day of the week or month of the year. A truly typical period should be used. For example, a strike in a small town's leading industry could give a misleading pattern.

ANALYSIS AND PROJECTION: TRIP GENERATION

Projection of Future Demand The collected data are analyzed to obtain a reasonably accurate picture of existing traffic volumes, trip purposes, trip generation by zones or neighborhoods, flow patterns, and distributions from any one zone to another zone of the trips generated by each zone. The problem of determining capacity excesses and deficits for existing routes and facilities, which must precede solutions both for present and future needs, will be deferred until later in this chapter.

---

[1]*Urban Mass Transportation Surveys,* prepared by Urban Transportation Systems Associates, Inc. for the U.S. Department of Transportation, Washington, D.C., August 1972.

FORM BHR 150A

ILLINOIS DIVISION OF HIGHWAYS
CHAMPAIGN-URBANA AREA TRAFFIC SURVEY

DWELLING UNIT SUMMARY

INTERVIEW ADDRESS _____   CARD _____ 1

IDENTIFICATION _____   DISTANCE FROM CBD _____

STREET ADDRESS _____   SOCIO-ECONOMIC TYPE _____

                                           SAMPLE NO. _____

DATE OF TRAVEL _____   ZONE _____

A. HOW MANY PASSENGER CARS ARE OWNED BY PERSONS LIVING AT THIS ADDRESS? _____

B. MAKE AND YEAR OF EACH CAR _____   M YEAR M YEAR M YEAR M YEAR

C. HOW MANY PERSONS LIVE HERE? _____

D. HOW MANY ARE 5 YEARS OF AGE OR OLDER? _____

E. HOW LONG HAVE YOU LIVED AT THIS ADDRESS? _____

F. HOUSEHOLD INFORMATION:

PERSON NO.	RACE SEX	PERSON IDENTIFICATION	AGE	CODE	OCCUPATION AND INDUSTRY	TRIPS YES	NO
01							
02							
03							
04							
05							
06							
07							
08							
09							
10							

G. TOTAL NUMBER OF TRIPS REPORTED AT THIS ADDRESS _____

1. NUMBER OF PERSONS (5 YEARS OF AGE OR OLDER) MAKING TRIPS _____

2. NUMBER OF PERSONS (5 YEARS OF AGE OR OLDER) MAKING NO TRIPS _____

3. NUMBER OF PERSONS (5 YEARS OF AGE OR OLDER) WITH TRIPS UNKNOWN _____

H. COMMENTS AND REASON IF COMPLETE INFORMATION WAS NOT OBTAINABLE _____

_____

I. FACTOR _____

ADMINISTRATIVE RECORD

INTERVIEWER _____

CALLS

	DATE	TIME
(1)		
(2)		
(3)		
(4)		

REPORT SUBMITTED INCOMPLETE

DATE _____

REASON _____

SUPERVISOR'S COMMENT _____

REMARKS: _____

_____

REPORT COMPLETED _____ (DATE) _____ (INITIAL)

INTERVIEWS CHECKED _____ (INITIAL)

CODED BY _____ (INITIAL)

CODING CHECKED BY _____ (INITIAL)

INTERNAL TRIP REPORT

SHEET ......OF.....

CARD [2] DISTANCE FROM CBD [ ][ ][ ] SOCIO-ECONOMIC TYPE [ ] SAMPLE NO. [ ][ ][ ][ ] ZONE NO. [ ][ ][ ] DAY OF TRAVEL [ ]

1	2	3	3a	3b	4	5	6	7	8	9	10	11
OCCUPATION AND INDUSTRY	PERSON NO.	TRIP NO.	SEX RACE	AGE	WHERE DID THIS TRIP BEGIN?	WHERE DID THIS TRIP END?	TIME OF TRIP WITHIN SURVEY AREA	PURPOSE OF TRIP (FROM / TO)	NO. IN CAR	SCREEN LINE	MODE OF TRAVEL	KIND OF PARKING
			1 2 3 4 5 6				A.M. P.M.	1 2 3 4 5 6 7 8 0 — WORK ...... 1 BUSINESS ...... 2 MED.-DEN ...... 3 SCHOOL ...... 4 SOCIAL-REC. .... 5 CH. TRAVEL MODE 6 EAT MEAL ...... 7 SHOPPING ...... 8 HOME 0			1 AUTO DRIVER 2 AUTO PASS. 3 STREETCAR-BUS 4 TAXI PASS. 5 TRUCK PASS.	1 STREET FREE 2 STREET METER 3 LOT FREE 4 LOT PAID 5 GARAGE FREE 6 GARAGE PAID 7 SERVICE OR REP. 8 RES. PROPERTY 9 CRUISED 0 NOT PARKED
			1 2 3 4 5 6				A.M. P.M.	1 2 3 4 5 6 7 8 0 — WORK ...... 1 BUSINESS ...... 2 MED.-DEN ...... 3 SCHOOL ...... 4 SOCIAL-REC. .... 5 CH. TRAVEL MODE 6 EAT MEAL ...... 7 SHOPPING ...... 8 HOME 0			1 AUTO DRIVER 2 AUTO PASS. 3 STREETCAR-BUS 4 TAXI PASS. 5 TRUCK PASS.	1 STREET FREE 2 STREET METER 3 LOT FREE 4 LOT PAID 5 GARAGE FREE 6 GARAGE PAID 7 SERVICE OR REP. 8 RES. PROPERTY 9 CRUISED 0 NOT PARKED
			1 2 3 4 5 6				A.M. P.M.	1 2 3 4 5 6 7 8 0 — WORK ...... 1 BUSINESS ...... 2 MED.-DEN ...... 3 SCHOOL ...... 4 SOCIAL-REC. .... 5 CH. TRAVEL MODE 6 EAT MEAL ...... 7 SHOPPING ...... 8 HOME 0			1 AUTO DRIVER 2 AUTO PASS. 3 STREETCAR-BUS 4 TAXI PASS. 5 TRUCK PASS.	1 STREET FREE 2 STREET METER 3 LOT FREE 4 LOT PAID 5 GARAGE FREE 6 GARAGE PAID 7 SERVICE OR REP. 8 RES. PROPERTY 9 CRUISED 0 NOT PARKED
			1 2 3 4 5 6				A.M. P.M.	1 2 3 4 5 6 7 8 0 — WORK ...... 1 BUSINESS ...... 2 MED.-DEN ...... 3 SCHOOL ...... 4 SOCIAL-REC. .... 5 CH. TRAVEL MODE 6 EAT MEAL ...... 7 SHOPPING ...... 8 HOME 0			1 AUTO DRIVER 2 AUTO PASS. 3 STREETCAR-BUS 4 TAXI PASS. 5 TRUCK PASS.	1 STREET FREE 2 STREET METER 3 LOT FREE 4 LOT PAID 5 GARAGE FREE 6 GARAGE PAID 7 SERVICE OR REP. 8 RES. PROPERTY 9 CRUISED 0 NOT PARKED
			1 2 3 4 5 6				A.M. P.M.	1 2 3 4 5 6 7 8 0 — WORK ...... 1 BUSINESS ...... 2 MED.-DEN ...... 3 SCHOOL ...... 4 SOCIAL-REC. .... 5 CH. TRAVEL MODE 6 EAT MEAL ...... 7 SHOPPING ...... 8 HOME 0			1 AUTO DRIVER 2 AUTO PASS. 3 STREETCAR-BUS 4 TAXI PASS. 5 TRUCK PASS.	1 STREET FREE 2 STREET METER 3 LOT FREE 4 LOT PAID 5 GARAGE FREE 6 GARAGE PAID 7 SERVICE OR REP. 8 RES. PROPERTY 9 CRUISED 0 NOT PARKED

FIGURE 14.4 QUESTIONNAIRE FORM FOR ORIGIN-DESTINATION STUDY. (COURTESY OF ILLINOIS DEPARTMENT OF HIGHWAYS, SPRINGFIELD, ILLINOIS.)

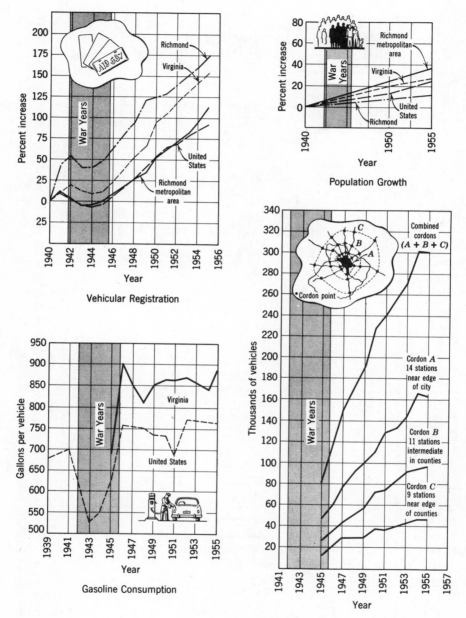

FIGURE 14.5 INDICES TO DETERMINE TRANSPORTATION GROWTH FACTORS. (COURTESY OF HARLAND BARTHOLOMEW AND ASSOCIATES, ST. LOUIS, MISSOURI.)

The next steps involve projecting trip generation and distribution to some later data to determine future demand, usually 20 years hence.[2] Historical data on transportation flow and demand can be extrapolated into the future. Such data are not always available in sufficient detail and accuracy. Other methods have been devised including the use of growth factors and of trip determinants. Land use models provide a means with which future land use, employment activity, population, and travel time can be interrelated.

Simple Growth Factors   Simple growth factors, useful for small communities, are obtained by extrapolating the projected increase in population, automobile ownership and use, level of industrial and commercial activity, or other factors that may be pertinent and computing the ratio of future to present activity or growth for each. These individual ratios are multiplied together to obtain an increase factor by which present demand and trip generation is multiplied to obtain demand and trip generation for the target date. See Figure 14.5.

For example a growth factor for automobile traffic may be based on population growth, automobile ownership, and gasoline consumption per capita. Assume present and projected populations are 20,000 and 28,000, respectively; car ownership increases from 12,000 to 18,000 vehicles, and fuel consumed per capita (a measure of vehicle use) grows from 700 to 800 gallons per person. The corresponding growth factors for each are 1.4, 1.5, and 1.1. The product of these gives an overall growth factor of 2.3. If the field survey has shown a present demand of 16,000 vehicle trips per day, the future trip demand will be $16,000 \times 2.3 = 36,800$. Present trip generation for each zone or neighborhood is then multiplied by the same growth factor, 2.3, to giving future trip generations for each. If the sum of these equals 36,800 approximately, one might have a reasonable degree of confidence in the initial projection. But a reasonable agreement may not be reached because of obvious inherent errors. Procedures described in a later section help in securing greater accuracy.

Trip Determinants and Regression Analysis   More recent procedures relate the level of trip generation to the type and intensity of land use. A reasonably accurate projection of future land use is required, but this can be established with some degree of confidence. Since trip generation (demand) is accepted as having a direct relation to land use, more realistic projections can be made through that approach than by trying to project generation and demand directly from present trip generation as is done

[2]The year 2000 is sometimes being used at the time of this writing.

with growth factors. The procedure requires that the relation between each particular land use and the trips generated thereby be determined. The characteristics of each type of land use that determine trip generation (or attraction) must also be identified as to the quantitative impact of each.

Equations are established that relate trip generation as a dependent variable of land use type and characteristics through linear regression analysis. Using current land use, population, and trip generation characteristics data, the independent variable, sometimes referred to as a trip determinant, is used to establish the number of trips that will be generated through the particular characteristic of land use being examined.

The regression equations so used assume a linear relation between the dependent variable and factors on which it depends.

$$Y = a_0 + a_1 X_1 + a_2 X_2 + \cdots + a_n X_n + U$$

Y is assumed to be a linear function (number of trips, for example) of independent variables X_1, X_2, X_n—such as household size, number of cars owned by the household, family income category, or distance to the CBD —with U representing a random error term. Coefficients a_0, a_1, a_n are selected so that the sum of squares of the error, U^2, between actual and predicted values is minimized. The coefficient, a, for any one explanatory variable, X, indicates the change to be expected in the dependent variable, Y, by a unit change in X. The accuracy of the procedure depends in part on introducing into the equation all the factors that can cause a significant variance in the dependent variable.

The traffic-generating potential of land and population are thus analyzed and established. Once the coefficients in the equation for a particular use of characteristic have been determined, one need only introduce the estimated extent of that use or characteristic at the future target date and compute the trips that will be generated or attracted. The weakness in the procedure is in assuming that the coefficients determined today will hold true for the target date as well. Changes in technology, social habits, or other factors can seriously affect the accuracy of these coefficients for later use.

Land use determines the purpose of trip making, but it is in the household or dwelling unit that a decision is made as to whether a trip for that purpose will or will not be made. The dwelling unit has a decision-making capability. More specifically, trip-making rates are found to be functions primarily of the number and age of those in a dwelling unit. Automobile ownership also directly affects trip making.[3] Car ownerhsip

[3]Walter Y. Oi and Paul W. Shuldiner, *An Analysis of Urban Travel*, The Transportation Center at Northwestern University, Northwestern University Press, Evanston, Illinois, 1962, p. 74.

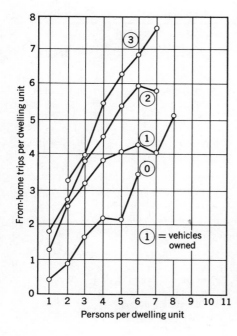

FIGURE 14.6 HOUSEHOLD COMPOSI-
TION AND CAR OWNERSHIP VERSUS
TRIPS GENERATED. (FROM *AN ANALY-
SIS OF URBAN TRAVEL DEMANDS* BY
WALTER Y. OI AND PAUL W. SHULDINER,
COURTESY OF THE NORTHWESTERN
UNIVERSITY PRESS, EVANSTON,
ILLINOIS, 1962, P. 91, FIGURE 10.)

reflects the size of household income.   The number of trips increases
generally, but at a decreasing rate, as car ownership increases from none to
three or more. Other factors such as length of trips, distance to the central
business district, and social and racial status have some relation to trip
production, but these in turn show a certain correlation to household
composition and car ownership. Household composition and car ownership
effects are shown in Figure 14.6.

A simple example of an equation developed by regression analysis is the
following based on unweighted averages for 58 traffic census zones in a
West Coast city of 80,000:

$$T = -0.627 + 1.216p$$

where

$T =$ average daily number of from-home motor vehicle trips per dwelling unit

$p =$ number of persons over five years of age per dwelling unit.[4]

---

[4]Ibid., p. 216.

TABLE 14.1 EFFECTS OF DENSITY ON TOTAL TRIP GENERATION: ILLUSTRATIVE EXAMPLES

Study	Year	Equation number	Equation[a]	Coefficient of correlation (r)
Between several cities:				
Future Highways and Urban Growth	1961	A	$Y_1 = 2.7 - 1.17x_1$	Hand fit
"Some Aspects of Future Transportation in Urban Areas"	1962	B	$Y_1 = 2.6 - 0.092x_6$	Hand fit
		C	$Y_1 = 2.6 - (0.092)x_1/x_3(10^{-3})$	Hand fit
Within cities:				
Detroit Area Transportation Study	1953	D	$Y_2' = 15.07 - 4.23\log x_2$	-0.75
		E	$Y_2' = 1.87 + 4.26\log x_4$ $- 1.60\log x_2$	0.83
"A Study of Factors Related to Urban Travel"[b,c]	1957	F	$Y_2 = 7.22 - 0.013x_2$	0.72
		G	$Y_2 = 4.33 + 3.89x_3 - 0.005x_2$ $- 0.128x_4 - 0.012x_5$	0.84
		H	$Y_2 = 3.80 + 3.79x_3 - 0.0033x_2$	0.84
St. Louis Metropolitan Area Transportation Study	1959	I	$Y_6 = 0.261 - 0.017x_7$	Not cited
Chicago Area Transportation Study	1956	J	$Y_4 = 6.64 - 2.43\log x_2$	-0.95
		K	$Y_5' = 4.32 - 1.90\log x_2$	-0.96
		L	$Y_2 = 11.80 - 4.246\log x_2$	-0.97
		M	$Y_3 = 7.34 - 3.29\log x_2$	-0.96
Pittsburgh Area Transportation Study	1962	N	$Y_2 = 9.62 - 4.19\log x_2$	-0.88
		O	$Y_3 = 5.55 - 2.64\log x_2$	-0.91
		P	$Y_4 = 5.02 - 2.17\log x_2$	-0.87
		Q	$Y_5 = 3.35 - 1.35\log x_2$	-0.90

[a]Dependent variables:

Y_1—Total internal person-trips per capita

Y_2—Person-trips per family

Y_2'—Person-trips per dwelling place

Y_3—Auto-trips per family

Y_4—Person-destinations per dwelling place

Y_5—Auto-destinations per dwelling place

Y_5'—Vehicle-destinations per dwelling place

Y_6—School trips per person

[b]Independent variables:

x_1—Gross urbanized-area density

x_2—Dwelling places per residential acre

x_3—Autos per dwelling unit

x_4—Distance from CBD

x_5—Family income

x_6—(Households per car) × urbanized area population density × 10^{-3}

x_7—Thousands of people per square mile

[c]*Public Roads*, Vol. 29, No. 7 (April, 1957), based on Washington, D.C. Source: Herbert S. Levinson Houston Wynn, "Effects of Density on Urban Transportation Requirements," *Community Val Affected by Transportation*, Highway Research Board, Washington, D.C., Highway Research Record (1963), p. 49.

If the average household 20 years hence is projected as having three persons over the age of five years, then daily from-home motor vehicle trips can be projected as:

$$T = -0.627 + 1.216 \times 3 = 3.021 \text{ trips}$$

Another example of the procedure is seen in an equation developed during the Chicago Area Transportation Study to show the number of mass transit trips in percent of total trips from a study zone:[5]

$$Y_{178} = 19.7331 + 0.0610X_7 + 0.0365X_8$$

where

Y_{178} = percent of mass-transit trips of the total number of trips from a particular zone

X_7 = net residential density

X_8 = automobiles per 1000 people

Table 14.1 presents several equations that relate various population characteristics, especially density, with trip generating potential.

Because each study area has its own peculiar characteristics of population, terrain, level of income, etc., the coefficients developed for one such area cannot be used safely in another study area (except, perhaps, to get a very general first-cut estimate).

Other major generators, additional to residential areas, include shopping centers, manufacturing plants, the CBD, and airports. A university campus can be a major generator in some communities; in others, a sports arena will fill that role. Whereas household composition, car ownership, family income, distance to CBD, for example serve as explanatory variables for residential areas, other factors must be utilized when considering other types of land use. Employment, floor space and retail sales have been used for commercial and retail activities. Employment and production have been used for manufacturing plants. These values can be divided by the land area occupied to gain the effect of density. Often a simple tabulation is used showing trip destinations per acre for various types of land uses. These can be further subdivided as to distance from a centroid, by sex, or type of transport used, etc. Table 14.2 shows such a list of generation rates. In any case, the goal is to determine a positive relation between the

[5]Robert Sharkey, "Mass Transit Usage," *C.A.T.A. News*, Vol. 3, No. 1, The Chicago Area Transportation Study, Chicago, Illinois, 9 January 1959.

TABLE 14.2 TRIP GENERATION RATES: CATS[a]

Ring	Average Distance from Loop	Trip Destination per Acre				
		Transpor-tation	Manu-facturing	Commer-cial	Public Buildings	Public Open Space
0	0.0	273.1	3,544.8	2,132.2	2,013.8	98.5
1	1.5	36.9	243.2	188.6	255.5	28.8
2	3.5	15.9	80.0	122.1	123.5	26.5
3	5.5	10.8	86.9	143.4	100.7	27.8
4	8.5	12.8	50.9	212.4	77.7	13.5
5	12.5	5.8	26.8	178.7	58.1	6.1
6	16.0	2.6	15.7	132.5	46.6	2.5
7	24.0	6.4	18.2	131.9	14.4	1.5

[a]*Guidelines for Trip Generation Analysis*, FHA, U.S. Department of Transportation, Washington, D.C., June 1967, p. 16.

amount, kind, and density of land use and present or future trip generation. The reader is referred to the footnoted items, especially to Oi and Shuldiner, and to Suggested Readings for further details on these procedures.

Land Use Models   There is evident usefulness in the ability to predict the type and extent of land use development and the demand created thereby for transportation. Transportation, in turn, can affect the rate and type of land development (urban sprawl, for example, could not have occurred without the automobile!), primarily through providing accessibility with reasonable constraints.

A variety of land use models have appeared to meet this need. These are of varying degrees of complexity and may require inputs that are difficult to evaluate quantitatively. The models include the operational suitable for use in ongoing site studies. Others are merely conceptual, a basis for further research, or are useful primarily in indicating relative trends and magnitudes but lacking expression in actual numbers. All models endeavor to express relationships between land used and available for population, land used and available for employment, and the adequacy of the connecting transport network and technology that provides accessibility between the two expressed in terms of the travel time. Certain models recognize that employment opportunities in commerce and industry and levels of population are affected over time periods by in-migration, out-migration, industrial growth and failure, and by births, deaths, and aging. Social factors, always difficult to evaluate, may also be considered as affecting availability of land in terms of race, income, family size, and place of employment. Some models treat separately the employment opportunities that vary endogenously, that is, from internal forces, and those that are dependent on exogenous factors and markets outside the area of study. Some models are recursive in time, using the output from one forecast of the model as the input for another forecast at some later time. Transportation enters the model usually in terms of spatial separation measured in travel time. Lesser travel times generally lead to more activity opportunities and to increased demand for transport capacity. The Suggested Readings for this chapter contain more detailed accounts that are beyond the scope of this book.

Chicago Area Transportation Study   The CATS effort represents a land mark in the transportation planning process through its use of techniques well beyond those previously in use and beyond the mere balancing of origin-destination-growth factor data. Emphasis is given to trip generation in terms of trip purpose and related land uses. There was no land use model as such but rather a system of land use accounting that provided a horizon year population and employment opportunities. Using a quarter square mile grid with lines at half mile intervals, land use was identified according to 10 categories: residential, manufacturing, commercial, public buildings, transportation, streets and alleys, public open space, parking, miscellaneous, and vacant land. The intensity of a particular use was measured by population density and floor area. Population growth was extrapolated from nation trends and distributed to residential land based

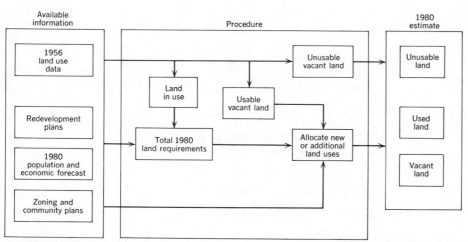

FIGURE 14.7 SIMPLIFIED DIAGRAM OF LAND USE ESTIMATING PROCEDURE. (CHICAGO AREA TRANSPORTATION STUDY, VOLUME I, 1958.)

on its holding capacity. Established areas were assumed to remain relatively constant with most of the growth occurring in outlying suburban areas and density decreasing as the distance from the CBD increased. Population density and the proportion of land in various uses was assumed to hold constant to the horizon date. Vacant land was distributed to those uses on that basis. Employment opportunities were based on capacities established from available density of land use and distance from the CBD. See Figure 14.7.

The CATS procedures have the virtue of proven operational capability. The reader is referred to the three-volume report from which the foregoing is summarized issued by the study group; see the Suggested Readings, No. 9.

TRIP DISTRIBUTION

Having established the traffic generated, by zones or other study area unit (probably through home interview surveys), and projected that traffic demand to some future date by use of growth factors or by trip determinants and regression analysis, the next step is to determine how those trips are distributed to individual zones or neighborhoods. We shall now describe procedures for making such distribution using an application of growth factors in one instance and, in another, the so-called gravity model.

Frater Method   The Frater method of trip distribution uses the arithemetic average of trips distributed from zones i and j, each use being weighted by the ratio of the projected zonal interchanges between zones i and j and the sum of all projected interchanges from each.[6]

$$T'_{ij} = T_{ij} \times F_i F_j (L_i + L_j)/2$$

where

$$T'_{ij} = \text{trips from zone } i \text{ to zone } j \text{ in target year}$$

$T_{ij} = $ trips presently made between zones i and j

F_i and $F_j = $ growth factors for zones i and j

$$F_i = \frac{\text{ratio of projected to present trip}}{\text{generation in zone } i}$$

$$F_j = \frac{\text{ratio of present to projected trip}}{\text{attraction to zone } j}$$

L_i and $L_j = $ factors of location

[6]"Traffic Assignment and Distribution for Small Urban Areas," Bureau of Public Roads, U.S. Department of Commerce, pp. IX-1 and IX-2, Washington, D.C., 1965.

where

$$L_i = \frac{T_{ij}}{\sum\limits_{j=1}^{j=n} T_{ij} \times F_i} \qquad \text{and} \qquad L_j = \frac{T_{ji}}{\sum\limits_{i=1}^{i=n} T_{ji} \times F_j}$$

The sum of the trips thus distributed from zone i to zone j should equal the number of trips estimated as being generated by zone i. Similarly, the sum of trips attracted to zone j should equal the number estimated for it. If these conditions are not met, a second (and even a third) iteration may be required using new values for F_i' and F_j' equal to the ratio between the original growth factors and the growth factor represented by the sums of the computed distributions and attractions.

Growth factors exhibit certain inherent inaccuracies. Their use assumes a uniform rate of growth for each zone, but rates of zonal growth are seldom uniform. A low density zone has a much greater potential for growth than one that is already fully developed. With the foregoing procedure, an empty zone would show zero growth over a 20-year period, yet its very emptiness makes it a likely place for extensive growth. A zone already developed would similarly show an excessive growth, perhaps beyond its areal capacity. There is some question as to whether the basic data used in any of these methods is sufficiently accurate to justify the additional manipulation needed to assure full agreement between projected and computed values.

Gravity Models   Because of a questionable similarity to Newton's law of gravity, certain trip distribution procedures have been termed "gravity models." Early models were used to compute trip interchanges between communities over intercity highways and had the form:

$$T_{xy} = \frac{(k)(P_x)(P_y)}{D^n}$$

where

$T_{xy} = $ trip interchanges between populations P_x and P_y at points x and y

$D = $ distance between population points

$n = $ an exponent of distance

$k = $ a coefficient

The terms n and k were determined by calibrating the model with known data.

In using the gravity model for urban trip distribution, consideration is given to the ability of various land uses to attract trips, of the effects of distance (and time) between origin and destination, and the effect these will have in determining the percent of the trips generated in a particular zone i that will terminate in some other zone j. Present generation and distribution are obtained by home interview surveys, future distribution is based on future trip generation as projected by growth factors or regression equations projections.

Bureau of Public Roads Model   The foregoing is generalized as:[7]

$$T_{ij} = T_i \times \dfrac{F_{ij} \times A_j \times K_{ij}}{\sum\limits_{j=1}^{j=n} A_j \times F_{ij} \times K_{ij}}$$

where

T_{ij} = trips for a given purpose—for example, work, shopping, recreation—generated in zone i, going to zone j

T_i = total number of trips generated in zone i for a specific purpose and equal to $\sum_{n=1}^{j=n} T_{ij}$

T_j = trip attractiveness of zone j, based on trips attracted per number of workers employed at j, the number of shopping trips per unit of commercial activity, or similar, depending on trip purpose under study

F_{ij} = a travel time factor expressing an average areawide effect of distance on trip interchanges

K_{ij} = a factor to permit zone to zone adjustments for the impact on travel patterns of specific economic or social conditions. If ignored, as it frequently is, this factor has a value of 1.

[7]Ibid., pp. IV-1 and IV-2 to IV-8.

The travel time factor expresses a measure of probability of trip making for each one-minute increment of travel time between the origin and destination zones. It varies inversely as the travel time raised to a power that varies with the travel time increment and trip purpose, that is, $F_{ij} = 1/t_{ij}^n$ where

t_{ij} = the travel time between i and j in minutes and

n = a factor to be determined by an adjustment or calibration process.

Unknowns are evaluated in the BPR model by a process of trial and adjustment. The calibration process uses known data from a survey sample of the study area or from data for a similar area taken elsewhere. By applying the model to the calibration data, one adjusts the unknowns to give the known trip distributions. The unknowns thus evaluated are then assumed to hold constant over the entire area and into the future.

With the BPR model, the calibration is principally concerned with developing a set of travel time factors for the trip purposes under study. A travel time factor is obtained by the trial and adjustment process for each minute of travel time in the study area.

Basic inputs to the gravity model are taken from the home interview origin-destination surveys. Zonal trip production, T_i, and trip attraction, A_j, are thus obtained. In defining trips, the linked trip concept is generally used, that is, a complete journey from origin zone to destination zone for a single purpose regardless of changes from one travel mode to another in the process. This is in contrast to the segment trip where each leg of a journey is considered a separate trip. Trips with either origin or destination at the trip maker's residence are considered trip productions for the zone of origin. Similarly, for trips with either origin or destination in the maker's residence zone, the trip that ends in the nonresident zone are trip attractions. Trips with neither end in the maker's residence zone are trip attractions of the destination zone.

The minimum path travel time, t_{ij}, is derived from field surveys that record the distance and speed of travel over major routes of the transportation system under study. The minimum travel time is further adjusted to include terminal time required at each end of the journey. From a point of origin (a zonal centroid), the travel time to all destinations along major routes is calculated. The process is that of building skim trees, a step in traffic distribution. This is most conveniently done by use of prepared computer programs.[8] A plot of trip length frequency then may be prepared showing the percent of total trips in the study area, for a given trip purpose, occurring for each incremental minute of travel time. The vehicle minutes of travel time for each time increment can then be summated for each trip purpose and mode of travel. By dividing the vehicle minutes of travel by the total number of trips in the study category, the mean trip length is obtained for use in determining the travel time factor, F_{ij}. In the calibrating process one must do the following:

1. One may either assume an initial travel time factor value of $F_{ij} = 1$ or use a set of factors from a study in a city of comparable size and characteristics.
2. These data are introduced into the gravity model formula from which a table of zone-to-zone movements, that is, values of T_{ij} is obtained.
3. The percent of trips for each trip purpose and mode are plotted on the previously prepared trip frequency chart. By comparing the actual (from O-D travel pattern inventories) with the estimated distribution, the degree of correctness of the initial travel time factors can be observed.
4. Both curves should be approximately the same with a difference between mean trip lengths (by time) for both sets of data within ±3 percent. If not within those limits, the travel time factors are adjusted for each travel time increment by the ratio of the surveyed trips to the percent of estimated trips for the respective time movements.

$$F_{adj} = F_{used} \times O\text{-}D\% / GM\%$$

where

F_{adj} = adjusted travel time factor for use in the next iteration of calibration

F_{used} = the initial travel time factor used in the first calibrating run of the model

$O\text{-}D\%$ = percentage of O-D survey trips of the appropriate time length

$GM\%$ = percentage of trips of the appropriate time length as determined by the gravity model run being analyzed

[8]Ibid.

5. Adjusted travel time factors can be "smoothed" by plotting against respective travel time increments on a log-log paper to obtain a "line of best fit" from which a new set of travel time factors are developed and used in a repeat run. Three iterations are usually the maximum number required to obtain reasonable agreement.

Problem Example

The general gravity model approach can best be understood by use of a problem example.

Figure 14.8 shows a simplified urban network having two residential zones, A and B, that produce respectively 700 and 600 work trips. Zones C and D are work opportunity zones attracting, respectively, 900 and 400 trips. Travel times between zones are shown on the network in circles and actual trip distributions as obtained from an origin-destination study are also shown. For this example, initial or trial travel time factors that correspond to the indicated travel times will be taken from the BPRs tabulation in Table V-2 of the foregoing reference. (One might also have taken a value of *one* as the initial value of F_{ij}.) The problem is to obtain a set of travel time factors that will duplicate approximately the trip distributions obtained in the O-D survey. These same factors may then be used to obtain trip distributions for some future horizon date.

A sample calculation using the gravity model is given for the trips between zones A and C, that is, T_{AC}.

$$T_{AC} = 700 \times \frac{900 \times 85 \times 1}{900 \times 85 \times 1 + 400 \times 10 \times 1} = 665 \text{ trips}$$

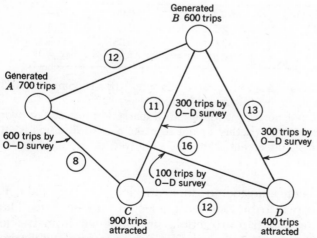

FIGURE 14.8 NETWORK FOR APPLICATION OF GRAVITY MODEL.

The following tabulation records the complete set of calculations for the first iteration of the network. As is frequently the practice, a value of $K_{ij}=1$ has been used throughout.

Path	Travel time (min)	Travel Time Factor (F_{ij}) (Initial)	(Adj)	K_{ij}	Computed Trips (Initial)	(Adj)	Observed (O-D) Trips
AC	8	85	(76.5)	1	665	(601)	600
AD	16	10	(28.5)	1	35	(99)	100
BC	11	61	(41.5)	1	439	(300)	300
BD	13	50	(93.5)	1	160	(300)	300
Total trips					1299	(1300)	1300

From this example it is seen that the computed trips do not approximate the observed trips of the O-D survey either individually or in total. There is evident need for a revised set of travel time factors; the model must be calibrated.

Using the BPR procedure, the percent the observed O-D trips ($O\text{-}D\%$) over each path are of the total and the percent the trips calculated by the gravity model over each path are of the total ($GM\%$) are computed and entered in the following tabulation. From these, an adjusted F_{ij} is computed where $F_{ij_{adj}} = F_{ij_{used}} \times (O\text{-}D\%/GM\%)$.

Path	Percent of Total O-D Trips	Percent of Total Computed Trips	$\dfrac{O\text{-}D\%}{GM\%}$	$F_{ij_{adj}}$
AC	46.2	51.2	0.90	76.5
AD	7.7	2.7	2.85	28.5
BC	23.1	33.8	0.68	41.5
BD	23.0	12.31	1.87	93.5

Using these adjusted F_{ij} values, a set of revised computed trips is computed and entered in the foregoing tabulation as figures in parentheses.

While the adjusted run is not perfect, it represents a marked improvement over the initial set of trip distributions. Whether or not to make a third iteration is, at this point, always a matter of judgement. Note the improved distribution agreement in this example.

Once the model has been adjusted to duplicate actual survey data, it may be used to obtain trip distributions for a selected horizon date. The values of trip productions and trip attractions are obtained from traffic generation projections. Horizon time travel times (skim trees of minimum travel times over each link in the network) must also be developed for anticipated conditions. The travel time factors already developed are assumed to remain constant to the horizon date, and trip length distributions are assumed to remain constant throughout the study area.

Some caution is needed in using and interpreting the results of the gravity models. The model represents primarily a mechanical process that does not account for the variability of human behavior. The assumed application of an average travel pattern to all zones is of questionable validity. Travel time can vary with the time of day, 10 a.m. vs. 10 p.m., for example, and with the day of the week. It may not remain constant for a given purpose throughout the study area. Mathematically, the number of trips between zones approaches infinity, at least becomes disproportionately large in practice, as the distance between zones approaches zero. One may also question the assumption that travel patterns, that is, impact on trip distributions by trip purpose and spatial separation, will remain constant to the projected target date.

Other Models   Variations on the gravity model utilize the concept of probability that trip purposes can be satisfied at some minimum travel time point other than the destination zone under consideration. The Intervening Opportunities Model was developed and used by the staff in the CATS program.[9] It is said to give greater reliability than the gravity models. The model is based on the probability that a trip purpose will be satisfied at an intervening point before the target zone is reached. The mathematical expression is:

$$T_{ij} = T_i \left[e^{-Lt_0} - e^{-L(t_0 + t_j)} \right]$$

where

T_{ij} = trips originating at i and going toward zone j

T_i = trips originated by zone i

T_j = trips presently attracted to zone j

t_0 = present (or future) volume of destinations closer in travel time to zone i than is zone j

L = probability per destination of the acceptability to satisfy the trip purpose of the destination at the zone under consideration; for example, one zone may offer more job opportunities than another

e = Napierian base of logarithms = 2.71828

[9]"Chicago Area Transportation Study," Vol. II, p. 111, 1960.

All possible destinations are listed in travel time sequence. L values are developed by a calibrating process applying to the equation values obtained in the original origin-destination survey.

Modal Split   In establishing demand, it is not enough to know trip generation and distribution and technoeconomic utility. Providing capacity is wasted effort if not used. Therefore the user's modal choice or preference must also be considered. The use of a particular mode is dependent on such factors as travel time, cost, comfort, walking distance, and lack of alternative (either no car, thus requiring public transit use, or no public transit, requiring thereby use of an automobile). The factor of choice is academic when no alternatives are available.

Trip distribution techniques have just been presented that can be used with or without a modal distinction. If the mode is not designated, the next step would be to establish the split between modes. Usually the distinction is between public transit and the use of private automobiles, but bicycles and walking should not be overlooked.

Numerous models and procedures have been developed for the purpose. One that is relatively simple to apply is a diversion curve procedure based on five related variables. It was developed by the Traffic Research Corporation for use by the National Capital Transportation Agency to estimate 1980 traffic requirements for Washington, D.C.[10]

(a)   Travel time ratio = door-to-door travel time by public transit ÷ door-to-door travel time by private car

$$= \frac{a+b+c+d+e}{f+g+h}$$

a = time on transit vehicle
b = time to transfer between transit vehicles
c = waiting time for transit vehicle
d = walking (or car driving time) to transit vehicle
e = walking (or driving time) from transit vehicle
f = automobile driving time
g = parking delay time at destination
h = walking time from parking place to final destination

(b)   Service ratio = $\dfrac{\text{Excess transit time}}{\text{Excess auto time}}$

$$= \frac{b+c+d+e}{g+h}$$

(c)   Cost ratio = $\dfrac{\text{Transit fare}}{\text{Auto cost per passenger}} = \dfrac{\text{Transit fare}}{i+j+(k/L)/L}$

[10]Sesslau, Hearne, and Balek, "Evolution of New Modal Split Procedure," Public Roads, U.S. Department of Commerce, Washington, D.C., Vol. 33, No. 1, pp. 5–17, April 1964.

where

i = gasoline cost = gal/mile × cost/gal × distance

j = oil change and lubrication cost per mile of distance

k = parking cost at destination

l = number of persons per vehicle

(d)   Economic status = median income of workers

(e)   Trip purpose = home-based work trips or all nonwork trips
(with the exception of school trips)
The home interview survey will provide the percents of travelers using
autos and public transit for each trip purpose and related to the foregoing
determinants. Cost ratios and excess time ratios are divided into four
ranges and income levels into five ranges. Five income levels were used in
the Washington, D.C. study: (1) $0 to $2499, (2) $2500 to $3999, (3) $4000
to $5499, (4) $5500 to $6999, (5) $7000 +. The four determinants and the
excess time ratios and income levels combine to give 80 time ratio and
diversion curves for each trip purpose. For each of these 80 combinations,
observations on percent of modal split can be plotted against the cost and
travel time ratios. See Figure 14.9. It is assumed that the same determinants and ratios will be effective at a future target date. Level-of-service
ratio seems to have a major importance while cost ratios have only a minor
role. The travel time ratio is reasonably sensitive throughout. See figure
14.9.

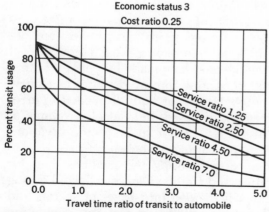

FIGURE   14.9 MODAL   SPLIT   RATIOS.
(COURTESY OF *PUBLIC ROADS*, UNITED
STATES DEPARTMENT OF TRANSPORTA-
TION, VOL. 33, NO. 1, APRIL 1964, PP. 5–17.)

This method neglects any evaluation of the tripmakers subjective response to comfort, amenities, or to the route followed. Neither does the method account for capital costs of owning a car. Only the out-of-pocket operating costs have been included. This ommssion can give a bias in favor of automotive travel.

A different approach is a procedure used in the Chicago Area Transportation Study that uses linear regression to show the relationships between residential density, automobile ownership, and use of mass transit. See page 513.

Traffic Assignment   Traffic assignment is a procedure for estimating the number of traffic units (persons, trips, vehicles) that will use each individual portion of a transportation system network, either in the present or at a future target date. The process attempts to align itself with the tripmakers demand for minimum travel time and cost and maximum comfort. Travel time frequently reflects cost and comfort and has been a central element in making trip assignments. The procedure was originally designed for highway networks but has application to all segments of a transportation system.

Assignment can be carried out on a vehicular basis, but this is a highway-oriented procedure and lacks modal flexibility. By working on a person-trip basis, trips can be assigned to any route or mode based upon existing divisions or by forecasts using modal choice models. The linked trip concept is required with the person-trip approach whereby a trip is defined as a person movement from point of origin to point of destination regardless of route taken or number of modes involved. Thus public transit and other than highway modes can enter into the assignment routine.

Present traffic flow, routing, peaking, and modal distribution are determined by surveys. Projections include determining what percentage of trips will be made by private automobile and how many by public or other transit. Assignments are then made to existing improved routes or to new routes, either highway or street, public transit, or both, if needed to meet demand for increased capacity. Demand is based on a selected volume of movement, probably during the rush hours or, for such specific situations as movement to a beach or sports arena, during a weekend.

A first step is preparation of a network using the facilities data for streets, bus, and rail routes, capacities, and traffic volumes collected during the inventory survey. The study area is divided into zones or sectors and the centroid of each located relative to population and trip generation density. The centroids are connected to the basic network of arterial streets and to public transit routes or stations. These connections serve in lieu of including local streets in the network. Nodes are numbered consecutively. The network should include all freeways, expressways, connector-distributor and arterial streets, and all streets protected by stop signs or traffic signals.

Intersections or nodes are indicated on the network and numbered suitably for computerized solution. See the section on networks in the next chapter for a suggested numbering system. Between each node, parameters for the link segments must be determined—distance, permissible speeds, delays such as traffic lights, "stop" intersections, and turning movements. Tables are prepared listing traffic volumes (present or projected) and travel time (including intersection delays) for each link in the network. If public transit is involved, gradients, junctions, interlockings, limiting curvatures, and stations and station halt and transfer times should be included; also running times for scheduled trains. Figure 14.10 and 14.11 show zonal centroids and network map, respectively.

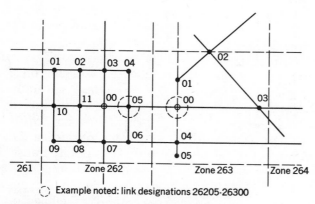

FIGURE 14.10 ZONAL CENTROIDS.

Minimum path "trees" are developed from destination to various nodes of origin. Principal "routes" are selected at a destination point and followed outward along the "trunk" and its branches to points of origins, tabulating (for computer use) the route characteristics, primarily capacity and related speed and accumulated travel time. After adjustments and checking for accuracy (no missing links or opposing one-way streets, for example) trip tables are prepared; they indicate trips originating or terminating at each centroid by origin or destination (and purpose). The goal is to determine the travel time over various routes for use in establishing the minimum travel time for various trip distributions.

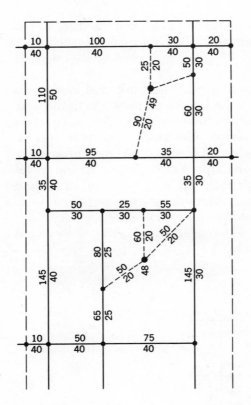

- **•** Nodes
- **●** Zonal centroids
- ---- Centroid—major route connectors
- $\frac{95}{40}$ Link distance—100th of miles
  speed—mph
- —— Study area boundary

FIGURE 14.11 TRAFFIC ASSIGNMENT NETWORK.

Traffic between any two points is then assigned by one of two methods: (*a*) all-or-nothing or minimum travel time or (*b*) diversion curve. Using the minimum travel time procedure, trips are assigned to the path or route that permits the shortest time of travel between the two points in question. Assignment work proceeds along the main route with traffic being fed from side branches. As the capacity of the route is reached, however, speed decreases and travel time increases so that the first route is no longer the minimum time path. An alternative route will then be selected and the remaining traffic assigned until it too is filled to its capacity when a possible third route may have to be selected.

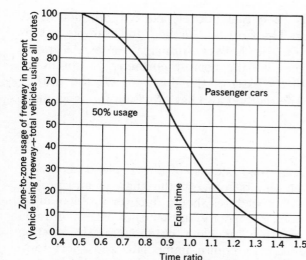

FIGURE 14.12 DIVERSION CURVE. (FROM *TRAFFIC ASSIGNMENT AND DISTRIBUTION FOR SMALL URBAN AREAS*, OFFICE OF PLANNING, BUREAU OF PUBLIC ROADS, SEPTEMBER 1965, FIGURE III-10, P. III-32.)

The relation between capacity and speed (Chapter 8) indicates the likelihood of tradeoffs developing between travel time (speed) and capacity so that a point is reached where there is no advantage timewise in taking one route over another. All routes become equally attractive on a time basis.

Assignment on the minimum time basis can be modified by resort to the Diversion Curve concept, which says that not all trips will take the shortest path, only those for which time is essential. The diversion curve indicates the percent of trips taking a more than minimum time path; this is presented in terms of the travel ratio between the shortest travel time and the alternative paths of longer travel time. Use of this concept permits more trips to be added to the minimum time route than would otherwise be possible or the entire process may follow the diversion process. See Figure 14.12.

The attractiveness of a route can be varied sometimes by increasing speed on a lightly loaded link to make it more attractive to the user. If the link is overloaded, speed can be decreased to reduce its attractiveness and direct the trip maker elsewhere.

Other assignment methods have been developed but are not in as extensive use as the foregoing. The Corridor method, for example, establishes dividing lines (that form a grid) between principal routes. All traffic north (or west) of the line goes via the nearest northerly (or southerly) route. All traffic to the south (or east) of the line goes via the nearest southerly (or easterly) route. The method also has provision for assignment to diagonal routes.[11]

FORMING SOLUTIONS

Whether or not planning is performed because of a recognized need or is conducted as a routine and continuing process, the analyses and projections and the traffic assignment procedures should uncover problem areas and identify deficits and excesses in capacity, instances of excessive travel time, and bottlenecks and points of accident potential in the existing system. Many of the problems will require immediate and short term solutions. The analysis should also identify a probable need for more extensive solutions at some target date $20 \pm$ years hence for which alternative solutions must be selected. Examples of the selection process, calling upon techniques discussed in this and in Chapter 13 are here given to show how solutions can be developed through the application of those techniques.

Problem Example

A capacity deficit on a 10-block street in a small community has been noted through the traffic assignment procedure. Traffic is congested at certain hours of the day, movement is slow, and numerous accidents have occurred. Parking is presently permitted on both sides of the street leaving only two 9-ft lanes for through movement.

1. The solution first suggested is to eliminate parking on the street, thereby gaining two additional lanes of traffic, a simple, effective, and low-cost solution.

But citizen input, invited or otherwise, voices strong protest at this solution. The street is lined with old homes in which low-rent apartments have been constructed. The tenants and landlords depend upon the street for parking spaces and the strongest protests come from them.

2. A second alternative would involve turning the street in question and an adjacent parallel street into a one-way couple. Protests from citizen input now develops from residents along the parallel street. Real estate dealers also oppose this alternative. They fear the loss of property values arising from the increased

[11]Wm. S. Pollard, Jr., *Corridor Capacity Determination: Procedural Outline*, an internal procedure paper, Harland Bartholomew and associates, Memphis, Tennessee.

traffic flow and from the noise, air, visual, and vibratory pollution and the potential for accidents.

3. A third alternative, street widening, meets with strong opposition from all quarters, including environmentalists and the local garden club, because of land taken from property frontage, the assessments for construction costs, and the need to cut down a row of curbside shade trees. An additional point of concern might arise if it develops that the proposed expansion is to accommodate part of a principal arterial that is planned by the traffic engineer of the city.

A selection between these alternative might be made using techniques set forth in Chapter 15. A cost evaluation could be made of each, using a weighting process to evaluate the preponderance of community concern for the several intangibles involved as well as the economic factors.

In this instance, to meet all objections, a fourth alternative might be devised whereby parking would be eliminated as in alternative 1 but spaces would be made available by the city, with the concurrence of the residents of that area, at the back of the properties on a little-used alleyway.

Example 2

The second example is that of a small city attempting to establish a bikeway network. The need for such a system has arisen through the local planning commission's projection of traffic growth and by the requests voiced before the commission, the two council, and in local newspapers by bike riders and motorists (the latter fearful of accidents and traffic delays). The local planning commission has therefore made an inventory of existing bikeway routes, volume of bicycle traffic, and origin and destination of that traffic. A major destination is a university campus where several Class I routes (completely separated bicycle path designated exclusively for bicycle use) have been constructed. The network lacks connections other than by Class III routes (roadway signs indicating that bicycles use the street) to certain on-campus residential areas and to off-campus areas used by students and faculty for residence.

First priority has been given, following a commission study and public meetings in which community and cyclists opinions and suggestions were heard, to connecting the campus network with the student and faculty residential areas. Problems arise in:

(a) Locating streets of sufficient width in the campus area to permit establishing Class II lanes (a lane designated for preferential or exclusive use by bicycles on a roadway used by other traffic). Note the preceding problem example.

(b) And in finding a way past or through or over a railroad carried on an earth viaduct standing about 12 ft above the surrounding terrain with only a few inconveniently spaced vehicular underpasses.

Alternative solutions to the railroad barrier include:

1. A bridge with long approaches costing half a million dollars.
2. A tunnel under the railroad costing only slightly less than the bridge.

3. A grade crossing with flashing lights and gate protection, with steep approaches, and costing approximately $100,000.

4. Use of one of the pedestrian sidewalks in an existing underpass, involving a reduction in pedestrian capacity, a bikeway of less than adequate width, and a Class III approach to the underpass.

5. Do nothing; let riders walk their bikes through the underpasses or take their chances in street traffic.

The alternatives and inputs from various groups and individuals are obviously many. The first three rail by-passing proposals can be evaluated on a cost effectiveness basis. The last two require evaluation of the accident potential and reduction in the use of bicycles.

The search for streets on which to locate bike lanes might find solution in a proposed university policy to prohibit all private autos from the campus area by substituting bus service from outlying parking lots.

The probable increase in cyclists and corresponding decrease in automobile traffic should be evaluated on the basis of cost effectiveness, benefit cost, relative accident potential, and effects on the environment. The economic cost study could indicate possibly a sufficient transfer of traffic from car to bicycle to warrant considerable expenditure for a railroad underpass. The cost effectiveness could determine the cost per 100 trips of moving those trips by each of the proposed methods.

Example 3

A third example might be that of a large urban area with traffic problems on its existing streets and expressways and a developing traffic demand from growing suburban areas. Six possible alternatives could arise for consideration.

1. Additional street and expressway routes and capacity involving new construction of roadways.

2. Increased rail commuter service requiring more schedules, train sets, the opening of several new stations, and rehabilitation of others. The railroads are loathe to share in any of these costs so it is anticipated some form of public support for the increase must be forthcoming.

3. A rail rapid transit system, partly in the median of existing expressways, partly underground, especially in the CBD.

4. Express bus service on existing highways and expressways, converting an existing lane in each direction for the purpose.

5. A combination of the foregoing.

6. Do nothing or status quo.

The problems engendered by these proposals and the various interests concerned are almost without limit. Some of the more obvious concerns can be cited by way of example as entering into the selection process.

1. Capital costs of each alternative.

2. Cost effectiveness of each alternative, that is, dollars per 1000 trips or trip miles, etc. of each one.

3. Benefit-cost ratio of each alternative.

4. Ability to secure funding, including city, state, and federal.

5. Relation of each plan to projected land uses and anticipated population densities.

6. Effects on aiding or retarding racial and social integration.

7. Reflection of community values and policy as regards city growth and expansion of the metropolitan area into the suburbs vs. developing high density land uses within the city.

8. Relative effects on the environment and on community values. These could include

   (a) Pollution: air, noise, visual, vibratory

   (b) Dwelling units lost

   (c) Work opportunities lost

   (d) Business displaced or lost

   (e) Disruption of mobility within the local neighborhood (Chinese Wall effect)

   (f) Loss of tax base; increased taxes

   (g) Destruction of parks, historic buildings, scenic vistas

   (h) Potential for accident; relative safety

   (i) Loss of business and inconvenience during construction

   (j) Effect on underground utilities

   (k) Gain in tax base, real estate values, business activity

9. Time saving for trip makers and cost of trip making

10. Reduction in accident incidence

11. Adaptability to physical terrain and topography

12. Extent to which all land uses are served

Many other factors, often local and peculiar to a particular community could be listed. In addition, it is only realistic to recognize that every community possesses a power structure that functions, usually unseen, to influence any community activity. The support of that structure can be vital to the success of a proposed goal or objective. Its opposition can involve costly delay and effort and, not infrequently, rejection of alternatives it does not favor.

QUESTIONS FOR STUDY

1. Outline the data to be secured in a transportation survey and explain what use is to be made of each item collected.

2. What is the role of the home interview survey? What data are secured and for what purpose?

3. Using the data of Figure 14.5, compute growth factors for the period 1940 to 1950 and for 1950 to 1970 and compare them with the projected study area totals. Explain sources of any difference.

4. Compute the number of trips 20 years hence going from zone A where the growth factor is 1.4 to zone B with a growth factor of 1.7. Present trips between A and B number 90 per day. Trip distributions and attractions obtained by the use of growth factors are:

Trip Distributions from Zone/A	*Trips Attracted to Zone/B*
$T_{AC} = 150$	$T_{CB} = 260$
$T_{AD} = 100$	$T_{DB} = 290$
$T_{AE} = 60$	$T_{BE} = 200$
$T_{AF} = 170$	$T_{BF} = 250$

Present trips generated by zone $A = 400$. Present trips attracted by zone $B = 400$.

5. What are trip determinants and what is their significance in transportation planning? How can these be used to project trip generation?

6. Using an equation from Table 14.2, how many person trips per dwelling unit will be made when a home interview survey indicates eight dwelling units per residential acre in a neighborhood containing 80 dwelling units located three miles from the central business district? In what city has this relation been found applicable? To what extent can it be applied to other cities?

7. An urban area containing residential zones E and F produces respectively 500 and 700 work trips. Zones G and H are work opportunity zones attracting, respectively, 650 and 550 trips. The travel times between zones are:

$$EG\text{---}6 \text{ min} \qquad FG\text{---}12 \text{ min}$$
$$EH\text{---}17 \text{ min} \qquad FH\text{---}10 \text{ min}$$

Actual observed trips are

$$EG\text{---}250 \qquad FG\text{---}400$$
$$EH\text{---}250 \qquad FH\text{---}300$$

Develop a set of travel time factors that will produce a reasonable approximation of the observed trips, that is, calibrate the model for the above set of data. Initial values of F_{ij} may be taken as 1; also take K_{ij} as 1 but explain the significance of so doing.

8. Select an origin and destination in the community in which you are located and develop: (a) travel time ratio, (b) service ratio, and (c) cost ratio for auto vs. bus or rail transit. Determine the modal split from the curves of Figure 14.8.

9. What are the general concepts and characteristics of most land use models? How does transportation technology enter into these concepts?

10. Identify a transportation problem or/and need in your community and outline (a) all possible conflicts between various concerned interests, (b) all possible impacts or consequences of alternative solutions, and (c) procedures by which alternatives can be compared and a choice made.

SUGGESTED READINGS

1. *Urban Mass Transportation Surveys,* prepared by Urban Transportation Systems Associates, Inc., for the U.S. Department of Transportation, Washington, D.C., August 1972.
2. A. M. Vorhees and R. Morris, "Estimating and Forecasting Travel for Baltimore by Use of a Mathematical Model," *Highway Research Bulletin 224,* Highway Research Board, Washington, D.C., 1959, pp. 105–114.
3. *Traffic Assignment and Distribution for Small Urban Areas,* Bureau of Public Roads, Office of Planning, U.S. Department of Commerce, Washington, D.C., September 1975.
4. Constantine Ben, Richard Bouchard, and Clyde E. Sweet, Jr., "An Evaluation of Simplified Procedures for Determining Travel Patterns in a Small Urban Area," *Highway Research Record No. 88,* Highway Research Board of the National Academy of Sciences, National Research Council, Washington, D.C., 1965, pp. 137–171.
5. Richard J. Bouchard and Clyde E. Peyers, "Use of Gravity Model for Describing Urban Travel," Highway Research Record No. 88, Highway Research Board, of the National Academy of Sciences–National Research Council, Washington, D.C., 1965, pp. 1–44.
6. Kevin E. Heanue and Clyde E. Peyers, "A Comparative Evaluation of Trip Distribution Procedures," *Highway Research Record No. 114,* Highway Research Board of the National Academy of Sciences–National Research Council, Washington, D.C., 1966, pp. 20–37.
7. B. V. Martin, F. W. Memmott, A. J. Bone, *Principles and Techniques of Predicting Future Demand for Urban Area Transportation,* Massachusetts Institute of Technology, R63–1, Research Report No. 38, Cambridge, Massachusetts, June 1961.
8. T. J. Frater, "Forecasting Distribution of Inter-Zonal Vehicular Trips by Successive Approximations," *Volume 33, Proceedings of the Highway Research Board,* Washington, D.C., 1954.
9. *Chicago Area Transportation Study, Volumes 1, 2, and 3,* Dr. J. Douglas Carroll, Director, Chicago, Illinois, 1959, 1960, 1962.
10. Walter Y. Oi and Paul W. Shuldiner, *An Analysis of Urban Travel Demands,* Transportation Center at Northwestern, Northwestern University Press, Evanston, Illinois, 1962.
11. *Detroit Metropolitan Traffic Study, Volumes I and II,* Speaker-Hines and Thomas, Inc., for the Michigan State Highway Department, Lansing, Michigan, 1955.
12. Alan M. Vorhees, "A General Theory of Traffic Movement," *Proceedings of the Institute of Traffic Engineers,* Washington, D.C., 1955, pp. 45–56.
13. A. R. Tomasinis, "A New Method of Trip Distribution in an Urban Area," *Bulletin No. 437,* Highway Research Board, Washington D.C., 1962.

14. W. R. McGrath, "Land Use Planning Related to Traffic Generation and Estimation," *Proceedings, Institute of Traffic Engineers,* Institute of Traffic Engineers, Washington, D.C., 1958.
15. A. B. Sosslau, E. K. Heanue, and A. J. Halek, "Evaluation of a New Modal Split Procedure," *Highway Research Record No. 88,* Highway Research Board of the National Academy of Sciences–National Research Council, Washington, D.C., 1965, pp. 44–63.
16. Robert H. Sharkey, "Mass Transit Usage," *C.A.T.S. Research News,* Vol. 3, No. 1, The Chicago Area Transportation Study, Chicago, Illinois, January 9, 1959.
17. N. A. Irwin, "Review of Existing Land-Use Forecasting Techniques," *Highway Research Record No. 88,* Highway Research Board of the National Academy of Sciences–National Research Council, Washington, D.C., 1965, pp. 182–216.
18. D. M. Hill, D. Brand, and W. B. Hansen, Prototype Development of Statistical Land-Use Prediction Model for Greater Boston Region," *Highway Research Record Number 114,* Highway Research Board of the National Academy of Sciences–National Research Council, 1966, pp. 51–71.
19. John W. Dickey, *Metropolitan Transportation Planning,* McGraw-Hill, New York, 1975.
20. Britton Harris, "Penn-Jersey Transportation Study," *PJ Papers,* No. 7, June 1961, No. 9, August 1961, No. 14, February 1962 and Revised Memos Nos. 1 and 2, February 1973.
21. Britton Harris, "Experiments in Projections of Transportation and Land Use," *Traffic Quarterly,* April 1962, pp. 305–319.
22. Lowdon Wingo, Jr., *Transportation and Urban Land,* Resources for the Future, Inc., Washington, D.C., 1964.
23. J. R. Meyer, J. F. Kain, and M. Wohl, *The Urban Transportation Problem,* A RAND Corporation Research Study, Harvard University Press, Cambridge Press, 1965.
24. *Modal Split—Documentation of Nine Methods for Estimating Usage,* U.S. Department of Commerce, Bureau of Public Roads, Office of Planning, December 1966.
25. Lyle G. Wermers, "Urban Mass Transit Planning," *Journal of the Urban Planning and Development Division,* Proceedings of the American Society of Civil Engineers, New York, March 1970.
26. *Transportation and Traffic Engineering Handbook,* John Baerwald, Editor, Institute of Traffic Engineers, Prentice-Hall, Englewood Cliffs, N. J., 1975.
27. Eugene V. Ryan, "Calibration of the Network Sensitive Mode Split Model," *C.A.T.S. Research News,* Chicago Area Transportation Study, Chicago, Illinois, Vol. 16, No. 3, December 1974, pp. 1–7.
28. William S. Pollard, "Operations Research Approach to the Reciprocal Impact of Transportation and Land Use," paper prepared for presentation at ASCE Transportation Engineering Conference, Minneapolis, Minnesota, May 1965.

29. Henry D. Quimby, "What is Involved in Transportation Project Planning?" *Public Works Magazine*, Ridgewood, New Jersey, April 1968.
30. "Transit Planning," *Transportation Research Record 559*, Transportation Research Board, National Research Council, Washington, D.C., 1976.
31. "Northeast Corridor Transportation Project Report," *NECTP—209*, Office of High Speed Ground Transportation, U.S. Department of Transportation, Washington, D.C., April 1970.
32. "Bus Use of Highways," *National Cooperative Highway Research Program Report 155*, Transportation Research Board, National Research Council, Washington, D.C., 1975.
33. "Analysis of High Speed Ground Transportation Alternatives," Office of High Speed Ground Transportation, U.S. Department of Transportation, January 1973 (PB 220079, National Technical Information Service).

Chapter 15

Evaluating Alternative Systems

THE ROLE OF EVALUATION

Transportation planning, especially on a regional or metro basis, deals with large quantities of diverse traffic and vehicle types, spread over a large land area, with complex and often obscure relationships, with some uncertainty as to the future, and usually with conflicting interests of groups and individuals. More than one solution is often possible; probably many. Decision makers must choose between alternative solutions and possibilities. Alternatives must be presented in a complete and orderly manner so that selections can be made on the merits of each alternative with a minimum of subjective, arbitrary, or uninformed opinion. Procedures for accomplishing such a selective process are examined in this section.

Systems Analysis A broad array of physical, social, aesthetic, economic, and political factors reflecting the user and community scales of values must be considered in choosing between alternatives. These interact for a common purpose. A change in one may have an impact or feedback effect upon one or all of the others. The relationships are often complex and not immediately obvious. A systems approach is required to evaluate alternative systems, to measure and predict how well a given alternative will meet the goals and criteria established.

Systems analysis offers a medium in which transportation alternatives can be analyzed and evaluated.

538

Steps in System Analysis   The general procedure involves the following:

1. Define the problem; possibly the most important step.
2. Develop a model of the system—graphical, mathematical, conceptual, or both—that shows the basic relationships between its general elements.
3. The model must be calibrated or quantified.
4. Criteria for performance evaluation are established.
5. Numbers are introduced and a given system is tested, feedback effects are noted, and evaluation performed.
6. Alternative solutions are tested and compared.

When several alternatives are being compared criteria must be available to guide judgement. For a transportation systems model these can be in terms of the degree to which objectives are obtained—capacity, travel time, accident reduction, pollution, quantity of land used, revenues, effect on community values, or ability to serve all areas of a region—to name a few. The costs of realizing the desired goals and objectives are always viable criteria.

A more general criterion relates to the level of the solution. The solution can be the optimum or it can be a good or acceptable solution. In terms of cost, optimization occurs when the proposed solution gives:

(a) The maximum output—speed, service frequency, capacity—at minimum cost, or
(b) The maximum capacity, speed, etc. for a fixed or given sum of available resources, financial or otherwise, or
(c) The minimum cost for a given requirement for capacity, speed, accessibility, etc.

To have an optimum solution, it is necessary to obtain all feasible solutions and all consequences of those solutions. It means that *all* related factors must be included in arriving at a solution. These requirements are not always easy or even possible to attain, and certainly not always financially worth the effort. By taking into account only feasible factors and data and considering only feasible solutions, a good solution, suitable for the intended purpose, can be obtained.

Evaluation Model   No fully adequate mathematical model has been developed to permit evaluation of alternative transportation systems. A

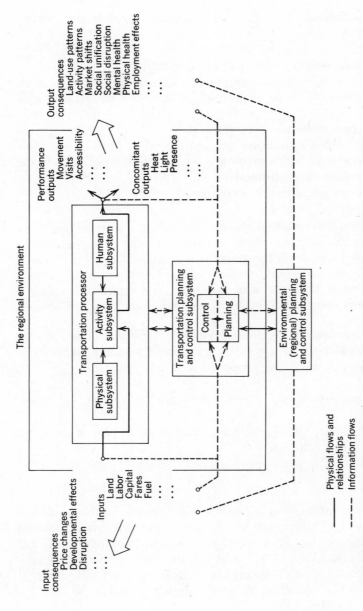

FIGURE 15.1 TRANSPORTATION SYSTEM MODEL. (FROM *STRATEGIES FOR THE EVALUATION OF ALTERNATIVE TRANSPORTATION PLANS*, NATIONAL COOPERATIVE HIGHWAY RESEARCH PROGRAM REPORT 96, HIGHWAY RESEARCH BOARD, WASHINGTON, D.C., 1970, FIGURE 10, P. 25.)

conceptual model presented in graphical form can be useful, indicating the various factors that should be considered and their relative relationships. Any such model should include two distinct but related features:

(a) A model of the transportation system that includes the physical elements (vehicles, roadway, terminals, control systems), people and groups (patrons, employees, builders), and activities (riding, driving, loading, unloading, building, dispatching, maintaining) with land, labor, capital, materials and equipment as inputs.
(b) A model of the region in which the transportation model is embedded. The regional model involves the physical elements (land, air, water, type of cover, manmade attachments, etc), the human subsystem (with physical, biological, social, and psychological characteristics) giving rise to aspirations and perceptions. The activity subsystem involves a response to the physical subsystem in terms of evaluation, satisfaction, or dissatisfaction.[1]

A complete transportation model would be similar to the one shown in Figure 15.1.

ECONOMIC CRITERIA

The capital and operating costs of a particular mode or alternative can determine the acceptability of that mode or alternative for a given situation. Too often, more attention is given to capital costs, not enough to the day-to-day costs of operation and maintenance. Both must be considered in the economics of transportation alternatives.

A primary responsibility of the engineer in the planning process is to determine whether or not a proposed system is economically justified. The ability to finance a project, to secure risk capital, or the commitment of public credit usually rests upon a favorable answer from some kind of initial estimate.

Has the project any economic justification for being in the first place? Will it earn a sufficient rate of return, or provide a sufficiently adequate and inexpensive service to be worth the money expanded? Will the benefits be more than the costs? Which of several alternatives will prove the most economically desirable? One alternative is always the status quo,

---

[1]*Strategies for the Evaluation of Alternative Transportation Plans*, National Cooperative Highway Research Report 96, by Edwin N. Thomas and Joseph L. Schofer, Highway Research Board, National Research Council, Washington, D.C., 1970, pp. 13–25.

continuing with the existing arrangement and (probable) mounting maintenance and operating expenditures. The new proposal must show a definite
advantage over the present practice; otherwise there is no point in making
the change.

The alternative is not only between transportation projects. Other
community and business opportunities compete for capital. Police and fire
protection, parks, schools, and hospitals, for example, compete for public
funds with transportation projects. A private source of capital faces num-
berless opportunities to invest its risk capital in other than transportation.

Economic costs can be determined by any of several well established
methods. If properly applied, all will give about the same "cost-profitabil-
ity" status of a project. These several techniques—rate of return, annual
cost, capital cost, benefit-cost ratio—are reviewed in following paragraphs.

Rate of Return   One of the oldest and soundest criteria of economic
feasibility is the rate-of-return method using the basic location formula.
This involves a comparison of the rates of return on investments from
several alternative systems, routes, locations, or designs using the formula
$(R - E)/C = p$, where R = anticipated revenues, E = operating expenses
including taxes and depreciation, C = investment in plant and equipment,
and p = rate of return. This method has application in the detailed comparison of alternative railroad locations where the effects of gradients,
curvature, and distance for each are under consideration. However, it has
application to almost any comparison that may arise in transportation
planning. It permits a determination of whether the rate of return is large
enough to attract capital, a necessary test for projects being financed in the
open money market and a conservative guide even for projects financed
directly by taxes. The method also permits evaluation of the revenue and
the costs of securing that revenue. For example, will enough revenue be
secured from a traffic source to justify a longer route and additional
operating expense to serve that traffic. It is useful in getting the most done
for the least money, that is, maximizing the rate of return and giving
priority to those projects yielding the highest rate.

The rate of return should be sufficient at least, even with public
projects, to amortize the investment and pay off all borrowed moneys with
interest. In addition, a commercial venture must show reasonable promise
of profit. It is possible that some or all schemes will be discarded as not
yielding an *attractive* rate of return.

A question will arise as to what value can be used for R, the revenue, in a
nonrevenue project such as a nontoll highway. Where revenue is constant or
where there is no direct revenue, consider the numerator of the formula as

the difference in total costs of operation of the old, or the base, project and the new. In this situation, the project, to be justified, must show a reduction in the overall costs of maintaining the roadway and of vehicle operation over it. Reduced gradients, flatter curves, shorter distances, faster speed—these should add to lower costs for fuel, maintenance, labor, and other pertinent items. The equation now actually shows the rate of return from the savings.

The use of this form of rate of return on investment is seen in the following simple illustration.

Given: a proposal to build a 20-mile (32.2-km) branch-line railroad to feed traffic to the main line. Three alternative locations are available: A, B, and C. Their revenues, construction costs, operating expenses, and, as a result of the rate-of-return calculations, rates of return are shown in tabulated form.

Formula Term:	A	B	C
R = revenue	$ 840,000	$ 840,000	$ 840,000
E = expense	720,000	780,000	660,000
C = construction cost	1,600,000	1,520,000	1,880,000
p = rate of return	7.5%	4.0%	9.6%

Sample calculation:

$$(R - E)/C = p; (840,000 - 720,000)/1,600,000 = 7.5\%$$

In spite of its higher construction cost, alternative C gives the most favorable rate of return because of lower operating expenses. B would hardly be acceptable, even if it were the highest, because it barely earns enough for fixed charges, let alone profit.

Another version of the rate of return determines the rate of interest at which two alternatives have equal annual costs. One alternative is taken as the base and the others compared with it. When improvements of existing facilities are contemplated, the status quo or do-nothing alternative must always be considered as a possibility and serves as a base for comparison. Each alternative is compared with that status quo, which will have no construction costs but will have its own operating costs, and the rate of return is computed. Those plans are discarded that do not show a rate of return sufficiently high to attract capital. The rate of return is next computed on the increase in investment between proposals of successively higher costs. Those not showing an attractive rate of return are discarded before computing the rate of return on the next increment. Economically speaking, the desirable alternative is the one having more than the attractive rate of return both on total and incremental investments.[2]

This method does not really avoid the problem of establishing a minimum acceptable rate of return and is not always an easy way to rank investment alternatives. The method has found principal application in highway studies, especially as a basis for making improvements. A straight rate-of-return comparison of alternative investments is the most direct, easily understood, and satisfactory procedure.

Annual Costs   With the annual-cost method, the capital or investment costs are reduced to an annual basis and added to the annual operating expenses to obtain total annual costs. The alternative giving the lowest annual cost is, logically, the one selected.

One form of the annual cost equation is:

$$C_a = (P - S) \times CRF + S \times I + E_a$$

where

C_a = total annual cost

P = initial capital cost or investment

S = salvage value at end of n years

I = interest on salvage

CRF = capital recovery factor for I rate of interest over n years

E_a = annual operating and/or maintenance expense

This method is adaptable to an industry planning its own transport system or to an evaluation of the effects of improvements on an existing route (in which the existing or status-quo situation is considered as a possible alternative). This method has the serious defect of giving widely differing results for various assumed interest rates and life-of-property elements, values for which engineers are not in full agreement. It does not give the rate of return.

Problem Example
Using the data of the preceding problem example, the annual cost would be compared in the following manner.

[2]L. I. Hewes and C. H. Oglesby, *Highway Engineering*, Wiley, New York, 1954, pp. 66–67, 70–71.

Formula Term:	A	B	C
Operating expenses	$720,000	$780,000	$660,000
Annual capital recovery (expense)[a]	80,832	76,790	94,978
Total annual cost	$800,832	$856,790	$754,978

[a]The construction or investment costs of the preceding problem example are reduced to an annual basis by multiplying the investment by the annual-capital-recovery factor (CRF). This procedure is given fuller explanation in the section entitled Capital Recovery. In the foregoing example, Project C, having the lowest annual cost, is preferable. Here, salvage has already been included in the capital cost figure.

Benefit-Cost Ratio The determining factor in benefit-cost studies is the ratio, R_b, of annual benefits, B_a, to the annual costs, C_a. Or, $R_b = (C_{u1} - C_{u2})/(C_{a1} - C_{a2})$ where C_{u1} and C_{u2} are the annual user costs before and after improvement, that is, the benefits, and C_{a1} and C_{a2} are the total annual costs before and after improvement, respectively. The total annual costs are computed as for the annual-cost method, using the CRF to include capital costs. A ratio greater than 1.0 indicates that the additional expenditure for the alternative over the base or equivalent cost is justified; a ratio less than 1.0, that the benefits are less than the costs. After taking the original or base condition, further ratios are computed between successively higher increments of investment cost. That alternative of maximum investment cost that reaches a satisfactory cost ratio on both total and incremental investment is the most acceptable.

Argument has arisen as to what benefits should be included. Should only the direct benefits to users of the system be counted or should such secondary benefits as enhanced property values, increased sales, and more industrial starts, be included? While secondary benefits are often quite real, they are difficult to evaluate accurately. The tendency is to overstate their importance, and there are possibilities of inadvertently counting these benefits more than once, that is, accounting for their effect in more than one situation. Conservative opinion holds to restricting estimated benefits to those enjoyed by the users of the system, that is, to the savings in transportation costs arising from reductions in fuel and wages, in road time, in accident costs, etc. User costs are more readily imposed. Secondary benefits will be recognized in more goods or capital available for other services or in lower prices of goods moved over the improved system.

The benefit-cost method has found extensive use in highway, waterway, and other public projects. The results obtained vary with the rate of interest assumed and with the values attached to various benefits. Adverse effects, negative benefits, should be included in the costs. The interest rate should be realistic and given uniform use throughout.

Capital Recovery   Not only must the capital investment be recovered by the project but the interest charges as well. On a yearly basis, the annual amount is obtained from the interest formula, $R = Pr\{(1+r)^n / [(1+r)^n - 1]\}$, where R = annual charge for n years to recover investment with interest, n = life of the property (or duration of the obligation), r = the rate of interest, and P = principal or initial investment. The value of $r(1+r)^n / [(1+r)^n - 1]$ is called the capital-recovery factor, CRF. Computations are facilitated by use of a table in which representative values for the CRF are given for various values of r and n. See Table 13.4. Intermediate values are found by interpolation. The capital-recovery cost per year for the annual-cost problem of the preceding section is computed as follows —assuming a 40-year life (n) and 4-percent rate of interest:

Alternative A: investment $\times$ CRF = \$1,600,000
         $\times 0.05052 = \$80,832$

Alternative B: $\$1,520,000 \times 0.05052 = \$76,790$

Alternative C: $\$1,880,000 \times 0.05052 = \$94,978$

Capital Costs and Recovery   The degree of accuracy of any method depends in part on the accuracy of the data used. For initial feasibility studies, average construction and equipment costs will serve for the investment. Eventually more accurate estimates will be required. Construction and equipment estimates present about the same problems and degree of accuracy as any other engineering estimate.

Problems arise in converting investment costs to annual costs. If the project has a finite life—as with a conveyor installed to move aggregates for dam construction—the total cost is simply divided by the anticipated duration of the project. Total cost in this case would include interest and any other financing costs. It could also include an estimate of the total operating cost for the life of the operation. If a finite life is not definitely known, one must be assumed. One assumption is to use the life of the financial obligations incurred. For example, if 50-year bonds have been used for financing, the interest and capital cost are spread over 50 years. This method may have advantages in setting up sinking funds to retire the obligations but does not square with the physical and economic facts in

TABLE 15.1 CAPITAL-RECOVERY FACTORS (CRF) FOR VARIOUS LIVES AND INTEREST RATES[a]

Life, Years	Interest Rate							
	0%	2%	3%	4%	5%	6%	8%	10%
5	0.20000	0.21216	0.21835	0.22463	0.23097	0.23740	0.25046	0.26380
10	0.10000	0.11133	0.11723	0.12329	0.12950	0.13587	0.14903	0.16275
15	0.06667	0.07783	0.08377	0.08994	0.09634	0.10296	0.11683	0.13147
20	0.05000	0.06116	0.06722	0.07358	0.08024	0.08718	0.10185	0.11746
25	0.04000	0.05122	0.05743	0.06401	0.07095	0.07823	0.09368	0.11017
30	0.03333	0.04465	0.05102	0.05783	0.06505	0.07265	0.08883	0.10608
35	0.02857	0.04000	0.04654	0.05358	0.06107	0.06897	0.08580	0.10369
40	0.02500	0.03656	0.04326	0.05052	0.05828	0.06646	0.08386	0.10226
50	0.02000	0.03182	0.03887	0.04655	0.05478	0.06344	0.08174	0.10086
60	0.01667	0.02877	0.03613	0.04420	0.05283	0.06188	0.08080	0.10033
80	0.01250	0.02516	0.03311	0.04181	0.05103	0.06057	0.08017	0.10005
100	0.01000	0.02320	0.03165	0.04081	0.05038	0.06018	0.08004	0.10001

[a]L. I. Hewes and C. H. Oglesby, *Highway Engineering*, Wiley, New York, 1954, p. 60, Table 5.

trying to establish a fair evaluation of a project.

More often, the economic life span of the plant is estimated. A new problem then arises in that the several elements of plant and equipment have different life spans. Trucks, tractors, and trailers may have a life of 6 to 10 years, airplanes 10 to 15 years, railroad cars and locomotives, ships, and barges 20 to 30 years. Structures are assumed by the AASHTO and the AREA as having a 40-year life. In that time, they are likely to become obsolete even if not completely worn out. Pavement life varies from 4 to 48 years, with an average life of 18.5 years being adopted by the AASHTO. Railroad track, in the United States is usually not depreciated but through continual piecemeal renewal, has an indefinite life. Lives of 50 to 100 years have been used in economic studies. The Bureau of Public Roads has adopted 100 years as the life for grading in highway. This would also be applicable to railroad and canal grading. Longer economic life is often assumed than the accuracy of predictions concerning wear and obsolescence justify. In computing capital-recovery values, each element is computed separately (or a weighted average is obtained through the use of interest tables). Research is needed to make more accurate evaluations of economic life for different classes of property.

The following problem illustrates the method of accounting for differential life spans.

Class of Property	Investment per mile	Economic Life	CRF, 4%	Annual Cost
Right of way	$ 3,000	80 yr	0.04181	$ 125
Grading	11,000	60 yr	0.04420	486
Track	72,000	80 yr	0.04181	3,010
Structures	20,000	40 yr	0.05052	1,010
Signals	8,000	30 yr	0.05783	463
Total	$94,000			$5,094

Capital versus Operating Costs   A definite relation exists between construction and operating costs. Construction costs generally occur only once, but their effects continue thereafter in interest and retirement charges, whereas operating expenses continue to accrue for the life of the route, even after construction costs have been amortized. A greater sum spent for construction or equipment can often reduce operating costs. However, undue refinement in the plant produces burdensome capital charges. Just how much additional or incremental construction cost should thus be incurred and the improvement in revenues and operating costs arising from them is a relation the engineer must determine, whether by rate of return or some other method. He must view each alternative in the light of the credit-income situation.

Operating Costs   Operating costs may be obtained from records of current and past experience or from generally accepted averages (as in Chapter 12), from national cost statistics published by regulatory agencies, or from the actual costs of a carrier similarly situated.

In assuming costs for a "fixed"-cost carrier, a principle of Chapter 12 should be recalled—only the variable percentage of total costs is used for costs because of incremental increases or decreases in traffic.

Units for measuring operating costs are the ton-mile (gross or revenue) or the 1000-ton mile and the vehicle mile (or train mile, truck mile, plane mile, ship mile, etc.) Ton-mile costs are usually used in railroad, conveyor, pipeline, and other commercial transportation studies, but vehicle miles are frequently used in highway comparisons.

Volume–Cost Relation   The utility of a system can be restricted by its costs in relation to volume of traffic handled. Rapid transit, for example, is generally considered economically feasible only with densely settled popu-

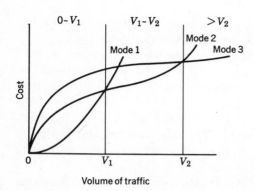

FIGURE 15.2 COST FACTORS IN MODAL CHOICE.

lations of one million or more. Automobiles and trucks offer low cost individual service useful economically in low density areas. Large volumes of traffic must be available to justify the high costs of pipelines.

An industry might evaluate its transport needs in terms of (a) one or more trucks for a small volume (V_1) of business, (b) contract with a contract carrier for a fleet of trucks as the volume of business increases (V_2), and (c) establish its own fleet of trucks when business has experienced a very large volume increase, (V_3). See Figure 15.2 and also the next section.

COST EFFECTIVENESS

Cost Effective Evaluation   Alternative systems must be evaluated through an orderly process with criteria that provide a means of identifying and comparing the consequences of each. Evaluation in the past was almost entirely based on dollar cost criteria such as capital costs, rate of return, present worth, or benefit-cost ratios. The last has had extensive use in evaluating public projects. These procedures, explained in more detail earlier in the chapter, lack the wide range of applicability that is necessary for systems analysis procedures.

A more recent method makes use of cost effectiveness methods, first developed for military planning but having suitable application to transportation planning. Cost effective evaluation methods provide a flexible framework for developing information that aids in selecting between alternatives. It permits the use of data from benefit-cost or other economic models but goes beyond those by permitting consideration of intangibles such as social and environmental consequences, separates costs from effectiveness, and, rather than developing a single figure such as a benefit-cost ratio or a rate of return to measure alternative values, permits a developing growth of objectives as a response to community knowledge about goals, objectives, and the consequences associated with them.

The term *cost* includes all of the materials, manpower, equipment, and effort that are required throughout the useful life of the system and the impact of these upon the regional environment in which the system functions.

Effectiveness is the degree to which an alternative achieves its objective. It is a measure of the system's utility. The decision maker must be informed as to the degree of effectiveness of alternative systems and costs of each level of effectiveness.

Consequences are the impact a proposed system or solution will have on:

(a) Those who use the system, operate, and maintain it (impacts of, for example, cost, comfort, convenience, safety, and wages).
(b) Those outside the system, the community and environment (for example, impacts of economic loss or gain, loss or gain in taxes, degree of pollution, land requirements, loss of dwelling units, and loss of job opportunities).

For a private, profit-seeking enterprise, cost effectiveness will usually be measured by one of the economic models—rate of return, annual cost, present worth, or cash flow. Alternatives are compared as to the increase in revenues reductions in cost, or increase in profits from route relocation, grade reductions, increased size of vehicle, redesign or relocation of terminals, new route construction, or from low volume–high profit vs. high volume–low profit pricing policies. Intangibles have played a less important role. With the close interrelation, especially financial, developing between all modes and the government, the distinction between the private and public interest is becoming blurred and there is need for more evaluation of intangibles to enter into private planning.

First Problem Example

An example of cost effective evaluation can be helpful at this point:

Consider the problem of providing transportation for grain movement in a Midwestern state. The state places a high value on assisting agricultural production. The initial objective is to serve the greatest number of farms at the least cost per farm. Because of road deterioration and the loss of rail branch lines in the Conrail reorganization, the Planning Group in the State's Transportation Department have developed five alternative networks of rural roads to be improved so that farmers can haul their grain to central on-rail elevators. The capital costs and number of farms served by each network are set forth in Table 15.2.

TABLE 15.2 NETWORK COSTS AND SERVICES

Alternatives	Capital Costs	Farms Served	Cost per Farm
A	$4 million	20	$200,000
B	6 million	25	260,000
C	10 million	40	250,000
D	14 million	60	233,000
E	20 million	100	200,000

These relations are better illustrated graphically as in Figure 15.3. The cost per farms served is given in the fourth column of Table 15.2 and show alternatives A and C to have the same capital costs. Alternative E would be preferable on this basis because more farms are served for that same cost. All other alternatives are less attractive.

One can go a step farther and suppose that the state budget permits only $14 million for this project. A new objective is thus introduced, that is, to stay within the budget constraints. Therefore the next highest number of farms served within that constraint will be chosen. Alternative D, serving only 60 farms but feasible from a budgetary standpoint, would meet that objective.

This is a highly simplified situation. Complexities arise when one considers the number of bushels of grain made available to the market by each alternative. The 40 farms omitted by choosing D over E could be large farms producing far more grain than the remaining 60. Also, only capital costs have been considered. There may be wide differences in the costs of maintaining the alternative routes and heavy grades or route circuity may incur high operating or hauling costs to the farmers. One or more of the routes may have to pass through and violate a forest preserve or follow a winding course that poses driving safety hazards. Not all of these factors are capable of economic evaluation.

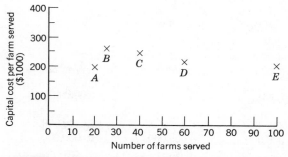

FIGURE 15.3 NETWORK COSTS AND SERVICES.

Second Problem Example

As a more elaborate (but still highly simplified) example, consider the proposal that a regional airport be built by a state that places a high value on industrial and aviation growth and development. The general location is near a city of 200,000 population with smaller communities of 30,000 to 80,000 population surrounding it at distances of 40 to 80 miles. The stated objectives of the planning effort are:

1. To satisfy the growing industrial needs of the region and to accommodate foreseeable traffic growth and technological development.
2. To remove traffic burden on airports in surrounding cities.
3. To be financially feasible.
4. To be compatible with the environment including safety to adjacent land uses.

Additional (and hypothetical) data regarding three alternative plans and sites are given in Table 15.3. The problem is to utilize cost-effective evaluation procedures for selecting one of the alternative sites.

TABLE 15.3 DATA FOR ALTERNATIVE PLANS AND SITES

Item	Alternative A	Alternative B	Alternative C
Land required	3000 acres	2500 acres	2000 acres
Estimated cost	$76,000,000	$66,000,000	$70,000,000
Projected on- and off-passengers	1,200,000	1,000,000	900,000
Projected landings and takeoffs	100,000	90,000	120,000
Traffic loss by adjacent airports	200,000	100,000	100,000
Crops lost from land taken (per year)	240,000 bushels	200,000 bushels	100,000 bushels
Dwellings lost by land taken	60	120	380
Proximity to centroid of traffic sources	20 miles	15 miles	10 miles
Nearest residential area to runway approaches	5 miles	2 miles	0.5 miles
Highway vehicle access per day	500,000	500,000	500,000
Operating costs (airport only)	$60,000,000	$94,000,000	$56,000,000
Operating revenues	$115,000,000	90,000,000	110,000,000
Interest rate	8%	8%	8%
Economic life	30 years	30 years	30 years

Some immediate evaluations can be made from the table. A prime consideration is the ability to meet the developing needs of the community, region, and state. Alternative A evidently serves a greater number of passengers; C the least. The loss of passengers by neighboring airports (or removal of traffic burden) is also greatest for site A and about equal at a much lower level for B and C. Thus site A is meeting the first two objectives. The relative amounts of land, construction costs, operating costs and revenues are also evident.

Because of its proximity to the large community, more takeoff and landings of private business and pleasure flying are projected for site C.

Social costs are related to the number of dwelling units lost at each site to make room for structures and runway approaches. Dwellings removed per million passengers are shown in Figure 15.4a. Alternative C has by far the worst impact in terms of lost dwelling space. A similar evaluation could be made in terms of takeoffs and landings but this would not be as significant.

Safety is always an important and often intangible factor. Economic costs have been associated with certain types of accidents if one has a reasonable estimate of the number of accidents likely to occur. The hazards connected with takeoff and landing are most pronounced within the immediate vicinity of the airport. The order of hazard could be in inverse relation to the distance to the nearest populated area. See Table 15.4.

TABLE 15.4 ACCIDENT HAZARD AS RELATED TO DISTANCE FROM AIRPORT

Alternatives	Distance to Populated Area (Miles)	Order of Hazard
A	5	1.0
B	2	2.5
C	0.5	10.0

That is, alternative A is indicated as being 2.5 times as safe as B and 10 times as safe as C. Another possibility would be to obtain records of accident frequency as a function of distance from point of takeoff and landing and use these probabilities as a measure of hazard. No dollar costs have here been developed, but a measure of relative safety has been obtained.

Noise pollution effects could be treated in the same manner as safety. Use could be made of sound intensity levels at various distances for the types of planes likely to be flown and in relation to their projected flight patterns.

Dollar costs can be analyzed first in terms of cost per passenger. The total capital cost divided by the total number of passengers on and off gives a series of costs as shown in Figure 15.4b. A similar cost per passenger based on estimated operating costs could also be obtained. Similarly costs per takeoff and landing based on capital costs could be computed as in Figure 15.4c. Revenues could be analyzed in the same way.

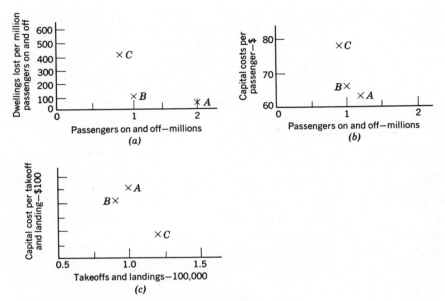

FIGURE 15.4 COST EFFECTIVE ANALYSIS.

The capital cost per takeoff and landing is found to be much less for C while A and B differ very little. Inquiry into these differences indicate that there would be more small private pleasure and business aircraft using alternative C because of its proximity to the supporting city. This raises the question of goals. Does the state wish to devote a significant amount of funding to provide flying facilities for private aircraft? The question might be answered either way depending on the values of the decision makers and the public they represent. The important thing is that analysis has disclosed the need to consider the question.

These procedures permit introduction of the benefit-cost ratio as one step in the evaluation. Detailed description of the procedures for computing the ratio and something of its significance and problems were given in an earlier section. For this example, the construction and operating costs of each have already been estimated. So have the revenues. One must compute in addition the saving in access mileage to the public by reason of the airport location. Access cost is the mileage distance from the centroid of traffic origination and destination × the number of passengers (assuming each on- and off-passenger represents a two-way trip by private car or taxi) × costs per vehicle mile (taken here as $0.12).

Many adjustments and additions to the foregoing can be suggested. Access distance cost could be further subdivided into private car, bus, and even rapid transit—if these be available. Travel time differences for the three alternatives could be computed with each minute of time lost or saved being based on the average wage for the community or similar measure. Costs for the airlines serving

TABLE 15.5 BENEFIT-COST RATIO

Alternative	Capital Cost	Operating Cost	Access Cost	Benefit-Cost Ratio
A	$76 million	$60 million	$115 million	1.65
B	66 million	54 million	90 million	1.61
C	70 million	56 million	110 million	1.74

the airport have not been included. The analysis in actual practice can become quite extended. The example will serve only to illustrate the procedure.

In any event, the benefit-cost ratio shows site *C* to have a greater return per dollar spent than the other two. That finding must be taken in conjunction with other criteria, however. These can be summarized from the foregoing in Table 15.6.

A number of conflicting factors are represented by the foregoing. Alternative *A* represents the highest capital investment and annual cost, but in spite of having the highest revenues, its benefit-cost ratio is the least. Site *C* has the highest benefit-cost ratio and has the least effect on farm productivity, but it has a disastrous impact on loss of dwelling units. It also rates poorly in terms of safety and pollution. Site *A* does, however, rank first in ability to satisfy demand.

TABLE 15.6 CRITERIA SUMMARY—COST EFFECTIVENESS EVALUATION

Criteria	Alternative A	Alternative B	Alternative C
Capital cost	$76 million	$66 million	$70 million
Annual operating cost	69.6 million	56 million	63 million
Annual revenue	115.0 million	70 million	110 million
Benefit-cost ratio	1.65	1.61	1.74
Safety-order of hazard	1.0	2.5	10.0
Dwellings lost per million passengers	50	120	380
Capital cost per passenger	$64.17	$61.00	$77.77
Capital cost per take off and landing	760	733	583
Pollution effects	1.0	2.5	10.0
Lost farm productivity—bushels per year	240,000	200,000	100,000
Bushels per year lost per passenger	0.2	0.2	0.11
Ability to satisfy traffic demand	1	3	2

The procedure does not give a specific answer. It defines, instead, the consequences of each alternative and indicates possible tradeoffs for each. The decision makers still have to make the decision, but it can now be a better informed decision.[3]

Additional features of analysis could be taken into account, such as the future effect on present land uses, the effects on financing costs and traffic demand of stage construction, and the costs and benefits to airlines serving the airport. Whether one location or another will serve to attract more of the desired industry to the state and locality would be a study in itself. Social and environmental inputs have been dealt with only superficially. This topic is given more discussion in a later section.

A weighting method to assist in making comparisons between the three alternatives with their varied criteria will be the subject of later paragraphs.

OTHER CRITERIA

Environmental Impact Statement An evaluation study relating to a broad sprectrum of environmental impacts is required by law for proposed major projects in which Federal funds are involved. Impact statements are required for Federal-Aid highways, for railroad projects that involve Federal money, and inland waterways. The statements are prepared by the supervising agency; the U.S. Army Corps of Engineers, for example, prepares impact statements for waterway projects. The impact statements are required by Section 102 (2) (6) of the National Environmental Policy Act of 1969. The statement requires: (1) a statement of the impact the proposed activity or project will have on the environment, (2) a statement of the known adverse effects occurring through implementation of the project, (3) a description of alternative courses of action, (4) a description of short-term vs. long-term effects on maintaining and enhancing the environment, and (5) a statement of irreversible and irretrievable commitments of resources needed to implement the project.

The report must be circulated and hearings held so that all elements of a community will be informed and given an opportunity to make their opposition or approval known. By so doing the possibility for citizen and community input to the planning process is made available. (See the section on citizen-community input later in this chapter.) The primary purpose of the requirement is, however, to protect the environment.

---

[3]See the National Cooperative Research Program Report 96, "Strategies for the Evaluation of Alternative Plans," for an extended discussion of the subject.

Weighting   Evaluating a set of criteria such as that in Table 15.6 presents difficulties. What importance is to be attached to one factor in contrast to another? How can tangible criteria such as a benefit-cost ratio be compared with a stated percent of the polluting substances in the atmosphere? Attaching weights to the various criteria aids in overcoming the problem. Weighting is usually a subjective process in which one or more groups or individuals, exercising judgement, attach numbers to each factor indicative of its perceived importance. Obviously such weighting will vary with the type of system or project, the goals and objectives, and the individuals and groups doing the weighting. Weights may be attached by technical experts on a planning commission or a consultant's staff or by a citizens advisory group. See the next section on citizen participation.

In Table 15.6, for example, the benefit-cost ratio could be considered the most important by a judge and given a ranking of 1. Loss of dwelling space could be considered of next importance and given a ranking of 2. Loss of agricultural production, considered not important, might receive a ranking of 7 or 8. Another judge might prepare an entirely different ranking.

To give the most important criterion the highest ranking, *converted rankings* are assigned on an $n-1$, $n-2$, etc. basis when n is the total number of items ranked. There are 12 such items in Table 15.6 of the preceding problem. The above raw rankings would have a converted ranking of 11, 10, and 5 or 4 respectively. To reduce subjective bias effects, several persons or groups should prepare rankings from which a composite rating can be developed.[4] The composite rank for criterion j is simply the sum of the ranks given to j by judge i when there are m judges ranking n criteria, that is, composite rank R_j is:

$$R_j = \sum_{i=1}^{i=m} R_{ij} \qquad \text{for } j = 1, 2, \ldots, n$$

and the normalized composite rank, weight, or utility value u_j is obtained by dividing the composite rank of criterion j by the sum of n such rankings:

$$u_j = \frac{R_j}{\displaystyle\sum_{j=1}^{j=n} R_j}$$

[4]Walter C. Vodrazka, Charles C. Schimpeler, Joseph C. Carradino, *Citizen Participation in Louisville Airport Site Selection*, Citizen's Role in Transportation, Transportation Research Record 555, Transportation Research Board, National Research Council, Washington, D.C., 1975, pp. 43-48.

Another technique is the use of a rating scale. Descriptive terms can be used—excellent, good, fair, poor, etc.—but these tend to create bias and are difficult to combine. A numerical rating reduces bias and permits combining the individual ratings. A list of criteria can be placed on one side of a sheet in a column and a rating scale from 0 to 100, 0 on the bottom, placed on the other side. A rating is assigned and indicated by drawing a line from an individual criterion to the appropriate point on the rating scale. These can then be rearranged in order of numerical rating.

As before, a numerical rating is obtained by summating the ratings of the several judges.

$$V_j = \sum_{i=1}^{i=m} V_{ij} \qquad \text{for } j = 1, 2, 3, \ldots, n$$

where

$$V_j = \text{composite rating for each criterion}$$

$$V_{ij} = \text{rating by judge } j \text{ of criterion } i$$

$$n = \text{number of criteria}$$

$$m = \text{number of judges}$$

The composite rating or utility for each criterion is obtained as before, that is,

$$u'_j = \frac{V_j}{\sum\limits_{j=1}^{j=n} V_j}$$

where

$$u'_j = \text{normalized rating}$$

The combined effects of the two composite ratings is obtained by taking their average whereby

$$u_{j(ave)} = (u_j + u'_j)/2$$

Having criteria established, one next applies the criteria to individual alternatives to determine an effectiveness value for each alternative. Each criterion is considered, one at a time, by each judge who assigns his own effectiveness values based on 1.0 when all aspects of the criterion can be met by a particular alternative, 0.5 when no advantage or disadvantage is evident; 0 indicates that the criterion cannot be met.

For any one of k alternatives, the effectiveness value $\times$ the utility value

for a particular criterion gives the alternative utility value for that criterion, u_j of n criteria:

$$U_t = \sum_{j=1}^{j=n} e_{ij} u_j$$

where

U_t = total effectiveness of an alternative

Again, the total effectiveness is normalized by:

$$U_t' = \frac{U_t}{\sum\limits_{j=1}^{j=n} U_t}$$

The overall utility for each choice can then be tabulated by alternatives and a 1- or 2-digit order of rank indicated. The foregoing procedures are based on a method given in the earlier reference to a study by Vodrazka, Schimpeler, and Corradino.[5]

Community-Citizen Input   A transportation system has considerable impact on the community and its individual citizens, whether the community be local, state, or national. The system is supposedly designed to serve the community and must look to it for traffic, financial, and, frequently, political support. Community-citizen imput to the planning process is a much needed element in the procedure.

Certain interests through their strong financial, industrial, or political power, may provide unpublicized input directly or indirectly to the planning and decision making process. Too often, citizen input from individuals and small neighborhoods or groups is lacking. Yet these individuals are most intimately affected, especially at the operational level, by adequacy of service, effects on the environment, disruption of land uses, loss of community values, loss of dwelling units, loss of work opportunities, costs, and taxes. The need exists to balance gains for one group against the losses for another. The process should bring all affected parties into the planning process rather than have a plan developed by a few technical experts (often for the benefit of a few powerful interests) and presented for implementation regardless of local desires. Public confidence in planning agencies whether state departments of transportation or regional, county,

[5]Ibid.

or municipal planning commissions—and acceptance of the plans developed—may well depend on the extent to which citizen input has been brought into the planning process.

Citizen-community input (hereinafter referred to simply as citizen input) is desirable then:

(a) To gain the benefit of insight and wishes of those most closely affected on a day-to-day basis, that is, of the users and neighbors of the proposed project.
(b) To win support for whatever plan or alternative is selected for implementation and for the agency promoting it.

Citizen input, when attempted, has generally been secured in one of two ways (or sometimes a combination):

(a) One or two alternative plans are developed by a state transportation department, a regional, metro, or city planning commission, or by a consultant acting in behalf of one of them. The alternatives, or perhaps only one plan, are presented at a public hearing with little or no advance preparation for citizen participants. On the basis of the one or perhaps two such presentations, citizens are asked to give their comments, approval or disapproval (approval is what is expected), and vote their tax money accordingly.
(b) The public is brought into the planning process. Their views and assistance are sought and made a part of an ongoing effort. Suggestions and arguments are acknowledged and the pros and cons of a particular position or positions are set forth. The citizen can see where his or her input, whether it be favorably acted upon or not, stands in relation to the goals, objectives, and constraints under which one or more alternatives have to be developed. There is greater likelihood for citizen support and less antagonism when such an approach to planning has been properly conducted.

Citizen input can come about in numerous ways.

(a) Meetings can be held at frequent times and in appropriately convenient locations, throughout the planning process. Those meetings, given widespread publicity, will, it is hoped, bring together those who have a concern about proposed planning and give them a means of expressing their views, objections, and suggestions. Citizen views can be expressed orally in such meetings, supplemented by questionnaires, letter forms, or other means of communication. Such expression has useful meaning only if the citizens have been briefed on progress and have had pertinent data made available to them.

(b) Citizen advisory committees (CAC) can be organized. Their membership should represent a broad spectrum of community population by age, sex, race, place of residence, and place and type of employment. Special interest groups—for example, merchants, real estate dealers, environmentalists, engineers, lawyers, and architects—can be included but many of these have their own means of voicing their viewpoints. The so-called average citizen should be well-represented.

Such groups can serve as a means of communication with the rest of the community, disseminating information and collecting critical comments, and as sources of ideas, and can bring into focus the needs of various groups and areas and the impacts of these needs on proposed alternatives. Much of the response will be at the operational level concerning routes, service frequency, schedules, noise, pollution, etc. Much response will also be received on the impacts the proposed plans might have on local land use, especially the land of the individual making the response. Citizens may participate in the weighting and rating processes earlier described. These could also be modified to obtain weights and ratings for goals and objectives in anticipation of plan formulations.

Community values and objectives so obtained can be quite different from those assumed by a planning commission. A highly useful outcome of such an input, experience has shown, can be the decision to abandon all efforts toward a particular proposal or objective. No single set of values will meet the needs or wishes of all parts of a community. Hence goals also will differ.

Certain pitfalls must be avoided in turning to citizen input.

(a) The groups or committees must be given realistic tasks to perform. Otherwise apathy, a feeling they are being used as window dressing and even antagonism, can arise.

(b) Membership in groups or committees must be so widespread and diversified as to belie a possible charge of "stacking" the membership to support a particular bias or portion of the community.

(c) Attention must be given to the acts and suggestions of the committee members. If a suggestion is not given positive action, the reason for its rejection must be made clear. Failure to respond and give consideration to inputs from citizen groups can have a worse effect than failure to invite citizen participation in the first place.

TABLE 15.7 MODAL PROFILE AND PROJECT ALTERNATIVE EVALUATION

Characteristic, Factor, or Impact	Alternative Evaluation (Poor to Excellent on Basis of 1 to 5 or Other Enumeration)			
	Alternative 1	Alternative 2	Alternative 3	Alternative 4
Adequacy of service				
Capacity				
Accessibility				
Availability				
Door-to-door capability				
All-weather capability				
Flexibility: Route				
Traffic type				
Quantity				
Transfer and interchange incidence				
Coordination with other modes				
User privacy				
Comfort and amenities				
Overall service evaluation				
Safety and dependability				
Guidance and control system				
Weather susceptibility				
Privacy of right of way				
Fail-safe capability				
Need to maintain motion				
Shock and impact incidence				
Accident rate				
Loss and damage rates				
Need for controlled environment—temperature, pressure, ventilation, etc.				
Overall safety evaluation				
Effects on community				
Ability to meet need				
Quantity of land required				
Work opportunities lost or gained				
Lost or gained housing opportunities				
Incidence of physical barriers				
Aesthetic impact				
Pollution effects				
Effect on community growth				
Relation to master plan				
Overall effect on community				

TABLE 15.7 (Continued)

Characteristic, Factor, or Impact	Alternative Evaluation (Poor to Excellent on Basis of 1 to 5 or Other Enumeration)			
	Alternative 1	Alternative 2	Alternative 3	Alternative 4
Environmental impact				
Impact on ecology: Plant				
Animal				
Soil				
Drainage				
Pollution effects: Air				
Water				
Noise				
Visual				
Vibratory				
Overall environmental impact				
Use of energy				
Efficiency (thermal) of use				
Availability of fuel type used				
Quantity used per ton mile or passenger mile				
Cost of energy used				
Overall use of energy				
Technoeconomic factors				
Propulsive resistance				
Payload–empty weight ratio				
Horsepower per ton or passenger				
Flexibility of movement: Overtake and pass, meet and pass, two-way movement				
Adaptability to terrain				
Productivity per hour				
Speed				
Overall technoeconomic				
Cost factors				
Project cost				
Cost effectiveness				
Rate of return				
Benefit–cost ratio				
Break-even load factor				
Fixed or variable cost type				
Need for subsidy				
Method and ease of financing				
Effects on community economy				
Overall cost factors				
Overall utility				

Useful information on techniques for citizen-community inputs and community interaction programs can be found in the NCHRP Report 156. Detailed procedures, organization and management guidelines, strategies, and catalogs of impacts, techniques, data sources, and conceptual models and project examples can be found there.[6]

A matrix format in which evaluation of various systems and alternatives can be entered is given in Table 15.7. Judgement values can be entered on some numerical basis after each characteristic or impact of the proposed alternatives.

QUESTIONS FOR STUDY

1. Prepare an outline of factors that warrant inclusion in a conceptual transportation systems model. Indicate the significance of each.
2. Route A has a possible revenue of $0.05 per net ton mile, operating expenses of $0.04 per net ton mile, construction costs of $150,000 per mile, available net tonnage of 1,800,000 annually. Alternative Route B has a possible revenue of $0.05 per net ton mile, operating expenses of $03.6 per net ton mile, and available net tonnage of 2,000,000 annually. Both routes are 500 miles long. The local attractive rate of interest is 9 percent. Determine which of these routes (if any) is preferable by using the rate of return on investment method.
3. Using the data of Question 2, determine which route is preferable using the annual cost method.
4. Using the data of Question 2, determine which route is preferable using the benefit cost method.
5. Given: Right of way at $6000 per mile, grading at $15,000 per mile, pavement at $45 per mile, structures at $60,000 per mile, maintenance of $3000 per mile, and interest at 9 percent; determine the average annual cost of the investment.
6. Four alternative bikeway networks have the following characteristics:

Alternative	Capital Cost	Travel time: Outer Edge to Centroid (Minute)	Estimated Two-Way Traffic (Trips)
A	$280,000	20	9000
B	200,000	30	7000
C	180,000	45	5500
D	100,000	49	5000

Assuming $4.00 an hour as the average value of time (based on the local going wage rate), compute and plot the cost effectiveness of each route on the basis of data given and select the desirable alternative.

[6]Marvin L. Manheim et al., "Transportation Decision Making: A Guide to Social and Environmental Considerations," *National Cooperative Highway Research Program Report 156*, Transportation Research Board, National Research Council, Washington, D.C., 1975.

7. Using the criteria data in Table 15.6, develop an overall utility figure for each of the three alternatives. (Each student in a class could prepare his or her own rank and rating; these could be combined to give a utility figure for the class as suggested in the section on weighting.)
8. Using a form similar to that of Table 15.7, prepare evaluations for (a) common types of intercity transport systems and/or (b) common types of urban transport systems.
9. Devise a program for including citizen input to the planning process where the project under consideration is:
   (a) A proposed rapid transit system for a metropolitan area of 900,000 population.
   (b) The creation of an urban mass transit district and operating system for a community of 200,000.
   (c) The creation of a regional airport located close to a community of 100,000 population.

SUGGESTED READINGS

1. Edwin N. Thomas and Joseph L. Schofer, *Strategies for the Evaluation of Alternative Transportation Plans*, National Cooperative Highway Research Program Report 96, Transportation Research Board, National Research Council, 1970.
2. Marvin L. Manheim et al., *Transportation Decision Making*, National Cooperative Highway Research Program No. 156, Transportation Research Board, Washington, D.C., 1975.
3. *Citizen's Role in Transportation*, Transportation Research Board, National Research Council, Washington, D.C., 1975.
4. *Issues in Public Transportation*, Special Report No. 144, Transportation Research Board, National Research Council, Washington, D.C., 1974.
5. Wm. S. Pollard, Jr., *Operations Research Approach to the Regional Impact of Transportation and Land Use*, paper prepared for American Society of Civil Engineers Transportation Engineering Conference, Minneapolis, Minnesota, May 1965.
6. *High Speed Ground Alternatives Study*, U.S. Department of Transportation, PB220079 National Technical Information, Washington, D.C., 1973.
7. *National Transportation Report (Present Status-Future Alternatives)*, Office of Assistant Secretary for Policy and International Affairs, U.S. Department of Transportation, Washington, D.C., July 1972.
8. *1974 Transportation Report (Current Performance and Future Prospect)*, Office of Assistant Secretary for Policy and International Affairs, U.S. Department of Transportation, Washington, D.C., 1974.
9. *Bus Use of Highways: Planning and Design Guidelines*, Wilbur Smith and Associates, National Cooperative Highway Research Program Report 155, Transportation Research Board, National Research Council, Washington, D.C., 1975.

10. *Transit Planning*, Transportation Research Record 559, Transportation Research Board, National Research Council, Washington, D.C., 1976.
11. *Financing Federal-Aid Highways*, Highway Planning Technical Report No. 34, Federal Highway Administration, U.S. Department of Transportation, Washington, D.C., January 1974.
12. R.F. Kirby et al., *Para-Transit: Neglected Options in Urban Mobility*, Final Report Volume II, Para-Transit Design, The Urban Institute.

Chapter 16

Regional, State, and National Transportation Planning

The preceding chapters have dealt with the planning process as such and with techniques and procedures primarily related to urban problems and with the evaluation of alternatives developed by those processes. This chapter emphasizes planning at the regional, state, and national levels, describing the designation of transportation corridors and the selecting of a detailed route location within those corridors.

STATE AND REGIONAL PLANNING

The multitude of problems arising in transportation have brought a recognition of the need to exercise some degree of state control over and participation in transportation planning and development as it affects the overall economy and society of the state. The result has been the creation between 1959 and 1975 of at least 23 state departments of transportation (frequently revamped state highway departments). Others can be expected in the future.

Just what should be the scope and authority of these departments has been a subject of lively discussion. As with urban areas, causal relations exist between population, land use, and transportation systems. In addition, the allocation of natural and other state resources and their effects upon the state's economy, society, and environment are closely associated factors. See Figure 16.1.

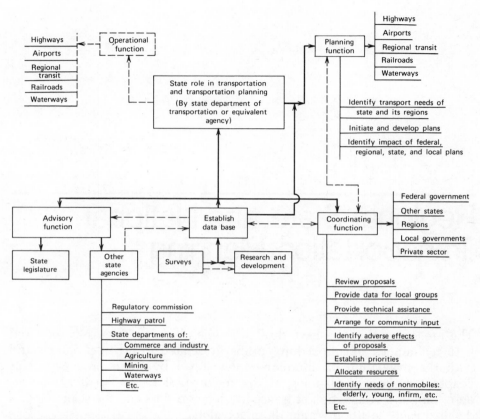

FIGURE 16.1 STATE'S ROLE IN TRANSPORTATION PLANNING.

State DOT Responsibilities Some of the recognized responsibilities and activities of a state department of transportation (DOT) include the following.

1. Define and formulate goals and policies for the state's multimodal approach to statewide transportation planning.

2. Propose alternative solutions and plans and set priorities.

3. Advise the legislature and regulatory commission on transportation policies, progress, and technical matters.

4. Establish guidelines for planning at all levels and for all modes and cooperate with local agencies in planning.

5. Establish guidelines for the allocation of resources—money, land, energy, and transportation systems.

6. Provide technical assistance and data to local planning agencies.

7. Assist in the transfer of private to public ownership of transportation facilities and of public to private ownership.

8. Allocate transportation services and facilities among the private and public sectors.

9. Assist in integrating privately owned transportation systems into the statewide system.

10. Be actively interested in problems of freight transportation and the appropriate modes.

11. Be actively interested in passenger transportation and the modes by which it is provided.

12. Foster citizen participation and evaluation.

13. Provide coordination with national and interstate planning programs.

14. Provide oversight on the use of funds granted to private transportation companies and to cities and regions.

Whether within a formal department of transportation or not, states should have planning units to carry out these functions. Among its specific duties such a unit could:

(a) Develop a data base for its own use and the use of other planning groups in the state.

(b) Prepare evaluations of alternatives.

(c) Determine the impact of alternatives on the life of the state.

(d) Prepare environmental impact statements and evaluate impact statements prepared by others.

(e) Keep abreast of changes in magnitude and character of demand and of ways to meet that demand.

(f) Participate in the active planning and development of alternatives.

(g) Plan for and build highways.

(h) Plan for and operate directly or indirectly needed rail lines and services that would otherwise be abandoned (Regional Railroad Reorganizational Act of 1973 and Amtrak legislation).

The planning unit may be staffed and financed to perform detail planning and designing or it may call upon consulting firms to provide or share with it these more specific task-related functions. It has been suggested that a liaison board that includes representatives from privately owned transportation companies and from local and regional operators be formed to advise and coordinate activities with those of the state.

There is an evident need for more research, experience, and improved techniques in the area of statewide planning. Policy-sensitive models that relate to modal allocations of resources and traffic and to subsidies and pricing are useful but not fully developed tools. The environmental impact statement is assuming greater importance as its scope and power are demonstrated with use. Improved models to determine environmental impact are needed. There should be more effective ties between planning and programming work and between state, local, and metropolitan comprehensive planning. Land use planning, on which the success of statewide planning may sometimes depend, is in its early stages and is viewed with suspicion by much of the citizenry, especially people in rural areas.

These matters and many more are discussed in some detail in *Special Report* 146 of the Transportation Research Board's "Issues in Statewide Transportation Planning."

Data Base   Because citizen input is more difficult to secure at a statewide level and because certain problems do not directly affect the entire state, planning may be performed in some instances at a regional level encompassing several zones or counties. The five-county Regional Transportation Authority in the Chicago area is an example.

Any state or region faces the problem of planning for the transportation needs of its population and economy. The scope of that planning should include all modes, all interests, all traffic (both freight and passenger), and all factors (social, environmental, economic) and the impacts that all of them will have on the economy, society, energy supply, and the environment.

Establishing a data base is one important step in the ability to include so wide a scope in the planning process. Establishing that data base can be a function of a state or regional planning unit. The base should include most or all of the following steps and items.

1. **Zoning.**   Division of the state or regions into zones, perhaps by counties or groups of counties. This permits "pinpointing" present conditions and future changes and gives more manageable quantities to manipulate.

2. **Population.**   Present location, numbers, and densities, both urban and rural, at the time of the survey. Population should be further broken into rural and urban by households, occupation and wage level categories, age groups and sexes. Population of towns and cities are quantified and SMSAs[1] located.

[1]Standard Metropolitan Statistical Area.

3. Land Use. Principal land uses by type, location, and extent (acres or square miles) devoted to each use. Land uses can be identified as residential, industrial, commercial, recreational, mining, and other. Major categories of use or type of production will be the set within each general category and the productivity of use secured for each general type and type category.

4. Physical Facilities. An inventory of physical facilities available in the state for transportation purposes. If concern is only with one mode, usually highways, the inventory can be limited to that mode. But all transportation needs of the state are interrelated. The inventory should be inclusive. Typical and pertinent data will include:

Airlines. Aircraft owned by type and owner, commercial airlines, routes flown and origination, destination, and intermediate stops of their flights, FRA route designations and navigational aids, and airports categorized by class, location, capacity, and ownership.

Highways. Miles of routes designated as urban or rural, as interstate, U. S. rural highway, state, county, or township, number of lanes, fully or partially limited access, divided lanes or not divided, type of surface, allowable loads, level of service—speed, travel time, capacity; all for individual segments of the system. Automobile and truck registration should be obtained by zones and type (for trucks) and average ownership by dwelling unit, farm, or other type of land use occupant.

Railroads. Miles of first main track and miles of second, third, fourth main track according to location, ownership, FRA, and DOT track classifications, physical condition, traffic density, system of operation and signaling, location and type of terminals, interchange facilities with other railroads and with other modes, and ownership of cars and locomotives by type and capacity.

5. Facility Use (Traffic Flow). Traffic patterns and flow, that is, the use made of the physical plant, are determined. Traffic flow can be synthesized (see page 509), but traffic counts will be needed to verify the synthetic processes. Volumes of traffic (by major commodities for freight), the origin and destination points, flight records and schedules, train schedules, passing reports, and train sheets, and locking and port records are prime sources of data.

Interstate traffic can be monitored at points of entry and exit by cordon stations and truckers weigh stations and classified as follows:

External-external.   Originating and terminating outside the state or region.
Internal-external.   Originating within the state but terminating outside the state.
External-internal.   Originating within the state but terminating outside the state.

Data on the kinds and types of traffic for specific routes and route segments and for specific commodities are usually difficult to secure from private companies, especially railroads. Private companies are loath to disclose any data as a matter of policy. Records are kept in terms of system totals, especially by railroads, rather than by specific route segments and commodities.

Service Factors   The need for improved or new transportation systems must first be recognized. Frequently that need establishes the general route or corridor with origin and destination points. The siting of a thermal power plant at Lake Powell and deposits of coal in the Black Mesa area of Arizona set the general location of the Black Mesa and Lake Powell rail line to haul coal to the power plant.

A different problem arose on a regional-national scale when the federal government through its agency, the U. S. Railway Association, undertook to form a viable rail network in the Northeastern United States. The problem was one of rationalization, a pulling together into one system, Conrail, those portions of a group of financially troubled properties that would be economically viable, possess suitable physical route characteristics, and provide a level of service commensurate with the needs of the region served. Certain little-used and unprofitable branch lines and duplicating parallel lines would be abandoned, but competition from other rail carriers was to be preserved. A complicating factor was the passenger-carrying Amtrak that operates over portions of the lines and that, for practical operating reasons, requires its own tracks in some areas.

Studies conducted by the Bureau of Public Roads in 1944 identified a pattern of highway movement corridors between major cities (of 50,000 population or more) that carried a major portion of all highway traffic. This led to the formation of a proposed network of 40,000 odd miles (64,360 km) of a federal interstate and defense highway system that would be financed and constructed starting in 1956. Then came the problem of locating the network in all its details where it would service the greatest volume of demand and an effort by cities omitted from the main network to gain access to it by high grade connecting spurs.

Corridor Identification   A corridor is basically defined by its origin and termination points. It can be further defined by certain intermediate controls—a metropolitan or industrial area that must be served, lakes or dams to be avoided, or favorable river crossing sites or mountain passes to traverse.

Traffic assignment procedures will have assessed the adequacy of existing rail, highway, and other routes and modes. The need for additional or new capacity will have been disclosed. Even with the intermediate constraints mentioned above, there will be wide latitude in the exact location of a line within the corridor. Details of terrain and of distance, curvature, and gradient will have an effect on detail location and will be discussed later.

Economic and environmental factors must also be taken into account. A location will be established that makes some compromise between direct movement and proximity to population centers, industrial areas, and other traffic generators sited between the origin and destination points.

In attempting to serve a particular traffic source, the alignment can be adjusted to pass through the source but at a cost. Alternatively, the alignment can be maintained and the traffic hauled from a distance by feeder service using rail or truck. The tradeoffs between main line deviation and feeder service, both capital and operating, enter into the decision. The rate of return method of cost evaluation forms a convenient way to evaluate the extent to which the length of the line can be introduced to attract additional revenue. Benefit-cost procedures can also be used but not so conveniently. A third alternative would envision a second corridor roughly parallel to the first. There is some right angle distance at which it is more economical to haul to the second corridor than to the first, introducing a problem in corridor spacing. Again, if two modes, say rail and highway, serve the same corridor, what is the most economical separation distance between the two from the standpoint of the shipper who takes into account the combined hauling costs and shipping charges? Also, how far should the two lines be separated in order for the traffic development of each to be realized? [Distances of from 300 to 2600 feet (91.44 to 792.5 m) have been suggested.] These and similar questions are not easy to answer but are highly pertinent to the overall impact of each alternative on the state's economy. Alternatives thus arise at an early stage in planning. Each must be subject to the several evaluation procedures.

Evaluation Procedures   The several alternative systems that arise from state and regional planning can be evaluated as described in Chapter 15. Some of the criteria to be used and the questions to be asked are listed below. Many more could be included depending on the particular goals, objectives, modes, regions, type of traffic, and land uses involved. By setting these in matrix form, questions and criteria down the page, alternatives across, an orderly evaluation procedure is maintained with the opportunity for individuals and groups to weigh and rate the intangible items.

1. Economic effects: Capital costs, operating costs, rates of return, and benefit cost ratios.

2. Cost effectiveness: In terms of trips produced, net ton miles produced, communities served, and vehicle miles covered.

3. User cost: Per mile, per ton mile, per vehicle mile.

4. Economic effects: On industry (including specific individual concerns), agriculture, commerce, mining, other transportation companies and systems.

5. Effect on land use: Will it present use increase, decrease, or change? Will undesirable land uses ensue? Will better land uses be obtained?

6. Does the alternative harmonize with state and regional goals and objectives?

7. Does the alternative fit into the comprehensive planning of the state? Region? Of the communities through which the route or system will pass?

8. Are all intended land uses served?

9. Could other land uses reduce the need for the proposed system?

10. Is the contemplated service adequate to the need it serves?

11. Is the alternative politically feasible?

12. What will be the effects on the environment?
   (a) Amount of pollution—air, water, noise, visual, vibratory.
   (b) Effect on land productivity—increase or decrease?
   (c) Effects on wild life.
   (d) Effects on plant life.
   (e) Will flooding, landslides, reduced water table, and fire hazards ensue?

13. What is the demand on resources—water, land, or energy?

14. Will job opportunities increase or decrease?

15. How will the project be funded? By the state? Will taxes increase? By how much?

The following examples of state and regional planning serve to put the process in focus.

Example 1 At the state level, planning is directed toward evaluating the needs of the entire state and relating existing and proposed networks to present and future requirements of a changing industrial and commercial population, agricultural productivity, and social patterns. Typical of this level of planning is a report prepared by a firm of consultants for the State of Illinois in cooperation with its Bureau of Planning, Illinois Division of Highways (now the Department of Transportation).

The studies were authorized by the Illinois General Assembly "...to provide a realistic appraisal of the impact of the expanding economy on highway requirements and on the sources of revenue for highway purposes,...(to) provide information and engineering data required to determine the type of highway systems needed, the cost of various improvement programs, and the most equitable means of financing highway needs...(including) studies of the economy, highway classification, highway needs, and fiscal requirements."[2]

This study fulfilled a directive of the Highway Study Commission authorized by the Illinois Legislature (in 1963 and 1965) to "...make a study of the Public Roads and Streets in Illinois and to classify and integrate the roads into a complete system to serve the highway transportation needs of the State...and make a complete study of the needs of the several highway systems as classified by the Commission and take into account the present and future needs for improvement, maintenance, and operation of these roads in accordance with their appropriate level of service."[3]

The scope of the study included:

(a)   Transportation economy: growth trends by major industrial sectors and shifts in location of activity. The study further identified 16 major industries, their productivity, and location, 11 standard metropolitan statistical areas and their industrial output and number of workers, agricultural production (9 commodities), and mineral production (14 groups), 24 recreational areas and attendance, distribution of urban and rural population, and projected urban and rural population increases and decreases by counties and SMSA.

(b)   Population growth and changes in rural and urban areas and the underlying reasons for them—by economic region, county, and metropolitan area. Population was further identified by income, number of households, number of passenger cars, and number of commercial vehicles. Correlations were developed between passenger car registration and age and sex of drivers.

(c)   Classification of highways according to standards of design consistent with changing service characteristics of the system in terms of anticipated use and travel volume 20 years hence. Highways were classified as state highways, county roads, township and road district roads and municipal streets. All traveled ways were inventoried to determine present conditions and needed construction or reconstruction for target date needs. Costs for such construction were established in terms of 10-, 15-, and 20-year programs.

A level-of-service classification defined roads as arterial, collectors, and access. Rural arterials were those connecting cities of over 25,000 population and involved trips of extended length and speeds of 70 mph (before the

[2]*Illinois Highway Needs and Fiscal Study: Final Report*, prepared by Wilbur Smith and Associates, New Haven, Connecticut, October 1967, p. 1.
[3]Ibid.

Federal 55-mph limit). Arterials included freeways and supplemental free-ways, major highways for interstate and regional traffic, some with partial control of access and serving all cities with populations of 5000 or more, area service highways (speeds of 60 to 70 mph (96.0 to 112.0 kph) providing service to all county seats and cities with populations of 1000 or more, collector highways serving smaller cities (50 to 70 mph or 80.0 to 112 kph), and land access roads to farms and low density homes. A similar type classification was made for urban areas.

(d)   Analyzed fiscal problems relating to the allocation of costs among users, the sources of funds, and the administration of funds.

(e)   Procedures involved synthesizing the existing travel patterns based on population, vehicle registration, and available origination-destination data. A total of 526 traffic analysis zones were established by grouping townships into approximately five zones per county plus seven zones for the City of Chicago. A network of U. S. numbered and major state highways was identified for traffic assignment purposes. It contained 11,500 miles (18,503.5 km) of arterial routes and 6500 miles (10,458.5 km) of centroid connectors. Each link was described by time, speed, and distance. Interchanges between zones were determined by the equation

$$T_{ij} = C\left(P_i P_j\right)^{1/2} e - KD$$

where

T_{ij} = trips originating in zone i and going to zone j

C = a coefficient having a value of 0.0026 work trips by auto, 0.0032 for other trips by auto, and 0.0023 for truck trips

K = a coefficient having values corre-sponding to C values of 0.035, 0.038, and 0.035, respectively

D = highway distance between i and j in miles

Screen lines (4) extending across the state were used as checks of the zonal interchanges. A zone-to-zone matrix was used to assign traffic to the network.

Future internal trips were projected using the above trip equations and projected populations for each zone; an adjustment factor was used for increased vehicle registration. External trips were estimated for 49 external stations using volume trends at each station and anticipated population growth of that particular region of the state. Traffic was assigned to the freeway and major highway network with an upward adjustment factor reflecting the influence of Interstate freeways that are established in the corridors. Similar studies for other modes have and will be made.

Example 2   Locating a portion of an Interstate route follows. The study was initiated by a request from the states of Illinois, Kentucky, Missouri, and Tennessee to extend Interstate Route 24 from Nashville to St. Louis. A consulting firm, employed to work jointly with the four state highway departments and the U. S. Bureau of Public Roads, conducted surveys and evaluated alternatives as here briefly described.[4]

The cities of St. Louis and Nashville defined the principal corridor through which the route would extend. Five subcorridor segments were identified as required by the ruggedness of the terrain, avoidance of the Kentucky Lake and the Barkley Dam, preferred river crossing sites, and connections to north-south freeways. Within the main and subcorridors, 21 alternative routes were projected and analyzed (Figure 16.2). The study areas were divided into zones composed of groups of counties. Population data for the study area and for adjacent external zones were secured from state records for 1960 and projected to 1975. Internal population changes were developed by counties and nonmetropolitan areas of 5000 or less and divided between agricultural, and nonagricultural employment. Future growth, of prime concern, was projected (1) as normal growth without the proposed highway, and (2) as growth impacted by the proposed highway. Industrial growth was anticipated because of the improved access and labor markets consequently expanded.

Recreational travel for swimming, boating, fishing, etc. in state parks, a National Wild Life Refuge, and the two dam areas was also analyzed and projected on the basis of population growth and improved accessibility. A 1960 network through the corridor was identified and described by travel time and distance over individual segments. The projected 1975 network included all anticipated changes, principally the addition of Interstate 24.

Because trip generation in rural areas was a relatively new effort, it was necessary to develop formulas by regression analysis for both trip generation and distribution in the study area. Data from 28 cities ranged from Chicago with a population of 4 million to Paris, Tennessee with 9000 people. Eighteen formulas were developed and tested with assembled data. Traffic generation for the 1960 network was thus synthesized using a form of the gravity model, one of the 18 developed and tested. It gave a squared

[4]*Location and Economic Impact Investigation for Interstate Route 24*, report submitted to the States of Illinois, Kentucky, Missouri, and Tennessee by Wilbur Smith and Associates, Columbia, South Carolina, 7 January 1963.

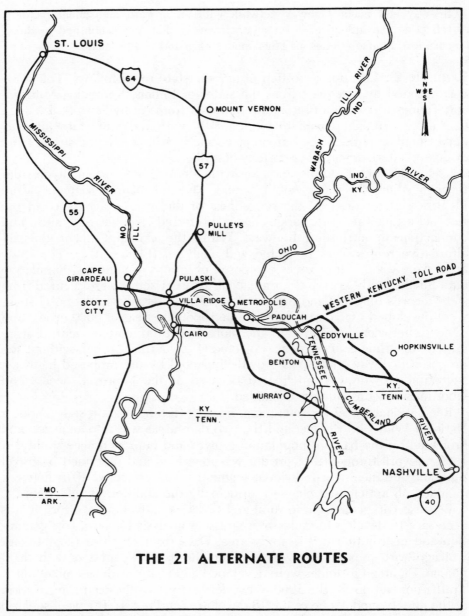

THE 21 ALTERNATE ROUTES

FIGURE 16.2 ALTERNATIVE ROUTE LOCATIONS (*LOCATION AND ECONOMIC IMPACT: INTERSTATE ROUTE 24* BY WILBUR SMITH AND ASSOCIATES, COLUMBIA, SOUTH CAROLINA, 1963, P. 15.)

correlation factor of 0.941 for the production of resident trips in individual zones.[5]

$$R_i = 1.03 P_i \left(P_i A_i / 1000 \right)^{-0.247}$$

where

R_i = total resident trips generated by zone i

P_i = population of zone i

A_i = area of zone i

1.03 and -0.247 are coefficient and exponent for the area under study

This gravity-type model, combined with the distribution curve, Figure 16.3, to make a computerized distribution of total resident and nonresident work, recreation, and other trips. An iterative balancing process was used to obtain agreement between resident and nonresident trip ends.

Traffic assignments were made using the minimum travel time path procedure. A 1960 network was prepared, and zones and zone centroids were identified and nodes numbered. Route segments or links were described by time and distance (using two runs to obtain average elapsed time then converted to speed) and by direction of travel—one-way or two-way. Assignment "trees" were built by which the computer selected the least travel time zone to zone route and accumulated total volumes over each link.

The 1960 network was updated to 1975 using the planning data and including the proposed new I-24 route. Trip ends for all 21 alternate routes required developing the model 21 times with an additional model being used as a base and projected trips distributed. Criteria used in comparing the 21 alternatives included:

(a) Travel distance between St. Louis and Nashville

(b) Construction length of proposed I-24

(c) Construction cost of proposed I-24

(d) Construction cost per mile of I-24

(e) Vehicles miles of freeway traffic on I-24

(f) Vehicles miles per mile over I-24

(g) Average daily traffic over I-24

(h) Construction costs per annual vehicle mile[6]

[5]Ibid., p. 72.
[6]Ibid., pp. 81–93.

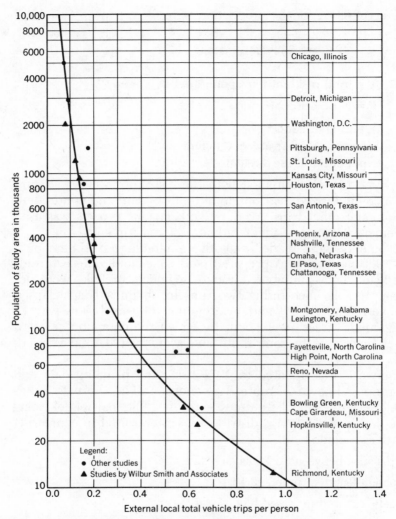

FIGURE 16.3 GENERATION OF DAILY EXTERNAL MOTOR VEHICLE TRIPS BY RESIDENTS. (*LOCATION AND ECONOMIC IMPACT: INTERSTATE ROUTE 24* BY WILBUR SMITH AND ASSOCIATES, COLUMBIA, SOUTH CAROLINA, 1963, FIGURE B-2, P. 175.)

The last item, it should be noted, is a type of cost effectiveness.

In computing the benefit-cost ratio, capital costs for the various items were divided into four life period groups (used also as amortization periods): items with 20-year, 40-year, 60-year, and 100-year lives. Annual maintenance costs combined with annual capital costs permitted computing the annual roadway costs for each route.

Annual road use costs by segments and routes were developed using ASSHTO data; these included data on fuel, oil, tires, and car maintenance, depreciation, gradient effects, speeds (of 40 and 52 mph), and whether the routes had free, restricted, or normal operation. Accident costs were computed as $0.0024 per vehicle mile.[7]

The 21 routes were also evaluated by the rate of return method. The route selected maximized both the benefit-cost ratio (of 3.24) and rate of return (17.4%). It gave 2.7 times more freeway service than the route giving the least, maximizing vehicle-miles. Construction costs were in an intermediate range. In length, it was the second longest.

Indirect benefits included qualitative comments on increased land values and increased potential for industrial location. Tourism and recreation were evaluated in dollars based on expenditures by tourists per 10,000 anticipated daily vehicles. Social benefits were said to include increased mobility, improved access to work, to recreational activities, and to relationships with business and governmental organizations. The number of towns with populations over 2500 served, total population served, and order rank by total population served were determined. Also considered were impacts on population growth and retail sales.

The final selection was based on first, second, third, and fourth order ranking for 12 factors applied to the most promising eight of the 21 routes: benefit-cost ratio, construction cost per vehicle-mile, average traffic volume, population served, cost of construction, alternative method benefit-cost ratio, travel distance—St. Louis to Nashville—length of new construction, total travel on the new route, and tourist expenditures. The alternative selected ranked first in five of the foregoing, second in two, third in two, and fourth in one. No other alternative fared as well.

Emphasis on environmental and social impacts was not being given, at the time of this study, the attention that those factors receive today. Consequently no evaluations were made of the relative demand for land among the alternatives, the extent of pollution, the invasion of parks and wilderness areas, or the loss of farmland, homes, work sites, historic buildings, and land marks that might characterize such a study today. The

[7]Ibid., p. 105.

extent of public input into the planning procedure was not indicated. Such input could arise through public hearings in conveniently located communities within the corridor and by requesting specific input on goals, objectives, and plans from public planning officials in the corridor's towns and cities.

An example of railroad location is given in Appendix 2.

National Level   At the national level, planning is primarily directed at establishing needs for the nation, state, or region. The impact on the economic, industrial, commercial, social, and racial status of the nation as a whole and of its several parts is a principal consideration. Specific planning and implementation are conducted, sometimes through the agency of the States as in constructing the Federal System of Interstate Highways. Transportation alternatives for the Northeast Corridor were evaluated by the Office of High Speed Ground Transportation for the Department of Transportation. In their report of 1970 nine combinations of auto, rail, bus, conventional air, vertical takeoff and landing, short takeoff and landing, and tracked air cushion vehicles were analyzed. Alternative modes and systems were described, patronage and travel patterns of the Northeast Corridor were established for each alternative, and estimates were made of capital expenditures, costs, and revenues. Various types of organizations were considered. A benefit-cost evaluation of each alternative was then performed including the effect that the proposed service between major points would have upon service to less densely settled areas, the extent of air pollution, the accident hazard, and the demand of each for land. Effects on the economics of the area traversed and on labor were noted.[8]

Rail Networks   Federal planning has already established a national highway system, a national system of airports and of waterways. Amtrak is setting a national network of rail passenger service. The identification of a national rail freight network is now underway. Section 503 of the Railroad Revitalization and Regulatory Act of 1976 (Public Law 94-210) provides a controversial process for classifying Class I railroads into categories. Individual lines have already received preliminary designations.[9]

Classifications are based on (1) traffic density in gross tons, (2) service to major markets, (3) levels of capacity, and (4) national defense. Lines thus

---

[8]"Northeast Corridor Transportation Report," *Northeast Corridor Transportation Project Report* 209, Office of High Speed Ground Transportation, U. S. Department of Transportation, Washington, D. C., April 1970.

[9]Preliminary Standards, Classification, and Designation of Lines of Class I Railroads in the United States, Vol. I & II, U. S. Department of Transportation, 3 August 1976.

categorized will receive priority in the allocation of Federal funds and probably in private investment as well. In general, excess capacity is considered to exist in a corridor having more than two competing lines; one or more will be termed "excess" and downgraded or abandoned. High category lines will be raised or kept to high standards of maintenance and efficiency. Many low density lines will thus be removed from the national network.

A problem exists for individual states that may have specific need for certain excess lines not recognized by Federal criteria. State transportation departments will probably have to determine which of such lines are required by their economies and plan their retention, funding, and management. The states have already funded a portion of the costs to keep certain Amtrak routes and trains that were deemed locally essential.

Branch line abandonment can lead to loss of income for farmers and others and may force some out of business into bankruptcy with loss of job opportunities. Longer grain hauls to central elevators (or goods from factory to market) by highway adds to the costs of those products and to the increased highway costs. The situation requires that each case be examined on its merits.

ENGINEERING FACTORS IN LOCATION

The preceding example with its 21 alternative routes is evidence that a great many individual routes and specific locations are possible, even within the relatively limiting confines of an established corridor. Service factors have a proper influence on location, but engineering factors of distance, gradient, curvature, and terrain can have significant construction and operating cost impacts on the specific location of a line and even on defining a corridor. These impacts can also be evaluated using rate of return, benefit-cost, and other criteria.

Route Elements The ideal location from an operating standpoint, one that minimizes travel time and operating costs, would be a straight (tangent) line between the origin and destination points with a level grade throughout. Such a location could be enormously expensive to construct (and might miss important intermediate traffic points), yet any departure from that ideal will increase costs of operation. The locating engineer has to select that combination of distance, curvature, and gradients, within the context of needed service, that will bring revenues, operating costs, and construction costs into economic balance. He must be alert that this effort to minimize capital costs doesn't burden the system with excessive operating costs throughout its active existence. He must also avoid the other extreme of designing a route with a degree of excellence that involves construction out of proportion to the service benefits that it will perform.

Distance   Other conditions being equal, a route should be as short as possible. In general, fixed costs are not affected by small changes in distance, but direct costs of construction will usually vary with distance. Consideration of costs due to distance should therefore be based primarily on variable or direct costs of construction and operation. For the latter, fuel, direct maintenance, and sometimes wages are the principal factors. Small or moderate increases in distance are of little consequence when a serious grade or curvature can thereby be improved or avoided or an additional source of traffic gained. In rail operation, the total operating cost does not vary in proportion but is only partly variable with a change in distance. Thus the added costs are only approximately 30 percent of the average costs per 1000 gross-ton-miles for changes in distance of less than 1 mile, 35 percent for 1 to 10 miles, and 48 percent for changes over 10 miles.[10]

Small changes in distance on highways can have an appreciable effect, especially when a heavy traffic flow is anticipated or experienced, because highway vehicles are variable-cost carriers. A three-mile reduction in distance over which operate 2600 vehicles per day at an average direct cost of $0.08 per vehicle mile would bring actual savings per vehicle of:

$$3 \times \$0.08 \times 2600 \times 365 = \$227,760 \text{ annually}$$

The many other factors entering into highway location make distance of secondary importance. Directness of route is, however, one of the characteristics that have recommended superhighways to the motorist, more from the standpoint of time saved, however, than from operating economy.

Distance is also important to pipelines. Flow resistance per mile accounts for a major portion of required pumping pressures except where excessive differences exist between the elevations of any two or more pumping stations. Since there is little restriction due to terrain on a pipeline location, the shortest route is usually the cheapest. Stations of low pumping pressure are relatively inexpensive but must be close together. Stations of higher pumping pressures are more expensive but can be placed farther apart, a certain economy thereby being achieved. An economic balance must be determined between pumping pressures and number of stations.

---

[10]*Proceedings of the A. R. E. A.*, Vol. 39, 1938, pp. 518–531, American Railway Engineering Association, Chicago 5, Illinois.

Canal costs are divided between route costs and locking costs. The latter depend on the differences in elevations rather than on route gradients (which should be level or nearly so), leaving distance as a vital route-cost factor. Where a lower elevation difference, leading to fewer lockings or smaller over-all lifts, can be had by increasing distance, the costs of the two alternatives must be compared to determine the one giving greater overall construction and operating economy.

Curvature Topography seldom permits an ideal straight line between two points. Curvature does not add materially either to highway construction costs or operating costs of highway motor vehicles. Some additional labor and material may be expended in widening and superelevating curves; also in painting center lines for safety. There is excess wear on tires and pavement due to tangential forces and side thrust but these are negligible if modern safety requirements are maintained. Today's practice is to flatten curves to the point where the effects of curvature are nominal. The theoretical layout of curves is a problem in geometric design.

The adverse effects of curvature on railway operating expenses (and construction costs to a lesser degree) are much more pronounced because of flange wear on rails and wheels, lateral thrust on the track structure, and consequent excess labor and material in maintaining gage, line, and surface. The total amount of curvature, the sum of the central angles of all curves in a route, is more important than the number of curves or degrees of curve. The same resistance is encountered in running through 528 degrees of central angle as is encountered by a train in running over 1 mile of tangent track (assuming an average value of 8 lb per ton for train resistance on tangent level track in still air for high modern train speeds). The costs of curvature can thereby be computed since it has been determined that the 528 degrees of an equivalent curve-mile will incur additional operating expenses equal to 30 percent of the average cost per 1000 gross-ton equivalent miles.

Curvature has a limiting effect on speed both for highways and railroads. This too is a matter of geometric design. In general, railroad curves for high-speed traffic—70 to 100 mph (113 to 161 kph)—are held to 1 to 2 degrees and 2 to 3 degrees for moderate speeds of 45 to 69 mph (72.4 to 111.0 kph). If sharper curves are required by terrain or construction costs, speed must be reduced accordingly. Minor-branch and slow-speed lines usually do not exceed 8 to 12 degrees. Industrial switching tracks may go as high as 30 to 40 degrees (but curvatures above 20 degrees are not recommended). Diesel-electric road and general-purpose locomotives are usually limited to curves not exceeding 20 to 22 degrees.

Curves of principal highways should not exceed 3 degrees, while super-highways and expressways will hold to the same standards as high-speed railroads. Secondary roads may safely go to 10 degrees. Sharper curves are limited to temporary and low-class roads and to city streets.

Curvature should be avoided in pipelines but presents no special problems. Bends of 20 to 30 degrees from the tangent are acceptable, but more sweeping changes in direction are preferred. The added flow resistance of such curves is considered only where precise calculations are required.

Curvature in canals and canalized rivers limits the size of vessels going around a bend and creates problems in maneuvering. The limiting length of vessels can best be determined by laying out curves and vessels to scale on paper with templates. There should be room for the passing of two large ships, barges, or tows plus a factor of safety and an allowance for increases in size of future equipment. Sharp curvatures are avoided wherever possible. Guiding a long tow around a sharp bend is always difficult, and sometimes disastrous when a heavy current is running.

Curvature is normally not encountered with conveyors. The changes in direction are made at the ends of the belt flights. Aerial tramways either change direction abruptly at an angle station with appropriate towers and guide rails, transferring from one cable to another, or direction changes are made by introducing long curves so slight as to be almost imperceptible.

Grades and Elevation The principles governing resistance to propulsion imposed by grades and differences in elevation have already been presented in Chapter 5. That chapter should be reviewed in studying these economic aspects of route location. In Chapter 5, it was shown that a resistance of approximately 20 lb per ton of vehicle and cargo weight per percent of grade must be overcome by all land-based vehicles—railroad, highway, monorail, aerial tramway, airway (during takeoff), and conveyor equipment.

For rail and highway operation, grades affect costs in two ways: (a) by adding to the costs (and road time) of operating any given train or motor vehicle (fuel, maintenance, wages, etc.) and (b) by limiting the size of load in any one train or motor vehicle and thereby determining how many trains or motor vehicles with all their attributable costs will be required.

The latter effect is by far the more important. For example, the locating engineer must select a ruling grade for railroads that will permit the maximum load per train (for the type of locomotive in use), and therefore the least number of trains, and the minimum of expense. His selection will probably involve a comparative study of several alternative routes and/or grades. In general, the more money invested in construction costs, the flatter will be the ruling grade that can be obtained and the lower the operating costs. The engineer must decide how much money to spend and which grade and route is the most economical to build and operate. He or she must apply one or more of the cost-study methods given in earlier chapters, preferably the rate of return through use of the basic location formula. In addition to selecting an economical ruling grade, he or she must also keep the minor grades, those less than the ruling grade, within economic limits. Since the ruling-grade location may be combined with a sequence of minor grades, these aspects of elevation form two individual elements in considering the overall capital or annual costs or rate of return for a route alternative.

Maximum grades on railroads should be kept to a maximum of two-percent, preferably to no more than one-percent, but to provide speed and increase tonnage capacity, few grades should exceed 0.50 percent. Highway grades are generally kept to 3 percent or less for high-speed super-highway-type construction but can go as high as 7 or 8 percent for secondary, slow-speed routes; city streets may go as high as 12 percent if need be. Ruling and minor grade effects should be given more consideration than they usually receive where truck traffic is a major part of the total. The limiting-grade concept is equally or even more important, at least for two-lane highways. The limiting grade reduces vehicle speed and, if speed is reduced below 25 to 30 mph, limits the number of vehicles that can negotiate that grade in a given time. The capacity of a highway is thereby limited. Heavy grades materially reduce the speed of trucks and buses, adding to the number of vehicle units required to move a given volume of traffic and to the costs of fuel and wages. With both railroads and highways, the length and total mileage of excessive grades are matters of concern. One long steep ruling grade may not be too serious, but a succession of such grades or of grades less than but approaching the ruling grade in magnitude can have a severely adverse effect on road times and operating costs.

Heavy grades introduce safety problems. Brakes must be in good condition and used judiciously to prevent runaways and jack-knifing. Jack-knifing is an especial hazard for trucks on icy roads. The trailer tends to swing onto the adjacent lane and crowd against the tractor. When dynamic brakes are applied on the locomotive consist at the head end of a long train, there is also danger of jack-knifing as the unbraked cars crowd against the locomotive units.

In highway location especially, man made works often have as important an effect on grades as does terrain. The necessity to provide grade separations, for example, in building a superhighway or other limited-access road may well establish the grade line without reference to other factors.

The material in Chapter 5 on momentum and floating grades should be reviewed in developing detailed aspects of costs of gradients.

The locating engineer must display much ingenuity and adopt many expedients to keep his grades as light as possible. In rugged, difficult terrain, light grades are usually constructed only at the cost of heavy excavation, bridging, tunneling, or development. These are all expensive. Cheaper but far less desirable from an operating standpoint are switchbacks (not unusual on mountain highways) and inclined planes. Several Alpine railways are so designed that a dual-drive locomotive can be used that operates by adhesion to the rails on normal grades but engages a driving cogwheel in a rack laid between the rails when grades of excessive rise (4 percent or more) are encountered.

Development consists of adding distance to the length of line to reduce the rate of grade. Thus a mountain route may head up a main valley but, to add distance and thereby reduce the gradient, it will follow a long loop along both sides of one or more side valleys, perhaps looping over or under itself.

Urban Locations Urban areas present difficult location problems. Access routes for railroads may be at street level, on viaducts, in open cuts, or in subways and tunnels. Street-level locations incur problems of traffic delay, accident hazards, and protection costs at street and railroad crossings. Viaducts, whether used for rail or highway, can act as "Chinese Walls," a barrier to movement from one side of the route to the other. Elevated structures are often unsightly and cast shadows on surrounding land uses. Open cut locations also physically divide the community. Expensive overpasses must be built, but many streets are dead-ended as with viaducts. Open cuts pose problems of drainage and collect debris and garbage. Whether to use a viaduct or open cut may revolve around the question of rail-over-street vs. street-over-rail. A street-over-rail requires at least 22 ft (6.71 m) of vertical clearance above the rails in contrast to 12 to 16 ft (3.66 to 4.88 m) of vertical clearance required above a pavement. The latter requires less earthwork, but the bridge structures to carry rail lines must be heavier than those for highways.

The problems of rail route location apply equally to highways, expressways, and freeways, except that the latter usually require more land area. Both can act as barriers to needed urban development.

Any route in the open adds to air, noise, and visual pollution. Costs of urban land acquisition are always high.

Tunnels, whether for intercity rail, rapid transit, or highways, face difficulties with subsurface utilities—power, sewer, water, gas, and communication and numerous other lines. A true tunnel location may be deep enough to avoid the problem, but in cut and cover work it is frequently necessary to relocate utilities temporarily, perform the subway construction, then restore utilities to their original location, an expensive operation. The interruption to street traffic and danger of settlement to adjacent buildings (also a problem for open cut work) during excavation offer additional problems and expense. Drainage, ventilation, lighting, and difficulties in performing maintenance add to tunnel costs.

Any location for rail lines other than ground level makes service to adjacent commercial and industrial patrons difficult if not impossible. Local service from expressways can be provided by parallel access roads. Approach gradients to cuts, tunnels, or viaducts should be kept below 1 percent for railroads, thereby introducing additional length to the land and construction required.

Urban routes often combine all four location possibilities within their total extent.

Ecological Factors   Only recently has the ecological impact of route location received deserved attention. Ecological concern is closely related to environmental problems. Maintaining a suitable ecological balance is a highly desirable goal for any planning process. The location of a line should be such that the ecology of the region traversed receives minimum disturbance. Animal life is susceptible to the proximity of human activity. Plant life too can suffer. Animals may withdraw from the vicinity or cease to reproduce. The land area required to support various wild species may be too greatly reduced by right of ways (often fenced). The pavement or track (or fence) can shut off access to water and feeding areas and interfere with migration. Many animals are killed each year by trains and highway vehicles. Game preserves, wild life refuges, and most certainly wilderness areas are not suitable locations for transportation routes. Even parks can lose their natural charm when traversed by a transport route. Provision for by-passing or, if the area *must* be traversed, fencing, with culvert underpasses are possible solutions.

Hillside cutting can lead to erosion and landslides. Filling or draining a swamp or marshy area may destroy the habitat of rare plant and aquatic life and the feeding grounds for birds. Air and water pollution may also lessen the quality and even the viability of life for farms and communities.

One useful technique for solving the problem is the use of a base map with overlays, a technique developed to a high degree by Ian McHarg to whose work the reader is referred.[11] The base map contains the general layout of the corridor and perhaps a trial location as well. As field data are

[11]Ian McHarg, *Design with Nature*, Natural History Press, Garden City, New York, 1969.

collected, transparent overlays are colored to represent traffic points, unstable ground, land marks and historic buildings, parks, wooded areas, marshlands, lakes, rivers, and other areas of environmental, ecological, and engineering sensitivity. The succession of overlays blocks out areas where the route should *not* be located. It also reveals those land areas suitable for the route. The best possible alignment is then traced between and around prohibited land types. Overlays may also include items of lower preservation priority if these are needed to make the route viable.

A route so chosen may be very different from one based on engineering excellence alone. It is at this point that public and ecological input is helpful in establishing goals, resolving conflicting goals and objectives, and effecting a reasonable compromise. Transportation should contribute to the quality of life rather than detract from the quality of life for those who happen to live in its vicinity.

QUESTIONS FOR STUDY

1. What are principal functions and responsibilities of a state Department of Transportation?
2. If a state planned to establish a principal highway route across the state, what steps would be taken to identify the corridor to be traversed?
3. What factors would enter into the selection of a specific route within the corridor of Question 2?
4. If the U. S. Railway Association decides that a branch line serving a series of rural grain elevators is to be abandoned, what considerations might cause the state's Department of Transportation to recommend to the legislature that the line be continued in operation at state expense?
5. Compare the operating costs of two proposed rail routes, each carrying 18,000 gross tons per day, one route containing 7890 degrees of central angle, the other including 6300 degrees of central angle. Assume operating costs per 1000 gross ton miles to be $5.00.
6. A route approaching a saddle or pass between two mountains is at an elevation of 980 ft above sea level. The saddle is 3 miles away and at an elevation of 1180 ft. What procedure can be used to achieve a maximum grade of 3 percent between the two points? What problems will that solution introduce?
7. What might be the effect of establishing classification categories of rail lines as required by recent legislation on (*a*) the railroad system, (*b*) the farmer, and (*c*) other modes of transport?
8. Explain the environmental and ecological factors that enter into the selection of a transportation route, and describe a technique for establishing an ecologically suitable location. Does this technique have economic and engineering applications as well?

9. In urban areas what special location problems are found in addition to those found in rural sites?
10. Outline all the factors that might enter into determining whether a rail-highway grade separation should be a rail over- or rail underpass.

SUGGESTED READINGS

1. A. M. Wellington, *The Economic Theory of the Location of Railways*, 1906 edition, Wiley, New York.
2. "Economics of Plant, Location, and Operation," chapter in the *Manual for Railway Engineering (Fixed Properties)*, American Railway Engineering Association, Chicago, Illinois.
3. W. W. Hay, *Railroad Engineering*, Volume I, Wiley, New York, 1953, Part I.
4. *Social, Economic, and Environmental Implications in Transportation Planning*, Transportation Research Record 583, Transportation Research Board-National Research Council, Washington, D. C., 1976.
5. L. I. Hewes and C. H. Oglesby, *Highway Engineering*, Wiley, New York, 1963, Chapter 3, "Highway Planning."
6. *Location and Economic Impact Investigation for Interstate Route 24*, report to the states of Illinois, Kentucky, Missouri, and Tennessee by Wilbur Smith and Associates, Columbia, South Carolina, 7 January 1963.
7. *Illinois Needs and Fiscal Study: Final Report*, prepared by Wilbur Smith and Associates, New Haven, Connecticut, October 1967.
8. 1972 *National Highway Needs Report*, Federal Highway Administration, U. S. Department of Transportation, Washington, D. C., May 1972.
9. *Preliminary Standards, Classification, and Designation of Lines*, Volumes I and II, U. S. Department of Transportation, Washington, D. C., 3 August 1976.
10. Ian McHarg, *Design with Nature*, Natural History Press, Garden City, New York, 1969.
11. Robert A. Snowbar, "Planning for Mass Rapid Transit," *Modern Government and National Development*, September 1969.
12. *Specifications for Development of a Railroad Network Model*, submitted by the Committee on Analytical Techniques, Association of American Railroads, Washington, D. C., 1969.
13. *The AAR Network Simulation System* (A Tool for the Analysis of Railroad Network Operations), developed by the Midwest Research Institute, Kansas City, Missouri, February 1971.
14. *A Model-Building Concept for Facilitating the Application of Existing Network Simulation Models*, PhD thesis by Seung Jai Kim, University of Illinois, Urbana, Illinois, February 1974.
15. *Issues in Statewide Transportation Planning*, Special Report 146, Transportation Research Board, National Research Council, Washington, D. C., 1974.

Chapter 17

Route Classification, Location, and Design

ROUTE CLASSIFICATION

For purposes of identification routes are classified into types or categories according to purpose, standards of use and construction, and for establishing sources and priorities in financing.

Highways The 1974 National Transportation Study identifies four categories of rural highways by service type: *Interstate, arterials (major and minor), collectors (major and minor),* and *local. These are further identified by location as rural, small urban (populations of 5000 to 25,000),* and *small urban (populations of 25,000 to 50,000).*[1] Note that interstate and arterial roads carry about 60 percent of the vehicle-miles of travel but constitute only 9.3 percent of the mileage; the reverse is true for collector and local roads.

Within urban areas are found divided lane routes with fully controlled access (freeways) and those with partially controlled access (expressways), also collector-distributor streets and principal main streets (and secondary), and local access streets without fully controlled access.

---

[1] *1974 National Transportation Report*, U. S. Department of Transportation, 1974, pp. VIII-1 to VIII-7.

TABLE 17.1 PRELIMINARY CATEGORIES OF RAIL LINES[a]

Category Title	Category Description	Percent of Route-Miles of Class I Network
1. A main line	20 million or more gross ton-miles per mile per year	15.5
	Three or more daily passenger operations in each direction	0.8
	Major transportation zone connectivity	0.8
2. Potential A main line	A temporary status for through lines located in corridors of excess capacity. They will be designated to another category upon resolution of the redundancy	11.6
3. B main line	Less than 20 million gross tons but at least 5 million	21.7
4. A branch line	Less than 5 million gross tons but at least 1 million	21.9
5. B branch line	Less than 1 million gross tons	25.6
6. Defense—essential branch line	Required for access of oversized military shipments	2.1

[a] *Preliminary Standards, Classification, and Designation of Lines of Class I Railroads in the United States*, U. S. Department of Transportation, Washington, D. C., August 3, 1976.

A frequently used design classification is that of the AASHTO that has been approved by every state and adopted by the Federal Highway Administration for the design of Federal-Aid Roads.[2]

Freeways. These include interstate and other multilane routes having fully controlled access and carrying a high volume of long distance traffic. Design speeds are generally 50 to 70 mph (80 to 113 kph) depending on terrain; grades are kept within 5 percent; curvatures seldom exceed 3 degrees.

Arterials. Highways other than freeways; for high volumes of long distance traffic. These form a part of the state's network but are without controlled

[2] *A Policy on the Geometric Design of Rural Highways*, American Association of State Highway and Transportation Officials, Washington, D. C., 1965, pp. 142–145.

access. Grades may be as much as 9 percent in mountainous regions and with design speeds of 50 to 70 mph (80 to 113 kph).

Collectors.   Secondary roads without controlled access move traffic to and from arterials to abutting properties. Design speeds are 30 to 50 mph (48 to 80 kph) with grades up to 9 percent.

Local Roads and Streets.   These are normally two-lane roads and give detailed access to individual land occupants; speeds are low; grades can be high, 12 percent at times; through traffic is discouraged.

Rail Routes   Railroad lines have traditionally been classified as main line, secondary main line, and branch line. The recently enacted Railroad Revitalization and Regulatory Reform Act of 1976 (the 4R Act), Section 503 (b) has led to the Secretary of Transportation issuing a report that establishes six preliminary categories of rail lines along with an orderly designation process for each line of the 193,500 miles of Class I railways. See Table 17.1.

Rail lines have been given a prelinimary designation based on (1) density (in gross tons moved or number of passenger trains), (2) service to market areas, lines that are important to connect major market areas, (3) appropriate level of capacity; high use of fixed plant and competition maintained, but more than two carriers on a through route must have sufficient traffic to avoid having one declared "excess," (4) defense essentiality.

LOCATION

Location Procedures   Field work and layout of routes are basically the same for all types of transport. Known terminal and intermediate points are first established, then a reconnaissance survey is made of a strip of land equal in width to about one third to one half the distance between any two fixed or control points. Reconnaissance in earlier days was made on foot or horseback or by small boat or canoe. These modes of travel can be supplemented today by motor car, airplane, and the use of maps and aerial photographs. These last can be used to delimit still further the width of the strip of land under study. Aerial photographs are especially useful in permitting an unhurried study and determination of critical locations where additional observations, perhaps on foot, in the field are required. The Map Information Office of the United States Geological Survey, Washington, D. C., makes available its U. S. G. S. maps, control data, and aerial photographs as well as furnishing lists of similar materials available from other government agencies. Steroscopic procedures enable a finding of elevations, contours, and earthwork estimates. Much of the reconnaissance can thus be performed in the office. Whether in office or field, the reconnaissance must secure information on all pertinent details of terrain, vegetation, climate, topography, works of man, and any other factors that might affect the choice of routes. The reconnaissance is the study of an area rather than an attempt to locate an actual route. Such a location within the area being reconnoitered comes later.

After the reconnaissance and paper survey, one or more partial or complete preliminary lines are set in the field. Tangents are run to their points of intersection but curves are not staked. Topography is noted in sufficient detail to permit comparison of the construction and operating costs of alternative locations. The effects of gradient, curvature, and distance on operation and on construction costs of earthwork, bridging, tunneling, etc. for each possible route are compared. Such comparisons are facilitated by the use of electronic computers. Several routes have thus been given detailed comparison in a fraction of the time that would have otherwise been required. Note that preliminary surveys can sometimes be avoided if adequate aerial mapping is available.

Once a selection of route has been made, the final location is staked complete with curves and offsets where those are needed. Complete topography (cross sections) is taken.

Construction surveys include a reestablishment of the final location and the setting of offset, grade, and slope stakes to mark the vertical and horizontal limits of cuts and fills.

Standards of Design and Construction   Certain standards of design and construction must be established, at least tentatively, before an adequate initial appraisal and location can be made. The quantity of traffic to be moved in a given period establishes one criterion for route design. Maximum permissible gradients, curvatures, sight distances, lifts per lock, number of lanes, tracks, belts, or cables, number of lines of pipe and the pipe diameters, and the number of cars, boats, trucks, or pumping stations are based on route and traffic factors. The horsepower required for tractive, propulsive, or pumping force derives from the foregoing. These are established in the light of traffic, terrain, soils, weather, works of man, etc. Many of these factors have been developed in earlier chapters.

Standards and recommended practices have been established by technical and government agencies. The *Manual for Railway Engineering of the American Railway Engineering Association*, design manuals of the Federal Aviation Administration, and *A Policy on Arterial Highways in Urban Areas* by the American Association of State Highway and Transportation

Officials, and the various safety standards for track, equipment, and signals of the Federal Railroad Administration are typical sources.

GEOMETRIC DESIGN

Geometrics involves the use of tangents and curvatures in various combinations to establish the horizontal trace or alignment of the route and vertical curves and grades to develop a profile of the route in a vertical plane. A few examples of typical geometric problems and elements are presented.

Alignment Horizontal curves are arcs of simple circles. The sharpness of curvature is the degree of curve, D, but the amount of curvature is measured by the central angle, I, equal to the deflection angle of the two intersecting tangents measured at their point of intersection, the PI. See Figure 17.1 in a preliminary survey, only the tangents are intersected, but in the final survey, the curves are staked. The equal distances from the PI to the PC and PT (point of curve and point of tangent respectively) are the tangent distances. By simple trigonometry, the tangent distance, $TD = R \tan I/2$, where R is the radius of the curve in feet.

The degree of curve is the amount of central angle subtended by a chord of 100 ft or, according to another definition, an arc of 100.007 ft. A relation

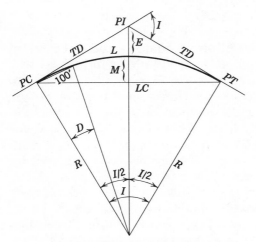

Tangent distance = $TD = R \tan I/2$
Long chord = $LC = 2R \sin I/2$
Midordinate = $M = R \text{ vers } I/2$
External distance = $E = R \text{ exsec } I/2$
Length of curve = $L = I/D \times 100$

FIGURE 17.1 FUNCTIONS OF SIMPLE CURVES.

between degree and radius is established by equating the circumference of the curve in feet to the circumference in unit of arc (or chord) subtended by D degrees, or $2\pi R = (360/D)(100.007)$. Hence, approximately, $D = 5730/R$ or, conversely, $R = 5730/D$. If the 100-ft chord definition is used, $R = 50/(\sin D/2)$.

The length of curve L is therefore equal to I/D in engineering stations of 100 ft each, or $L = I/D \times 100$, in feet.

Curves are laid out in the field by intersecting the tangent lines and measuring the exterior deflection angle, I. The tangent distance is computed and the point of curve, PC, and point of tangent, PT, are established by measuring the tangent distance to each from the point of intersection, PI. Points on the curve are set by turning deflection angles that differ by $D/2$ degrees. For a four-degree curve with an intersection angle of 12 degrees, deflection angle $a_0 = 0°$, $a_1 = 2°$, $a_3 = 4°$, and $a_4 = 6° = I/2$. As the deflection angles are turned, points on the curve are set by intersecting the line of sight with a 100-ft chord originating at the previously set station. The final deflection angle should coincide with the PT, a check on accuracy.

Simple curves of different degrees and radii may adjoin to form compound curves where terrain or other obstructions are severely restrictive. Compound designs are undesirable (and even dangerous, especially in highway design) and should be used only where all other measures fail.

Superelevation The outer rail of a track and the outer edge of a highway slab are elevated over the inner rail or edge to compensate for the centrifugal force that acts on a vehicle as it traverses a curve. In Figure 17.2 the weight W, acting downward through the center of gravity, CG, and the centrifugal force, F_c, combine in the resultant, R. For equilibrium, that is, the weight equally distributed by each wheel, the resultant should pass through the midpoint of the interwheel distance. By use of similar triangles for weight and distance

$$\frac{e}{F_c} = \frac{4.9}{W} \quad \text{and} \quad e = \frac{4.9 F_c}{W}$$

where 4.9 is the horizontal projection of the 60-inch (152.4 cm) distance between the *bearing points* of the wheels on the rails for standard gauge track of 4.71 ft (1.44 m). From physics, the equation for centrifugal force is

$$F_c = \left(\frac{w}{g}\right)\left(\frac{v^2}{R}\right)$$

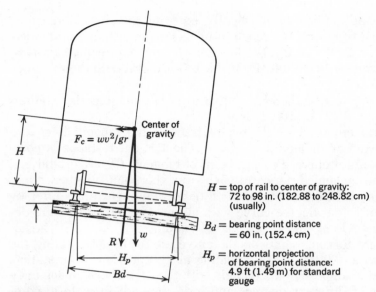

H = top of rail to center of gravity: 72 to 98 in. (182.88 to 248.82 cm) (usually)

B_d = bearing point distance = 60 in. (152.4 cm)

H_p = horizontal projection of bearing point distance: 4.9 ft (1.49 m) for standard gauge

FIGURE 17.2 SUPERELEVATION FOR RAILROAD TRACK.

where

w = vehicle weight in lb

v = speed in ft/sec = $1.47V$ where V is in mph

R = radius of curvature in ft = $5730/D$

g = acceleration due to gravity = 32.2 ft/sec²(980cm/sec²)

By introducing these factors in the equation for superelevation and converting to mph,

$$e = 0.0007\ DV^2 \dots \text{in inches} \qquad (0.001778\ DV^2 \dots \text{in cm})$$

Trains will run at different speeds on the same track with a change in centrifugal force that causes the resultant to move toward the high or low rail depending on whether the speed is greater or less than for equilibrium. Safe and comfortable riding occurs when the resultant falls within the middle third of the interrail space. This is roughly equivalent to 3 in. (7.62cm) of unbalanced superelevation whereby

$$e_c = 0.0007DV^2 - 3 \dots \text{inches}$$

The maximum permissible speed becomes:

$$V_m = \sqrt{\frac{e_a + 3}{0.0007D}}$$

where

$$V_m = \text{maximum permissible speed in mph}$$

$$e_a = \text{actual superelevation in inches}$$

A further restraint is placed on V_m by restricting the amount of supereleva-tion that may be imposed because of the possibility of derailing slow moving trains or starting ones. The Federal Railroad Administration's *Track Safety Standards*, paragraph 213.57, limits e_a to no more than 6 in. (15.24cm).

In applying superelevation to track, the low or inside rail of a curve is normally held to the profile elevation and the outside rail raised to the full amount of the superelevation.

If we follow a similar approach, the equilibrium superelevation for highway slabs becomes $e = 0.0000117DV^2$ ft per foot of slab width, where D and V have the same values as before. The weight W and centrifugal force, F_c, again combine to form a resultant R.

Highway vehicles, unlike rail-flange guided rail vehicles, are guided only by the driver's steering and the lateral friction between the pavement and tires, that is, centrifugal force must be balanced by weight and the resistance to lateral friction. In Figure 17.3a, an automotive vehicle is shown on a flat curve having no superelevation. There is an overturning moment equal to the centrifugal force × the moment arm of the height of the center of gravity (h) above the pavement. This is restrained in part by the weight of the vehicle × the moment arm consisting of half the distance between the wheels. With the low center of gravity of passenger cars, the vehicle tends to slide sideways rather than to overturn. This is not always

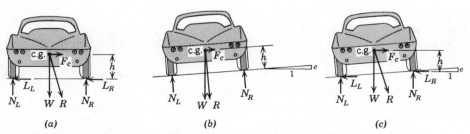

FIGURE 17.3 CURVILINEAR FORCES ON A HIGHWAY VEHICLE.

true for trucks that often have a high center of gravity. The restraint of friction between tires and pavement is needed to act against outward slip or skid.

To balance centrifugal force completely,

$$Wf = (N_L + N_R)f = \frac{wv^2}{gr}$$

where N_L and N_R are the normal forces of tires on the pavement for the inside and outside wheels and f is the coefficient of friction between tire and pavement. When centrifugal force is balanced $f = V^2/15R$. Values of f range from 0.30 to 0.50 for smooth dry pavement but drop to 0.20 for wet pavement, and 0.15 or lower for snow and ice.

At equilibrium speed the vehicle weight is evenly distributed on all wheels and there is no lateral or sliding force. See Figure 17.3b. But all cars do not travel at equilibrium speed. At speeds greater than equilibrium there is a tendency for the vehicle to slide outward and/or overturn. For stability, there must be restraining lateral forces L_R and L_L at the tire rim, as shown in Figure 17.3c. The effective coefficient of friction becomes

$$f_e = \frac{V^2}{15R} - e_e$$

Substituting the value of $5730/D$ for radius R and solving for V,

$$V = 293.2\sqrt{\frac{e_e + f}{D}}$$

This equation may also be rearranged to give the maximum or design degree of curve for a given combination of speed, superelevation, and coefficient of friction:

$$D_{max} = 85,950\frac{(e + f)}{V^2}$$

Because much of the centrifugal force effect will be overcome by super-elevation, there is need for less dependence on friction to prevent side slip. Values have been recommended that vary from 0.16 at 30 mph (48.3 kph) to 0.12 at 70 mph (113 kph).[3] Dependence on higher values than these can lead to driver apprehension and discomfort.

There are limits as well on the amount of superelevation. Too much superelevation may cause sliding toward the inside if ice or snow reduce the coefficient of friction. To protect against the worst conditions likely to

[3]C. H. Oglesby and L. I. Hewes, *Highway Engineering*, Wiley, New York, 1963, p. 220.

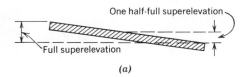

One half-full superelevation

Full superelevation

(a)

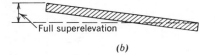

Full superelevation

(b)

FIGURE 17.4 DISTRIBUTION OF SUPERELEVATION. (a) SLAB REVOLVED AROUND CENTER LINE. (b) SLAB REVOLVED AROUND INSIDE EDGE.

be encountered, the maximum superelevation permitted by AASHTO is 0.12 ft (0.037m) per ft for normal conditions but is reduced to 0.08 ft (0,024m) per ft for snow and ice.[4] Based on $f = 0.027$, no superelevation is recommended on curves of $1°30'$ or less at speeds up to 30 mph (48 kph) or for speeds of 70 mph (113 kph) on curves of $0°15'$ or less.

Superelevation further acts to assist in "steering" a car around a curve, the slope of the pavement acting against the tangential tendency of the wheels.

The center line of the roadway is usually held to grade and the outside edge raised by half the total superelevation, the inside edge is depressed by the same amount, (Figure 17.4). When needed to facilitate drainage, all the superelevation can be placed on the outer edge of the pavement.

Transition Curves   To provide a smooth transition from tangent to full curvature (change in lateral acceleration) and from zero to full super elevation, use is made of an easement curve. Drivers of highway vehicles may effect their own easements by varying the vehicle's path, but this is an unsafe practice that can lead to encroachment on adjoining lanes. For railroads, the track circumscribes the train's path. Transition curves are a necessity for the smooth, safe operation of high-speed railroad and highway curves.

To connect the tangent to the simple curve, frequent use is made of the cubic parabola, although other spirals have been used. Use of the cubic parabola must be anticpated during the design procedures and the simple curve offset inward a distance from the tangent to allow room for the spiral. See Figure 17.5 for the geometric relations of this spiral.

[4]A *Policy on the Geometric Design of Rural Highways*, Association of American State Highway and Transportation Officials, Washington, D. C. 1965, Chapter 3, "Elements of Design," Table III-5, p. 158.

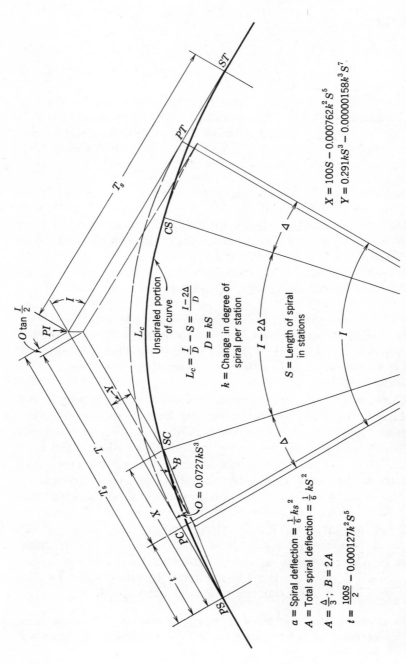

FIGURE 17.5 *FUNCTIONS OF A SPIRALED CURVE. (FROM RAILROAD ENGINEERING, VOL. I, BY W.W. HAY, 1953, FIGURE 5-2, P. 49. (COURTESY OF WILEY, NEW YORK.)*

The following labels and equations appear within the figure:

$O \tan \frac{I}{2}$

PI

I

T_s

T

t

X

Y

PS

PC

SC

B

$O = 0.0727 kS^3$

L_c

Unspiraled portion of curve

$$L_c = \frac{I}{D} - S = \frac{I - 2\Delta}{D}$$

$$D = kS$$

k = Change in degree of spiral per station

S = Length of spiral in stations

$I - 2\Delta$

Δ

I

CS

PT

ST

$X = 100S - 0.000762 k^2 S^5$

$Y = 0.291 kS^3 - 0.00000158 k^3 S^7$

a = Spiral deflection = $\frac{1}{6} kS^2$

A = Total spiral deflection = $\frac{1}{6} kS^2$

$A = \frac{\Delta}{3}$; $B = 2A$

$t = \frac{100S}{2} - 0.000127 k^2 S^5$

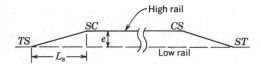

FIGURE 17.6 LENGTH OF SPIRAL.

For the cubic parabola, the offset o and the length S (in 100-ft (30.48 m) stations) are mutually bisecting. The spiral can be laid out by offsets from the tangent that vary approximately as the cube of the distance along the tangent or by deflection angles that vary as the square of the distance. The degree of spiral increases linearly at a given rate of $k°$ per 100 ft station so that

$$S = \frac{D}{k}$$

where D is the degree of curves and S, again, is the length of spiral in stations. Having the value of k and the degree of curve, the length of spiral can be ascertained. More likely, the length of spiral in ft, L_s, will be determined and k computed for use in curve layout.

L_s can be based on the rate of superelevation run-in. In Figure 17.6 the spiral length $L_s = vt$ and $t = L_s/v$ where v is the speed in fps and t is the time in seconds to travel the distance L_s. During the same time, t, the outer wheels of the vehicle are rising a distance e, which equals the superelevation so that $e = v't$ and $t = e/v'$. Since the time t is the same for both, $L_s/v = e/v'$ and $L_s = (1/v')ev$. Various rates of run-in have been used— $1\frac{1}{4}$ in. per sec, $1\frac{1}{6}$ in. per sec (recommended by AREA), and $1\frac{1}{8}$ in. per sec. By changing V to mph and introducing these rates with appropriate changes in units,

$$L_s = 1.17 \ eV \text{ for a run-in rate of } 1\tfrac{1}{4} \text{ in. per sec}$$

$$L_s = 1.26 \ eV \text{ for a run-in rate of } 1\tfrac{1}{6} \text{ in. per sec}$$

$$L_s = 1.301 \ eV \text{ for a run-in rate of } 1\tfrac{1}{8} \text{ in. per sec}$$

The lowest value would be used on low-speed lines, 50 mph or less, the intermediate is suitable for speeds up to 79 mph, and the highest value is desirable for speeds greater than 79 mph.

For highways, the value of k is related to the rate of increase in lateral (centripetal) acceleration C with a value of 2 being frequently used. From this relation,

$$L_{s(min)} = 0.0055 \ V^3 D/C \quad \text{and} \quad k_{max} = 173,000C/V^3$$

The AASHTO has set minimum standards for easement curves based on design speeds (Table 17.2).

TABLE 17.2 MINIMUM EASEMENT CURVE STANDARDS[a]

Design Speed					
(mph)	30	40	50	60	70
(kph)	48	64	80	96	112
Minimum degree of curve requiring spirals	3°30′	2°00′	1°30′	1°00′	1°00′
Minimum length of spiral and superelevation runoff in feet	100	120	150	175	200
runoff in meters	30.48	36.58	45.72	53.34	60.96

[a]*A Policy on Geometric Design of Rural Highways*, Association of American State Highway and Transportation Officials, Washington, D. C., 1964, p. 171, Table III-10.

Horizontal Sight Distance   Horizontal sight distance should be such that a driver has ample stopping distance at the design speed before reaching an obstruction on the road ahead. The obstruction and the driver's eye are assumed to be on the pavement centerline. The driver's eye is assumed to be 3.75 ft (1.14 m) above the pavement and the obstruction to be 0.5 ft in height.[5]

As explained in Chapter 5, stopping distance, L_s, is the sum of the driver's perception-reaction time distance, L_r, and actual braking distance, L_b, that is,

$$L_s = L_r + L_b$$

where

$$L_r = 1.47 V_i t_r \ldots \text{ (or } 0.278 V_i t_r \text{ with } V_i \text{ in kph)}$$

and

$$L_b = V_i^2 / 30(f \pm g) \ldots \text{(or } V_i^2 / 25.5(f \pm g) \text{ metric)}$$

[5]*Transportation and Traffic Engineering Handbook*, Institute of Traffic Engineers, John Baerwald, Editor, Prentice-Hall, Englewood Cliffs, New Jersey, 1976, p. 613.

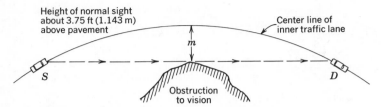

FIGURE 17.7 HORIZONTAL SIGHT DISTANCE.

where

$$V_i = \text{initial speed in mph (or kph)}$$

$$t_r = \text{driver's perception-reaction time, taken as 2.5 sec}$$

$$f = \text{coefficient of friction, 0.40 to 0.27 for wet pavements, decreasing with speed }^{6}$$

$$g = \text{percent of grade divided by 100}$$

On curves the obstruction to sight must be clear of the line of sight by the distance between the midordinate and the line of sight, m. See Figure 17.7. From the functions of simple curves,

$$m = R\,\text{Vers}\,(I/2); \qquad \text{also } m = 0.125\,\overline{SD}^2/R$$

where m is the midordinate for the long chord, the line of sight $\overline{SD}$. With a curve length S in 100-ft stations to the pavement obstruction and expressing R in terms of D,

$$m = (5730/D)\,\text{Vers}\,(\overline{SD}/200)$$

The AASHTO recommends stopping distances of 200 ft at 30 mph, 275 ft at 40 mph, 350 ft at 50 mph, 475 ft at 60 mph, and 600 ft at 70 mph.[7]

[6]*A Policy on Design of Urban Highways and Arterial Streets*, American Association of State Transportation and Highway Officials, 1973, p. 136.

[7]*A Policy on Geometric Design of Rural Highways*, Association of American State Highway and Transportation Officials, Washington, D. C., 1965, Figure III-13, p. 188.

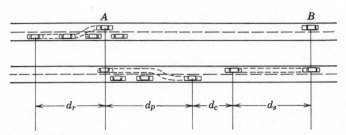

FIGURE 17.8 PASSING SIGHT DISTANCE. (AFTER AASHTO, *A POLICY ON GEOMETRIC DESIGN OF RURAL HIGHWAYS*, 1965, FIGURE III-2, P. 143.)

Passing Sight Distance   On two-lane highways there must be sufficient sight distance so that one vehicle can pass another vehicle and return to its proper lane before meeting an opposing vehicle appearing after the passing maneuver is underway. To promote safety and the flow of traffic, as much of a two-lane highway as possible should be available for passing.

The sight distance is composed of four distance elements.[8]

1.   The perception-reaction distance, d_r, in which the driver of the car making the maneuver perceives the passing opportunity and initiates the pass; d_r, is composed of two elements, $v_i t_r$, plus the acceleration distance (including entry into the opposing lane), which from physics is $\frac{1}{2} a t_r^2$ where v_i is the initial speed, t_r, is the perception-reaction time, taken as 3 to 4 seconds, and the acceleration a has an approximate value of 1.4 ft/sec^2.

2.   Distance in the passing lane, $d_p = v_2 t_2$ where v_2 = average passing speed and t_2 is time in seconds in the left lane, varying from 9 to 11 seconds.

3.   Distance to return to original lane, d_c, equal to the clearance time, about 4 seconds times the average speed at which the two vehicles are approaching each other, that is, $d_c = t_3(v_3 + v_4/2$ where v_3 is the speed of the passing car in fps and v_4 is the speed of the opposing car in fps.

4.   Distance traveled by the opposing car from the time it is first perceived until the passing maneuver is successfully completed plus a safe clearance distance of 100 to 300+ ft (30.4 to 91.44 m), depending on the speed, or $d_s = v_4 t_4$ where v_4 is again the speed of the approaching car and t_4 is the time needed to complete the maneuver.

---

[8]After *A Policy on the Geometc Design of Rural Highways*, American Association of State Highway Officials and Transportation, 1965, pp. 142-145.

TABLE 17.3 MINIMUM DESIGN PASSING SIGHT DISTANCES[a]

Design Speed		Minimum Passing Sight Distance	
mph	kph	Feet	Meters
30	48	1100	335
40	64	1500	457
50	80	1800	549
60	97	2100	640
65	105	2300	701
70	113	2500	762
75	121	2600	793
80	129	2700	823

[a] *A Policy on Design of Urban Highways and Arterial Streets*, American Association of State Transportation and Highway Officials, Washington, D. C., 1973, pp. 268-276.

The total passing sight distance is the sum of the foregoing. See Figure 17.8.

The many variables of speed and acceleration and perception-reaction time make these terms difficult to evaluate. Recourse can be had to values recommended for design purposes as given in Table 17.3.

Roadway   A lane width of 12 ft (3.66 m) is usual for freeways and arterials. Widths of 11 ft (3.35 m) may be used in collectors and local roads while roads with very low traffic density may have 10 ft (3.05 m) lanes.

The number of lanes will depend on the level of service (speed and volume of traffic) desired. See Chapter 8. Design is usually based on daily hourly volumes (DHV) from which the thirtieth highest hour of the year is taken for the design volume. Freeway and multilane highways may be based on the design daily volume by direction (DDHV). The average daily volume (ADV) is also sometimes used.

Auxiliary lanes supplement the traveled lanes for speed changes in entering or leaving a highway, for parking, hill climbing by commercial vehicles, turning, and queuing for turning.

Access may be established as a feature of the right-of-way acquisition. It can also be secured by providing paralleling frontage roads that give detailed access to adjacent land uses. Rights of way vary in width from 50 ft (15.24 m) for local roads to 350 ft (106.68 m) or more for multilane highways with frontage roads. Access to freeways is obtained through interchanges, the subject of a later section.

Channelization   To improve safety and promote flow of traffic, various channelization procedures are used. Channelization has several purposes.

1. The number of choices a driver must make in passing through an intersection are reduced. The driver is confined to a specific and evident path for the maneuver he or she is making; his confusion is reduced.
2. Other drivers are informed as to where the car is going thereby avoiding their confusion that a free choice of routes would entail.
3. Right and left turns can be facilitated or prohibited as desired.
4. Safety islands or refuges are provided for pedestrians and turning and storage lanes for cars making turning movements.
5. The angle at which lines of traffic intersect can be controlled. Cars approaching at a flat angle are less likely to experience severe collisions at intersections than at wide and 90-degree angles.

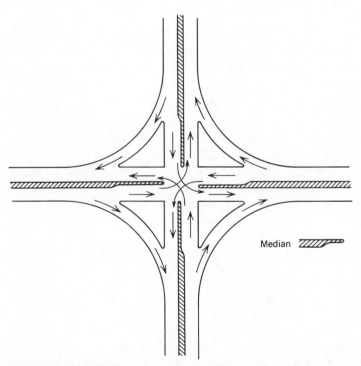

Median ⊿⊿⊿⊐

FIGURE 17.9 EXAMPLES OF CHANNELIZATION.

Various means have been employed to effect channelization.

1. Painted lane lines and arrows are inexpensive and quickly applied but are easily ignored by the motorist, wear away with time, and can be obliterated by ice and snow.
2. Sand bags can be used temporarily to delineate lanes and restricted areas. The bags, inexpensive and portable, are useful when experimenting with several possible arrangements.
3. Portable concrete "turtles" are sometimes used instead of sandbags.
4. Raised, drive-over, median strips, preferably corrugated and marked with reflector buttons, are in use, primarily as median strips or separators.
5. Raised curbs and islands, usually of concrete or similar material and often supplemented by turning lights, offer a positive and permanent medium for channelizing an intersection. Lane lights and divided lane highways might also be thought of as types of channelization. Two examples, applied to intersections, of the many possible channelized designs are shown in Figure 17.9.

Grade Separations   At-grade intersections are not a part of freeway or other limited access road design. One roadway, usually the minor route, is carried over the other on a bridge. If access between the two roads is required, it is provided by one of many forms of ramp arrangements. Simple sloping ramps give access in one direction from a parallel roadway. If lanes in the opposite direction are to be reached, a bridge over the lower level route is required giving the diamond arrangement of Figure 17.10e. Access onto the minor route is protected by stop signs and left-hand turns required for movement in the left-hand direction. The diamond takes a relatively small amount of land space, but the stop signs can be sources of delay if there is heavy traffic on the minor route.

Partial cloverleafs with ramps in only two quadrants are seen in Figure 17.10a. Here, again, the minor route must be protected by stop signs and traffic may be delayed.

A full cloverleaf giving greater flexibility is shown in Figure 17.10d. It takes a large amount of land, especially if the approach ramps are designed for high speeds. Speeds of 25 mph (40 kph) are possible with curves of 45 degrees or less. In any event, the length of ramp must be sufficient to overcome the vertical rise based on 16 ft (4.88 m) of clearance between the lower top of pavement and the bottom chord of the overhead structure.

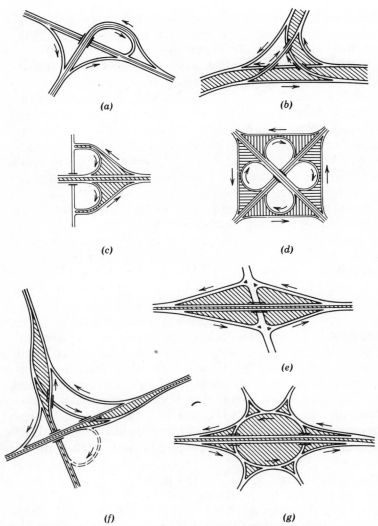

FIGURE 17.10 GRADE SEPARATED INTERCHANGES. (*a*) T OR TRUMPET. (*b*) Y. (*c*) PARTIAL CLOVERLEAF (RAMPS IN TWO QUADRANTS). (*d*) CLOVERLEAF. (*e*) DIAMOND. (*f*) DIRECTIONAL. (*g*) ROTARY. (COURTESY OF THE AMERICAN ASSOCIATION OF STATE HIGHWAY AND TRANSPORTATION OFFICIALS, *A POLICY ON GEOMETRIC DESIGN OF RURAL HIGHWAYS*, 1965, FIGURE IX-1, P. 494.)

Intersections other than right angle ones may call for special and more complex arrangements. The wye intersection shown in Figure 17.10*b* is such an example.

In any of the foregoing designs, the entry or exit from or to the principal roadway should always be from the right rather than from the left.

Where there are large differences between the roadway and ramp speeds, acceleration and deceleration lanes are placed adjacent to the outer lane of the freeway to permit the necessary acceleration and deceleration outside the line of free flowing traffic. Such lanes are full for most of their length with a taper toward the freeway end. The AASHTO recommendations vary from 175 to 250 ft (53.34 to 76.20 m) for freeway speeds of 40 to 70 mph (64 to 112.6 kph). For 30-mph ramp speeds, the total lengths of lanes are recommended as 250, 500, 800, and 1000 ft (76.2, 152.4, 243.8, and 304.8 m) for corresponding speeds of 40, 50, 60, and 70 mph (64.4, 80.5, 96.5, and 112.6 kph). Deceleration lanes maintain the same taper length but have overall lengths of 175, 250, 350, and 400 ft (53.34, 76.2, 106.68, and 121.92 m).

Railroad Intersections   Railroad lines intersect (or diverge) by means of a turnout comprised of a frog, which permits wheel flanges to pass through opposing lines of rails, a split switch that guides the flanged wheels to the desired path, and the closure rails between. The amount of diversion is a function of the frog angle, F, but frogs are usually designated by the frog number, N. In Figure 17.11, N is seen as the ratio between unit spread and the distance to the point of the frog from where the unit spread is measured. The actual point of the frog is located NP ft from the theoretical point; this point is normally used because the theoretical point is blunted to reduce damage from wheel impact. Guard rails placed opposite the frog

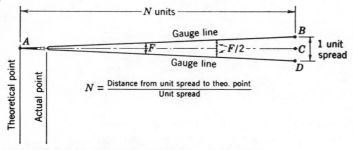

FIGURE 17.11 FROG ANGLE VS. FROG NUMBER.

point reduce batter on the point and guide the wheels. The portion of frog nearest the switch is the toe of the frog, that farthest away is the heel. Heel and toe distances to the point of the frog vary with the frog angle and can be obtained from tables of turnout data or from the manufacturer. For a number 10 turnout, the heel distance is 6 ft 5 in. (1.98 m) and the toe distance is 10 ft 1 in. (3.07 m). Coresponding distances for a number 20 frog are 11 ft $\frac{7}{8}$ in. (3.375 m) and 19 ft 10 in. (6.045 m). The relation between the frog angle and frog number is given in Figure 17.11 as

$$N = \tfrac{1}{2}\cot{(F/2)}$$

The switch angle, S, is usually taken as approximately one-fourth of the frog angle, that is, $S = F/4\pm$. From the switch angle and the gauge to gauge distance of $6\frac{1}{4}$ in. (15.88 cm) at the heel of the switch (Figure 17.12) the length of the switch point is computed as in Figure 17.12:

$$l = h\text{-}t/\sin S$$

where

$l = $ length of switch point rail in feet

$h = $ heel spread in inches, usually $6\frac{1}{4}$ in. (15.88cm)

$t = $ thickness of point at point of switch, $\frac{1}{8}$ to $\frac{1}{4}$ in. (0.32 to 0.64 cm)

Turnouts are also designated and installed by the lead distance from the point of the switch to the actual point of frog. The radius of the connecting

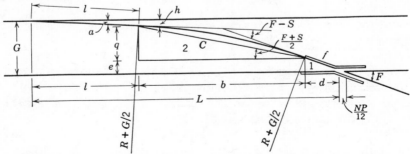

FIGURE 17.12 SPLIT SWITCH TURNOUT. (FROM *RAILROAD EN-GINEERING*, VOL. I, BY W.W. HAY, WILEY, NEW YORK, 1953, FIGURE 27.8, P. 438.)

curve between the heel of the switch and the toe of frog is

$$R = \frac{c}{2\sin(F-S)/2} - G/2$$

where

R = center line radius in feet of lead
curve connecting the heel of
switch with the toe of the frog

F = frog angle

S = switch angle

c = long chord of the connecting curve

G = gauge distance between the inside heads of rails with a value
of 4.708 ft (1.44 m) for U. S.
standard gauge

Where railroad tracks cross at grade, four 90-degree crossing frogs are used. For other angles of intersection, two end frogs with angles equal to the degrees of divergence are used and two side frogs that have angles supplementary to the end frogs.

Profiles   The profile of a line is its position in the verticle plane. It is also composed of tangents (grades) and connecting curves. In the United States, grades are usually measured in percent, the rise in feet per 100 ft (30.48 m).

Vertical curves form a transition from one grade to another, easing the sharp break at the point of intersection or vertex. A parabola having among others the following properties is commonly used. See Figure 17.13.

1. A line drawn to the intersection of the two tangents from the midpoint of the long chord bisects the curve and is bisected by the curve.
2. The offsets from the grade line to the curve vary as the square of the distance along the tangent (grade line).

The length of the curve is a function of the rate of change in grade per station in easing from one grade tangent to the adjoining tangent. Expressed mathematically, $L = (G_1 - G_2)/r$, where r = rate of change in feet

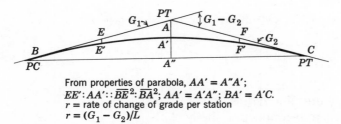

From properties of parabola, $AA' = A''A'$;
$EE':AA'::\overline{BE}^2:\overline{BA}^2; AA' = A'A''; BA' = A'C.$
r = rate of change of grade per station
$r = (G_1 - G_2)/L$

FIGURE 17.13 PROPERTIES OF A PARABOLIC CURVE.

per 100-ft station, G_1 and G_2 are the two intersection grades used with proper signs indicating whether the grade is ascending $(+)$ or descending $(-)$, and L is the length of curve in stations. To provide for smooth slack action in the train, the American Railway Engineering Association recommends a value of $r = 0.05$ ft per 100-ft station in sags and not more than 0.10 ft per 100-ft station on summits. For secondary, low-speed tracks, the values may be twice the foregoing.[9] These values also permit adequate sight distances for the engineman in going over the crests of hills.

Sight distance is an important factor for safety in highway design. A curve long enough to produce 1000 ft of sight distance is generally recommended for high-speed modern highways. Adequate sight distance si based on an assumed average height of the driver's eye of 3.75 ft (1.143 m) above the pavement (but may be as much as 6 ft or 1.830 m for large trucks) and should permit him to see an object 6 in. (15.24 cm) at the far range of vision. See Figure 17.14. According to AASHTO recommendations, with A = the algebraic difference in grades in percentage $\div$ 100, S = sight distance in feet, and L = length of vertical curve in feet, then when S is greater than L, $L = 2S - (3295/A)$ and when S is less than $L, L = AS^2/3295.$[10]

Aerial-tramway design is based on a parabola of the form $x = 2c/y^2$, which is a smooth curve representing approximately the funicular polygon formed by weights hung equidistantly along the chord between two supporting points. For uniformly spaced loads, the cable deflection at any point is practically the same as for a uniformly loaded cable with a weight per foot taken as equal to the weight per foot of the cable alone plus that of one load divided by the load spacing in feet. Using this approximation

[9]*Manual for Railway Engineering (Fixed Plant)*, 1953 edition, American Railway Engineering Association, Chicago, Illinois, p. 5-3-13.
[10]A *Policy on Geometric Design of Rural Highways*, American Association of State Highway and Transportation Officials, Washington, D. C., 1965, p. 207.

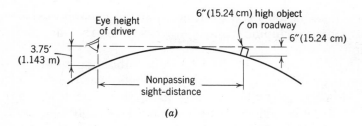

FIGURE 17.14 SIGHT DISTANCE OVER CRESTS.

and taking moments about the midpoint of c of a level span between two supports, (Figure 17.15), equations for the deflection, based on the loaded side, may be obtained:

$$h_L = s^2(L/d + r)/8t$$

and

$$h_L' = (L/d + r)mn/2t$$

where h_L = the midpoint deflection, h_L' = deflection of the cable at any point, S = span in feet between the two towers, L = weight of the loaded cars spaced d feet apart, r = weight per foot of the cable, and m and n are the respective horizontal distances to the point where the deflection is being found, and t is the midpoint tension in pounds. Cable deflection in

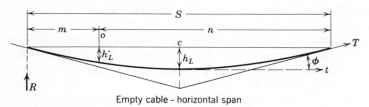

FIGURE 17.15 AERIAL TRAMWAY CABLE.

aerial-tramway design must be checked to insure that the cars and cable clear all ground obstructions including manmade works, trees, rock out-juttings, and drifted snow.

To permit the track cable to lie snugly in the saddle without an excessive angle between tangents at the crest, the deflection angle over the tower saddle is held to 2 degrees, 52 minutes, using more than one tower if necessary with the total angle divided between them (although German practice has used a total crest angle at a tower as high as 15 degrees). Heights of towers are determined by fitting a simple circular curve best suited to the profile between the two tangents. For a complete discussion of geometric design for aerial tramways, the reader is referred to "Aerial Tramways and Cableways" by Edward B. Durham in Peele's *Mining Engineers' Handbook*, Volume II, 3rd edition, Wiley, New York, 1952, pp. 1-51, from which the foregoing equations are derived.

QUESTIONS FOR STUDY

1. What specfic pruposes are served by the classifying of (*a*) highways and (*b*) rail lines?
2. Derive tangent distance, long chord, degree, and length for a simple curve.
3. Explain the environmental and ecological factors that enter into the selection of a transportation route. Explain a technique for establishing an ecologically suitable location. Does this technique have economic and engineering applications as well?
4. What superelevation will be required on a 3-degree curve for a 2-lane (12-ft width) pavement with an operating speed of 70 mph? What would be the effective coefficient of friction?
5. What maximum speed can be operated on a railroad's four-degree curve when the curve is given equilibrium superelevation for 40 mph unit coal trains?
6. Using AASHTO stopping distances, what is the minimum clearance required of a sight obstruction for a highway speed of 60 mph on a 4-degree curve?
7. What length of spiral is required for a 2-degree railroad curve to be operated at a maximum speed of 79 mph? What rate of change in degree of curve should be used in making the field layout?
8. What length of curve would be required to give a vertical sight distance of 1200 ft at a crest where an ascending 2.00-percent grade intersects a descending 1.00-percent grade?
9. What is the overall length, point of spiral to spiral to tangent for a 3-degree highway curve designed for 60 mph with an intersection angle of 60 degrees?
10. What area of land will be required for a clover leaf connecting two interstate highways with two 12-ft lanes in each direction and a 30-ft median strip and 6-ft shoulders, using a 70-mph design speed, 30-mph ramp speed?

11. Determine the frog angle for a number 12 turnout, the switch angle, the degree of connecting curve, and the lead distance when the toe distance is 7 ft $9\frac{1}{2}$ in. and the overall length of frog is 20 ft 4 in.
12. Using the turnout of Problem 11, what would be the total distance from point of switch to where the turnout track becomes parallel to the main track when the track centers are 14 ft, assuming a degree of return or connecting curve to be equal to or less than the connecting curve within the turnout?
13. What length of ramp would be necessary to give access to an interstate highway from an overhead city street if the grade is to be held to 4 percent and the street pavement is 18 ft above the interstate pavement?

SUGGESTED READINGS

1. A. M. Wellington, *The Economic Theory of the Location of Railways*, 1906 edition, Wiley, New York.
2. "Economics of Railway Plant Location and Operation," *Manual for Railway Engineering (Fixed Plant) of the A.R.E.A.*, 1976 edition, American Railway Engineering Association, Chicago, Illinois, Chapter 16.
3. W. W. Hay, *Railroad Engineering*, Volume I, Wiley, New York, 1953, Part I.
4. L. I. Hewes and C. H. Oglesby, *Highway Engineering*, Wiley, New York, 1963 Chapters 6 and 7, "Highway Surveys and Plans" and "Highway Design."
5. Highway and Bridge Surveys, *Journel of the Surveying and Mapping Division, Proceedings of the A.S.C.E.*:
   Reconnaissance, Paper 1593, April 1958.
   Introduction to Bridge Surveys and Reconnaissance, Paper 1713, July 1958.
   Preliminary Surveys, Paper 1697, July 1958.
   Location Surveys, Paper 1698, July 1958.
   Preliminary Bridge Surveys, Paper 1842, November 1958.
6. *A Policy on Urban Highways and Arterial Streets*, American Association of State Transportation and Highway Officials, Washington, D.C., 1973.
7. *A Policy on Geometric Design of Rural Highways*, American Association of State Transportation and Highway Officials, Washington, D.C., 1965.
8. *Transportation and Traffic Engineering Handbook*, Institute of Traffic Engineers, John Baerwald, Editor, Prentice-Hall, Englewood Cliffs, N.J., 1976.
9. George E. MacDonald, "Survey and Maps for Pipelines," Separate No. 393, January 1954, American Society of Civil Engineers, New York.
10. Ian McHarg, *Design with Nature*, Natural History Press, Garden City, New York, 1969.
11. G. W. Pickel, and C. C. Wiley, *Route Surveying*, 3rd edition, Wiley, New York, 1949.

Appendix 1

Typical Transport Units

TYPICAL RAILROAD TRANSPORT UNIT

Type: **Diesel-Electric Locomotive**
Service: Freight and passenger
Model: SD-40-2
Builder: GMC Electro-motive Division
Classification: C-C
Overall length: 68 ft, 10 in. (inside knuckles)
Overall width: 10 ft, $3\frac{1}{8}$ in.
Overall height: 15 ft, $7\frac{3}{16}$ in.
Distance between bolster centers: 43 ft 6 in.
Motors: Six D77, dc series wound
Driving wheels: 6 pairs, 40-in. wheels
Weight on drivers (loaded): 368,000 lb
Fuel capacity: 3200 gal
Maximum curvature radius: 193 ft (29.7 degrees)
Prime mover: 16-cylinder, turbocharged GM diesel, series 645E3
   Horsepower: (manufacturer's rating): 3000
Tractive effort at 25-percent adhesion: 92,000 lb

TYPICAL HIGHWAY TRANSPORT UNIT

Type: **Highway-Motor-Carrier Tractor**
Model: Astro 95
Service: Hauling highway-motor-freight trailers

FIGURE A1.1 DIESEL-ELECTRIC LOCOMOTIVE. (COURTESY OF THE GMC ELEC-TROMOTIVE DIVISION, LA GRANGE, ILLINOIS.)

FIGURE A1.2 HEAVY DUTY HIGHWAY TRACTOR. (COURTESY OF THE TRUCK AND COACH DIVISION, GENERAL MOTORS CORPORATION, DETROIT, MICHI-GAN.)

Builder: Truck and Coach Division, General Motors Corporation
Prime Mover:
   Engine: Detroit diesel, 12-cylinder, 71 N, 60 mm injectors, 2-cycle, $4\frac{1}{4}$-in.
      bore, 5-in. stroke, 851.0 in.^{3} displacement, 18.7-to-1 compression ratio
   Horsepower: 390 SAE net at 2100 rpm
   Torque: 1078-lb-ft SAE net at 1200 rpm
Transmission: 13-speed synchromesh overdrive
Tires: 10.00|20.f
Dimensions—tractor:
   Wheelbase: 195 in.
   Bumper to end of frame: 308 in.
   Width: 96 in.
   Height: 111 in.
Capacities:
   GVW: 50,500 lb
   GCW: 76,800 lb
   Axle load: 12,000 lb front and rear tandem: 38,000 lb

TYPICAL GREAT LAKES TRANSPORT UNIT

Type: **Great Lakes Bulk-Cargo Carrier**
Name: *Roger Blough*
Owned by: Great Lakes Fleet: United States Steel Corporation
Built: 1974
Trade: Iron ore pellets—Two Harbors, Minnesota and lake ports of South
   Chicago, Gary, and Conneaut
Motive power: Two 2-16 cylinder diesel engines—14,000 SHP
   Propeller: 4 stainless steel blades
Dimensions:
   Length overall: 858 ft, 0 in.
   Breadth, molded: 105 ft, 0 in.
   Depth, molded: 41 ft, 6 in.
   Designed draft: 28 ft, 0 in.
   Block ratio (approximate): 0.92
Tonnage:
   Empty plus fuel: 15,000 long tons (approximate)
   Cargo: 45,000 long tons
Hatches:
   Number: 21
   Center-to-center spacings: 24 ft, 0 in.

FIGURE A1.3 GREAT LAKES BULK CARGO CARRIER. (COURTESY OF THE GREAT LAKES FLEET, UNITED STATES CORPORATION, PITTSBURGH, PENNSYLVANIA.)

Self-unloading rate: 10,000 long tons per hour
Speed, loaded: 16.5 mph

TYPICAL RIVER TRANSPORT UNIT

Type: **Towboat** (Pusher)
Name: *A. D. Haynes II*
Service: Inland-river "towing" for Mississippi Valley Barge Line
Builder: Dravo Corporation
Prime mover:
   Engines: diesel-electric Nordbergs
   Horsepower: 4200 at 514 rpm
Propellers: 2 stainless-steel, 10-ft diameter with Kort nozzles

FIGURE A1.4 RIVER TOWBOAT. (COURTESY OF DRAVO CORPORATION, PITTS-BURGH, PENNSYLVANIA.)

Dimensions:
   Length: 200 ft
   Depth: 12 ft
   Width: 45 ft
   Draft: 9 ft (approximate)

TYPICAL AERIAL TRANSPORT UNIT

Type: **Jet Air Transport**
Model: Boeing 747
Builder: Boeing Aircraft Corporation
Prime Movers:
   Engines: 4 Pratt and Whitney JT9D–3W turbofans
   Thrust: 43,500 lb
Cruising speed: Mach 0.86 600.89
Dimensions: Wing area: 5,500 ft^2
   Cabin width: 20 ft
Weight:
   Takeoff gross weight: 710,000 lb
   Maximum landing weight: 564,000 lb

FIGURE A1.5 TRANSPORT AIRCRAFT (COURTESY OF AIR TRANSPORT ASSOCIATION OF AMERICA, WASHINGTON, D.C.)

Fuel capacity: 47,210 U.S. gal
Capacity:
Mixed class: 374 to 405 seats
All economy: 446 to 490 seats

HIGHWAY MOTOR COACH

Builder: GMC Truck and Coach Division of General Motors Corporation
Service: Urban and high-speed expressway
Seating capacities: 47 to 53 seats
Engines: Detroit Diesel Allison, 6V- and 8V-71, two cycle
        Rated brake horsepower (6V-71): 172 at 2000 rpm
        Rated brake horsepower (8V-71): 239 at 2000 rpm
Fuel capacity: 95 gal
Brakes: Service, 4-wheel, air, internal expanding, 2-shoe
        Emergency, 2-shoe, internally expanding, hand brake

Appendix 2

Problem Examples

The multiplicity of situations and individual problems that are involved in transportation planning make the presentation of a problem example that includes all elements of planning difficult. The problem that follows contains a goodly number of the planning processes and will present patterns that can be applied to other situations. They are not, however, a comprehensive statement of the contents of this book. The reader is warned not to be misled by the oversimplification.

PROBLEM EXAMPLE This example gives some of the problems and procedures in planning a new transport route in an undeveloped country.

The Need for Transportation A large ore-producing organization wishes to add 500,000 tons annually to its output by opening a new mine in a remote and inaccessible part of an undeveloped country. The mine property is located in a valley surrounded by rugged mountain ranges rising precipitously to elevations of 6000 ft. The mine capacity is estimated at 500,000 tons annually for a 20-year period.

The undeveloped character of the land in which the mine is located and the type of ore determine that this product must be transported to overseas markets via ocean shipping.

Terminal Points The mine location has established the originating terminus of the route. Three ocean ports are available. Port X is so far away that simple inspection rules it out. Port Y is just beginning to develop but is found to have ample room for expansion and a good harbor. Port Z is more fully developed but is becoming congested, with little room for expansion, and has adverse tidal currents that lead to frequent ship delays. Port Z also involves 50 additional route miles of distance. Port Y, a location called Welcome Bay, is therefore selected as the ocean terminus. (*Note*: this problem will not be concerned with transportation problems beyond Welcome Bay.)

Selection of Carrier   Ore, a granular-mass commodity, is most efficiently moved by rail, water, or conveyor transportation. A rail route is available part of the distance, but the best mode and route giving access to the rail line are not readily ascertainable. Consideration is given to highway trucking, a conveyor, an aerial tramway, a suspension-flow pipeline, or a combination of these to reach the existing rail line.

The pipeline and conveyor systems are ruled out because the transport used must permit inbound movements of fuel and supplies to the mine. Also, the region is too arid to provide the water necessary for a pipeline flow. A selection among the other possibilities can be made only on the basis of field-reconnaissance data.

The Reconnaissance   Maps of the area show no topographic detail. A field reconnaissance by car, rail, helicopter, and on foot is conducted that involves a study of terrain, drainage, elevation, and other pertinent features of the intervening land between the mine and the railroad. A map of the area is developed from this reconnaissance. See Figure A2.1. On the basis of the reconnaissance, three principal routes and modes are selected for further consideration.

Route A   A 30-mile highway haul over existing 8-percent mountain grades to a lightly constructed and little-used rail line owned by the mining company, and thence by rail (on a ruling grade of 2 percent) 200 miles to Welcome Bay. The existing portion of the route would require new and heavier rail, strengthened bridges, new ties and ballast, and additional cars and locomotives. The estimated cost of rail rehabilitation is $50,000 per mile. The cost of improving the "highway" by resurfacing, bank and cut widening, and improving drainage is estimated at $30,000 per mile. The total length of the route is 230 miles.

Route B   Ten miles of aerial tramway across a mountain range and 20 miles of new railroad construction through rolling country to the same

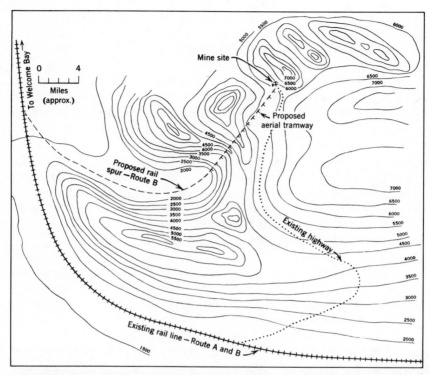

FIGURE A2.1 PORTION OF RECONNAISSANCE MAP FOR PROPOSED
NEW TRANSPORTATION ROUTE.

railroad with the same 2-percent ruling grade as in Route A but at a point
only 160 miles from Welcome Bay. The total length of Route B is 190
miles.

Route C   An aerial tramway 140 miles long over rough mountainous
terrain directly to Welcome Bay.

Preliminary Analysis of Cost   The following data are next computed in
order to have a preliminary estimate of cost for economic justification.

Common Data:
Tonnage—annual mine production = 500,000 tons
            daily mine production (310 days per year) = 1612 tons
Cost of mine and mine equipment = $20,000,000
Estimated daily cost of mine operation = $3.50 per ton
Estimated sales price of ore at Welcome Bay = $20 per ton

Route A

Truck Requirements A truck is considered that weighs 18 tons empty and has a 260-hp (net) engine operating at 3000 rpm, with gear ratios of 6:1, 4:1, 2:1, 1.7:1 and 1:1 for the transmission and 6.5:1 and 5.7:1 for the differential. Because of steep, 8-percent, ascents and descents, this vehicle is assumed to operate in third gear (2:1), leaving the lower gear combinations for emergency situations. Drive-line mechanical losses are 10 percent. Tires are 48×12 in.

Torque $= T = \text{hp}/0.00019N = (260/0.00019) \times 3000 = 456$

Tractive effort $= TE = T \times G_1 \times G_2 \times e \times r/12$ (as explained in Chapter 5)
$$= 456 \times 2.1 \times 5.7 \times 0.9 \times 24/12$$
$$= 9828 \text{ lb} \qquad TE = 375 \times \text{hp}/V$$

$V = 375 \times 260/9828$

$\quad = 10$ mph

Road resistance $= R_r = \left(17.9 + \dfrac{1.39V - 10.2}{W}\right)r_r + \dfrac{CAV^2}{W} W$

where $r_r =$ ratio of the fair-road-surface resistance, 40 lb per ton, to resistance for good surface conditions $= 2$
$W =$ empty weight of truck unit $= 18$ tons
$A =$ cross-sectional area $= 90$ ft^2

$R_r = \left[\left(17.9 + \dfrac{1.39 \times 10 - 10.2}{18}\right)2 + \dfrac{0.002 \times 90 \times 100}{18}\right]18$

$\quad = 670$ lb $\qquad$ (in unit terms, $R_r = 670/18 = 37.2$ lb per ton)

Total empty-truck resistance on 8-percent grade $= 670 + 8 \times 20 \times 18 = 3550$ lb

Net tractive resistance $= DBP = 9828 - 3550 = 6278$ lb

Pay load $= 6278/(37.2 + 8 \times 20) = 31.3$ tons (Use 30 tons.)

Truck loads per day required $= 1612/30 = 54$

Running time per trip $=$ round-trip distance$/V$
$$= 60/10 = 6 \text{ hours}$$
(The truck can probably run faster on the return trip because of a lighter load, but many return trips will be under load, bringing in supplies, so, as a factor of safety, the same speed is assumed throughout.)

Loading, unloading, and delay time = 1 hour
Total round-trip time = 7 hours
Round trips per truck per day = 24/7 = 3.4
Number of trucks required (truck loads per day/trips per truck)
　　1.10—to include 10 percent standby and in-shop equipment—
　　= (54/3.4)1.10 = 18 trucks
Two other types of trucks are available and offer the following operating and cost situation, which would be computed in the same way as for the 18-ton truck:

Truck Number	Pay load	Price of Truck (dollars)	Cruis- ing Speed (mph)	Loads per Day	Round- Trip Time (hours)	Round Trips (24 hours per truck)	Vehicles Required	Purchase Cost (dollars)
1	20	20,000	12	81	6	4.0	23	460,000
2	30	30,000	10	54	7	3.4	18	540,000
3	40	40,000	8	40	$8\frac{1}{2}$	2.8	16	640,000

Truck Number	Annual- Capital- Recovery Factor[a]	Total Annual Cost (dollars)	Truck miles per Year[b]	Operating Cost per Vehicle Mile (cents)	Annual Operating Cost (dollars)	Total Annual Cost (dollars)
1	0.12950	59,570	1,656,000	32	529,920	589,490
2	0.12950	69,693	1,011,840	36	364,262	433,955
3	0.12950	82,880	806,400	48	374,976	457,856

[a]Based on a 10-year life for the trucks at 5 percent. See Table 15.1.
[b]Round trips per day × length of round trip × number of vehicles × 310 days.

This tabulation of comparative annual costs shows that truck No. 2, the 30-ton-pay-load truck in the computation example, gives the lowest total annual cost and will therefore be selected for use in the rest of this problem.

Rail-Equipment Requirements This part of the study is based on a 2-percent ruling grade, an average train resistance of 6 lb per ton, 60-ton cars (40-ton pay load), and an 1800-hp 100-ton diesel-electric locomotive.
Locomotive resistance = R_{loco} = tractive resistance and grade resistance × locomotive weight
$$= (6 + 2 \times 20)100 = 4600 \text{ lb}$$

Drawbar pull (DBP) at 10 mph and 25-percent adhesion
$$= \text{tractive effort} - \text{resistance}$$
$$= (308 \times 1800/10) - 4600 = 50{,}840 \text{ lb}$$
Car resistance $= (6 + 2 \times 20)60 = 2760$ lb
Cars per train $= 50{,}840/2760 = 18$
Net tons per train $= 40 \times 18 = 720$ tons
Number of trainloads per day $= 1612/720 = 2.2$
Total turn-around time (average speed of 10 mph) is computed from the following items:
  Round-trip running time $= 400$ miles$/10 = 40$ hours
  Delays en route $= 4$ hours
  Terminal time (loading, unloading, inspection, etc.) $= 16$ hours
Total elapsed round-trip (turn-around) time $= 60$ hours
Equipment for the foregoing:
  6 complete sets of trains (cars and locomotives) plus one standby (or in-shop) set $= 7$ trains
  Cars $7 \times 18 = 126$
  Locomotives $= 7$
  Purchase price:

Cars $= 126$ at $6000 =$	$756,000
Locomotives $= 7$ at $300,000 =$	$2,100,000
Total equipment $=$	$2,856,000

Route B

Aerial-Tramway Requirements Tons per hour (24-hour operation) $= 1612/24 = 67$ tons
  (Use 70 tons per hour.)
Maximum average cost $= \$120{,}000$ per route mile
  (Taken because of the rugged terrain, and because $50,000 is needed for a power plant to operate the 10-mile aerial-tramway route in one section)
Operating costs $= 8\cent$ per ton-mile

Railroad Requirements

(As in Route A, 2.2- 18-car trainloads per day will be necessary to move the 1612-ton daily output of the mine.)

Round-trip running time $= 360$ miles$/10 =$	36 hours
Delays en route $=$	4 hours
Terminal time (loading, unloading, inspection, etc.) $=$	16 hours
Total elapsed round-trip time $=$	56 hours

Equipment for the foregoing $= (56/48)(2.2) = 5$ sets

  5 complete sets of trains (cars and locomotives plus one set of standby
    equipment) $= 6$ trains

  Cars $= 6 \times 18 = 108$

  Locomotives $= 6$

  Purchase price:

$$
\begin{array}{lr}
\text{Cars} = 108 \times \$6000 = & \$648{,}000 \\
\text{Locomotives} = 6 \times \$300{,}000 = & \$1{,}800{,}000 \\
\hline
\text{Total equipment} = & \$2{,}448{,}000
\end{array}
$$

Traffic Capacity of Rail Line   The existing rail line is already carrying 4 trains per day (2 each way). The addition of 4.4 trains (2.2 each way) makes

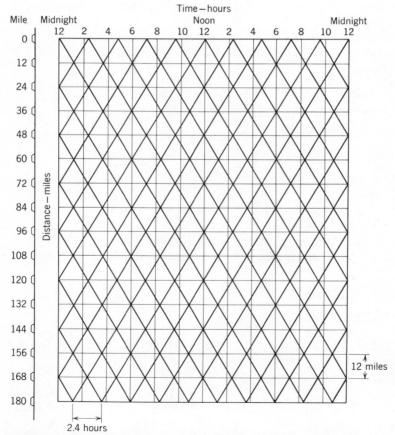

FIGURE A2.2 THEORETICAL TRAIN-HOUR GRAPHICAL SCHED-ULE.

6.6 trains total. Sidings are spaced approximately 12 miles apart (1.2 hours apart if trains run at 10 mph). As the spacing between trains on a single-track line (disregarding fleet movements) must be twice the siding spacing in hours, Figure A2-2, 10 trains can be started in each direction in a 24-hour period, a total of 20 trains per day. Even though theoretical, this can be seen by inspection to be adequate for the proposed 7 trains per day. The 7 trains are well below the possible 20 to 30 trains that a single-track line can handle in 24 hours. A closer siding spacing would, however, reduce delays. It can be assumed that the cost of rehabilitation per mile will include intermediate sidings at several locations.

Using the formulas of Chapter 8, train hours $= TH =$ number of sidings $\times$ 24. If the sidings are spaced 12 miles apart, the 180-mile distance includes 15 sidings, each 1 mile long. The total theoretical train hours will be $15 \times 24 = 360\,TH$. As each train is running at 10 mph, a running time of 18 hours per train, the total number of trains will be $360/18 = 20$ trains.

Comparative Cost Analysis   A tabulation of route and equipment (construction and purchase) costs and of operating costs for the several routes is as follows:

Route costs

Route	Improve Present Highway	Construct New Railroad	Rehabilitate Present Railroad	Construct New Cableway	Total Route Construction
A	(30 miles @ $30,000) $900,000	—	(200 miles @ $50,000) $10,000,000	—	$10,900,000
B	—	(20 miles @ $80,000) $1,600,000	(160 miles @ $50,000) $8,000,000	(10 miles @ $120,000) $1,200,000	$10,800,000
C	—	—	—	(140 miles @ $120,000) $16,800,000	$16,800,000

Equipment Costs

Route	Highway Trucks	Railroad Locomotives	Cars	Aerial-Tramway Power-Plants	Total Equipment Cost
A	(18 trucks @ $30,000) $540,000	$2,600,000	$756,000	—	$3,396,000
B	—	$1,600,000	$648,000	$50,000	$2,498,000
C	—	—	—	14 × $50,000 = $7,000,000	$7,000,000

Operating Costs

Route	16 Trucks @ 36¢ per Vehicle Mile	Trains @ $7.00 per Train Mile	Aerial Tramway @ 8¢ per Ton-Mile	Total Operating Costs
A	$364,262	(272,800 train miles) $1,909,600	—	$2,273,862
B	—	—	$400,000	$2,118,640
C	—	—	$5,600,000	$5,600,000

In computing the rate of return, only the effects of transportation are needed at this stage. The evaluation of the mine property is not a part of the problem. The ore, on being transported to tidewater, has a sales-price value of $20.00 per ton. Thus the original revenue will be $500,000 \times \$20.00$, or $10,000,000. Using this foregoing figure in the location formula $(R - E)/c = p$, the tabulated results are the following:

Rate of Return

Route	Revenue	Operating Costs	Capital Costs, R & E (Route Costs)	Equipment Costs	Total R & E	Rate of Return
A	$10,000,000	$2,273,862	$10,900,000	$3,396,000	$14,296,000	54.1%
B	$10,000,000	$2,118,640	$10,800,000	$2,498,000	$13,298,000	59.2%
C	$10,000,000	$5,600,000	$16,800,000	$7,000,000	$23,800,000	18.5%

Route C, the all-cableway route, is definitely out of the picture at this point. Route B, combined rail and aerial tramway, has a slight edge over the track operation. It would probably be wise to give more extensive study to each to determine if a better operating grade or shorter route could be had for the truck, tramway, or railroad route.

Economic Justification Assuming that Route B continues to be best after further study, the next step is economic justification. Here the justification for the route is dependent upon the justification for its purpose. If the mine plus transportation is not justified, then there is no need to consider the transport problem at all.

Assuming the mine to have a total capital cost of $20,000,000 and its annual total operating costs to be $3.50 per ton, the rate of return on the entire project (and its calculation) is given herewith [the value added by transportation would be the value of the mine per ton at tidewater ($20) less the operating costs at the mine ($20 − $3.50 = $16.50) less capital costs of the mine operation]:

Revenue:			$10,000,000
Operating costs: Mine ($3.50 per ton)	$1,750,000		
Transportation	$2,118,640		
Total	$3,868,640	$3,868,640	

Capital costs: Mine	$20,000,000		
Transportation	$13,298,000		
Total	$33,298,000	$33,298,000	

Rate of return = ($10,000,000 − $3,868,640)/$133,298,000 = 18.4%

The entire project, including Route B, is thus seen to give a justifiable rate of return even if a 10-percent allowance for contingencies is made throughout. There is no need to test Routes A and C as those have already been eliminated in the rate-of-return comparison. The revenue and the mine costs are the same in all three alternatives.

In making the final location for the route, considerations of detailed gradient, curvature, and distance factors over the general route selected will govern its final design. The same tabulation procedure, this time determining detailed costs instead of general costs, will again be carried out. As suggested in a preceding paragraph, a detailed survey might well be made for Route A as well as for Route B. However, another comparison needs to be made here, quickly. Can the possible saving deriving from a detailed study of a Route A revision offset the added cost of making the study?

One other possible alternative situation should be considered at this point. The railroad might have had different ownership and offered to build and operate the 20-mile rail spur and 10-mile aerial tramway in return for a guaranteed rate to Welcome Bay of $5.00 per ton. In this analysis the rate of return is based on the saving. The R term in the rate-of-return equation now becomes the cost of railroad freight rates, or $5.00 × 500,000$ tons = $2,500,000 per year. The E term is the expense of operating Route B, and C is the cost of constructing and equipping Route B. Then $(\$2,500,000 - 2,118,640)/13,298,000 = 2.9$ percent. Under these circumstances it would be hard to justify financially the investment in a private transportation system. The mining company should be able to earn more than a 2.9-percent return on this money by putting it in some other enterprise. Nevertheless, a desire to control the ore movement to the coast or doubt regarding the railroad's ability to maintain their agreement might carry sufficient weight to warrant a private transport system even with the foregoing rate of return.

Name Index

Subject Index

(Bold face folios indicate principal references)